Measurement Essentials

Measurement Essentials

Edited by **Miles Eron**

Clanrye International

New Jersey

Published by Clanrye International,
55 Van Reypen Street,
Jersey City, NJ 07306, USA
www.clanryeinternational.com

Measurement Essentials
Edited by Miles Eron

International Standard Book Number: 978-1-63240-346-9 (Hardback)

Printed in China.

Contents

Preface

This book was inspired by the evolution of our times; to answer the curiosity of inquisitive minds. Many developments have occurred across the globe in the recent past which has transformed the progress in the field.

With a qualitative introduction to the field of measurements, this book presents the wide ranging implications of this subject. Measurement is an interdisciplinary investigational science. Measurement systems synergistically combine science, engineering and arithmetic methods to offer primary data for study, design and expansion, control of processes and operations, and make possible safe and economic performance of systems. In the past few years, measuring techniques have grown rapidly and have gained a certain position through extensive research and hard work. This book consists of a compilation of researches and theories based on measurements.

This book was developed from a mere concept to drafts to chapters and finally compiled together as a complete text to benefit the readers across all nations. To ensure the quality of the content we instilled two significant steps in our procedure. The first was to appoint an editorial team that would verify the data and statistics provided in the book and also select the most appropriate and valuable contributions from the plentiful contributions we received from authors worldwide. The next step was to appoint an expert of the topic as the Editor-in-Chief, who would head the project and finally make the necessary amendments and modifications to make the text reader-friendly. I was then commissioned to examine all the material to present the topics in the most comprehensible and productive format.

I would like to take this opportunity to thank all the contributing authors who were supportive enough to contribute their time and knowledge to this project. I also wish to convey my regards to my family who have been extremely supportive during the entire project.

Editor

1

Precise Measurement System for Knee Joint Motion During the Pendulum Test Using Two Linear Accelerometers

Yoshitake Yamamoto[1], Kazuaki Jikuya[2], Toshimasa Kusuhara[3], Takao Nakamura[3], Hiroyuki Michinishi[4] and Takuji Okamoto[3]
[1]Himeji Dokkyo University
[2]Kawasaki University of Medical Welfare
[3]Okayama University
[4]Okayama University of Science
Japan

1. Introduction

The pendulum test is a means to evaluate the knee joint reflex from the pendulum motion induced by letting the lower leg drop freely after it has been lifted up (Watenberg, 1951). Many researchers have attempted to quantify the spinal cord stretch reflex from this pendulum motion in order to diagnose spasticity (Fowler et al., 2000; Kaeser et al., 1998; Lin & Rymer, 1991; Nordmark & Andersson, 2002; Stillman & McMeeken, 1995; Vodovnik et al., 1984). However, even today, much remains unknown about the relationship between this pendulum motion and the mechanism that produces the stretch reflex. For this reason, quantification studies on the stretch reflex have progressed slowly.

One method to advance the quantification of the stretch reflex may be to implement the following items in order:

1. Analyze the unknown behaviors in the pendulum test from various view points by trial and error, using existing physiological, clinical, and control engineering knowledge and theory as appropriate.
2. Modify the existing pendulum test model (Jikuya et al., 1991) based on the results of 1.
3. Elucidate the detailed mechanism of the stretch reflex using the model in 2, and investigate quantification methods.

We have already elucidated various phenomena following this procedure, but in this process, it has often been necessary to know angle, angular velocity, and angular acceleration values at arbitrary times during knee joint motion as initial and boundary conditions to solve nonlinear differential equations. Obtaining this kind of waveform with existing simple methods is difficult, as described below.

In principle, various existing sensors can be used to detect knee joint motion. However, several such sensors are not practical because of the knee joint's unique structural complexity. In addition, all existing sensors can measure only one of angle, angular velocity, or angular acceleration. Because of this, the only method that we can produce more than one type of waveform using such sensors is to differentiate and integrate the measured waveforms. As a result, it is difficult to ensure sufficient amplitude accuracy for waveforms obtained in this way and precise synchronization with measured waveforms.

For these reasons, we have recently begun investigating sensors that are suitable for the pendulum test. We have developed a new sensor that can precisely measure knee joint motion using two linear accelerometers. This article provides a comprehensive description of this sensor and related matters.

Section 2 briefly explains basic matters related to the pendulum test, such as the skeletal structure of the knee joint and the kinesiology of the stretch reflex. section 3 explains the measurement principle, assessment of accuracy in the laboratory, and the precision estimates when measuring subjects with the knee joint motion measurement system that is the main topic of this article. section 4 examines the results with the knee joint motion measurement system using these sensors; that is, the angle waveform and angular acceleration waveform of the knee joint in the pendulum test. We then touch briefly on a pendulum test simulator and an inverse simulation of measured waveform to more effectively utilize the results of the measurements, including the future outlook. section 5 provides a brief summary.

2. Biomechanics of the knee joint

2.1 Structure of the knee joint

The general motion of the knee joint is flexion and extension in the sagittal plane, caused consciously (actively) or unconsciously (passively). The leg structure that contributes to this motion is shown in Fig. 1.

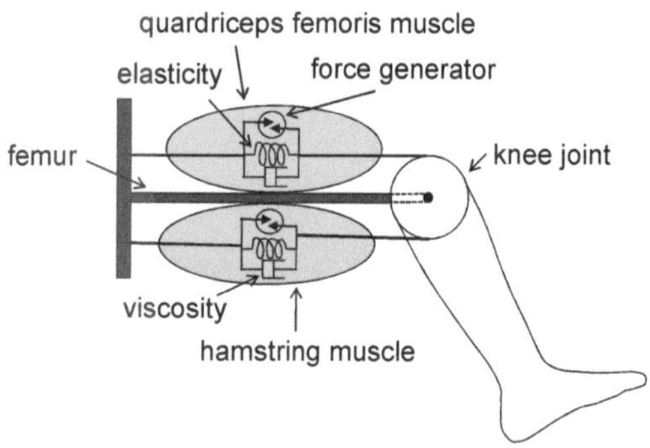

Fig. 1. Mechanism of extension and flexion.

The disc located at the end of the femur represents the knee joint, and the center of the disc is the rotation axis of the knee joint. The lower leg is fixed to the disc. The upper and lower ellipses are the quadriceps femoris muscle and hamstring muscle, respectively. One end of each muscle is fixed on the circumference of the disc. The spring and dashpot drawn in each of these ellipses are the respective elasticity and viscosity of the muscle. The force generator is the source that generates muscle contraction. The efferent fiber that controls it is not drawn. The knee joint oscillatory system consists of elasticity, viscosity, muscle contraction, and lower leg mass.

The above-mentioned quadriceps femoris muscle and hamstring muscle are the agonist and antagonist, respectively. When contractile force occurs in the agonist, the agonist shortens regardless of whether it is triggered consciously or unconsciously (when it is conscious, the antagonist also extends simultaneously), and consequently the knee extends. Similarly, the knee flexes when contractile force occurs in the antagonist. When conscious contractile force disappears or external forces that flex or extend the knee passively are eliminated, the lower leg will subsequently have damped oscillation with repeated flexion and extension unless it is resting in a stable position. In the following, we call such a dumped oscillation free one.

Next, let us look at the movement of the knee joint rotation axis. In general the knee joint is classified as a uniaxial joint that performs flexion and extension movement, but strictly speaking its rotation axis, as described below, moves according to a complex mechanism in which the lower end of the femur slides while rolling along the top of the tibia (Kapandji, 1970). That is, though the position of the knee joint rotation axis seems as if it is fixed to the center of the disk, it slightly moves together with flexion or extension. The rotation axis that moves based on this kind of phenomenon is called the axis of motion.

The skeletal structure of the knee joint is shown in Fig. 2(a). The axis of motion during flexion and extension corresponds to the imaginary point where the collateral ligament and cruciate ligament intersect (shown with a black dot (•)). Fig. 2(b) shows the migration of the intersection. The uppermost and lowermost black dots are the positions of the axis of motion in full extension and full flexion, respectively. When the knee joint rotates from full extension toward flexion, the condyle of the femur moves by rolling only up to a certain angle, beyond which an element of incremental sliding begins to apply. At the vicinity of the maximum flexion, there is only sliding movement. The relationship between the amount of movement and the angle of the knee joint is therefore mechanically complex, and analyzing it quantitatively is not an easy task. Similarly, neither the position nor the trajectory is easy to estimate by any simple means.

For the above reasons, unlike the elbow and other joints, it is not easy to measure exactly the knee joint motion in the pendulum test.

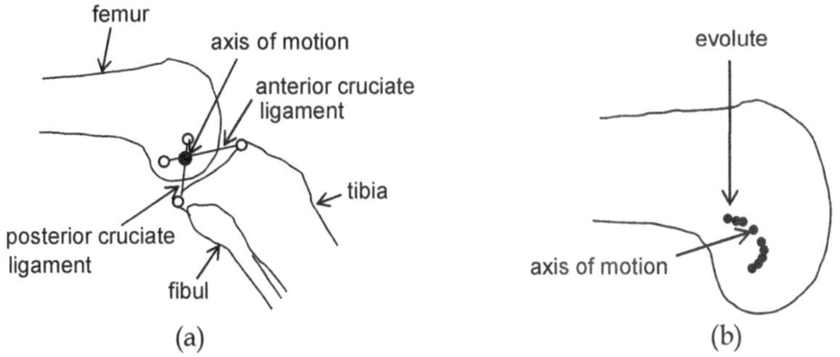

Fig. 2. Skeletal structure and locus of axis of motion of knee joint. (a) Skeletal structure of knee joint; (b) Locus of axis of motion.

2.2 Physiology of the stretch reflex

2.2.1 Principle of the stretch reflex

When muscle is stretched, it reflexively contracts in response. This kind of muscle response is called the stretch reflex. The stretch reflex is the target of the pendulum test. Fig. 3 shows the conceptual pathway of the stretch reflex. When muscle is stretched by some factor, receptor (called muscle spindle) detects it as a stimulus and transmit it as an afferent signal up to the spinal cord i.e., the α-motoneuron. The spinal cord receives the signal and sends a command (efferent signal) to effector (muscle) to restore this stretched state to the original state. These processes are executed unconsciously. Afferent fiber and efferent fiber function respectively as the transmission pathways for the afferent signal and efferent signal, which are both transmitted as impulses.

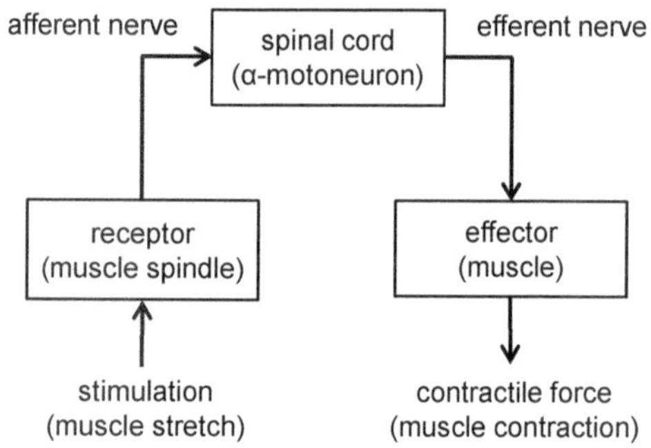

Fig. 3. Path of stretch reflex.

The reflex pathway described above (receptor → spinal cord → effector) is called a reflex arc. The stretch reflex is generated not only by passive muscle stretching, but also in response to conscious stretching of a muscle. The pendulum test is a test to estimate the sensitivity with which the reflex arc responds to the stimulus of knee flexion (extension of the quadriceps femoris muscle). The knee joint motion in this case is induced unconsciously by adding external force with the subject in a resting state for ease of estimation.

2.2.2 Structure of the spindle and its functions

Muscle is made up of many extrafusal muscle fibers arranged in parallel. Both ends of each muscle spindle are attached to one of these extrafusal muscle fibers. The muscle spindle is covered with a capsule, as shown in Fig. 4. In the capsule, there exist two types of intrafusal muscle fibers, called nuclear bag intrafusal muscle fiber and nuclear chain intrafusal muscle fiber. Stretching of the extrafusal muscle fiber affects the nuclear bag intrafusal muscle fiber and nuclear chain intrafusal muscle fiber, and stretch velocity and displacement, respectively, are detected. The detection sensitivities of the stretch velocity and displacement are regulated by efferent commands that are sent from phasic γ-motoneuron and tonic γ-motoneuron present in the spinal cord, respectively. The two kinds of detected information are consolidated into the afferent signal within the muscle spindle and transmitted to the spinal cord through Group Ia afferent fiber. Group II afferent fiber that send only nuclear chain intrafusal muscle fiber information to the spinal cord is also present, but they have a little influence on the stretch reflex in the pendulum test, and so it is not shown in the figure.

The afferent signal of Group Ia fiber is given as follows as the impulse frequency f_s (primary approximation) (Harvey & Matthews, 1961).

$$f_s = k_{1s}x + k_{2s}f_{\gamma s} + k_{1d}\dot{x} + k_{2d}f_{\gamma d} \tag{1}$$

Here, x is extrafusal muscle fiber (muscle) displacement, $f_{\gamma d}$ and $f_{\gamma s}$ are the respective impulse frequencies from the brain to phasic and tonic γ-motoneurons, and k_{1s}, k_{1d}, k_{2s}, and k_{2d} are constants. As shown in the above equation, there are two types of components in stimuli detected by the muscle spindle in the stretch reflex: a stretch velocity component expressed by the first and second terms, and a muscle displacement component expressed by the third and fourth terms.

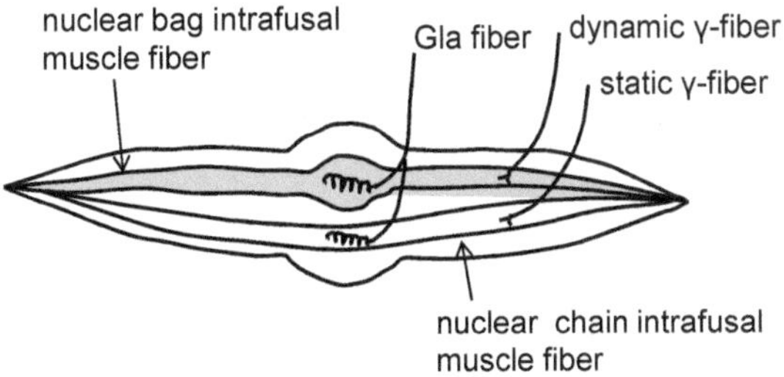

Fig. 4. Structure of muscle spindle.

2.2.3 Phasic stretch reflex and tonic stretch reflex

Commands to control the lower extremities are transmitted from the brain to muscle via the spinal cord. They are broadly divided into commands for flexion and extension, commands for maintaining of posture and commands for adjusting of the muscle spindle sensitivity. The first commands are generated only consciously, the second and third ones are generated consciously and/or unconsciously. Measurements of knee joint motion in the pendulum test are however done under the unconscious state of the subjects, and so the commands in this case are only unconscious ones to maintain posture and adjust the muscle spindle sensitivity. Consequently, the presence or absence of the efferent command toward the muscle and its strength during the pendulum test are determined only by these unconscious commands.

Fig. 5 shows the reflex arcs in the pendulum test schematically with a focus on the quadriceps femoris muscle. It includes phasic γ-motoneuron, tonic γ-motoneuron and α-motoneuron that play principal roles in the stretch reflex. The upper part enclosed by the solid line is the spinal cord. Signals f_e and f_i are commands to determine the posture, and represent frequencies of the impulses from the brain to the α-motoneuron and presynaptic inhibition part, respectively. The presynaptic inhibition part usually suppresses afferent signal from the muscle spindle so that it does not reach the α-motoneuron. Signals $f_{\gamma d}$ and $f_{\gamma s}$ are commands to adjust the muscle spindle sensitivity, and represent frequencies of the impulses from the brain to the phasic γ-motoneuron and tonic γ-motoneuron, respectively.

In normal subjects, f_e, $f_{\gamma d}$, $f_{\gamma s}$ have rather small values and f_i has rather large value, so that the α-motoneuron does not fire and no reflex occurs. Consequently, the knee joint motion at pendulum test becomes a free oscillation. On the contrary, in subjects having injuries of central nervous system, more than one of f_e, $f_{\gamma d}$, $f_{\gamma s}$ have rather large values and/or f_i has rather small value, so that the α-motoneuron fires and the stretch reflex occurs in the knee joint. Consequently, the knee joint motion is forced to disturb from free oscillation by the contractile force. In the following, we call such an oscillation forced oscillation.

The forced oscillation is classified into two types (William, 1998). One is a forced oscillation that is caused by stretch velocity component included in the afferent signal from the muscle spindle to the α-motoneuron. The value of the contractile force induced by such a component becomes maximum at the time when stretch velocity of the quadriceps femoris muscle reaches about maximum value. We call the reflex caused by such a component phasic reflex. The other is a forced oscillation that is caused by displacement component in the afferent signal from the muscle spindle. The contractile force in this case has maximum value at the time when the displacement of the muscle is about maximum. We call the reflex caused by such a component tonic reflex.

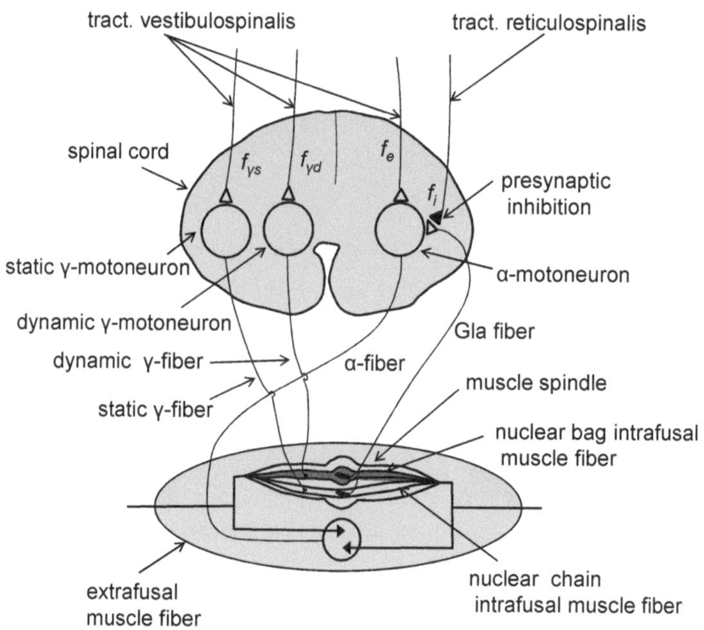

Fig. 5. Reflex arc.

As mentioned above, two types of reflexes can occur in the reflex arc composed of spindle, GIa fiber, presynaptic inhibition, α-motoneuron, α fiber and muscle. Evaluation of these reflexes therefore requires consideration of not only the size of the reflex but also the timing of their generation. Naturally, therefore, measurements of knee joint motion used in analyzing these reflexes demand high accuracy.

3. Detection of knee joint motion using acceleration sensors

3.1 Accelerometers as biosensors of knee joint motion

Measurements of physical movement have long been done focused on gait analysis. Recently, various types of advanced measurement technology are used in the field of sports science. Among them, sensors and measurement systems thought to be applicable to measurements of lower extremity motion, including sensors for pendulums described in 3.2-3.3, may be listed as follows.

a. Electrogoniometer (commonly called potentiometer)
This is a fixed rotation axis sensor that uses a rotating variable resistor. The rotation angle is detected as an electrical potential proportional to it. It has high reliability. On the other hand, it is unsuitable for measurement of high-speed movement, because large torque is required to drive contact points and they are abrasive.
b. Magnetic-type goniometer
This is a fixed rotation sensor with multiple magnetic pole and magnetic elements arranged along its circumference. It detects an electrical potential proportional to the rotation angle. It has high reliability and high accuracy. It requires a little torque since it is a non-contact-type device, and has no wearable parts.
c. Distribution constant-type electrogoniometer ("flexible goniometer")
This sensor was developed for angle measurements of complex joints (Nicol, 1989). It is not affected by movement of the joint axes, with the basic axis and movement axis set on either end of a bar-shaped resistor that changes electrical resistance with changes in shape. The angle between the two axes is measured as change in the resistance value. It has both rather large non-linearity and hysteresis.
d. Marking point measurement (or motion capture system)
Many marking points are attached to the surface of the subject's body, and images are made while the subject is moving (Fong et al., 2011). The subject is completely unrestricted. The angles at multiple points can be measured simultaneously. Its application is limited to experimental use for reasons of large filming space requirement, low time resolution, large scale of the system, etc.
e. Accelerometer
This sensor detects the movement of an object along a single axis as an acceleration signal, using a built-in strain gauge or similar element set. It is applicable to detection of accelerations in a wide range of fields, and various types have been developed from perspectives such as model type, accuracy, and stability. It does not restrict the movement of subjects, because a sensor only needs to be attached to one side of a joint even for joint movement measurements. It can also measure angle and angular acceleration simultaneously.
f. Gyroscope
Ultra-small devices have been developed using the Coriolis force and piezoelectricity based on micro-electro-mechanical systems (MEMS) technology (Tong & Granat, 1999). Currently, however, stability and reliability remain problematic.

To summarize, the requirements for knee joint motion measurement systems suitable for the pendulum test in clinical practice include: (1) sufficient accuracy; (2) low susceptibility to effects from the motion of the knee joint axis; (3) no restriction of the knee joint when worn; (4) ability to be attached simply and stably; and (5) ability to obtain waveforms of angle, angular velocity, and angular acceleration simply and with high accuracy. In the light of the above, the following conclusions may be reached with regard to the suitability of these sensors or measurement systems.

First, potentiometers are the most basic kind of angle sensor, and they have been used by Vodovnik et al. (1984), Lin & Rymer (1991), and others in studies of the pendulum test. However, when measuring knee joint angle using one potentiometer, accurate measurements cannot be made because of the axis of motion of the knee joint. Furthermore, it is not easy to attach and maintain the axis of the potentiometer in alignment with the rotation axis of the knee joint, and knee joint movement is restricted. Moreover, when seeking angular velocity and angular acceleration, one must depend on the differential, which is problematic in terms of accuracy. Magnetic goniometers perform well as angle meters, but they have the same problems as potentiometers with respect to the motion of the knee joint axis. Flexible goniometers have good properties with respect to the motion of the knee joint axis and ease of use, but the sensor itself has inadequate accuracy. Moreover, for the optical motion capture system that measures score, it is expected that the angle of knee joint motion (in some cases, angular velocity) will be detected faithfully with no contact mode, but the construction of the apparatus is too large for measurements of knee joint angle only with the body at rest, making it difficult to apply clinically. In recent years, many types of small and lightweight gyroscopes have been developed, and they have many features, such as ease of attachment, that make them suitable for measuring knee joint motion. However, stability and reliability are lacking in ultra-small types. In addition, the values detected are basically limited to angular velocity or one of the angles.

From the above, one can conclude that accelerometers fulfill nearly all of the preceding requirements, and, overall, they are the best option.

3.2 Principle of the knee joint motion measurement system using two accelerometers

We developed a method that can detect knee joint angle and angular acceleration simultaneously using two linear accelerometers in accordance with the conclusions stated in 3.1 (Kusuhara et al., 2011).

The fundamental configuration for the detection of knee joint pendulum motion is shown in Fig. 6(a). Accelerometers 1 and 2 are fixed on an accelerometer mounting bar separated by a certain distance (L_1, L_2) from point A on the rotation axis. The sensing direction of the accelerometer is the direction orthogonal to the bar on the paper. At this time, the direction of sensor attachment must be accurately fixed. However, attachment of the bar when measuring knee joint motion only needs to be fixed freely in a position within the plane of rotation of the knee joint and along the fibula as shown in Fig. 6(b). The lower leg is lifted until the bar reaches a certain angle θ (left on paper), and the pendulum motion is generated by letting the leg drop freely.

The outputs of accelerometers 1 and 2 with respect to this pendulum motion are taken as a_1 and a_2, respectively. a_1 and a_2 are given as follows.

$$\alpha_1 = L_1\ddot{\theta} + g\sin\theta \tag{2}$$

$$\alpha_2 = L_2\ddot{\theta} + g\sin\theta \tag{3}$$

Here, g is the gravity acceleration.

For both equations (2) and (3), the first term on the right side is angular acceleration from the pendulum motion, and the second term is the sensing direction component of the accelerometer, influenced by the gravity acceleration.

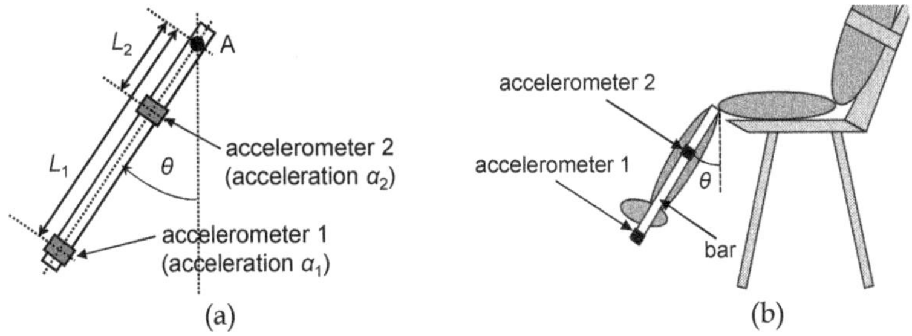

Fig. 6. Rotary motion detection by two linear accelerometers. (a) Accelerometer-bar with two linear accelerometers; (b) Attachment of accelerometer-bar on the lower leg.

When the first term on the right side disappears from both equations, the above-mentioned sensing direction component $g\sin\theta$ of the acceleration due to gravity and the angle θ of the knee joint are obtained in equations (4) and (5), respectively.

$$g\sin\theta = \frac{L_2\alpha_1 - L_1\alpha_2}{L_2 - L_1} \tag{4}$$

$$\theta = \sin^{-1}\frac{L_2\alpha_1 - L_1\alpha_2}{g(L_2 - L_1)} \tag{5}$$

When the second term on the right side disappears from equations (2) and (3), the angular acceleration $\ddot{\theta}$ of the pendulum motion unaffected by the acceleration due to gravity is given as follows.

$$\ddot{\theta} = \frac{\alpha_1 - \alpha_2}{L_1 - L_2} \tag{6}$$

In addition, for the angular velocity $\dot{\theta}$, the temporal differential of values on the right side of equation (5) and temporal integration of values on the right side of equation (6) can be obtained from the following equation.

$$\dot{\theta} = \frac{d\theta}{dt} = \int \ddot{\theta}\, dt \tag{7}$$

From the above, according to the proposed method, waveforms for angle, angular velocity, and angular acceleration that are unaffected by the acceleration due to gravity and synchronized are obtained with the addition of a single differentiation or integration.

3.3 Evaluation of the measurement system in the laboratory

3.3.1 Generation of simple pendulum motion

When evaluating the performance of the knee joint motion measurement system constructed in accordance with the principles described in 3.2, error can arise from the movement of the knee joint axis, imperfect attachment of the bar when evaluation is done, etc. This makes it difficult to accurately grasp the performance of the instrumentation body unit. In the following, therefore, we evaluate the instrumentation body unit by generating pendulum motion in a simulation.

The prototype performance evaluation system made for this purpose is shown in Fig. 7. The reference angle gauge is a high-accuracy, non-contact type, rotation angle gauge (CP-45H, Midori Precisions, Japan) used for comparison and evaluation of detector performance in the proposed method. An aluminum bar corresponds to the bar in Fig. 6(a), to which a weight is attached midway to make the period of the pendulum about the same as the lower leg. The fulcrum point A is set on the rotation axis of the reference angle gauge as in the figure for accurate comparison of the detection results. Accelerometers 1 and 2 (AS-2GA, Kyowa Electronic Instruments, Japan) are located in positions separated by only L_1 (60 cm) and L_2 (15 cm), respectively, from the rotation axis on the aluminum bar. Acceleration a_1 and a_2 detected by the accelerometers are input to a computer via matching amplifier and an A/D converter (PCD-300B, Kyowa Electronic Instruments). The output of the other rotation angle gauge is input to a computer via an A/D converter.

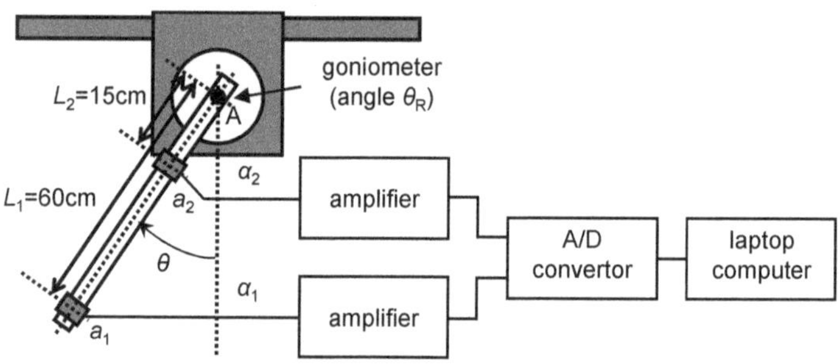

Fig. 7. Construction of performance evaluation system.

Here, the aluminum bar of the apparatus in the figure was moved as a rigid pendulum, and performance was evaluated from the results of simultaneous measurements of the pendulum motion with the detector of the proposed method and the reference angle gauge.

3.3.2 Results of evaluation

Pendulum motion was induced by freely dropping the aluminum bar after tilting it to about 40 deg. This pendulum motion had damped oscillation of a sinusoidal waveform with a period of 1.14 s, nearly the same as knee joint motion.

Output waveform examples of accelerometers a_1 and a_2 when the amplitude of damped oscillation is about 30 deg are shown in Fig. 8. a_1 and a_2 are in opposite phases because, with a_1, acceleration from pendulum motion is greatly affected by acceleration due to gravity, whereas the opposite is true with a_2.

The $g\sin\theta$ waveform obtained from equation (4) and the angle θ waveform obtained from equation (5) using these waveforms are shown in Figs. 9 and 10, respectively. In Fig. 10, the reference angle gauge output θ_R (broken line) and the $error_\theta$ between θ and θ_R (thin solid line) are added. From Figs. 8 and 9 it is seen that the values for the $g\sin\theta$ component included in a_1 and a_2 are large enough that they cannot be ignored. In addition, as understood from the example in Fig. 10, the θ and θ_R waveforms are in excellent agreement. Moreover, the component of acceleration due to gravity remaining in angle waveform θ is small enough to be indistinguishable from noise (see $error_\theta$). The correlation coefficient of θ and θ_R for 10 periods, including the 2 periods shown in this figure, was 0.999, and RMSE was 0.992 deg.

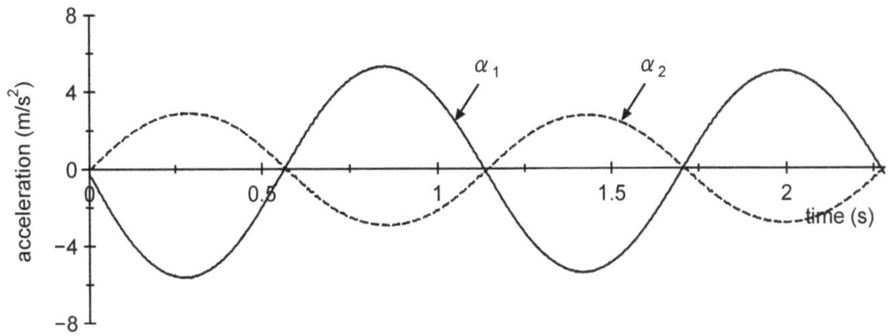

Fig. 8. Output waveforms of linear accelerometers (a_1 and a_2).

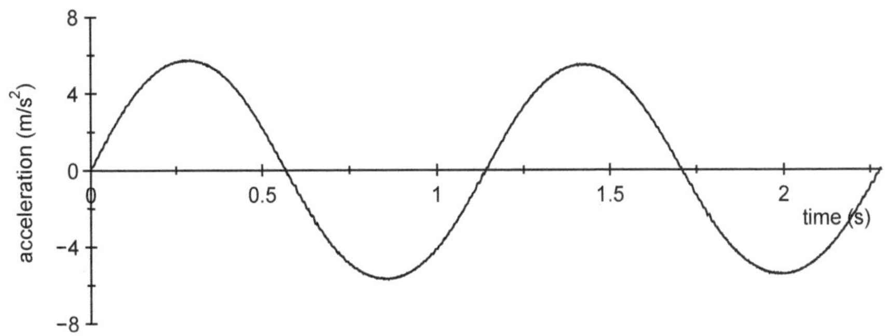

Fig. 9. Gravity acceleration components ($g\sin\theta$) in a_1 and a_2.

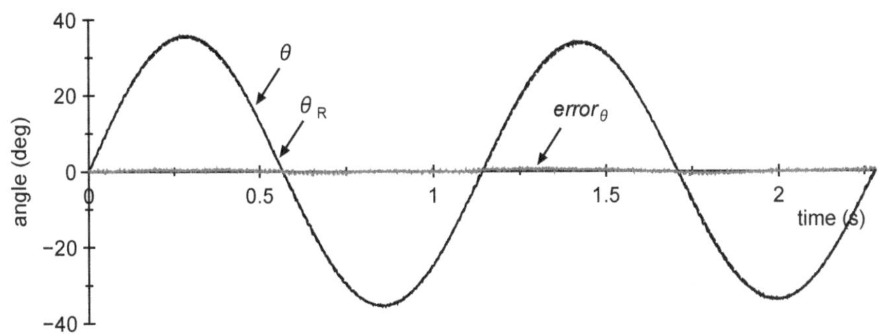

Fig. 10. Angle waveforms (θ and θ_R).

The detection results for angular acceleration $\ddot{\theta}$ were as follows. The angular acceleration $\ddot{\theta}$ obtained by substituting acceleration waveforms a_1 and a_2 from Fig. 8 into equation (6), and the angular acceleration $\ddot{\theta}_R$ obtained by twice differentiating θ_R in Fig. 10, are shown in Fig. 11. The solid and broken lines are $\ddot{\theta}$ and $\ddot{\theta}_R$, respectively, and the thin solid line is the $error_{\ddot{\theta}}$ of the two. Noise is superimposed in $\ddot{\theta}_R$ obtained with a differential, but there is good agreement between the two. The correlation coefficient of $\ddot{\theta}$ and $\ddot{\theta}_R$ and RMSE for 10 periods, including the 2 periods shown in Fig. 11, was 0.998 and 0.749 rad/s^2, respectively.

Next, let us look at the angular velocity waveform. The angular velocity waveform $\dot{\theta}_1$ obtained by differentiating angle waveform (θ) in Fig. 10 (solid line) and the waveform $\dot{\theta}_2$ obtained by integrating angular acceleration waveform $\ddot{\theta}$ in Fig. 11 (broken line) are shown in Fig. 12. There is good agreement between the two, although this is due partly to the fact that these values were obtained in a simulation trial done in a laboratory with little noise.

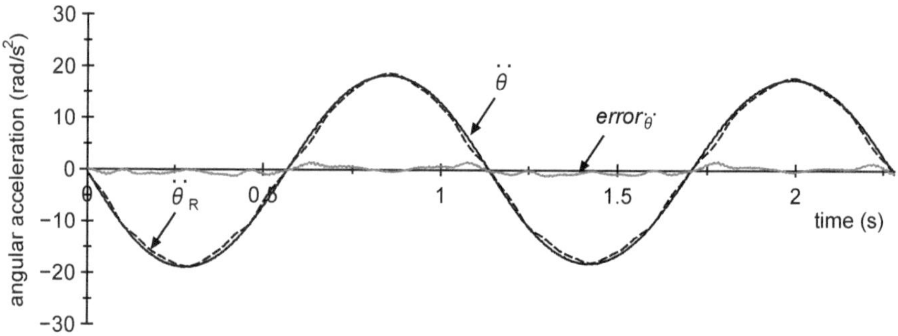

Fig. 11. Angular acceleration waveforms ($\ddot{\theta}$ and $\ddot{\theta}_R$).

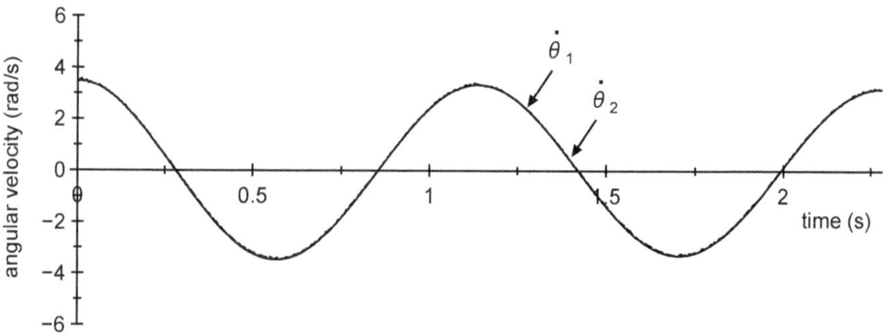

Fig. 12. Angular velocity waveform ($\dot{\theta}_1$ and $\dot{\theta}_2$).

The accuracy of the above-mentioned knee joint motion measurement system itself was obtained using two accelerometers of the same type purchased with no special conditions. This accuracy, when compared with the accuracy of detecting uniaxial arm motion with a gyroscope, goniometer, and potentiometer (correlation coefficients (0.9997-0.9999) and RMSE (1.37-1.47 deg (Furuse et al., 2005)) for similarity between these measured values), had a similar correlation coefficient and RMSE of about 30% smaller. Attaching the aluminum bar to the subjects is easier than these sensors. The next subsection discusses the effect on knee joint motion.

3.4 Estimation for the accuracy of the proposed system

For the principles given in 3.2, when using this knee joint motion measurement system created for the pendulum test, there is the problem of axis of motion mentioned in 2.1, in addition to the unique aspects of biological measurements, such as the state of attachment of the aluminum bar to the knee joint and slight changes of posture by the subject during the test. Therefore, one would predict that the decrease in accuracy due to these factors cannot be ignored. However, we have found no method that can directly and precisely evaluate the decrease in accuracy resulting from these factors. In this study, therefore, the following indirect method was used to evaluate the decrease in accuracy when measuring subjects.

First, the subject sitting in a chair for measurement and the evaluation system used in 3.3 were arranged as shown in Fig. 13. The chair and subject seen on the plane of the paper are located just in front of the evaluation system. The height of the evaluation system from the floor and the left-right positions seen on the plane of the paper were adjusted so that the rotation axis of the evaluation system and the rotation axis of the knee were on about the same line. The aluminum bar was fixed to the lower leg with rubber bands as shown in the figure so that the motion of the knee joint would be restricted as little as possible.

Next, the pendulum test was done by freely dropping the lower leg after it had been lifted about 50 deg.

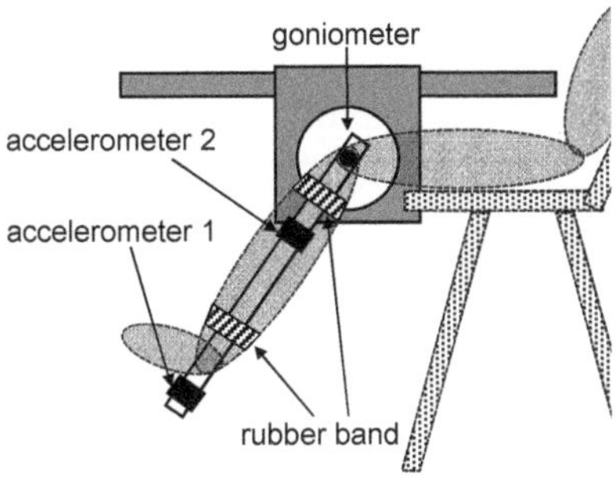

Fig. 13. Evaluation system for knee joint motion detector.

Angle waveforms obtained in this way are shown in Fig. 14 (a). θ and θ_R are the angle measured with the present method and the angle measured with the reference angle gauge, respectively. It is clear from this figure that the agreement is so close that it is difficult to distinguish the two angles. To examine the angle detection accuracy with this method in greater detail, Fig. 15 shows a window display of one section of the waveform in Fig. 14 (a). The correlation coefficient in this part was 1.000, and RMSE was 0.672 deg. The RMSE value, compared with the value for simple pendulum motion (Fig. 10), was 1.84-fold, equivalent to 1.94% with respect to the maximum amplitude (34.7 deg) of θ_R. Fig. 14 (b) shows the angular acceleration waveforms measured with this method. Good agreement between this waveform and the waveform when the reference angle gauge output was differentiated twice was also confirmed.

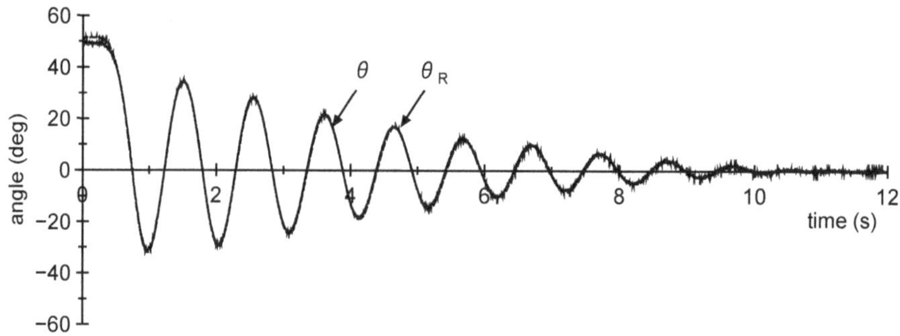

(a)

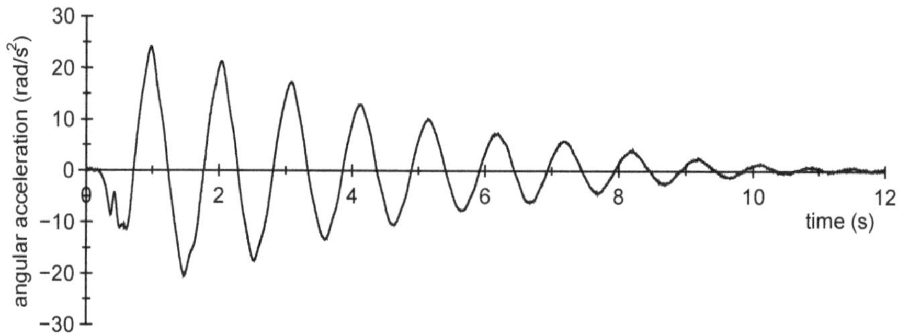

(b)

Fig. 14. Angle and angular acceleration waveforms measured from a normal subject by pendulum test. (a) Angle; (b) Angular acceleration.

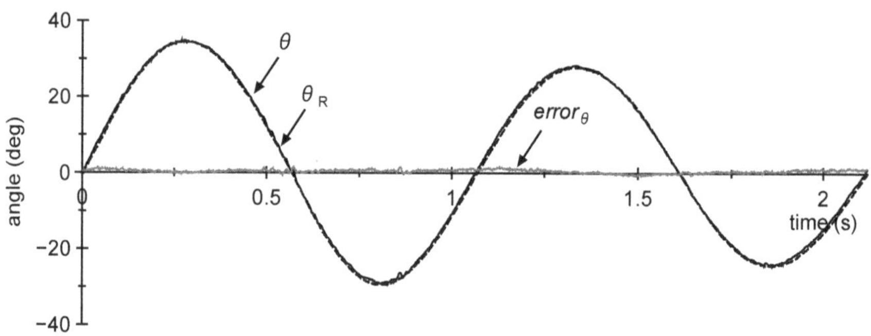

Fig. 15. A window display of the angle waveforms in the Fig.14(a).

From the above, the accuracy of the knee joint motion measurement system based on the present method was assured to be comparable to that of a reference angle gauge when applied to the pendulum test.

This knee joint motion measurement system is also thought to have sufficiently high accuracy to be applicable to the pendulum test for the following reasons.

The evaluation system used for application to the pendulum test is the same as the system used in the waveform measurements in Fig. 10. Therefore, the difference in RMSE for the waveforms in Fig. 10 and the waveforms in Fig. 15 is thought to have been produced by the difference in damped oscillation that is generated artificially and damped oscillation that occurs in the biological body, by whether or not there was positional displacement or distortion of the aluminum bar with shaking of the lower leg, or by the rotation axis movement described in 2.1. This result means that, when the fulcrum point of the aluminum bar is nearly matched to the knee joint rotation axis, both of RMSEs for the angles with this method and with the angle gauge worsen by only about 0.3 deg compared with the instrumentation body unit.

Finally, let us briefly consider the decrease in accuracy with the addition of the aluminum bar. When the aluminum bar (85 g) and two accelerometers (15 g × 2) are added, the lower leg mass of an average normal subject (about 3 kg) increases by roughly 4%. However, in the accuracy evaluation results mentioned up to this point, the descriptions have shown that there is almost no effect. Even so, the effect on measured knee joint motion in subjects cannot be ignored. When the center of the gravity of the lower leg changes with the addition of the aluminum bar and accelerometers, the moment of inertia changes in proportion to the square of the distance to the rotation axis, affecting the period of the oscillation and the damping coefficient in the pendulum motion. In both θ and $\ddot{\theta}$, the effect of the aluminum bar and two accelerometers on the knee joint motion in an average normal subject is an increase of about 3% in the time for one period and a decrease of about 6% in the damping coefficient in the results of rough trial calculations. When more precise measurements are required, the increase in the moment of inertia can be suppressed, and the influence on the period and the damping coefficient can be decreased, by shifting the attachment position of the aluminum bar upward.

4. Analysis of knee joint motion using the simulator with waveforms measured

In this section, we will deal with spastic patients as the subject of the analysis. Such patients have high phasic reflex.

4.1 Examples of waveforms measured

The knee joint motion of a normal subject was measured in the pendulum test using the measurement system shown in Fig. 6 in section 3. Fig.16 shows examples of the waveforms measured. In the figure, (a) and (b) show the angle waveform and angular acceleration waveform, respectively. There is absolutely no restriction on motion of the lower leg, since only two accelerometers were attached to it. It is difficult to estimate the error in this measurement result quantitatively, but the measurement was probably made with about the same accuracy as obtained in the investigation in the preceding section. The collapse of the waveform that appears in the early stage of oscillation is noise produced by the state of contact between the hand of the investigator and the lower leg in the instant when the lifted leg was released. If this portion is eliminated, the angle waveform and angular acceleration waveform have typical damped oscillation with nearly the same periods, although the phases differ. These waveforms are the free oscillations mentioned in subsection 2.1. Both waveforms are described theoretically by the following differential equation derived by Vodovnik et al. (1984).

$$J\ddot{\theta} + B\dot{\theta} + K\theta + \frac{mgl}{2}\sin\theta = 0 \quad (8)$$

Where J, B, K, m, ℓ, and g are the moment of inertia, viscosity, elasticity, mass, length of the lower leg and gravity acceleration.

Even if each coefficient value is taken as a constant, this is not simple to solve analytically. However, according to the result of numerical analysis, the waveforms have similar damped oscillations as in Fig. 16.

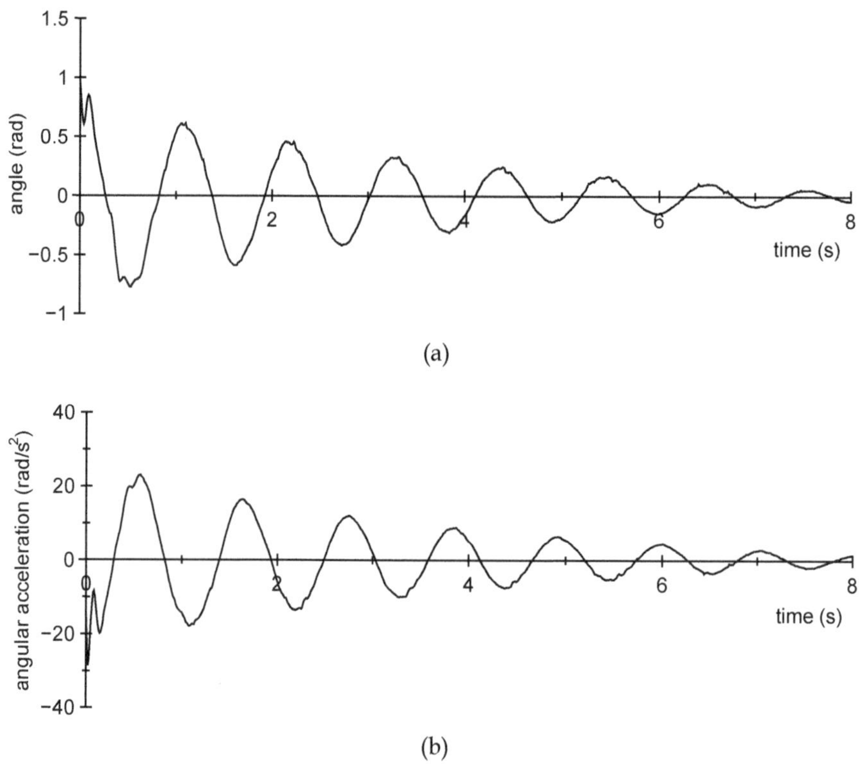

(b)

Fig. 16. Waveforms of knee-joint motion measured from a normal subject by pendulum test. (a) Angle; (b) Angular acceleration.

Next, we will look at the waveforms of spastic patients. Sample waveforms are shown in Fig. 17. The subject is a spastic patient with a moderately increased phasic reflex. Comparing them with the waveforms in Fig. 16, the first peak of the angular acceleration waveform is larger. This is because the muscle stretch velocity reaches a maximum in the vicinity where the acceleration first intersects the time axis and a phasic reflex is produced, and the resulting contractile force in the quadriceps femoris muscle acts to extend the lower leg (see 2.2.3). The knee joint motion in this case is not free oscillation, but restricted one by the contractile force. Thus, as shown in the following equation, such a knee joint motion is given with an equation that is obtained by adding this contractile force Q_h to the right side of equation (8).

$$J\ddot{\theta} + B\dot{\theta} + K\theta + \frac{mgl}{2}\sin\theta = Q_h \tag{9}$$

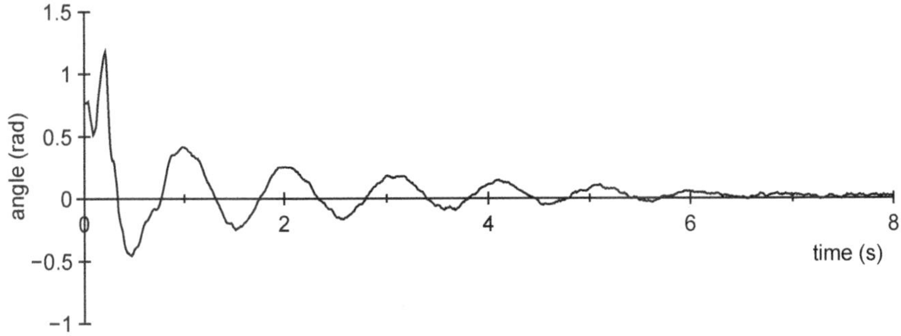

(a)

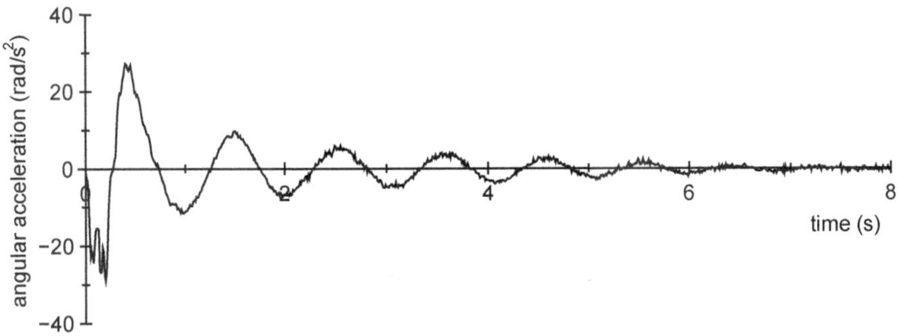

(b)

Fig. 17. Waveforms of knee-joint motion measured from a moderate spastic patient by pendulum test. (a) Angle; (b) Angular acceleration.

The angle waveform and the angular acceleration waveform are accurately synchronized, because they are calculated directly from the outputs of the same linear accelerometers. Consequently, the results of measurement with the above-mentioned measurement system are suitable for the purpose of investigating behavior while rigorously referencing both waveforms.

As seen from the above examples, according to the measurement system shown in Fig. 6 of section 3, the angle and angular acceleration of the knee joint motion in the pendulum test can be measured simultaneously with high accuracy. However, investigating in detail the phasic reflex generation mechanism of each subject or estimating the degree of increase in the reflex often requires values of physical quantities and/or waveforms for various sections. Constructing a simulator for the pendulum test using the model described in the next subsection, the values of physical quantities and waveforms appropriate for an arbitrary purpose will be able to be calculated freely.

4.2 Pendulum test simulator

Since equation (9) described in 4.1 is a basic model expressing knee joint motion, a simulator can be constructed using it. However, to obtain a simulator with good accuracy, the nonlinearity of B and K needs to be considered together with a detailed mathematical formulation of Q_h. The following briefly describes these implementation methods.

Q_h is muscle contraction that occurs in the quadriceps femoris muscle during the stretch reflex, expressed by equations (10) and (11) (Jikuya et al., 2001).

$$Q_{\mathrm{h}} = \exp(-T_{\mathrm{d}}\mathrm{s})\frac{F_{\alpha}}{(1+T_{\mathrm{m}}\mathrm{s})^2} \tag{10}$$

$$F_{\alpha} = \frac{F_{\gamma\mathrm{d}} - k_{\mathrm{f}}\theta\mathrm{s}}{F_{\mathrm{i}}} + F_{\mathrm{e}} \tag{11}$$

$\exp(-T_d s)$ and $1/(1+T_m s)^2$ are transfer functions that express the sum of the transmission delay times in Group Ia fiber and α-fiber (T_d: time required for an impulse to pass through Group Ia fiber and α-fiber) and the characteristic of excitation contraction coupling (T_m: twitch contraction time of muscle), respectively. s is a Laplace operator, and k_f is a coefficient to convert knee joint angle to muscle length. F_α is the output of α-motoneuron. $F_{\gamma d}$, F_i and F_e are normalized command frequencies of $f_{\gamma d}$, f_i and $f_e - V_{th\alpha}$ ($V_{th\alpha}$: α-motoneuron threshold), respectively. Equation (11) expresses that the phasic afferent information $F_{\gamma d} - k_f\theta s$ sent from muscle spindles, after being inhibited by presynaptic inhibition F_i, is added to command F_e from the brain and becomes the output of α-motoneuron.

Next is a description of the modeling of B and K with large nonlinearity. Each extrafusal muscle fiber that make up muscle contain many actin and myosin molecules. It is known that, in a resting state, the vast majority of these molecules are in a gel state, but when the muscle starts to flex or extend movement, these molecules solate in accordance with the velocity of the flexion or extension (Lakie et al., 1984). Using the properties of these actin and myosin molecules, the temporal changes in B and K values are described by the following differential equations (Jikuya et al., 1995).

$$\dot{B} = a(B_{\mathrm{M}} - B) - b(B - B_{\mathrm{m}}) \mid \dot{\theta} \mid \tag{12}$$

$$\dot{K} = c(K_{\mathrm{M}} - K) - d(K - K_{\mathrm{m}}) \mid \dot{\theta} \mid \tag{13}$$

Here, $B_M(K_M)$ and $B_m(K_m)$ are the maximum and minimum values, respectively, of $B(K)$, and a, b, c, and d are all constants. These equations express equality of the differential of viscosity and elasticity coefficients to the value when the proportion that is solated ($b(B-B_m) \mid \dot{\theta} \mid$) is subtracted from the portion of actin and myosin molecules that is gelated ($a(B_M-B)$).

The pendulum test simulator is a program specifically for analyzing knee joint motion, prepared according to the mathematical models in equations (9)–(13) above. The inputs to the program are values of the constants that appear in these equations, and the output is angle waveform (θ), angular velocity waveform ($\dot{\theta}$), and angular acceleration waveform ($\ddot{\theta}$).We created this program in C language for execution on a personal computer.

4.3 Applications of pendulum test simulator

Using the simulator described in the previous section, knee joint motion can be freely analyzed for arbitrary combination of input values of the simulator that cannot be directly measured from subjects. In addition, by slightly modifying the program as needed, the waveforms for arbitrary section of the model can be easily analyzed. Moreover, obtaining a model for each subject by inverse simulation using the simulator, it becomes possible to analyze the knee joint motion of that subject with the model only. The following 4.3.1, shows a simple analysis of knee joint motion using the simulator, and 4.3.2 shows a case of high-level analysis of the stretch reflex using inverse simulation.

4.3.1 Examples of waveforms obtained from simulation

Figures 18, 19, 20, and 21 are waveform examples of knee joint motion in a normal subject and patients with mild, moderate, and severe spasticity, respectively. In all of the figures, (a), (b), (c) and (d) show angle waveform, angular velocity waveform, angular acceleration waveform, and muscle contraction waveform, respectively. It is clearly understood from the figures that there is a relationship between knee joint motion and muscle contraction that changes as spasticity increases. Although not shown in the figure, output waveforms such as for muscle spindles and α-motoneurons can also be analyzed simply.

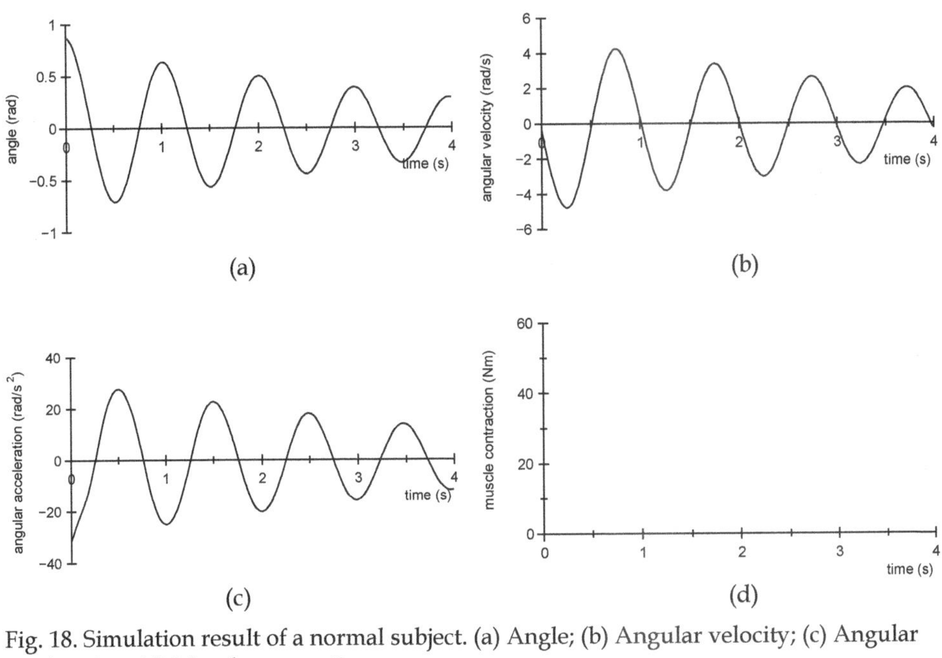

Fig. 18. Simulation result of a normal subject. (a) Angle; (b) Angular velocity; (c) Angular acceleration; (d) Muscle contraction.

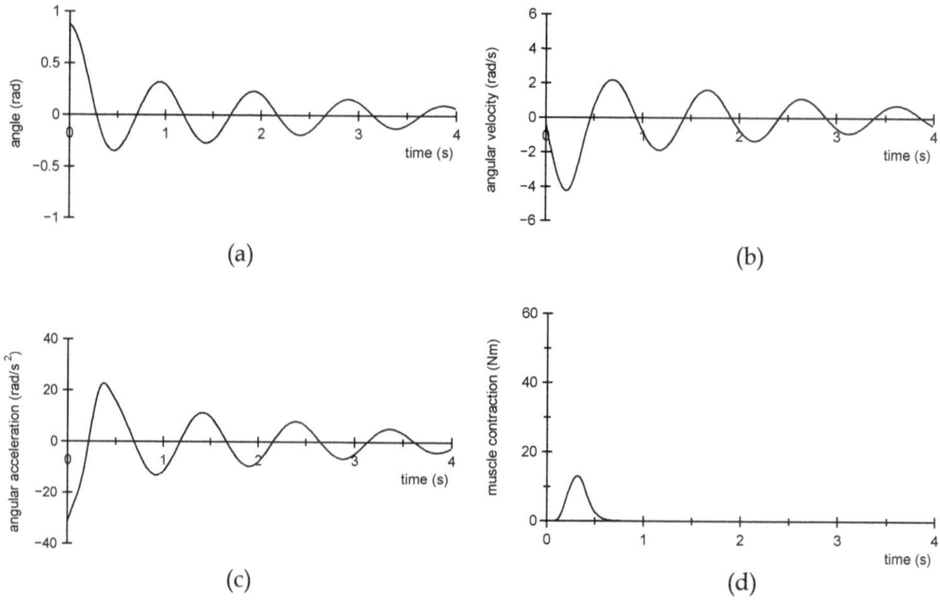

Fig. 19. Simulation result of a mild spastic patient. (a) Angle; (b) Angular velocity; (c) Angular acceleration; (d) Muscle contraction.

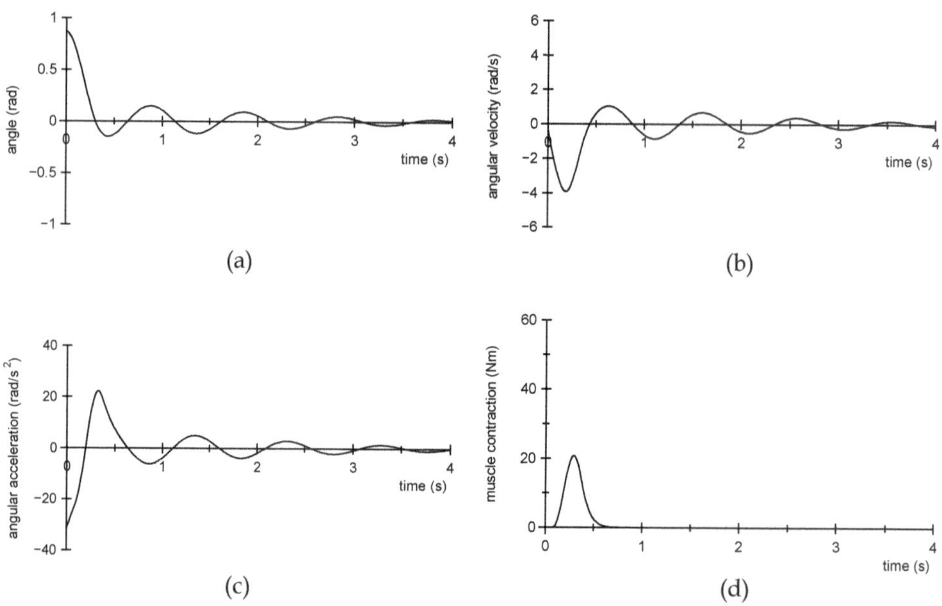

Fig. 20. Simulation result of a moderate spastic patient. (a) Angle; (b) Angular velocity; (c) Angular acceleration; (d) Muscle contraction.

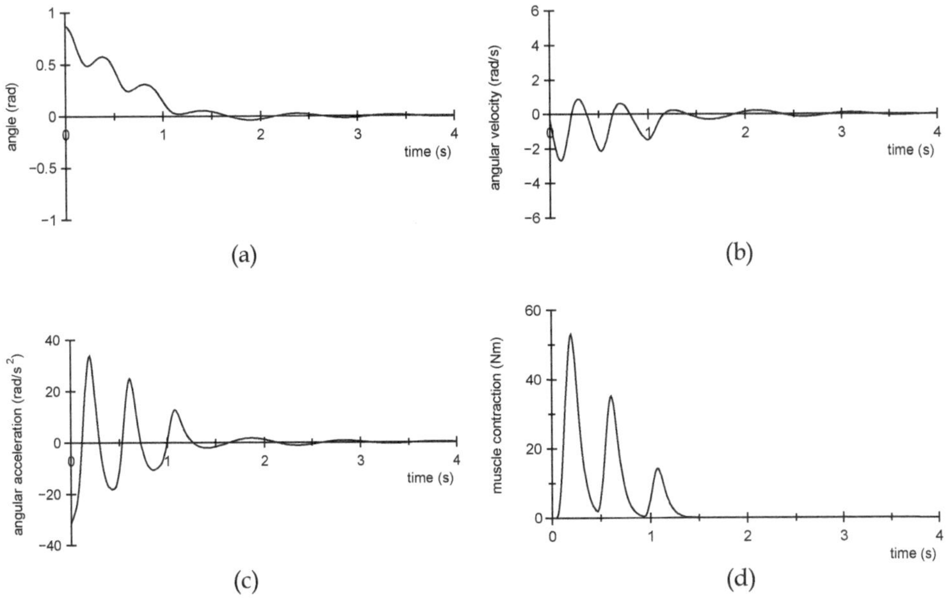

Fig. 21. Simulation result of a severe spastic patient. (a) Angle; (b) Angular velocity; (c) Angular acceleration; (d) Muscle contraction.

4.3.2 High-level analysis of the stretch reflex by inverse simulation

Determination of the input values for the simulator, so that the waveforms obtained with the pendulum test simulator agree as closely as possible with the waveforms for actual knee joint motion measurements, is called inverse simulation. If inverse simulation is conducted for knee joint motion measured with the pendulum test, the waveforms generated in the simulator, as already mentioned, are nearly the same as the measured waveforms for that subject. The constant and command frequency values at this time are values that characterize the individual subject. Therefore, if simulation is conducted based on these constant and command frequency values and waveforms and information for each part of the reflex arc are analyzed, a detailed understanding of the enhancement of the reflex for that subject can be gained.

The results of inverse simulation for one patient with spasticity are shown in Fig. 22. The solid line shows the result of actual measurement, and the broken line shows the result of simulation. There is extremely close agreement between the two results. Considering the sufficient accuracy of the knee joint motion measurement system and this kind of good agreement between the two with this method, the simulator is also assumed to have sufficient accuracy. At the current stage, however, some problems remain in aspects such as the accuracy of constant and command frequency values obtained from the inverse simulation and the time required to implement inverse simulation.

If these problems are solved, it is expected that the following issues can be resolved with application of inverse simulation.

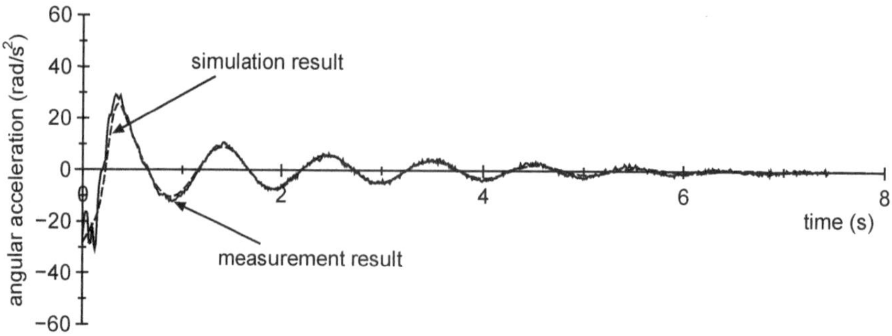

Fig. 22. Result of inverse simulation of a spastic patient.

1. Regular estimation of the status of the increase in spasticity in specific patients with $F_{\gamma d}$ and F_i values.
2. Reduction of individual differences and quantitative assessments of spasticity based on a uniform scale by standardizing inputs values of the pendulum test model for all subjects, under the premise that subject's body type and geometric structure of internal tissue are considered to be similar.
3. Verification of the effect of antispasticity drugs using this system.

5. Conclusions

This article thoroughly discusses a new knee joint motion measurement system constructed using two linear accelerometers, from the basic stretch reflex to the analysis of measurement results, focused on achievements obtained up to this time.

In section 2, we first explained the mechanical structure of knee joint flexion and extension as background knowledge to understand the discussion in section 3 and subsequent subsection, briefly touching also on the movement of the knee joint axis. Next, we looked at the source of the generation of the stretch reflex, which is the subject of measurement of the knee joint motion measurement system. We also showed the phasic and tonic reflex loops centered on muscle spindle function.

In section 3, we summarized the principles and results of performance evaluation of the knee joint motion measurement system. First, we looked comprehensively at accelerometers, which are the best option among sensors that can be used to measure knee joint motion, and then discussed the principles of the knee joint motion measurement system using two linear accelerometers. Next, we showed that the detection error of this device for simple pendulum motion and the pendulum motion of subjects is about the same as with high-accuracy, rotation angle gauges.

In section 4, we showed that the angle and angular acceleration of the knee joint could be simultaneously synchronized and measured in patients showing spasticity with enhanced phasic reflexes, and that the timing at which reflexes are produced could be easily estimated. Next, we showed the principles of a simulator to analyze measured waveforms and examples of analysis using this simulator, together with an additional statement on the outlook for high-level analysis of reflexes in patients with spasticity.

As a general conclusion, we demonstrated that the developed knee joint motion measurement system does not restrict subjects' movement as other systems do, and that it has many features that other systems do not, such as simple system configuration and the ability to acquire large amounts of information with simple data processing.

Future issues are the accumulation of clinical data using the features of this knee joint motion measurement system and quantification estimates of abnormal stretch reflexes such as spasticity, rigidity, and rigidospasticity based on those data.

6. Acknowledgment

We sincerely thank the late honorary professor Ken Akashi (Kawasaki Medical School and Kawasaki University of Medical Welfare) for his invariable support to this project.

7. References

Fong, C.M., Blackbum, J.T., Norcross, M.F., McGrath, M., & Padua, D.A. (2011). Ankle-dorsiflexion range of motion and biomechanics. *J. Athl. Train.*, Vol.46, No.1, pp.5-10

Fowler, E.G., Nwigwe, A.I., & Ho, T.W. (2000). Sensitivity of the pendulum test for assessing spasticity in persons with cerebral palsy. *Developmental Medicine & Child Neurology*, Vol.42, pp.182-189

Furuse, N., Watanabe, T., & Hoshimiya, N. (2005). Simplified measurement method for lower limb joint angles using piezoelectric gyroscopes. *Transactions of the Japanese Society for Medical and Biological Engineering*, Vol.43, No.4, pp.538-543

Harvey, R.J., & Matthews, P.B.C. (1961). The response of de-efferented muscle spindle endings in the cat's soleus to slow extension of the muscle. *J. Physiol.*, Vol.157, pp.370-392

Jikuya, K., Yokohira, T., Okamoto, T., Tsubahara, A., & Akashi, K. (2001). Quantitative analysis of spasticity by pendulum test - Improvement in measurement system. *Proceedings of the 40th conference the Japan Society of Medical Electronics & Biological Engineering*, p.292, Nagoya, Japan, May, 2001

Jikuya, K., Okamoto, T., Yokohira, T., & Akashi, K. (1995). Improvement in accuracy of pendulum test model for spastic patients. (Japanese), *Transactions of Institute of Electronics, Information and Communication Engineers D-II*, Vol.J78-D-II, No.4, pp.650-660

Jikuya, K., Okamoto, T., Yokohira, T., & Akashi, K. (1991). Mechanism of knee joint motions in a pendulum test. (Japanese) *Transactions of Institute of Electronics, Information and Communication Engineers D-II*, Vol.J74-D-II, No.9, pp.1289-1300

Kaeser, H.E. et al. (1998). Testing an antispasticity drug (Tetrazepam) with the pendulum test: a monocentric pilot study. *L. Neuro. Rehabil.*, Vol.12, pp.169-177

Kapandji, I. A. (1970). *The physiology of joints*, Vol.II, Churchill Livingstone Inc., N.Y.

Kusuhara, T., Jikuya, K., Nakamura, T., Michinishi, H., Yamamoto, Y., & Okamoto, T. (2011). Detection method of knee-joint motion using linear accelerometers for pendulum test. (Japanese) *IEEJ Transactions on Electronics, Information and Systems*, Vol.131, No.9, pp.1564-1569

Lakie, M., Walsh, E.G., & Wright, G.W. (1984). Resonance at the wrist demonstrated by the use of a torque motor: An instrumental analysis of the muscle tone in man. *J. Physiol.*, Vol.353, pp.265-285

Lin, D.C., & Rymer, W.Z. (1991). A quantitative analysis of pendular motion of the lower leg in spastic human subjects. *IEEE Trans. Biomed. Eng.*, Vol.38, pp.906-918

Nicol, A.C. (1989). Measurement of joint motion. *Clin. Rehabil.*, Vol.3, pp.1-9

Nordmark, E., & Andersson, G. (2002). Watenberg pendulum test: objective quantification of muscle tone in children with spastic diplegia undergoing selective dorsal rhizotomy. *Developmental Medicine and Child Neurology*, Vol.44, pp.26-33

Stillman, B., & McMeeken, J. (1995). A video-based version of pendulum test: Technique and normal response. *Arch. Phys. Rehabil.*, Vol.76, pp.166-176

Tong, K., & Granat, M.H. (1999). A practical gait analysis system using gyroscopes. *Med. Eng. Phys.*, Vol.21, pp.87-94

Vodovnik, L., Bowman, B. R., & Bajd, T. (1984). Dynamics of spastic knee joint. *Med. & Biol. Eng. & Comput.*, Vol.22, No.1, pp.63-69

Watenberg, R. (1951). Pendulousness of the legs as a diagnostic test. *Neurology*, Vol.1, pp.18-24

William, D.W., Jr. (1998). Spinal organization of motor function. *Physiology 4th*, Robert, M.B. & Matthew, N. L., Mosby, Inc. U.S.A., pp.186-199

2

XAFS Measurement System in the Soft X-Ray Region for Various Sample Conditions and Multipurpose Measurements

Koji Nakanishi and Toshiaki Ohta
The SR Center, Ritsumeikan University, Japan

1. Introduction

An X-ray absorption fine structure (XAFS) spectroscopy is a powerful and useful technique to probe the local electronic structure and the local atomic structure around an absorbing atom in an unknown material [Stöhr, 1996, Ohta, 2002]. A highly bright X-ray source, synchrotron radiation (SR) is usually used for XAFS measurements to obtain reliable spectra, even for elements of very low content in a sample.

For visible/ultraviolet (UV) and infrared absorption spectroscopies, the transmission mode is generally used, where incident and transmitted photon intensities are monitored. This is also the most fundamental technique for XAFS measurements in the hard X-ray region. However, it is hard to apply it in the soft X-ray region because of very low transmission. Instead, other techniques equivalent to the transmission mode have been developed; total/partial electron yield (EY) and fluorescent yield (FY) modes. The former is a widely adopted mode in the soft X-ray region, where the yield of Auger electrons and/or secondary electrons is proportional to the X-ray absorption coefficient. Since the electron escape depth is very short, the EY mode is surface sensitive. The latter is useful for XAFS measurement of heavy elements of low concentration in the hard X-ray region, and it is also useful as a bulk sensitive method in the soft X-ray region, although the probability of radiative decay is much smaller than that of Auger decay. It is often the case that an appropriate mode is chosen for sample conditions.

We have developed a practical and useful XAFS measurement system in the soft X-ray region applicable for various sample conditions and multipurpose measurements. In this system, it is possible to measure not only solid samples (such as powder, grain, sheet and thin film samples) but also liquid and gel samples. It is also applicable to in-situ measurements of anaerobic samples. In addition, it provides us some information of depth profiles with combined use of the EY and FY modes.

2. XAFS measurement system in the soft X-ray region

The XAFS measurement system is an assembly of several components; a soft X-ray beamline, sample chambers, detection systems, and a sample transfer system. Details of each component are described follow.

2.1 Soft X-ray double-crystal monochromator beamline

For a beamline below 1000 eV, a grating monochromator is generally used, while a crystal monochromator is advantageous above 1000 eV, since several high quality crystals are available which have proper lattice spacings. For XAFS measurements in the higher-energy soft X-ray region (above 1000 eV), the double-crystal monochromator (DCM) beamline (BL-10) was constructed in the SR center, Ritsumeikan University in Japan [Iwasaki, et al., 1998, Handa, et al., 1999]. Then BL-10 has been developed and upgraded, and many measurements have been performed (see Fig. 1) [Nakanishi & Ohta, 2009]. It consists of a 5.1 μm thick Be foil, a Ni-coated Si toroidal mirror, a Golovchenko-type DCM [Golovchenko et al., 1981], an I_0 monitor made of either Cu or Al mesh, a high-vacuum (HV) sample chamber kept below 2×10^{-5} Pa, an atmospheric-pressure (AP) sample chamber, and some masks and slits (see Fig. 2 (a)). The Be foil and the toroidal mirror at the front line are cooled by water in order to reduce a heat load by direct irradiation of white X-rays. The Be foil functions to cut visible and vacuum-ultra-violet photons which cause a background of an XAFS spectrum. The SR beam with 6 mrad (horizontal) and 2 mrad (vertical) is deflected upward by 1.4 ° and focused at the sample position in the AP sample chamber about 9 m apart from the source point with the 1:1 geometry by the toroidal mirror. The beam shapes and sizes are in good agreement with those simulated by the ray trace analysis, as shown in Fig. 2 (b)-(d). The available photon energy covers a range from about 1000 to 4500 eV by choosing a pair of monochromatizing crystals, such as beryl(10-10), KTP(011), InSb(111), Ge(111), Si(111) and Si(220) whose 2d lattice spacings are 1.5965, 1.0954, 0.7481, 0.6532, 0.6270 and 0.3840 nm, respectively. The incident angle to the monochromatizing crystal, θ is read in high accuracy with an angle encoder. The photon energy is determined with the 2d lattice spacing of a monochromatizing crystal and the incident angle using Bragg's law. Several masks and slits were inserted to minimize stray lights.

2.2 Tandem-type high-vacuum and atmospheric-pressure sample chambers

It is challenging to obtain reliable spectra from highly reactive compounds. In the soft X-ray region, XAFS spectra are usually measured in vacuum because of the low transmittance of X-rays in air (see Fig. 3). A vacuum environment is also necessary to measure reliable spectra from highly hygroscopic samples. In contrast, some compounds change their structures in vacuum. A typical case is hydrated compounds, in which hydrated water molecules are easily desorbed in vacuum. In addition, liquid solutions or nano-particles suspended in liquid cannot be introduced in vacuum without a special cell. For such samples, XAFS measurements in AP are necessary.

Another compact AP sample chamber, made of an ICF70 six-way cross nipple, was installed at the downstream of the HV sample chamber (see Fig. 2, 4 (a)). Two chambers are separated by a thin Be window, which should be tolerable against 1 atm pressure difference and whose thickness should be as thin as possible to minimize the intensity loss. The beam size at the sample position in the HV sample chamber is about 2.5 mm (vertical) × 6 mm (horizontal) (not focused), while that in the AP chamber is about 2 mm (vertical) × 5 mm (horizontal) (focused). Thus, we chose the diameter of 10 mm and thickness of 15 μm, respectively (see Fig. 4 (b)). It has been working without any trouble for more than one year [Nakanishi et al., 2010].

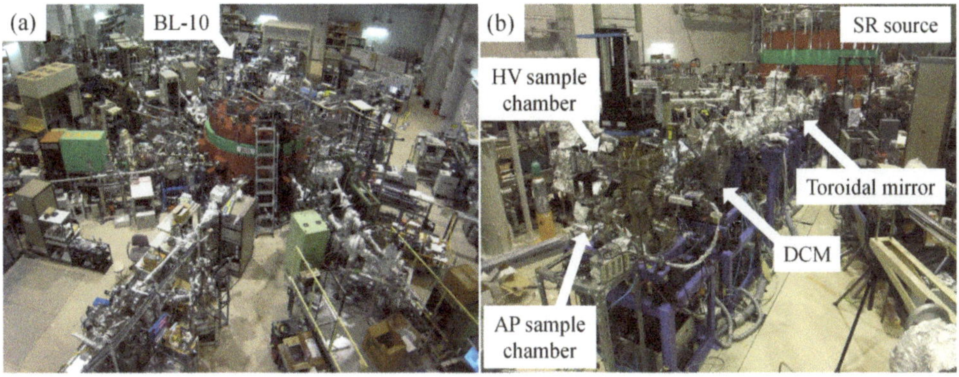

Fig. 1. (a) Bird-eye view of the SR center, Ritsumeikan University; (b) Photo of BL-10. Fourteen beamlines have been installed in the center; five beamlines for XAFS, three for X-ray lithography, three for photoelectron spectroscopy, and one for soft X-ray microscopy, infrared microscopy and X-ray reflectivity.

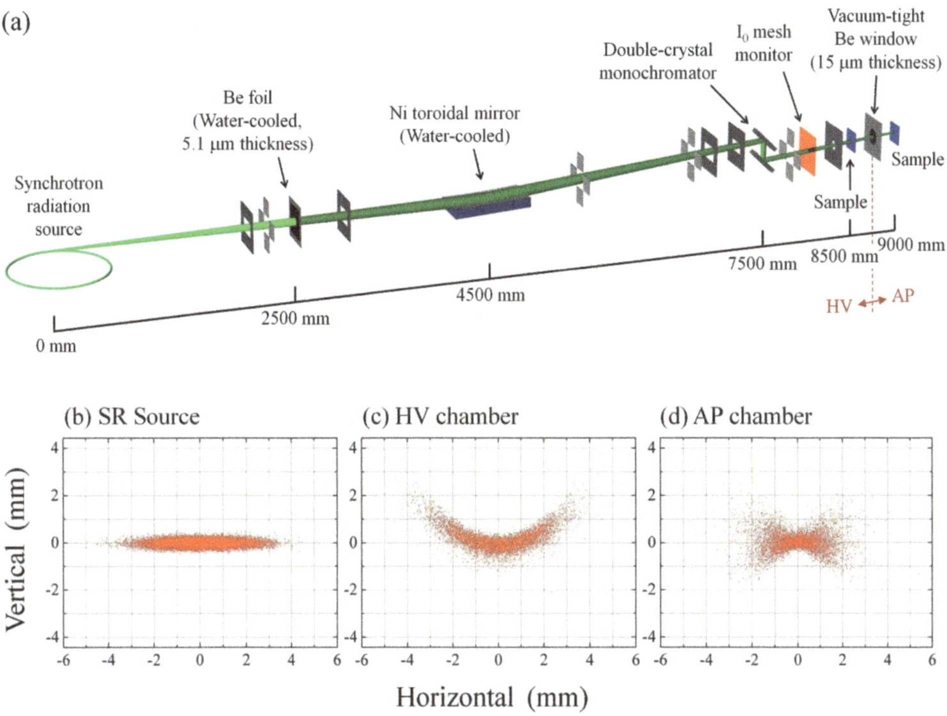

Fig. 2. (a) Schematic configuration of BL-10, and simulated X-ray beam profiles at (b) the source point; (c) the sample position of the HV chamber and (d) the sample position of AP chamber. The SR ray-tracing program "SHADOW" [Lai & Cerrina, 1986, Welnak et al., 1996] was used for the simulations.

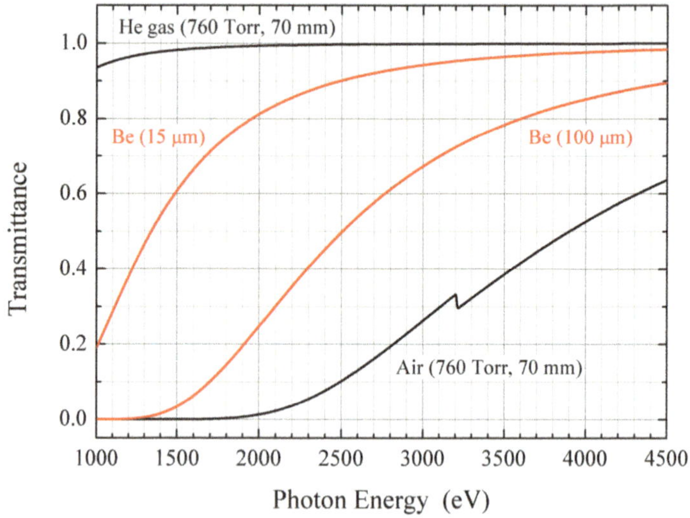

Fig. 3. Simulated transmittances of Air and He gas at 70 mm, which is the distance from the vacuum-tight window to the sample position in the AP sample chamber, and Be of 15 and 100 μm thickness. These simulations were used "X-ray Interaction with Matter Calculator", the Center for X-ray Optics, Lawrence Berkeley National Laboratory, USA. Available from "http://henke.lbl.gov/optical_constants/".

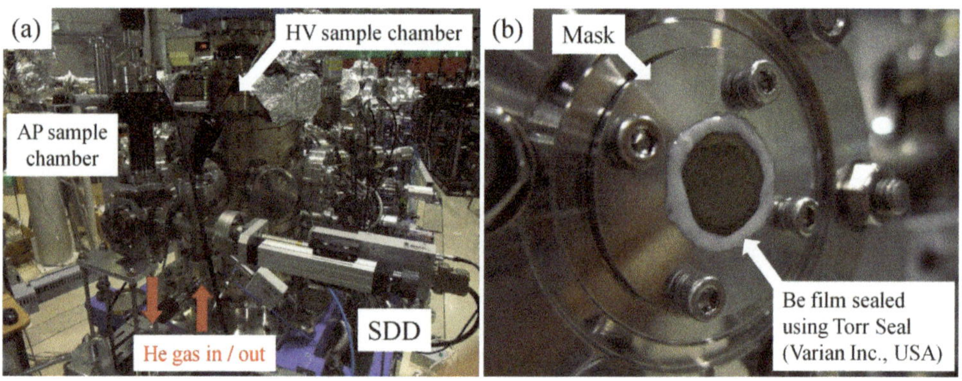

Fig. 4. Photos of the HV and AP sample chambers (a), and the vacuum-tight Be window (b).

Prior to the measurement, the AP chamber was filled with He gas to increase the transmittance of X-rays (see Fig. 3). It takes about 10 minutes to replace the air inside with He gas completely with the flow rate of 8.45×10^{-1} Pa•m^3/s (500 sccm). The He gas flow rate was kept constant, typically at 3.38×10^{-2} Pa•m^3/s (20 sccm) during measurements.

Now, the total EY (TEY) with specimen current, partial EY (PEY) using a PEY detector (as will hereinafter be described in detail), and partial FY (PFY) using a silicon drift detector (SDD) can be carried out in the HV sample chamber, while PFY can be carried out in the AP sample chamber.

For the performance test of BL-10, the photon flux was estimated from 1000 to 4500 eV using a Si PN photodiode (AXUV-SP2, International Radiation Detectors Inc., USA). In general, the photon flux Φ (photons/s) is given by the following formulae,

$$\Phi = \frac{i_p}{\eta e}. \quad (1)$$

where i_p (A) is the short-circuit current of a photodiode. η is a quantum efficiency and e is a charge (A·s) [Saleh & Teich, 1991].
η is in the range ($0 \leq \eta \leq 1$), and is approximately proportional to the photon energy $h\nu$,

$$R = \frac{\eta e}{h\nu} \quad (2)$$

where R is called as 'resposibility', whose typical values are available from the WEB site of International Radiation Detectors Inc. (http://www.ird-inc.com/). The equation (3) is changed by,

$$\eta e = R h \nu \quad (3)$$

where ηe is called a device quantum yield. Substituting this equation (3) into the equation (1), we obtain the following,

$$\Phi = \frac{i_p}{R h \nu} \quad (4)$$

Fig. 5 shows the photon fluxes at the sample positions in the HV and AP sample chambers. All photon fluxes are normalized by the SR ring currents (200 mA). The difference of photon fluxes between in HV and AP is larger in the lower photon energy because of the transmittance for the Be window (see Fig. 3). However, the difference for the KTP crystal was relatively smaller than predicted that. This might be due to the radiation damage for the crystal, since the measurement was performed first in AP and later in HV sample chambers.

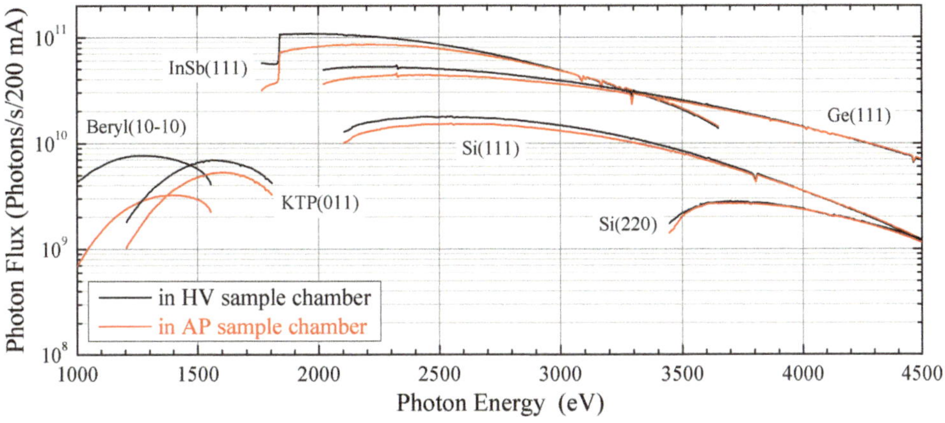

Fig. 5. Photon fluxes at the sample position in HV (black line) and AP (red line) sample chambers using each monochromatizing crystal.

To demonstrate the availability of the HV and AP chambers, XAFS measurements were carried out for anhydrous $MgCl_2$ and $MgCl_2 \cdot 6H_2O$. The crystal structures of $MgCl_2$ and $MgCl_2 \cdot 6H_2O$ are schematically shown in fig. 6. It is known that both samples are highly deliquescent. Since the local structures around Mg and Cl atoms are different from each other, it is expected different XAFS spectra are observed between $MgCl_2$ and $MgCl_2 \cdot 6H_2O$.

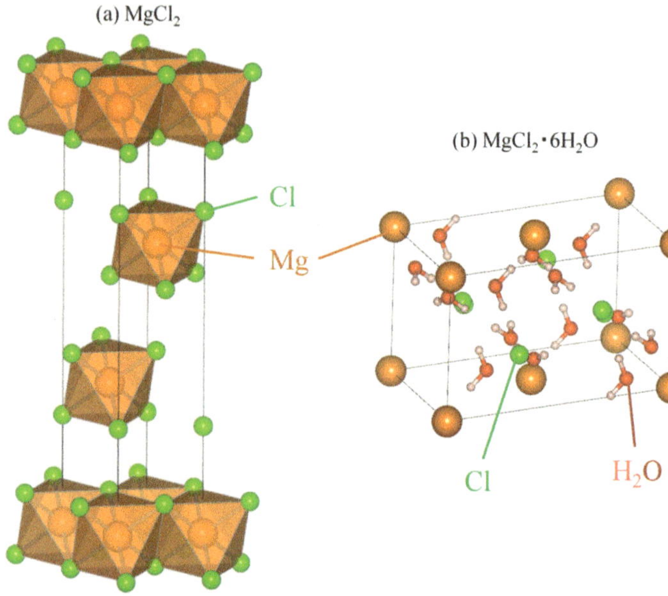

Fig. 6. Crystal structures of $MgCl_2$ (a) and $MgCl_2 \cdot 6H_2O$ (b) drawn by VESTA program [Momma & Izumi, 2008]. Each crystal information is referred from [Wyckoff, 1963] and [Agron et al., 1969].

Observed Mg and Cl K-edge X-ray absorption near edge structure (K-XANES) spectra which were measured both in HV and AP sample chambers are shown in Fig. 7. They are compared with theoretical XANES spectra simulated with the FEFF-8.4 program based on the real-space full multiple-scattering theory [Rehr et al., 2000]. Note that the white lines of PFY spectra in Fig. 7 are heavily suppressed compared with those of TEY spectra. This is due to the self-absorption effect in the PFY spectrum. In Mg K-XANES spectra of $MgCl_2$ (fig. 7 (a)), we can observe characteristic peaks; a white line at 1309.5 eV and a shoulder at 1311.5 eV. The spectral profile of TEY (HV) is well reproduced by the FEFF simulation. Those of PFY (HV), TEY (HV) are similar to that of PFY (AP), though the shoulder at 1311.5 eV is more enhanced in the spectrum of PFY(AP). In contrast, Mg K-XANES spectrum from $MgCl_2 \cdot 6H_2O$ in AP is distinctly different from those in HV, as shown in fig. 7 (b). The simulated spectrum is very similar to that of PFY (AP). This clearly indicates that the sample in vacuum is not $MgCl_2 \cdot 6H_2O$ anymore, but changed to anhydrous $MgCl_2$, desorbing crystalline water molecules. In fact, the spectra from $MgCl_2 \cdot 6H_2O$ in HV (both PFY and TEY) are close to those of $MgCl_2$ in fig. 7 (a). Careful examination revealed that the spectrum from $MgCl_2 \cdot 6H_2O$ in HV (PFY) can be interpreted as a superposition of the spectra from

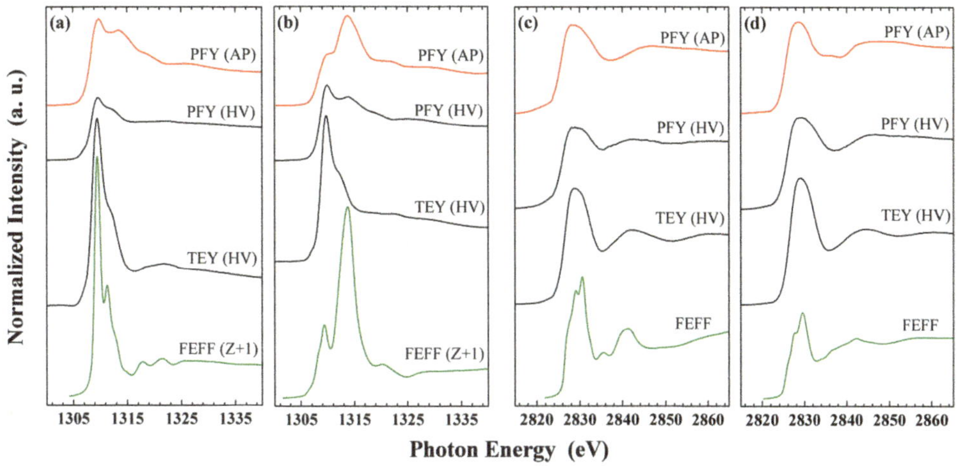

Fig. 7. Mg K-XANES spectra of $MgCl_2$ (a) and $MgCl_2 \cdot 6H_2O$ (b), and Cl K-XANES spectra of $MgCl_2$ (c) and $MgCl_2 \cdot 6H_2O$ (d) [Nakanishi et al., 2010]. Theoretical XANES spectrum with FEFF-8.4 is also shown in the bottom of each figure, where the Z+1 approach was used for (a) and (b) [Nakanishi & Ohta, 2009].

$MgCl_2$ and $MgCl_2 \cdot 6H_2O$. A part of $MgCl_2 \cdot 6H_2O$ might remain in bulk and be detected with the bulk-sensitive FY method. In fact, the spectrum was quickly measured in HV after evacuation.

As described above, an additional peak appears at 1313.5 eV in the spectrum from $MgCl_2$ of PFY (AP) in fig. 7 (a). This feature can be interpreted as the contribution from $MgCl_2 \cdot 6H_2O$, since highly deliquescent $MgCl_2$ adsorbed water during the sample preparation in air. This 'surface layer' is estimated to be about several μm orders. On the other hand, in the Cl K-XANES spectra from $MgCl_2$ and $MgCl_2 \cdot 6H_2O$, no significant difference was observed between the spectra measured in AP and HV, as shown in Fig. 7 (c), (d). This shows that hydrations occur exclusively around the Mg ions.

These results clearly demonstrate the necessity and the importance of XAFS measurements both in AP and HV in the soft X-ray region to obtain reliable spectra from hygroscopic and hydrated compounds.

2.3 Multiple Soft X-ray XAFS measurement system

The local structure and chemical composition of a sample surface are sometimes different from those of the bulk. Such differences play a critical role in functions of materials, such as catalytic activities, electronic properties of semiconductors, etc. Thus, there is a strong demand of depth-profiling techniques. As described in Introduction, the EY mode is generally used in the soft X-ray region, which is surface sensitive. The FY mode is sometimes used to probe bulk structures, although the intensity of the FY mode is one or two orders of magnitude lower than that of the EY mode. Combined use of these modes is known to be a powerful XAFS method for the depth profiling, as published so far [Yoon, et al., 2004].

There are some types in the EY mode. The TEY mode can be performed by only monitoring a specimen current without selecting electron energies using any detector or analyzer. Therefore it is adopted in many soft X-ray XAFS. The EY mode is surface sensitive compared with the FY mode, but especially in the higher-energy soft X-ray region, the sampling depth is not so small as expected [Frazer et al., 2003, Kasrai et al., 1996]. Others are needs to select electron energies with an electron detector or analyzer. The Auger electron yield (AEY) mode is the most surface sensitive and a high signal to background (S/B) ratio by collecting only Auger electrons using an electron analyzer [Gao et al., 2009]. However it is difficult sometimes to obtain a high signal to noise (S/N) ratio spectrum because of the low signal of detectable electrons for low concentration elements. The partial electron yield (PEY) mode collects only high energy electrons by filtering out secondary electrons and Auger electrons from other lower-energy absorption edges. The PEY mode is more surface-sensitive than the TEY mode, which is dominantly contributed by secondary electrons. A typical electron detector for the PEY mode is composed of two metal mesh grids for ground and retarding voltages, a chevron microchannel plates (MCPs) assembly with double MCPs and a metal collector [Stöhr, 1996]. In this detector, a suitable retarding voltage excludes low-energy electrons emerged from a deep bulk and extra signals from other lower-energy absorption edges. Hence, it enables us to obtain a spectrum with higher surface sensitivity and higher signal to background (S/B) ratio. The PEY method provides us with high quality spectra, and has been used in many XAFS studies, especially in the lower-energy soft X-ray region [Sako et al., 2005]. However, it should be noted that soft X-ray XAFS spectra by the PEY mode may sometimes cause a serious problem. It is well known that MCPs can detect not only electrons but also X-rays [Wiza, 1979]. When one uses the conventional MCP detector as an electron detector in soft X-ray XAFS experiments, one would get a spectrum deformed by unexpected inclusion of fluorescent X-rays.

In the lower-energy soft X-ray region below 1000 eV, the influence of fluorescent X-rays is very small and negligible, but it is not negligible in the higher-energy soft X-ray region, since the radiative core hole decay channel, i.e. fluorescent X-ray emission starts to open though the Auger decay process is still dominant [Krause, 1979]. Thus, we should be careful whether the PEY mode with an MCP detector provides reliable spectra or not.

A typical example is shown in Fig. 8 (a), which exhibits Si K-XANES spectra of a commercial thermally oxidized Si (SiO_2/Si) wafer with a 100 nm oxide overlayer (100 nm-SiO_2/Si wafer) taken with a conventional MCP detector. Observed TEY spectrum with specimen current is also shown for comparison, which gives the typical spectrum of SiO_2. This result is reasonable, because the sampling depth of SiO_2/Si wafer by the TEY mode with specimen current is estimated to be about 70 nm, as reported by Kasrai et al. [Kasrai et al., 1996] and also confirmed by our experiments shown in fig. 11 (b). On the other hand, the XANES spectrum recorded by the conventional MCP detector at the retarding voltage of 0 V shows not only the feature of SiO_2 but also a weak feature of bulk Si at 1840 eV. In addition, by increasing the retarding voltage, the feature of bulk Si is more enhanced. In other words, increasing the retarding voltage appears to give bulk sensitive spectra. This result contradicts with a general tendency of the relation between a retarding voltage and sampling depth by the PEY method.

In order to explain this incongruous phenomenon, we show unnormalized spectra in fig. 8 (b). It shows how the retarding voltages change the spectra. As the retarding voltage

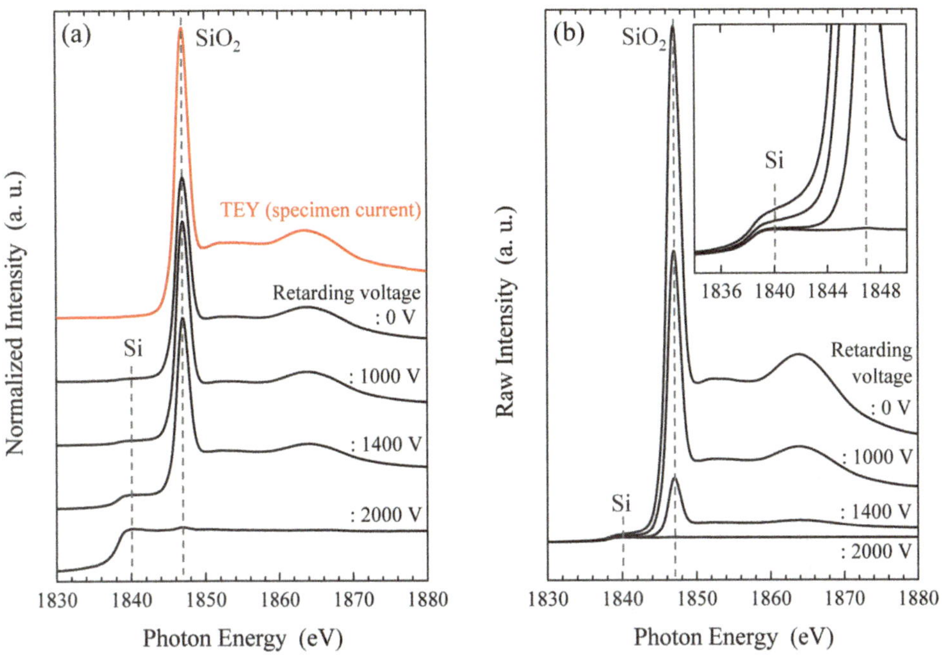

Fig. 8. Si K-XANES spectra of 100 nm-SiO_2/Si wafer obtained with the conventional MCP detector as a function of the retarding voltage. (a) Normalized spectra. The TEY spectrum with specimen current is also shown for comparison; (b) Unnormalized spectra.

increases, the peak intensity associated with SiO_2 decreases drastically, but the peak intensity at 1840 eV does not change at all (see inset of Fig. 8 (b)), even when the high retarding voltage (2000 V) is applied enough to eliminate all emitted electrons from the sample. It turns out that the origin of the peak at 1840 eV is fluorescent X-rays from bulk Si. This indicates that the XANES spectra by the conventional MCP detector at the retarding voltages of 0, 1000, and 1400 V are mixtures of both the PEY and total FY (TFY) spectra. By increasing the retarding voltage, number of electrons decreases, but that of fluorescent X-rays does not change. As the result, the TFY signal is relatively enhanced and the MCP detection gives a bulk sensitive spectrum apparently when the high retarding voltage is applied. In other words, the PEY spectrum is deformed by inclusion of unexpected TFY spectrum. Thus, in the soft X-ray region, one should use the conventional MCP detector as a PEY detector carefully, especially in the case that fluorescent X-rays from the bulk are not negligible.

Above results suggest that it is necessary to make the influence of fluorescent X-rays as small as possible for the PEY mode. Bearing it in mind, we designed and fabricated a new PEY detector using a MCP assembly. The schematics and photograph are shown in Fig. 9. There are two major modifications from the conventional MCP detector. The first is to bend the trajectory of emitted electrons from a sample so as to collect only electrons in the MCPs.

For this purpose, three cylindrical austenitic stainless steel (ASS) grids, whose the transmission rate is about 77.8 %, were used. The bending voltage of 3000 V was applied between

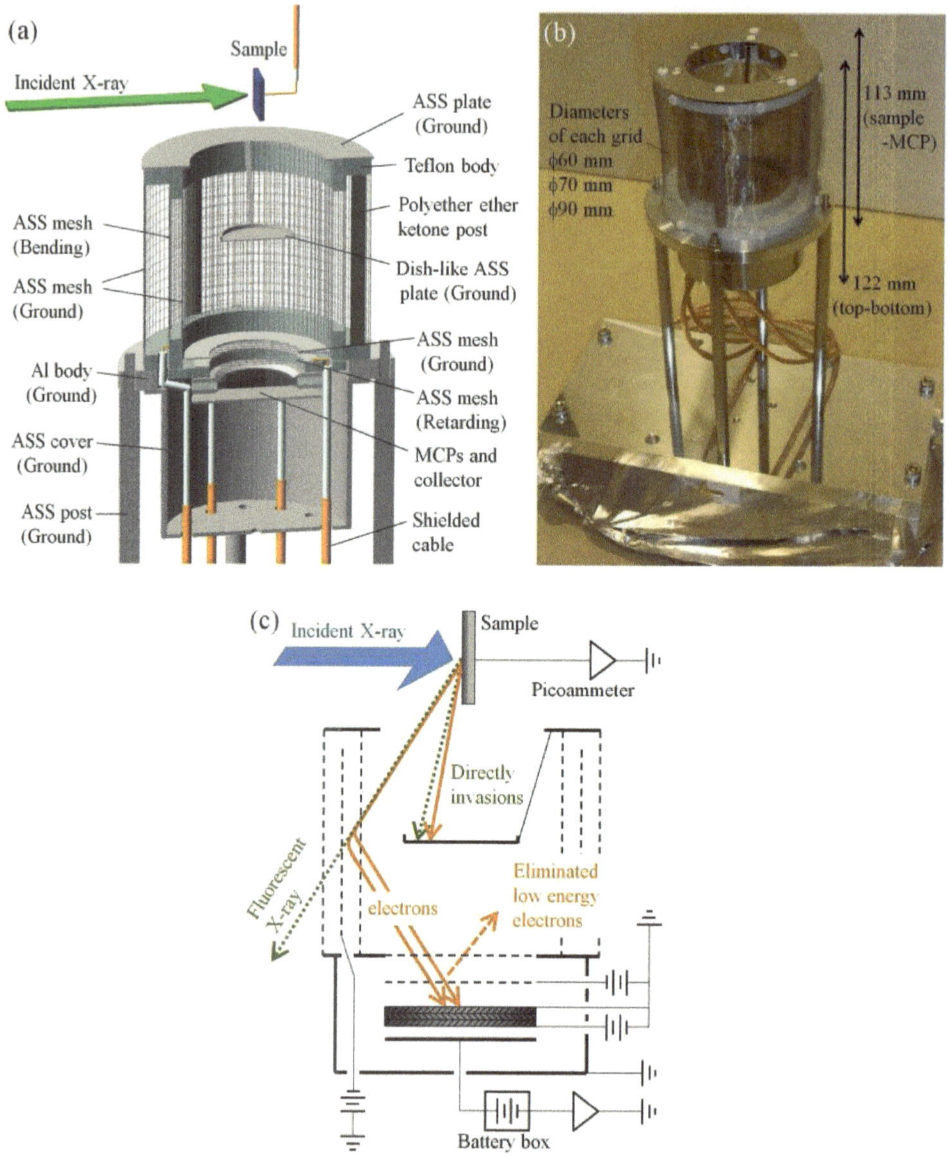

Fig. 9. Newly-developed PEY detector for the soft X-ray region [Nakanihshi & Ohta, In press]. (a) The experimental layout. The cross section of the detector is shown for the details; (b) Photograph; (c) The operating principle for the PEY measurement.

the inner and the intermediate grids, and the outer grid works to prevent from a leak voltage. The second is to place the dish-like ASS plate between a sample and MCPs so as to avoid direct invasions of fluorescent X-rays (and electrons) into MCPs. These two modifications make the PEY measurements work effectively in the higher-energy soft X-ray region as same as the conventional MCP detector in the lower-energy soft X-ray region. For the detector, a Z-stack MCP assembly, whose MCPs have the diameter of 25 mm and the aspect ratio of 60:1 (Long-Life MCPs, Photonis USA Inc., USA) was used.

To examine the performance of the detector, we prepared a Si wafer etched in 1.0 % aqueous solution of HF (HF-Si wafer) and several SiO_2/Si wafers with different oxide overlayer thickness by heating at 950 K in an electric furnace. The oxide overlayer thickness of each sample was controlled by the heating time and estimated by ellipsometric measurements (λ = 633 nm), in which the refractive index was fixed to 1.457 [Malitson, 1965]. The photon energy was calibrated by setting the first peak of the first derivative in the Si K-edge XAFS (K-XAFS) spectrum of a Si wafer to 1839 eV [Nakanishi, 2009]. PFY spectra were using the SDD with the selected energy window for Si-Kα X-rays.

Fig. 10 (a) shows observed Si K-edge PEY spectra of 25.3 nm-SiO_2/Si wafer as a function of the retarding voltage, together with the TEY and PFY spectra for comparison. Compared with the TEY spectrum, the PEY spectrum at the retarding voltage of 0 V is slightly more surface sensitive. This is because the detector collects electrons emitted in the oblique angle, as shown in Fig. 9 (c).

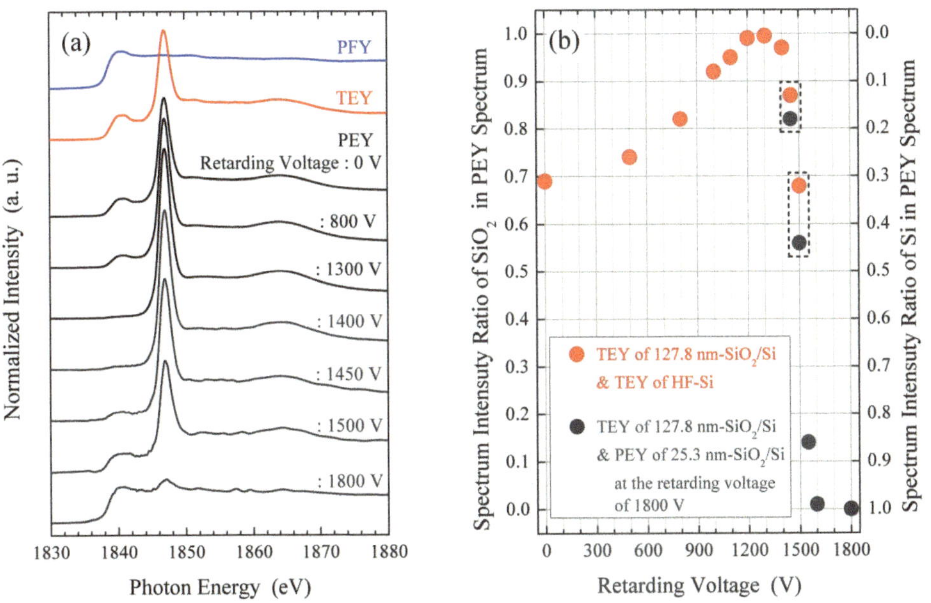

Fig. 10. (a) Observed Si K-XANES spectra of 25.3 nm-SiO_2/Si using the PEY detector at several retarding voltages. The PFY spectrum using the SDD, and TEY spectrum with specimen current are also shown for comparison. The self-absorption effect of the PFY spectrum is not corrected; (b) The distribution of SiO_2 and Si intensity ratio in PEY spectra of 25.3 nm-SiO_2/Si wafer as a function of the retarding voltage.

Fig. 10 (b) shows how surface SiO_2 and bulk Si contribute to each PEY spectrum as a function of the retarding voltage. Spectrum intensity ratios in each PEY spectrum are analyzed as the superposition of these two spectra: SiO_2 and bulk Si. Here, we adopt the TEY spectrum of 127.8 nm-SiO_2/Si wafer to stand for SiO_2, and the TEY spectrum of HF-Si wafer to stand for bulk Si. Note the PEY spectrum of 25.3 nm-SiO_2/Si at the retarding voltage of 1800 V is also used as the spectrum of bulk Si for the analysis, when the retarding voltage of PEY spectra is higher than 1400 V. As the retarding voltage increases from 0 to 1300 V, the SiO_2 intensity increases and the Si intensity decreases. This is the reasonable tendency of the PEY measurements unlike that of the conventional MCP detector in the previous section. However, as the retarding voltage increases further from 1300 to 1800 V, the SiO_2 intensity decreases dramatically and the bulk Si intensity increases, relatively. At the retarding voltage of 1800 V, we could obtain a PEY spectrum which was close to that of bulk Si, even though Si KLL Auger electrons could not reach the MCP at the voltage.

About the above phenomenon, we think the following reason. A part of fluorescent X-rays or electrons emitted from the sample were scattered on the surface of parts of the detector, then they are entered the MCP on unexpected trajectories. In addition, A part of fluorescent X-rays and electrons excited atoms in parts of the detector as a probe, then generated newly fluorescent X-rays and electrons also entered the MCP on unexpected trajectories. Here, most of newly-generated electrons were excluded by the retarding voltage, but electrons emitted from the intermediate grid for bending electrons of the detector could reach the MCP because they were accelerated by the voltage between the intermediate grid and the inner grid. These effects could be neglected when the retarding voltage was below 1300 V, since the intensities of intrinsic electrons are dominant. Above 1300 V, These effects could not be neglected, since the intensity of intrinsic electrons suddenly dropped. However, the MCP gain had to be enhanced by increasing the MCP voltage from 2400 V to 3200 V in order to get spectra when the retarding voltage was above 1300 V. These indicate that retarding voltages above 1300 V are not suitable for the PEY detection at Si K-XAFS measurements using this detector.

It was determined to be 1300 V for Si K-edge from the above results. Here, the S/B ratio was confirmed for the PEY spectrum with the optimum retarding voltage. At the photon energy of 2000 eV, the S/B ratio of the PEY spectrum of 25.3 nm-SiO_2/Si wafer was 4.95. This was superior to that of the PEY spectrum without retarding voltage (3.40) and the TEY spectrum with specimen current (1.64).

Then, we estimated the sampling depth of the PEY detection. Fig. 11 (a) shows the PEY spectra of SiO_2/Si as a function of the oxide overlayer thickness at the retarding voltage of 1300 V. For comparison, the result of TEY spectra is also shown in fig. 11 (b). As the oxide overlayer thickness increases, the spectrum intensity ratio of SiO_2 increases and saturates in fig. 11 (c). From this spectrum intensity ratio profile, the sampling depth at Si K-edge of the PEY was estimated to be about 30 nm in the SiO_2/Si system. This is less than half of that of the TEY.

Combining the PEY detector (PEY) with the specimen current (TEY) and the SDD (PFY), we can get multiple information about sampling depths; surface, interface and bulk. It is greatly valuable and efficient to obtain the depth profile of an unknown sample. Fig. 12 shows the detection system and demonstrative Si K-XANES spectra of 25.3 nm-SiO_2/Si. The three spectra were observed with different spectral features depending on each sampling depth (shown in fig. 12 (b)). This simultaneous soft X-ray XAFS system is not only useful to obtain detailed information about a sample but also important for XAFS users with limited available beam time in a synchrotron radiation facility.

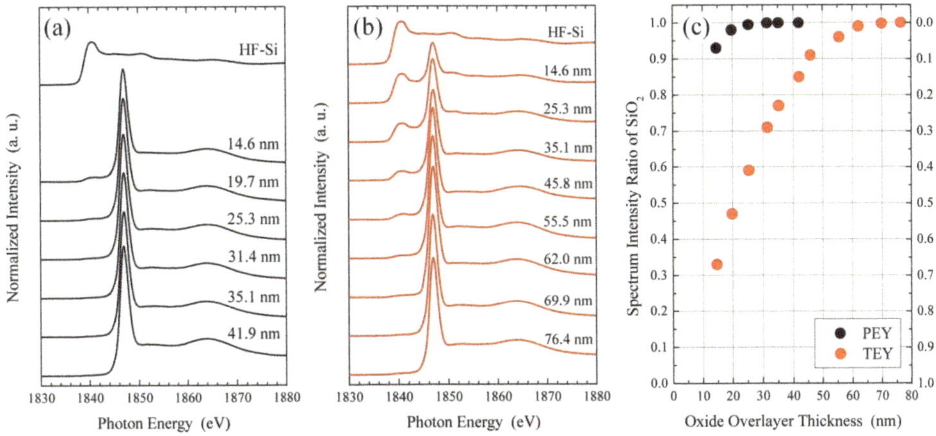

Fig. 11. Si K-XANES spectra of several SiO_2/Si samples with the PEY mode (a) and the TEY mode (b) as a function of the oxide overlayer thickness; (c) The distribution of SiO_2 and Si intensity ratio in PEY and TEY spectra of several SiO_2/Si samples as a function of the oxide overlayer thickness.

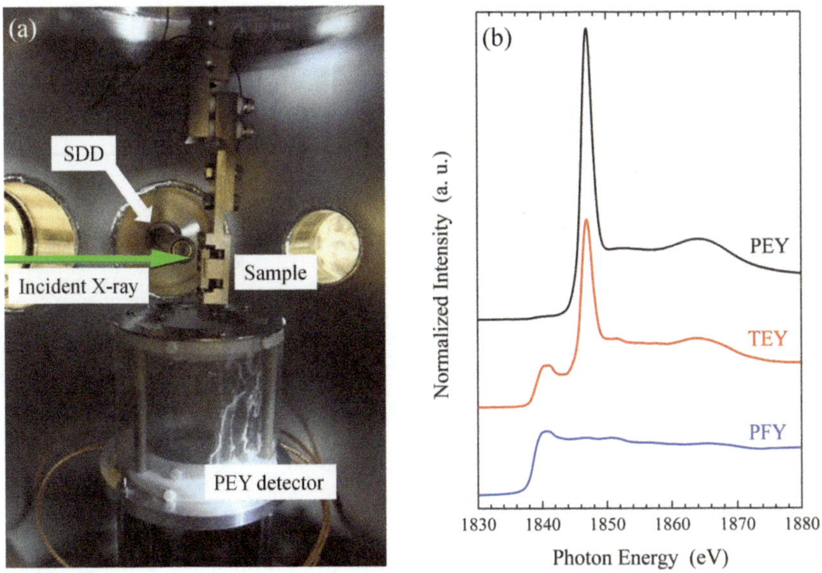

Fig. 12. Multiple soft X-ray XAFS measurement system. (a) Photograph of the experimental setup in the HV sample chamber; (b) Si K-XANES spectra of 25.3 nm-SiO_2/Si detected by the PEY detector at the retarding voltage of 1300 V (PEY), the specimen current (TEY), and the SDD (PFY). The self-absorption effect is not corrected in the PFY spectrum.

2.4 Transfer vessel system for anaerobic samples

We sometimes encounter a problem to measure anaerobic samples, such as highly deliquescent materials, e.g. $MgCl_2$ and $MgCl_2{\cdot}6H_2O$ as described in section 2.2 and highly hydrolytic materials, e.g. Li-ion battery (LIB) materials. One of the difficulties is how to carry such a sample from a laboratory to an SR facility and how to set it up in an XAFS equipment without exposing to air. An aluminum-laminated bag is often used to seal the sample with a high-purity Ar gas in a glove box. It is possible to measure the sample in the hard X-ray region, but is impossible in the soft X-ray region because of the low transmittance for an aluminum-laminated bag. In order to solve the problem, we developed a compact transfer vessel system (see fig. 13) [Nakanishi & Ohta, 2010].

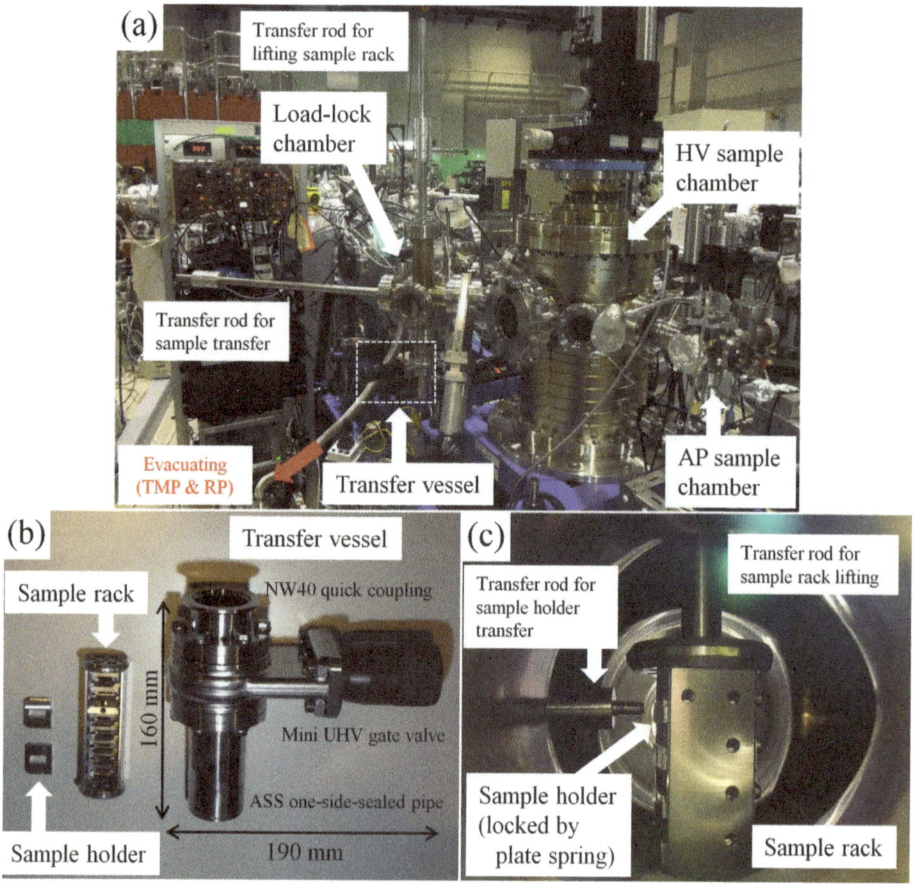

Fig. 13. Photos of developed transfer vessel system. (a) The load-lock chamber and the HV sample chamber; (b) the transfer vessel, sample rack and sample holders; (c) inside of the load-lock chamber.

It consists of an ICF70 UHV gate valve (Mini UHV gate valve, VAT Vacuumvalves AG, Switzerland), an ASS one-side-sealed pipe, an ICF70-NW40 flange. The vessel is compact enough to be handled in a commercial glove box. Normally two samples are mounted on the sample holder with carbon tapes. Eight sample holders can be set in the sample rack. After the sample rack with sample holders is loaded into the vessel, the UHV gate valve of the vessel is closed tightly together with a high-purity Ar gas in a glove box. Then the vessel is taken out from the glove box. The sealed vessel can be carried to the SR center and is connected to the load-lock chamber, as shown in Fig. 13 (a). After evacuating the chamber using a turbomolecular pump (TMP) and rotary pump (RP), the gate valve of the vessel is opened. The rack with sample holders is pulled out from the vessel and lifted to the transfer position in the load-rock chamber by the transfer rod. Each sample holder in the rack is transferred and loaded into the HV sample chamber for XAFS measurements. The HV sample chamber and load-lock chamber are kept in HV until all XAFS measurements are over.

For evaluation of the sealing capacity of the vessel, we monitored dew points in the vessel. The transfer vessel with a dew point temperature sensor probe (Moisture Target Series 5, GE Measurement & Control Solutions, USA) was prepared (see fig.14 (a)) and monitored dew point temperatures enclosed sixteen LIB electrode samples, $LiCoO_2$ powders with acetylene black and poly-vinylidene difluoride (PVDF) coated on Al films, with an Ar gas in a glove box (see fig. 14 (b)). The sealed vessel was ejected from the glove box as soon as closing the valve. It is shown that the dew point temperature increased gradually. After 24 hours, the dew point temperature is about -80 °C. This is the sufficient value to carry LIB samples or others prepared in a glove box without changing the condition.

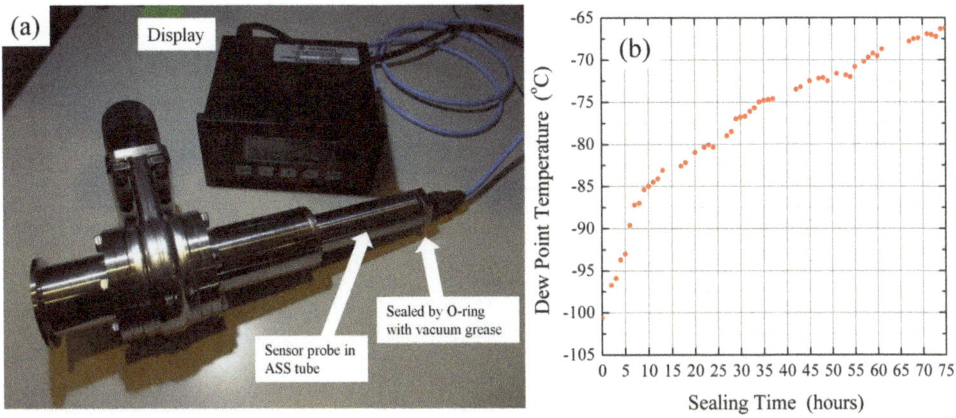

Fig. 14. (a) Photograph of the developed transfer vessel with a dew point temperature sensor probe; (b) Monitored dew point temperature plots in the sealed vessel with LIB samples.

XAFS measurements of $LiPF_6$ known as an LIB electrolyte material were demonstrated. Fig. 15 (b) shows P K-XANES spectra of $LiPF_6$ powders carried by the sealed vessel without air exposure and air exposure for 1 day by the TEY and PFY modes. For comparison, the simulated spectrum of $LiPF_6$, the experimental spectra of Li_3PO_4 powder and H_3PO_4 solution are also shown in fig. 15 (b). Experimental spectra of $LiPF_6$ without air exposure are good agreement between the TEY and PFY spectrum, and are also reproduced by simulated spectrum. However, the small difference at the energy position of the pre-edge peak is confirmed between the TEY and PFY spectrum. Meanwhile both experimental spectra of $LiPF_6$ with air exposure have similar features to Li_3PO_4 and H_3PO_4, such as the white line at 2152.9 eV (black dashed line) and the broad peak around 2170 eV. This indicates most local structures of P atoms in $LiPF_6$ changed the octahedral coordination with F atoms shown in fig. 15 (a) into the tetrahedral coordination with O atoms (i.e. phosphates). Hence we identified surface P atoms in $LiPF_6$ without air exposure has been also changed into phosphates, because the energy position of the pre-edge peak of the TEY spectrum closes to that of the white line of phosphates. We think that only the surface of $LiPF_6$ samples has been changed by negligible moisture in a glove box over the course of several weeks. About $LiPF_6$ with air exposure, both the TEY and PFY spectrum shape are very similar to that of Li_3PO_4, but the shoulder peak at 2154.5 eV (red dashed line) can be seen only spectra of $LiPF_6$ with air exposure. This origin is not clear yet, but may originate from POF_3, $POF_2(OH)$ or other materials generated by hydrolyzed $LiPF_6$ [Kawamura et al., 2006].

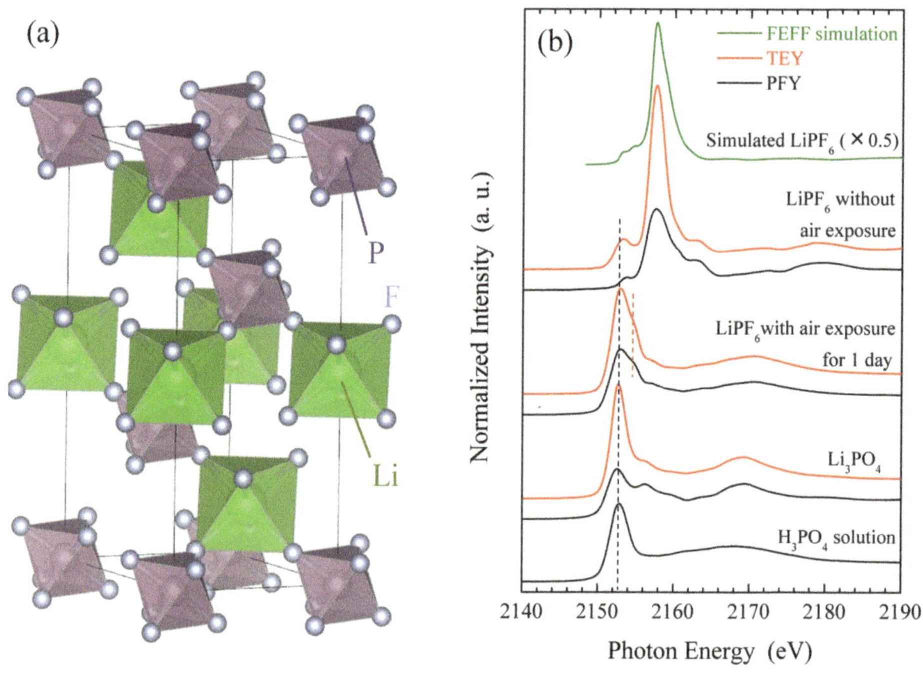

Fig. 15. (a) Crystal structure of $LiPF_6$ drawn by VESTA program [Momma & Izumi, 2008]. The crystal information is referred by [Röhr & Kniep, 1994]; (b) Simulated spectra of $LiPF_6$ by FEFF program and experimental P K-XANES spectra of $LiPF_6$ with and without air exposure. TEY spectra of Li_3PO_4 and H_3PO_4 are also shown as a reference sample. The self-absorption effect is not corrected in PFY spectra.

From these results, it is clearly indicated the efficiency of the vessel to carry the highly hydrolytic materials without changing the chemical condition. This transfer vessel system is using in many case, and especially in LIB materials, the system is absolutely essential.

3. Conclusion

We developed the unique and efficient soft X-ray XAFS measurement system at BL-10 of the SR center, Ritsumeikan University. As well as being able to accept stable solid samples in the HV sample chamber, hygroscopic and hydrated compounds, which are changed the condition in vacuum, and liquid samples are also able to accept in the AP chamber with the vacuum-tight window of a thin Be foil and in He gas atmosphere. In the HV chamber, a multiple measurement combined with the TEY, PEY, and PFY methods can give us the depth profile of samples. This is very useful for chemical analysis of practical samples. In addition, the transfer vessel system is an efficient tool for carrying anaerobic samples without changing chemical conditions. These setups will satisfy with many demands for various sample conditions. We hope that this XAFS system will be used by many users and stimulate further soft X-ray XAFS studies.

4. Acknowledgment

The authors acknowledge Prof. Shinya Yagi (Nagoya University, Japan) for his invaluable advices in development of the atmospheric-pressure sample chamber, Dr. Yasuhiro Abe (Ritsumeikan University, Japan) for his technical support in ellipsometric measurements, Dr. Masatsugu Oishi (Kyoto University, Japan), Mr. Takahiro Kakei (Kyoto University, Japan), Dr. Tomonari Takeuchi (National Institute of Advanced Industrial Science and Technology, Japan) and Dr. Hiroyuki Kageyama (National Institute of Advanced Industrial Science and Technology, Japan) for their assistances in providing samples to evaluate the transfer vessel.

These works were partially supported by Nanotechnology Network (Nanonet) Project from Ministry of Education, Culture, Sports, Science and Technology (MEXT) in Japan, and by Research and Development Initiative for Scientific Innovation of New Generation Batteries (RISING) project from New Energy and Industrial Technology Development Organization (NEDO) in Japan.

5. References

Agron, P. A. & Busing, W. R. (1969). A neutron diffraction study of magnesium chloride hexahydrate, *Acta Cryst. A*, Vol. 25, pp. 118-119.

Frazer, B. H.; Gilbert, B.; Sonderegger, B. R. & G. D. Stasio, (2003), The probing depth of total electron yield in the sub-keV range : TEY-XAS and X-PEEM, *Surf. Sci.*, Vol. 537, pp. 161-167.

Gao, X.; Chen, S.; Liu, T.; Chen, W.; Wee, A. T. S.; Nomoto, T.; Yagi, S.; Soda, K. & Yuhara, J. (2009). Si clusters on reconstructed SiC (0001) revealed by surface extended x-ray absorption fine structure, *App. Phys. Lett.*, Vol. 95, pp. 144102-144105.

Golovchenko, J. A.; Levesque, R. A. & Cowan, P. L. (1981). X-ray monochromator system for use with synchrotron radiation sources, *Rev. Sci. Instrum.*, Vol. 52, pp.509-516.

Handa, K.; Sakai, I.; Izuhara, O.; Iwasaki, H.; Yoshimura, Y.; Masui, S. & Murata, T.(1999). Soft X-ray XAFS Beamline for Advanced Materials Research at Compact Superconducting Ring, *Jpn. J. Appl. Phys.*, Vol. 38, Suppl. 38-1, pp. 654-657.

Iwasaki, H.; Nakayama, Y.; Ozutsumi, K.; Yamamoto, Y.; Tokunaga, Y.; Saisho, H.; Matsubara, T. & Ikeda, S. (1998). Compact Superconducting Ring at Ritsumeikan University, *J. Synchro. Rad.*, Vol. 5, pp. 1162-1165.

Kasrai, M.; Lennard, W. N.; Brunner, R. W.; Bancroft, G. M.; Bardwell, J. A. & Tan, K. H. (1996). Sampling depth of total electron and fluorescence measurements in Si L- and K-edge absorption spectroscopy, *Appl. Surf. Sci.*, Vol. 99, pp. 303-312.

Kawamura, T.; Okada, S. & Yamaki, J. (2006). Decomposition reaction of $LiPF_6$ based electrolytes for lithium ion cells, *J. Power Sources*, Vol. 156, pp. 547-554.

Krause, M. O. (1979). Atomic Radiative and Radiationless Yields for *K* and *L* Shells, *J. Phys. Chem. Ref. Data*, Vol. 8, pp. 307-327.

Lai, B. & Cerrina, F. (2002). SHADOW: A synchrotron radiation ray tracing program, *Nucl. Instrum. Methods A*, Vol. 246, pp. 337-341.

Malitson, I. H. (1965). Interspecimen Comparison of the Refractive Index of Fused Silica, *J. Opt. Soc. Am.*, Vol. 55, pp. 1205-1208.

Momma, K. & Izumi, F. (2008). VESTA: a three-dimensional visualization system for electronic and structural analysis, *J. Appl. Crystallogr.*, Vol. 41, pp. 653-658.

Nakanishi, K. & Ohta, T. (2009). Verification of the FEFF simulations to K-edge XANES spectra of the third row elements, *J. Phys. : Condens. Matter*, Vol. 21, pp. 104214-104219.

Nakanishi, K. & Ohta, T. (In press). Improvement of the Detection System in the Soft X-ray Absorption Spectroscopy, *Surf. Interface Anal.*

Nakanishi, K.; Yagi, S. & Ohta, T. (2010). Development of a XAFS Measurement System in the Soft X-ray Region for Various Sample Conditions (in Japanese), *IEEJ Trans. EIS*, Vol. 130, pp.1762-1767.

Nakanishi, K.; Yagi, S. & Ohta, T. (2010). XAFS Measurements under Atmospheric Pressure in the Soft X-ray Region, *AIP Conf. Proc.*, Vol. 1234, pp. 931-934.

Ohta, T. (Ed.), (2002). *X-ray AbsroptionSpectrooscopy* (in Japanese), Industrial Publishing & Consulting, Inc., Japan, ISBN: 978-4901493215.

Rehr, J. J. & Albers, R. C. (2000). Theoretical approaches to x-ray absorption fine structure, *Rev. Mod. Phys.* Vol. 72, pp. 621-654.

Röhr, C. & Kniep, R. (1994). The crystal structures of $Li(PF_6)$ and $Li(AsF_6)$: on the crystal chemistry of compounds $A(E^VF_6)$, *Zeitschrift für Naturforschung. B, A journal of chemical sciences*, Vol. 49, pp. 650-654.

Sako, E. O.; Kondoh, H.; Nakai, I.; Nambu, A.; Nakamura, T. & Ohta, T. (2005). Reactive adsorption of thiophene on Au(111) from solution, Chem. Phys. Lett., Vol. 413, pp. 267-271.

Saleh, B. E. A. & Teich, M. C. (1991). *Fundamentals of Photonics, Wiley Series in Pure and Applied Optics*, Wiley-Interscience, ISBN 978-0471358329.

Stöhr, J. (1996). *NEXAFS Spectroscopy, Springer Series in Surface Science,* Vol. 25, Springer-Verlag, Berlin, ISBN: 978-3642081132.

Welnak, C.; Chen, G. J. & Cerrina, F. (1994). SHADOW: A synchrotron radiation and X-ray optics simulation tool, *Nucl. Instrum. Methods A*, Vol. 347, pp. 344-347.

Wiza, J. L. (1979). Microchannel plate detectors, *Nucl. Instr. and Meth.*, Vol. 162, p. 587-601.

Wyckoff, R. W. G. (1963). *Crystal Structures,* Vol. 1, Interscience Publishers, New York, ISBN: 978-0471968696.

Yoon, W-S.; Balasubramanian, M.; Yang, X-Q.; Fu, Z.; Fischer, D.A. & McBreen, J. (2004). Soft X-Ray Absorption Spectroscopic Study of a $LiNi_{0.5}Mn_{0.5}O_2$ Cathode during Charge, J. Electrochem. Soc., Vol. 151, pp. A246-A251.

3

Laser-Induced Damage Density Measurements of Optical Materials

Laurent Lamaignère
CEA, DAM, CESTA, BP N°2, F-33114 Le Barp, France

1. Introduction

The prediction of the lifetime of optics in high power fusion lasers is a key point for mastering the facilities (Bercegol et al., 2008). The laser damage scenario is seen as occurring in two distinct steps. The first one concerns the damage occurrence due to the first optic irradiation: the initiation step (Feit et al., 1999). Afterwards, damage sites are likely to grow with successive new shots: the growth step (Norton et al., 2006 ; Negres et al., 2010). The damage growth study requires the use of large beams due to the exponential nature of the process, leading to centimetre damage sites. At the same time, damage density measurements on large optics are mainly performed off-line by raster scanning the whole component with small Gaussian beams (ϕ~1mm) (Lamaignère et al., 2007), except for a wide range of results using larger beams which permit also a comparison between procedures (DeMange et al., 2004). Tests are currently done at a given control fluence. The goal is to irradiate a known area in order to reveal all the defects which could create damage. This procedure, named *rasterscan procedure*, gives access to the optical damage densities (Lamaignère et al., 2007). This measurement is accurate and reproducible (Lamaignère et al., 2010). In standard tests (ISO standards, 2011), results are given in terms of damage probability. This data treatment is size dependent; on the contrary, it is important to focus the attention rather on the density of sites that damage at a given fluence.

The purpose of the present chapter is to explain how the knowledge of the entire test parameters leads to a comparable and reproducible metrology whatever beam and test characteristics. To this end, a specific mathematical treatment is implemented which takes into account beam shape and overlap. This procedure, which leads to very low damage densities with a good accuracy, is then compared with 1/1 tests using small beams. It is also presented with a peculiar attention on data reduction. Indeed an appropriate treatment of these tests results into damage densities gives at once good complementarity of the several procedures and permits the use of one procedure or the other, depending on the need. This procedure is also compared with tests realized with large beams (centimetre sized as used on high power facilities). That permits to compare results given by specific table-top test facilities with those really obtained on laser lines (Lamaignère et al., 2011).

In section 2, facilities using small and large beams are described. The several procedures used are presented in section 3. Developments of data treatment and analysis are given in section 4. Section 5 is devoted to the determination of error bars on both the fluence measurement and the damage density. Few results presented in section 6 illustrate the complementarity of procedures, the repeatability, the reproducibility and the representativeness.

2. Tests facilities

2.1 Small beam facilities

Measurements are currently performed on Q-switched Nd:YAG lasers supplying three wavelengths, the fundamental (1.064µm, 1ω), the second (0.532µm, 2ω) and third (0.355µm, 3ω) harmonics. Laser injection seeding ensures a longitudinal monomode beam and a stable temporal profile. The lasers deliver approximately 1 J at a nominal repetition rate of 10Hz at 1ω. Beam is focused into the sample by a convex lens which focal length is few meters. It induces a depth of focus (DOF) bigger than the sample thickness, ensuring the beam shape to be constant along the DOF.

Two parameters are essential in damage tests: the beam profile and the energy of each shot. Equivalent surface S_{eq} (i.e. defined as the surface given at 1/e for a Gaussian beam) of the beam is determined at the equivalent plane corresponding to distance between the focusing lens and the sample (see Fig. 1). At the focus point, beam is millimetric Gaussian shaped and diameter (given at 1/e) is around 0.5 to 1 mm. Shot-to-shot laser fluences fluctuations (about 15% at one standard deviation) are mainly due to fluctuations of the equivalent surface of the beam. This is the reason why fluence has to be determined for each shot:

- Energy measurements are sampled by pyroelectric cell at 10 Hz: it is a relative data; this cell has to be systematically compared with full-beam calorimeter calibrations before each test.
- A photoelectric cell records the temporal profile (rise time of about 70 ps). This parameter is rather stable (standard deviation of less than 2% of the average value). Nevertheless, it is important to check that the laser goes on working with a single longitudinal mode.

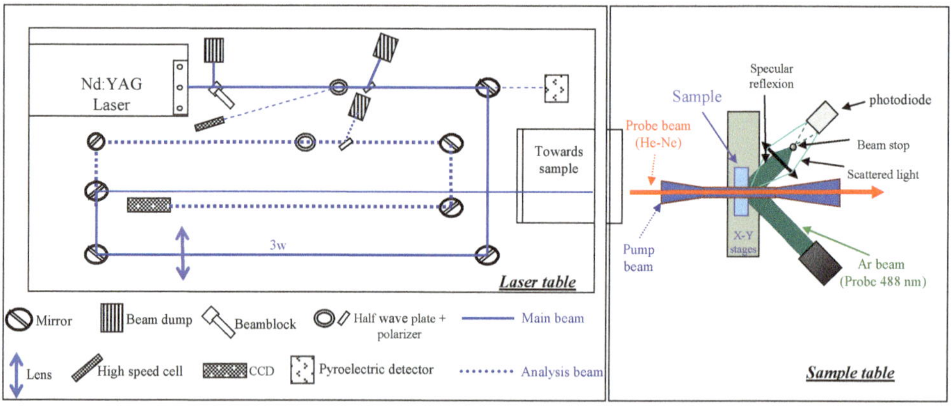

Fig. 1. Typical experimental setup for laser damage testing (Nd:Yag facility) with small beam.

- A CCD camera records the spatial profile. This tool is positioned at a position optically equivalent to that of the sample. The two main parameters obtained from this set-up are the maximum value (pixel) of the beam which is directly connected to the maximum fluence, and the pointing position which permits to build up the fluence map. Beam wings integrate a large fraction of the energy (up to 50%) and are responsible for a large part of the fluctuations. Above a diameter of 1/e, the shape is very nearly Gaussian.

2.2 Large beam facilities

The advantage of tests with large beams is that they are representative of real shots on high power lasers. They are carried out on high power facilities where parametric studies are conducted: pulse duration or phase modulation (FM) effects. With energies about 100J at 1053 nm and 50J at 351nm and after size reduction of the beam on the sample, high fluences are delivered with beam diameters about 16 mm. Due to a contrast inside the beam itself (peak to average) of about few units, a shot at a given average fluence covers a large range of local fluences (Fig. 2). The laser front-end capability makes possible parametric studies like the effect of FM, temporal shapes and pulse durations on laser damage phenomenology. The characteristics of this kind of laser are quite similar to high-power lasers for fusion research: the front-end, the amplification stage, the spatial filters, the frequency converters crystals. Then laser damage measurements performed with this system should be representative of the damage phenomenon on high-power lasers.

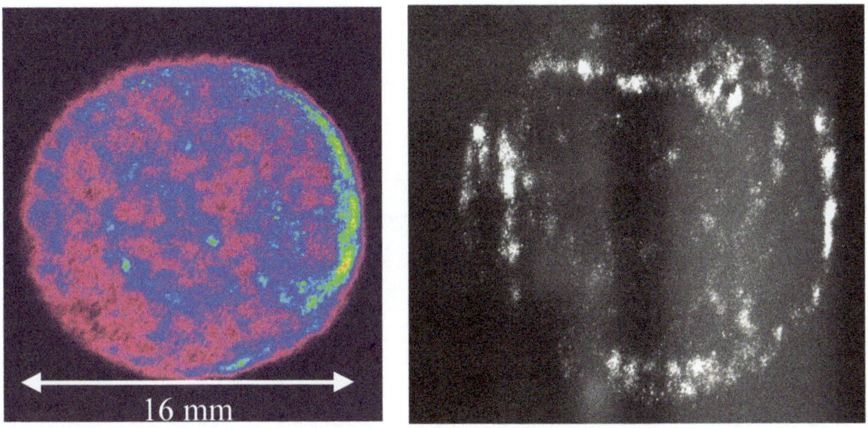

Fig. 2. Damage test with a centimetre-sized beam. On the left, spatial profile of the 16mm-diametre beam at the sample plane as measured on CCD camera. On the right, the corresponding damage photography is reported. Matching the two maps allows to extract the fluence for each damage site.

3. Procedures

3.1 1on1 procedure

The 1/1 test is made on a limited number of sites (ISO standards, 2011). Results are generally given in terms of damage probability as a function of fluence. Since the beam sizes of several benches are different, it is compulsory to convert probability into damage density. 1/1 tests are also done because with a small number of tested sites, a rapid result is achieved. Another advantage is that a relative comparison of several optical components is possible if the test parameters are unchanged. Generally, the 1/1 test is used instead of S/1 to avoid material ageing or conditioning effects.

3.1.1 Small beam

In the 1/1 procedure:

- 1 shot is fired on 1 pristine site. Energy, spatial and temporal profiles are recorded and lead to beam fluence and intensity for each shot.
- Damage detection is achieved by means of a probe laser beam co-linear to the pump laser beam.
- Many shots are realized at different fluences (about twenty sites are tested per fluence, on the basis of a ten of fluences set); gathering shots by fluence group, probability plots versus fluence are determined.

The next step is to convert the probability into damage density. This data treatment is presented on paragraph 4.1.

3.1.2 Large beam

The metrology is very close to small beam metrology: energy, temporal and spatial profiles are recorded. For the latest, a CCD camera is positioned at a plane optically equivalent to that of the sample. Beam profile and absolute energy measurement give access to energy density $F_{(x,y)}$ locally in the beam:

$$F(x,y) = \frac{E_{tot}}{S_{pix} \times \sum_{i=\min}^{i=\max} (n_i \times i)} \times i(x,y) \tag{1}$$

E_{tot} : total energy
S_{pix} : CCD pixel area
$i_{(x,y)}$: pixel grey level

Figure 2 shows fluence spatial distribution of a large beam shot (beam diameter of about 16 mm). Due to a beam contrast (peak to average) of about 40%, a shot at a requested fluence (mean value) covers a large range of fluences. This make compulsory the exact correlation between local fluence and local event (damage or not). Data treatment is then completed by matching fluence and damage maps (map superposition is realized by means of reference points in the beam, hot spots). The two maps are compared pixel to pixel to connect a local fluence with each damage site detected. Then data are arranged to determine a damage density for each 1 J/cm^2 wide class of fluence. Damage density versus fluence is then given by the following relation:

$$D(f) = -\ln(1 - P) / S \tag{2}$$

Where P is the damage probability: rate between the damage sites number (N_d) over the irradiated sites number (N_t) and S is the size of the pixel camera.

3.2 Rasterscan procedure

In order to scan a large area, the sample to be tested is translated continuously along a first direction and stepped along a second direction (Fig.3), the laser working at predetermined control fluence. Repeating this test at several fluences on different areas permits to determine the number of damage sites versus fluence thus the damage density. But in the case of Nd:Yag tripled frequency, shot to shot fluence fluctuations have to be taken into account. Fluence can display a standard deviation of up to 15% for this kind of laser (Fig. 5 of reference (Lamaignère et al., 2007)). When thousands of shots are fired, modulation of a factor 2 in fluence is obtained. Thus, it is important to monitor the specific fluence of each shot in order to get a precise correlation between damage occurrence and laser fluence. During the test, energy, spatial and temporal profiles, beam position on the sample are recorded for each shot (at 10 Hz) in order to build up an accurate map of peak fluences corresponding to the scan.

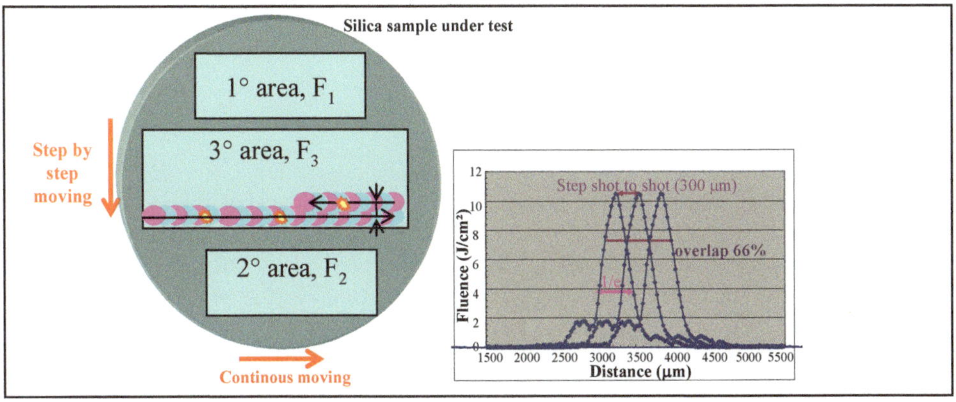

Fig. 3. On the left, the successive laser pulses overlap spatially to achieve an uniform scanning. On the right, the beam overlap.

Much attention has to be paid to the position of the zero level in the beam image, because the value of fluence is very sensitive to small errors in that level. The determination of that position has to be checked very often, for example by verifying that the total energy integrated on the CCD is proportional to the pyroelectric cell measurement. The uncertainty on the zero level is responsible for the largest part of the total uncertainty on fluence measurement. This point is treated on paragraph 5.1.

3.3 Damage detection system

Depending on facilities, damage detection is realized in-situ after each shot during scans or after the irradiation with a "postmortem" observation. The "post-mortem" observation of irradiated areas is realized by means of a long working microscope (Fig. 6 of ref. (Lamaignère et al., 2007)). The minimum damage size detected is about 10 µm whatever the morphology of damage. Below this value, it is difficult to discriminate between a damage site and a defect of the optic that did not evolve due to the shots. Damage sites might be of different types: some are rather deep with fractures, others are shallower.

In the case of the in-situ detection, a probe beam is focused at the same position as the pump laser on the sample. The specular reflexion is stopped by a pellet set at the centre of a collecting lens. The probe beam is scattered when damage occurs and the collecting lens redirects the stray light onto a photocell: that corresponds to a Schlieren diagnostic. Since the optic moves during the scan, the signal is recorded just before and just after the shot. The comparison of the two signals allows making a decision on damage occurrence or not. The smallest damage detected is again about 10 μm.

Matching the maps of fluence and damage sites allows one to extract the peak fluence F_p for each damage site. A first data treatment is then realized by gathering damage sites in several fluence groups [(F_P-$\Delta F_P/2$) to (F_p+$\Delta F_P/2$)]. Knowing damage number and shot number, and attributing an area to each shot, the damage density is determined for each group (Fig. 4). Then with only one predetermined control fluence, damage density is then determined on a large fluence range, due to fluence fluctuations during scan. At this point, however, only the relation between damage sites and peak fluences is available. In next section, damage density is calculated as a function of local fluence.

4. Data treatment

4.1 1/1 procedure

The damage probability is converted into damage density. Note that interaction between materials defects is neglected. Thus, if defects damaging at a given energy density F are randomly distributed, the defect density $D(F)$ follows a Poisson law. Then, on a given area S, damage probability $P(F,S)$ and defect density $D(F)$ are related by (Feit et al., 1999):

$$P(F,S) = 1 - \exp(-D(F).S) \tag{3}$$

The treatment is the same considering surface or volume damage densities. The total volume illuminated V is the product of the beam area by the optical component thickness θ (Lamaignère al al., 2009). $V_{eq.g}$ is defined as the Gaussian beam equivalent volume ($S_{eq.g}$ being the Gaussian beam equivalent area):

$$V_{eq.g} = S_{eq.g}.\theta \tag{4}$$

Defining δ_m as the measured damage density which is then obtained from damage probability:

$$P = 1 - \exp\left(-\delta_m.S_{eq.g}\right) \tag{5}$$

The measured damage density is then:

$$\delta_m = -\frac{\ln(1-P)}{S_{eq.g}} \tag{6}$$

When a spatially Gaussian beam is used, the absolute damage density can be expressed as the logarithmic derivative of the measured density δ_m:

$$D(F) = F.\frac{d\delta_m(F)}{dF} \tag{7}$$

The experimental curve of $\delta_m(F)$ can often be fitted by a power law of the energy density:

$$\delta_m(F) = \alpha.(F)^\beta \tag{8}$$

Where α and β are calculated from the best fit of the measurements.
From equations (7) and (8), the absolute damage density is obtained:

$$D(F) = \beta.\delta_m(F) \tag{9}$$

Or again:

$$D(F) = \beta.\left(-\frac{\ln(1-P)}{S_{eq.g}}\right) \tag{10}$$

Note that if P<<1, then relation (10) is equivalent to:

$$D(F) = \beta.\frac{P}{S_{eq.g}} \tag{11}$$

For each energy density level F, the observed damage probability P is given by the equation:

$$P = \frac{n^D}{n^D + n^{ND}} \tag{12}$$

where $(n^D + n^{ND})$ is the total number of exposed sites and n^D the number of damage sites. Then, if $n^D << n^D + n^{ND}$, the measured damage density is:

$$\delta_m(F) = \frac{n^D}{\left(n^D + n^{ND}\right).S_{eq.g}} \tag{13}$$

Finally, the absolute damage density given in (11) is:

$$D(F) = \beta.\frac{n^D}{\left(n^D + n^{ND}\right).S_{eq.g}} \tag{14}$$

In order to determine the absolute damage density, the knowledge of the number of damage sites and the Gaussian beam equivalent area is sufficient. The final step consists in determining the exponent β from power law fitting of experimental data.

For a top-hat beam, the absolute damage density is directly given by equation (13), the same equations can be used than for Gaussian beams taking $\beta=1$

4.2 Rasterscan procedure

In this part, the calculations used to analyze the experimental data are described. When tests are done with small quasi-Gaussian beams, the overlap is not perfect (see the right insert figure 3): a large area is irradiated at a fluence smaller than peak fluence. Then the precise beam shape and more precisely the Gaussian equivalent area have to be taken into account. Each shot is identified with its maximum fluence F_p. On the high fluence side, it is observed experimentally that the damage density varied rapidly with fluence, approximately as a power law (Fig. 4). The shape of the beam at the top is then the relevant function. Shots are thus characterized by the equivalent area of the Gaussian peak. When this is done, it appears that, schematically, the resulting curve could be divided in two parts (Fig. 4).

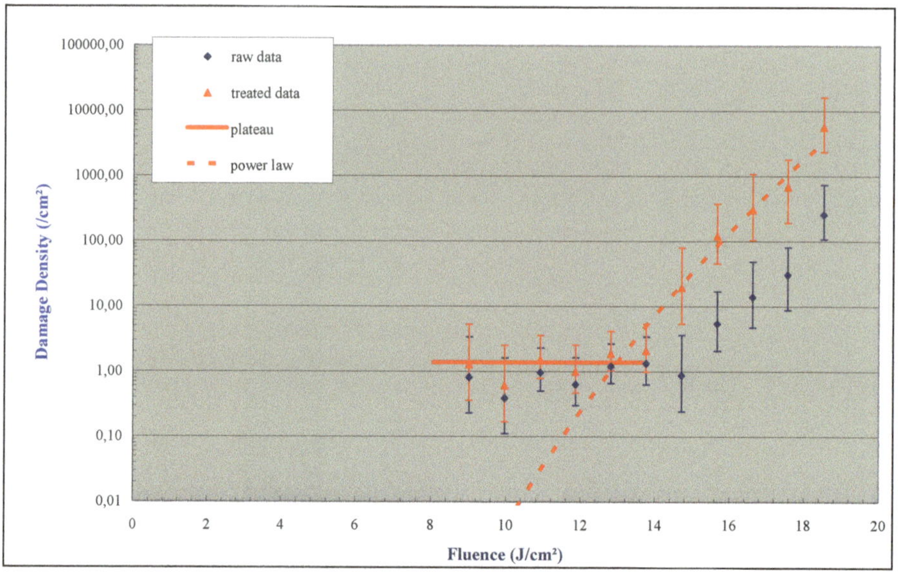

Fig. 4. Damage density versus fluence, after treatment taking care of beam shape and to derive experimental uncertainty. Diamonds are the raw data. Triangles represent treated data; in this case fluence is the local fluence. Data on plateau are issue from relation (23) and those in the high fluence range from relation (22). Error bars calculations are explained in §5.2.

4.2.1 At high fluences, above F_{cut}, (see Fig. 4)

Damage density increases quickly with fluence. Calculations in this section are made easy by considering spatially Gaussian shapes, for which fluence distribution can be expressed as a function of r, the distance to the peak. Each shot is identified with its peak fluence F_p.

$$F(r) = F_p \exp\left(-\frac{\pi r^2}{S_{eq.g}}\right) \tag{15}$$

becomes, when derived

$$\frac{dF}{F} = \frac{2\pi r dr}{S_{eq.g}} \tag{16}$$

where we defined:

$S_{eq.g}$ = Gaussian beam equivalent area.
Let us also define, for the following equations
S(total): Total scanned surface area
F_{cut}: Transition fluence between "plateau" and power law behaviors
N: Number of damage sites
n: Number of shots

Let D(F) be the damage density distribution that we are trying to measure, a function of local fluence F. We call $\delta_m(F_p)$ the experimentally measured density at F_p. The number of damage sites created by a shot of maximum fluence F_p for a Gaussian beam is given by the relation:

$$N\left(F_p\right)=\int_0^{+\infty} 2\pi r.dr.D[F(r)]=\int_0^{F_p} S_{eq.g}.dF.\frac{D[F]}{F} \tag{17}$$

The third member of eq. (17) is obtained by changing variables in the integral and using eq. (16). As shown in figure 4, we defined the experimentally measured density by:

$$\delta_m\left(F_p\right)=\frac{N\left(F_p\right)}{n(F_p).S_{eq.g}} \tag{18}$$

This permits to express easily the measured density as a function of the true damage density D(F):

$$\delta_m\left(F_p\right)=\int_0^{F_p}\frac{D(F)}{F}.dF \tag{19}$$

Thus D(F) can be easily expressed as the logarithmic derivative of the measured density δ_m.

$$D(f)=f\cdot\frac{d\delta_m(f)}{df} \tag{20}$$

Since the experimental curve of $\delta_m(F_p)$ is rather dispersed, it is better to use a functional approximation of $\delta_m(F_p)$ to derive D(F). At high fluences (above F_{cut}), damage density is well fitted by a fluence power law:

$$\delta_m\left(F_p\right)=\alpha.\left(F_p\right)^{\beta} \tag{21}$$

α and β can be calculated from the best fit of the measurements. From eq. (20) and (21), one obtains the absolute damage density:

$$D(f)=\beta.\delta_m(f)=\beta.\frac{N(f)}{n(f).S_{eq.g}} \tag{22}$$

Practically, above F_{cut}, to determine the absolute damage density for each fluence group, we have to know the number of damage sites and Gaussian effective area from the experiment, and then we have to determine the exponent β from power law fitting of experimental data.

4.2.2 At low damage fluences, below F_{cut}

The measured damage density $\delta_m(F_p)$ is nearly constant for fluences lower than F_{cut}. This plateau must be treated differently from the high fluence part. First of all, in this fluence range, and especially when damage sites are attributed to low fluence shots, F_p is not a local fluence maximum. Since $\delta_m(F_p)$ shows a very weak variation with F_p, it is reasonable to assume that damage density D(F) is really a constant on this fluence range. Thus D(F) is simply obtained by taking the ratio of the total number of damage sites in this fluence range, by the total area covered by these shots (with $F_p < F_{cut}$). This area is proportional to the number of these shots.

$$D(F < F_{cut}) = \frac{N(F < F_{cut})}{S(F < F_{cut})} = \frac{N(F < F_{cut})}{\dfrac{n(F < F_{cut})}{n(total)} \cdot S} \tag{23}$$

Damage density D is then once again almost δ_m multiplied by a numerical factor $\dfrac{n(total).S_{eq.g}}{S}$, which is not far from 1. The result of this treatment can be visualized in figure 4. Since fluence is low, the number of damage sites is low too, and the result is dominated by the error bar, as we are going to see in paragraph 5.2.

5. Error bars

5.1 Fluence error

Synthesis of error margins is reported in Table 1. The first error (contributor 1) is due to the acquisition of the energy with the pyroelectric cell. This measurement is systematically compared with an absolute calorimeter calibrated yearly: the measurement uncertainty is given at 2% at 1σ (contributor 1).

A special attention has to be paid to the "qualification" of the camera used on each facility. For example series of 1000 shots at different positions around the waist position are realized in order to measure the mean equivalent area. It appears that two identical cameras (same model) gave similar results within 4% (contributor 2). The same deviation is found between three different cameras of different providers (two being 8-bit analog cameras with rectangular pixels (13*11.5 μm^2), the third one being a 12-bit digital camera with square pixels (9.9*9.9 μm^2)). The fluctuations of cameras parameters (amplification, transmission factors) and the laser instability must be taken into account; they certainly play a part in these deviations. Thus, the different cameras give a rather similar result. These verifications were completed by checking that the effective area determined on the equivalent plane was the same as on the sample. It can be done by measuring around the waist position the areas on the two paths "camera" and "sample" (Fig. 5). For the latter, a camera is placed instead of the sample. Series of 1000 shots can then been realized for each position, each camera recording spatial profiles at the same time. It appears, in Fig. 5, a good correspondence between the two paths. From analysis, in the vicinity of the Rayleigh length, a deviation between the two paths of less than 4% was obtained (contributor 3). This includes the deviation coming from the cameras. It appears, in the same figure, that the deviation and fluctuations shot-to-shot are drastically reduced close to the waist position. Last, it is also important to check that the same fluctuations are recorded on the two paths. A good correlation between the two sets of data (of 1000 shots each) is found with a relative variation of about 6% at 1σ, contributor 4 (see Fig. 4 of ref (Lamaignère et al., 2010)).

After reducing the total uncertainty on equivalent area measurement, a comparison between the total energy integrated on the CCD camera and the measurement from the pyroelectric cell has to be performed. The two signals should be proportional. The comparison gives a random error of about 5% (due to the determination of the zero level on the camera, determined at each shot and repeatability of the pyroelectric cell), contributor 5. Thus, considering the whole analysis of error measurements (the different contributors are summed up in Table I with the hypothesis that there are not correlated), it is appropriate to consider for the different facilities, that the absolute fluence values are known with an accuracy around 10%.

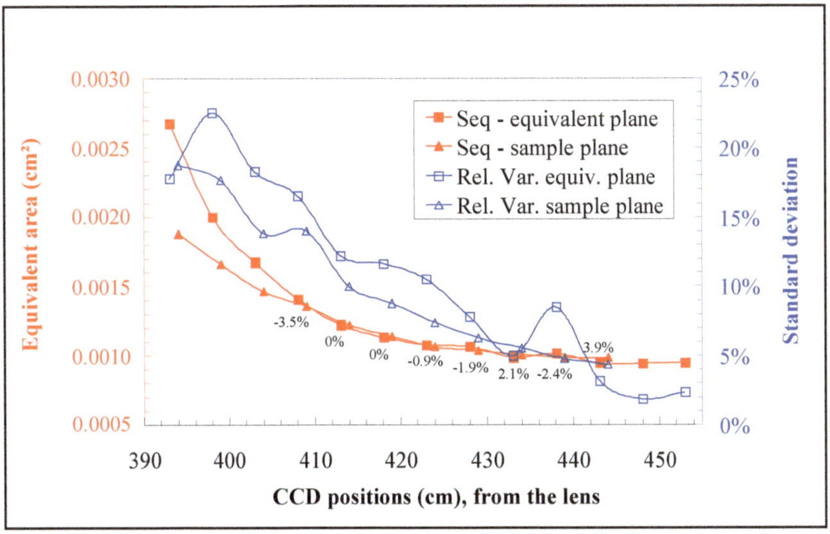

Fig. 5. Optical paths comparison. Measurements of the beam equivalent areas (full symbols) at the sample and equivalent planes with the same camera, at several positions close to the focal point. At each position, the deviations between the two optical paths are indicated.

Contributor		Error bar at 1σ (%)
1	Calorimeter	2
2	Deviation between cameras	4
3	Deviation between optical paths	4
4	Shot-to-shot correlation between optical paths	6
5	Deviation between camera and pyroelectric cell	5

Table 1. Synthesis of error margins for identified contributors [Error budget]. A quadratic summation provides an accuracy around 10% for the determination of fluences.

5.2 Intervals of confidence

Unlike the measure of fluence, values of damage densities are rarely presented with any notion of accuracy. However, it is not possible to compare measures in one's lab or with other installations without error bars. These error bars are especially needed in the low fluence range were rare damage events are observed. The uncertainty is not on the measurement of this particular sample, because this one is now destroyed. The interesting statistical figure is the uncertainty that makes possible the comparison of two similar but physically distinct samples.

To determine the statistical error on the measurement of D(F), one assumption on the nature of the distribution of damaging defects is made: it is supposed that defects are randomly distributed over the area. A specificity of this hypothesis is that there is no interaction between defects. This supposition is common in laser damage research, and it probably holds for optical components tested with millimeter sized beams. However, it is possible to make the hypothesis that defects collaborate in damage, and draw some useful conclusions, for the case of optical multilayers or that of bulk damage of KDP crystals.

The assumption of randomly distributed and independent defects is applied to any set of defects: For example a set of defects damaging at a given fluence is supposed to be distributed that way in the sample. So a set of defects is damaging in a given fluence range. A damage test is an experimental sampling of a distribution of defects that characterizes a very large area, for example the area of all the optical components and samples produced with the same recipes. The error made will be calculated as a possible discrepancy between density obtained at the sample and the "true" density that would be measured if the whole production was damage tested.

The uncertainty depends on the number of damage sites generated within each fluence group. By considering a Poisson distribution, it is rather straightforward to determine the interval of possible measurements when the true characteristic is known. Let us define:

Σ: The surface area of the total production
N: number of potential damage sites on the total production
k: number of detected damage sites on S
$\nu = N\,S/\Sigma$, number of expected damage sites on S
k is not equal to ν, but rather follows Poisson law, when Σ is very large compared to S, the area of the tested surface. Thus the probability of finding a given k value is:

$$P_{\nu}(k)=\frac{\nu^{k}e^{-\nu}}{k!} \tag{24}$$

This formula (24) is only the first step. The problem is then inversed since we want to know ν from the measurement k. The main difficulty lies in the fact that we are interested in small values of k (or ν). When a law of probability is determined with a high enough number of events, then one can use the law of large numbers to express neatly the error in terms of erf function. In case of rare events, especially in the low fluence range, we had to make a special derivation (the full demonstration is given in reference (Lamaignère et al., 2007)).

Let k be the known number of detected damage sites. The interval of values of ν, for the confidence to be better than 1-ε can be written:

$$\nu \in [\nu_{min};\nu_{max}]$$

with

$$\int_{\nu=0}^{\nu_{min}} \frac{\nu^k e^{-\nu}}{k!}.d\nu = \frac{\varepsilon}{2} \text{when k} \neq 0 \quad ; \quad \nu_{min} = 0 \text{ when k} = 0 \tag{25}$$

and

$$\int_{\nu=\nu_{max}}^{+\infty} \frac{\nu^k e^{-\nu}}{k!}.d\nu = \frac{\varepsilon}{2} \tag{26}$$

This means that we calculate a probability 1-ε for ν to lie in the interval between ν_{min} and ν_{max}. In this section, we know use specifically the confidence limits that corresponds with 2 standard deviation (2σ) of a Gaussian variable ε = .0455, or ε/2 = .02275.

The confidence limits are very far apart when the measured number of damage sites is low. Table 2 gives a numerical derivation of these limits for low k values. At k=0, when no damage site is detected, we can only say that the average number of sites is lower than 3.7 with an error rate of 2.3%. This number of sites must be translated into a density.

One should notice that these error bars are only given by the statistical variations due to the limited number of data (connected to the size of the sample). Potential errors due to inaccurate damage detection are not taken into account.

k	0	1	2	3	4	5	6	7	8	9	10	11	12	13	14	15	16	17	18	19	20
ν_{min}	0	0,2	0,6	1,1	1,6	2,2	2,8	3,4	4	4,7	5,4	6,1	6,8	7,6	8,3	9	9,8	11	11	12	13
ν_{max}	3,7	5,6	7,2	8,8	10	12	13	15	16	17	19	20	21	22	24	25	26	27	29	30	31

Table 2. Interval of confidence of ν for a given measured value of k.

6. Results

This chapter is dedicated to experimental results and illustrates:

- The complementarity of the several procedures which can be indifferently used depending on the information asked;
- The repeatability;
- The reproducibility; these two notions permit to validate the whole procedures;
- The representativeness of tests realized with small beams compared to large and real beams on high laser facilities.

6.1 Complementarity

Figure 6 shows results obtained on the same optical component tested with the 1/1 and rasterscan procedures, and on the same facility. Results are directly reported in terms of damage density with the presented formalism in §4.1 and 4.2. During rasterscan, about 6000 shots (this corresponds to a scanned area of about 6 cm^2) are fired with fluences between 2.5 and 4.5 J/cm^2. During the 1/1 tests, only 200 shots have been fired with fluences between 4.5 and 6.5 J/cm^2. These results, which were presented, in the past, in terms of damage probabilities, are translated in terms of damage densities. The good complementarity of the two test results leads to validate the developed formalisms (for clarity, fluence error bars are not reported). We remark that due to a large number of tested sites, low damage densities are available with the rasterscan technique. On the opposite, 1/1 test covers damage probabilities between 0 and 1, and due to the small illuminated area, damage densities available are higher. We observe that the intervals of confidence are smaller for rasterscan procedures, due to a large number of damage sites in spite of small damage densities, than for 1/1 where a limited number of sites are tested. Nevertheless, the overlap of damage densities indicates a good measurement complement and that data reduction permits to determine repeatable damage densities; in spite of a non 100% overlap during rasterscan and a small number of tested sites during 1/1. Identical results were also obtained with different pulse durations and spatial beam profiles.

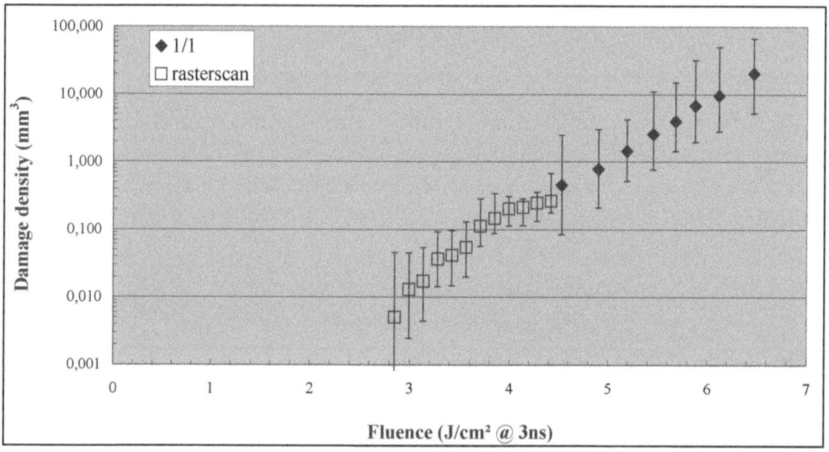

Fig. 6. Comparison of 1/1 and rasterscan procedures. Damage densities vs fluence measured on the same optical component with the 1/1 test procedure (diamonds) and 3 rasterscans at 3 different fluences (squares).

6.2 Repeatability

It is necessary to check that the procedures applied on similar optics give always similar results on a unique set-up. This should be done regularly. Fig. 7 shows results obtained on 3 samples from the same vendor, tested on the same facility on a one year period. Damage densities lay inside confidence limits. This result and those obtained with other lasers (not presented here) show that it is possible to achieve a good repeatability.

6.3 Beam overlap

A non-negligible parameter of the procedure is the shot-to-shot step i.e. the beam overlap. In order to scan a large area and to be sure to irradiate all the defects, a good overlap is necessary. That is possible with top-hat beams but not with Gaussian ones. In the latter case, it is preferable not to use too small a step, in order to avoid that the defects experience a long irradiation ramp, that the scan duration is too long and that damage grows due to successive shots on the same site. Moreover, a good correspondence between damage and fluence maps requires the step not to be too small. On the contrary, a large step implies a large area to be scanned or a low statistic. At last, whatever steps and overlaps, data

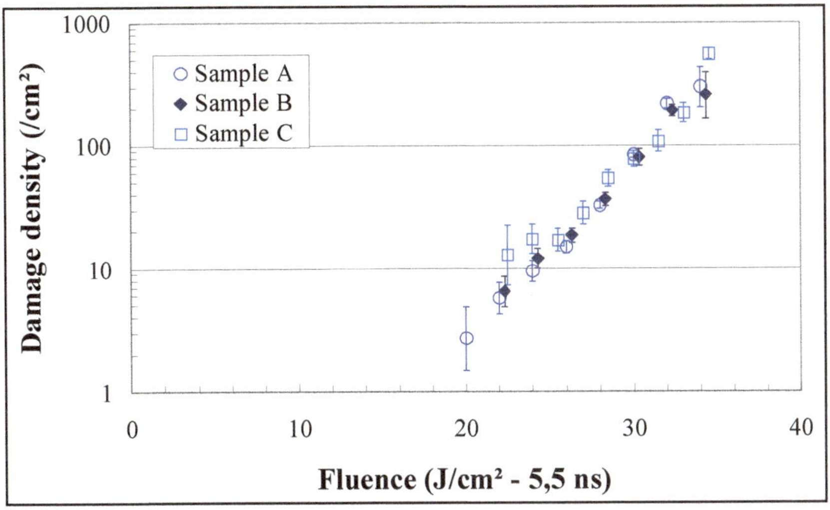

Fig. 7. Repeatability. Damage densities D vs fluence F for three different optical components (from the same vendor) tested on the same facility within a one year period with the rasterscan procedure.

treatment must provide the same and accurate results. In Fig. 8 results obtained on the same optic (330*330 mm²) for 5 different steps are reported (0.15; 0.3; 0.6; 1; 2 mm corresponding to an overlap of 95; 92; 66; 31 and 3%). For a step smaller than 0.15 mm, catastrophic damage growth was observed. No tests were conducted under this value. As can be seen in

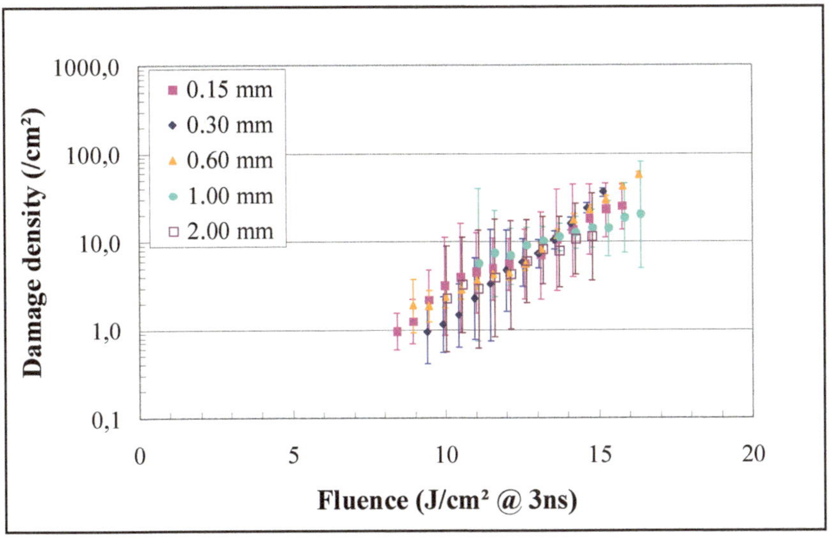

Fig. 8. Steps. Damage densities vs fluence D(F) for the same optic tested on the same facility with 5 different steps on 5 different zones. Data have been processed following data reduction presented in § 4.2, this example is extracted from a Gaussian test.

Fig. 8, from 9 to 15 J/cm², results are gathered inside the largest interval of confidence (given at 2σ). These results indicate that this procedure, with its data treatment, is able to provide comparable measurements for a wide range of beam overlap. Consequently, comparison is possible between several facilities where the shot-to-shot step inevitably varies from one test to another.

6.4 Reproducibility

The comparison is on 4 optical components from the same batch. Due to the limited available area on each sample, data exploration is realized for fluences between 10 and 20 J/cm². In this range, damage density increases quickly with fluence and values are sufficiently high for the number of damage sites to be high enough. For lower fluences and lower damage densities, the areas to be scanned become too large and the error bars could be too high to make any conclusion. For each sample, an area about 40 cm² was irradiated.

Data treatment presented in §4.2 is first applied (see insert of Fig. 9). Next pulselength scaling is used. The best correspondence is obtained when α is equal to 0.6. (Fig. 9), value slightly different from the usual scaling $\tau^{1/2}$. Up to 18 J/cm², measurements are gathered inside the largest interval of confidence (given at 2σ). Fluence error bars are not reported for clarity (values are provided with an accuracy around 10%) but taking into account both fluence error bars and level of confidence, results are quite comparable. Above a few hundred damage per cm², damage sites can aggregate and the comparison is no more feasible.

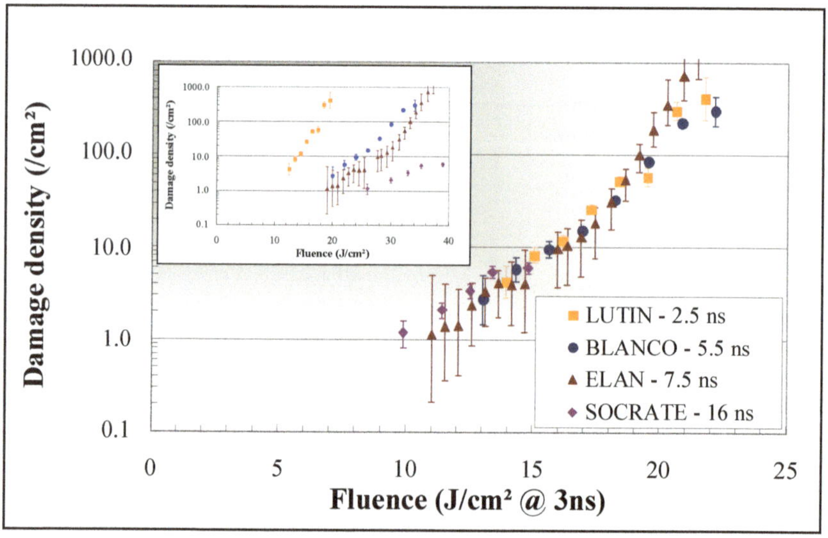

Fig. 9. Reproducibility. Damage densities vs fluence for several components (with the same polishing process) tested on 4 different facilities. Fluences are scaled using a temporal scaling law (τ^{α}) with an α exponent of 0.6. In the insert, the raw data are reported.

6.5 Representativeness

For the question of representatives, an area of 40 cm² has been tested on a large optic with the small beam rasterscan procedure, i.e. illumination, damage detection and data processing. A second optics has been tested with the large beam. For this comparison, the number of parameters, that are different, is reduced selecting shots that are quite similar: the pulses are single mode longitudinally; the temporal profiles are Gaussian with equivalent pulse durations τ=2.5 and 2.3 ns, respectively. Data treatments are first applied. Next pulselength scaling is applied with the help of a temporal scaling law ($F\sim\tau^{\alpha}$) and α = 0.6, this value being previously determined (see §6.4). Nevertheless, the two pulse durations being close, the comparison can also be made with the usual scaling $\tau^{1/2}$ or without the use of any scaling law. The absolute fluence values are known with an accuracy of about 10% (Lamaignère et al., 2010); the sources of errors on the two facilities being different, it is compulsory to take into account these errors. Figure 10 shows that measurements are gathered inside the largest interval of confidence (given at 2σ) and thus the two results are quite comparable. This confirms the reproducibility that was noted in the previous paragraph.

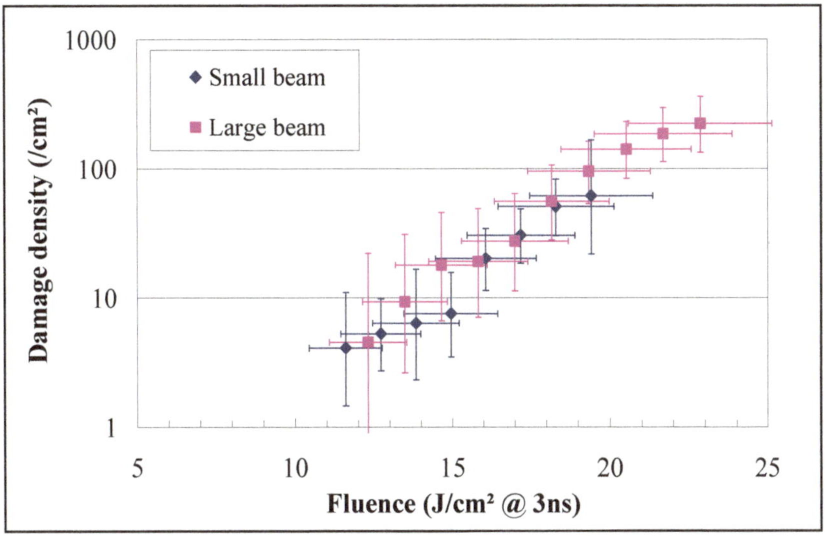

Fig. 10. Damage densities versus fluence measured on two optics from the same batch with small and large beams.

7. Conclusion

With rigorous data analysis and treatment, it is possible to measure damage density on an optical component with high accuracy and repeatability, whatever the beam overlap and the beam shape. Since tests are destructive, the same area cannot be measured on two different instruments. However the consideration of error bars on damage density allows comparing results from several components assumed to be comparable.

A particular attention has to be devoted to the error budget on fluence determination, more precisely on the measurement of beam equivalent area. CCD cameras have to be carefully qualified. Thus the error on calculation could be estimated. It is vital to ensure that this equivalent area determined on the reference path is equal to that on the sample controlling the CCD position and by verifying that the optics in front of the camera do not alter the beam profile. This measurement has to be recorded at the frequency laser to monitor shot-to-shot laser fluctuations. The calculations on error bars not only allow comparing results from several samples tested on several facilities but also give an upper limit of damage density, particularly useful when a small area is scanned or in a low damage density range.

Depending on the available area on the sample to be tested and the level of damage density, it appears that the 1/1 and rasterscan procedures are comparable and complementary with the use of an appropriate data reduction. More, these procedures give access to representative measurements when compared to results and behaviours observed on high laser facilities irradiated with very large beams.

8. References

Bercegol, H., Boscheron, A., DiNicola, J.M., Journot, E., Lamaignère, L., Néauport J. & Razé, G. (2008). Laser damage phenomena relevant to the design and operation of an ICF laser driver, *J. Phys. Conf. Ser.*, 112, 032013-032016.

DeMange, P., Carr, C.W., Radousky, H.B., & Demos, S.G. (2004). System for evaluation of laser-induced damage performance of optical materials for large aperture lasers, *Rev. Sci. Instr.* 75, 3298.

ISO standards: (ISO 21254-1:2011) - (ISO 21254-2:2011) - (ISO 21254-3:2011).

Feit, M.D., Rubenchik, A.M., Kozlowski, M.R., Génin, F.Y., Schwartz, S. & Sheehan, L.M. (1999). Extrapolation of damage test data to predict performance of large-area NIF optics at 355 nm, *Proc. SPIE* 3578, 226-234.

Lamaignère, L., Bouillet, S., Courchinoux, R., Donval, T., Josse, M., Poncetta, J.C. & Bercegol, H. (2007). An accurate, repeatable, and well characterized measurement of laser damage density of optical materials, *Rev. Sci. Instr.* 78, 103105.

Lamaignère, L., Donval, T., Loiseau, M., Poncetta, J.C., Raze, G., Meslin, C., Bertussi, B. & Bercegol, H. (2009). Accurate measurements of laser-induced bulk damage density, *Meas. Sci. Technol.* 20, 095701.

Lamaignère, L., Balas, M., Courchinoux, R., Donval, T., Poncetta, J.C., Reyné, S., Bertussi, B. & Bercegol, H. (2010). Parametric study of laser-induced surface damage density measurements: Toward reproducibility, *J. Appl. Phys.* 107, 023105.

Lamaignère, L., Dupuy, G., Donval, T., Grua, P. & Bercegol, H. (2011). Comparison of laser-induced surface damage density measurements with small and large beams: toward representativness, *Applied Optics* 50, 442.

Negres, R.A., Norton, M.A., Cross, D.A. & Carr, C.W., (2010). Growth behavior of laser-induced damage on fused silica optics under UV, ns laser irradiation, *Opt. Express* 18, 19966-19976.

Norton, M.A., Donohue, E.E., Feit, M.D., Hackel, R.P., Hollingsworth, W.G., Rubenchik, A.M. & Spaeth, M.L. Growth of laser damage in $Si0_2$ under multiple wavelength irradiation, *Proc.SPIE* 5991, 599108 (2006).

4

System for High Speed Measurement of Head-Related Transfer Function

Andrzej B. Dobrucki, Przemysław Plaskota and Piotr Pruchnicki
Wroclaw University of Technology
Poland

1. Introduction

Recently surround-sound systems have become popular. The effect of "surrounding" the listener in sound is achieved by employing acoustic phenomena which influence localizing the source of the sound. Similarly to stereophony system the time, volume and phase interrelations in signals coming from each sound source are taken into account. Additionally the influence of the acoustic system created by the pinna, head and torso on the frequency characteristic of the sound is taken into consideration. This influence is described by the Head-Related Transfer Function (HRTF). The knowledge of the human physical body characteristics' influence on the perception of the sound source location in space is being more and more frequently applied to building sound systems.

So far the best method of including the influence of the human body on the frequency characteristic of the sound is the HRTF measurement for different locations of the sound in relation to the listener. Then the achieved measurement results are used for creating a database meant for the sound reproduction. Creating a proper HRTF database is a difficult problem - every human exhibits individual body characteristics therefore it is not possible to create one universal database for all the listeners. For this reason applying the knowledge of the human body influence on the frequency characteristic of the sound is impeded. In order to include these parameters it is necessary to conduct all these laborious measurements for each individual.

2. Head-Related Transfer Function

The HRTF is a representation of the influence of the acoustic system formed by the pinna, head and human torso on the deformation of the acoustic signal spectrum reaching the listener's ear. The head's shape and tissue structure have a bearing on acoustic signal spectrum distortion (Batteau, 1967; Blauert, 1997; Hartmann, 1999; Moore, 1997). The changes in the spectrum enable the listener is able to more accurately localize the sound source in the space which surrounds her/him. In case of headphone listening the influence of the acoustic system formed by the pinna, head and human torso is eliminated and the acoustic signal received by the listener is unnatural - the listener localizes the sound source inside her/his head. Through the use of HRTF measurement results the signal can be so deformed that the listener subjectively identifies the spatial properties of the sound whereby the location of the sound source in the space surrounding the listener is reproduced (Hartmann & Wittenberg, 1996; Horbach et al., 1999; Hen et al., 2008, Plaskota & Kin, 2002; Plaskota et al. 2003). Since there are many sound source locations in the space surrounding the listener many HRTFs are needed to accurately reproduce the location of the sound source in this space.

The function describing the direction-dependent acoustic filtering of sounds in a free field by the head, torso and pinna is called HRTF. Although it is obvious that the linear dependence between Interaural Time Difference (ITD), Interaural Level Difference (ILD) and the perceived location in space needs to be predicted, it is less obvious how the spectral structure and the location in space can be mathematically interrelated (Cheng & Wakefield, 2001). The first step towards understanding the significance of the signal spectrum in directional hearing was an attempt at physical modeling and empirical measurement followed by computer simulations of the ear's frequency response depending on the direction. The measured frequency response of the ear is subject to further analysis.

Formally a single HRTF function is defined as an individual right and left ear frequency response measured in a given point of the middle ear canal. The measurement is conducted in a far-field of the source placed in a free-field. Typical HRTFs are measured for both ears in a particular distance from the head of the listener for several different points in space. Thus the transmittance function related to the head depends on the azimuth angle, elevation angle and the frequency, and apart from that it has a different value for the left ear (L) and for the right one (R): $HRTF_{L,R}(\theta,\phi,f)$. The HRTF's time-domain equivalent is the Head-Related Impulse Response (HRIR).

In fact a measured transmittance function includes also a certain constant factor. This factor characterizes the measurement conditions - the measurement chamber characteristic and the measurement path. This is a reference characteristic, and the value of this parameter is determined by measuring the impulse response without the presence of the measured subject. Therefore by additionally taking into consideration the reference characteristic the result of the transmittance function can be presented as

$$h_{L,R}(\theta,\phi,t)=s(t)*c(t)*HRIR_{L,R}(\theta,\phi,t), \tag{1}$$

where: $h_{L,R}(\theta,\phi,t)$ - impulse response by the entrance to the ear canal,

$s(t)$ - measurement signal,

$c(t)$ - impulse response of the measurement system,

$HRIR_{L,R}(\theta,\phi,t)$ - impulse response related to the head.

In some conditions it can be assumed that $c(t)$ is constant and not influenced by the measurement point's position in space. Then the $c(t)$ value is a mean measurement result for several different azimuth angles and elevation angles. But if the measurement chamber does not fulfill the conditions of the anechoic chamber or in the room are present some elements which cause generating undesirable reflections, the $c(t)$ factor is influenced not only by the time, but also by the position of the measurement point in the space surrounding the listener, and it differs for the left and right ear: $c_{L,R}(\theta,\phi,t)$. In order to increase the accuracy of the measurement the $c_{L,R}(\theta,\phi,t)$ value can be measured for every measurement point and then these values can be applied while processing the results of the measurement.

Formula (1) can be also written in the frequency domain:

$$H_{L,R}(\theta,\phi,f) = S(f)\,C(f)\,HRTF_{L,R}(\theta,\phi,f)\,. \tag{2}$$

Further the HRTF value is calculated according to the following interrelations:

$$\left|HRTF_{L,R}(\theta,\phi,f)\right| = \frac{\left|H_{L,R}(\theta,\phi,f)\right|}{\left|S(f)\right|\left|C(f)\right|}\,, \tag{3}$$

$$\arg HRTF_{L,R}(\theta,\phi,f) = \arg H_{L,R}(\theta,\phi,f) - \arg S(f) - \arg C(f)\,, \tag{4}$$

$$HRTF_{L,R}(\theta,\phi,f) = \left|HRTF_{L,R}(\theta,\phi,f)\right| \exp\left[j \arg HRTF_{L,R}(\theta,\phi,f)\right], \tag{5}$$

and the HRIR value:

$$HRIR_{L,R}(\theta,\phi,t) = \mathcal{F}^{-1}\left[HRTF_{L,R}(\theta,\phi,t)\right], \tag{6}$$

where: $\mathcal{F}^{-1}$ - inverse Fourier transform.

It has been empirically proven that HRTFs are minimum-phase, therefore minimum-phase FIR filters are used to simplify the HRTF description interrelated (Cheng & Wakefield, 2001). Firstly, minimum-phase requirement allows to explicitly define the phase on the basis of the amplitude response. This is a consequence of the fact that the logarithm of the amplitude response and phase response in a casual system are related by the Hilbert transform. Secondly, the minimum-phase requirement allows to isolate the information about the ITD from the FIR characteristic describing the HRTF. When the minimum-phase filter has the minimum group-delay property and the minimum energy delay, most of the energy is accumulated at the beginning of the impulse response and the appropriate for the left and right ears minimum-phase HRTFs have zero delay.

In order to achieve the characteristic of the hearing impression related to a particular point in space there are three values to be measured: the left ear amplitude response, right ear amplitude response, and ITD. The characteristics of the filter include both the ITD and ILD information: time differences are included in the phase characteristic of the filter, whereas the level differences correspond with the total power of the signal transmitted through the filter interrelated (Cheng & Wakefield, 2001). The interaural time difference can be calculated by many various measurement methods: as a result of measurement with the participation of people, a result of the dummy-head measurement, simulations performed on the spherical and elliptical models, calculation based on Woodsworth-Slosberg formula (Minnaar et al., 2000; Weinrich, 1992).

Conducting the measurements for a big number of people is a complicated issue (Møller et al., 1995; Møller et al., 1992). The head-related transmittance functions show a great individual variability: the discrepancy between the measurement results reaches about 3 dB for the frequency to 1 kHz, 5 dB for the frequency to 8 kHz and about 10 dB for the higher frequencies. The first reason is an obvious dependence on individual physical body differences. Other reason are the measurement errors which are hard to be calculated in the final results - e.g. the error resulting from the differences in positioning the head in relation to the sound source or the differences in placing the measurement microphone in the ear canal. The individual HRTF variable is lower for the measurements conducted with a closed ear canal than for the measurements with an open ear canal.

3. HRTF measurement requirements

In general, the HRTF parameters are measured in anechoic chamber, e.g. Møller et al., 1995. During measurement it must be possible to place the sound source in a distance of minimum 1 m from the middle of the listener's head in each direction. Especially the direction above the listener's head is important because of chamber size. Taking into account the listener's height and minimal distance between the loudspeaker and the human head it can be assumed that the minimal height of the measurement room is ca. 3 m. The intermediate solution is to place the listener sitting on a chair, although in this case reflections from knees can be observed (Møller et al., 1996). The reflections from measurement device placed into the measurement room have more significant influence on the result of the measurement in comparison with the reflections from body parts (Møller et al., 1995), so these last can be omitted.

The HRTF measurement can be provided in ordinary room, e.g. auditorium (Bovbjerg et al., 2000; Møller et al., 1996). Measurements in non-anechoic chamber are convenient because of availability of this kind of room. Usually, when measurements with people go on a few days, there is a necessity to leave measurement devices in a fixed setup for long time. To make measurements in an ordinary room a noise-gate must be used for eliminating the reverberation signals (Plaskota & Dobrucki, 2004).

In the measurement room it is necessary to place the video devices for controlling and eventually recording the head position and head movements. Head movements are a significant source of errors. Verifying a head position allows to increase the measurement accuracy (Algazi et al., 1999; Gardner & Martin, 1995).

For measurements in many points in space around the listener it is needed to use many sound sources in fixed positions or use movable set of loudspeakers. Generally, it is possible to apply two methods of changing the position of the loudspeaker relatively to the listener's pinna. One of them is a movement of the sound source (one loudspeaker or set of loudspeakers) around the listener's head (Algazi et al, 2001; Bovbjerg et al., 2000; Grassi et al., 2003). The listener can improve measurement's accuracy by a visual control of head position. In the case of changing the listener's position relatively to the loudspeaker set (e.g. by chair rotation) it is needed to use an additional equipment for monitoring the head position (e.g. video camera) (Møller et al., 1995). A convenient situation is when the position of the listener and positions of the loudspeakers are fixed. In this situation very good control of measurement setup is obtained, but the number of measurement points is limited (Møller et al., 1996).

The next important parameter of the measurement system is a placement of measurement microphone in an ear canal. In publications four main positions are considered: a few millimeters over an ear entrance, an ear entrance, a few millimeters under an ear entrance, directly over the tympanic membrane (Pralong & Carlile, 1994). Additionally, the ear entrance closing influence on the measurement result is considered. It was found out that a smaller individual variation is obtained in measurements with closed ear entrance (Møller et al., 1995). It was also determined that the ear canal transfer function is independent of sound source position in the space around the listener (Bovbjerg et al., 2000).

The parameters of electroacoustic transducers have a great influence on the measurement result, especially a frequency response. The frequency responses of microphones are more important than the frequency responses of loudspeakers (Plaskota, 2003). It is suggested to use loudspeakers with a frequency response without large deeps (Møller et al., 1995).

In the studies there are informations available about used signals during the HRTF measurements. One of the applied signals is the Maximum Length Sequence (MLS) (Møller et al., 1995). It is possible to use Golay codes (Algazi et al., 2001), but difficulties in results interpretation are known (Zahorik, 2000). In anechoic chamber, the use of chirp signal is adequate to measurement conditions. It can be supposed that in a non-anechoic chamber the impulse signal is applied. It comes from a necessity of providing good measurement conditions.

4. Measuring system

4.1 Conception of measuring system

The HRTF measuring device is built for a special group of test participants. It is assumed that the measurement will be made for people with severe vision problems (Bujacz & Strumiłło, 2006; Dobrucki et al. 2010). Therefore, the device is designed to reach many demands such as the highest automation of measurements which assures a short measurement time (ca 10 minutes) and offers great ease of manipulation. The participant of the test should feel comfortable during the measurement process and should be given sufficient information on each part of the measurement. To reach these demands, the device is equipped with a bidirectional communication system allowing the participant to report the problem at any time. In addition to voice communication, a visual control of the room is provided. It is possible to monitor the test room using a camera mounted on an arc with loudspeakers.

To provide a short measurement time the HRTFs are measured for both ears simultaneously. The way sound sources are configured significantly shortens this time too. The loudspeakers are mounted on vertically positioned arc (see Fig. 1). It allows to measure the range of vertical angles from -45° to +90° in one chair position. In certain points in the space of the room the measurement is made by switching the measurement signals to subsequent loudspeakers by an electronic switch.

The number of measurement points for elevation angles is adjusted by changing the number and position of the loudspeakers. On the other hand, the number of measurement points for horizontal angles depends on the size of the rotation step of the chair. The rotation of the chair is controlled by a stepper motor which assures high horizontal resolution. Default vertical resolution is 9° in regular sound source positions. Assuming the same horizontal resolution the number of measurement points is 640. The measurement in 16 points for one horizontal angle and simultaneous measurements for both ears allows to conduct the whole measurement in less than 10 minutes. Obviously, the number of measurement points can be modified. Changing the resolution in a vertical plane means changing the position of the loudspeakers. In a horizontal plane, changing the resolution means changing the rotation step of the chair.

Fig. 1. Overview of the HRTF measurement equipment.

The HRTF measurement can be done within the range of frequencies from 200 Hz to 8 kHz. The lowest frequency depends on the test room parameters. The device works in an anechoic chamber, therefore the cut-off frequency of the chamber limits the operational range of the device. The high cut-off frequency of the device is on the one hand confined by the set of loudspeakers, and on the other - by the set of microphones. Miniature microphones used in hearing aids, but with an untypical flat frequency response, are used in the device (Fig. 2). Another factor limiting the high cut-off frequency are the dimensions of microphone fixing elements. For 5-mm tubes the wave phenomena are significant for the frequencies above 10 kHz.

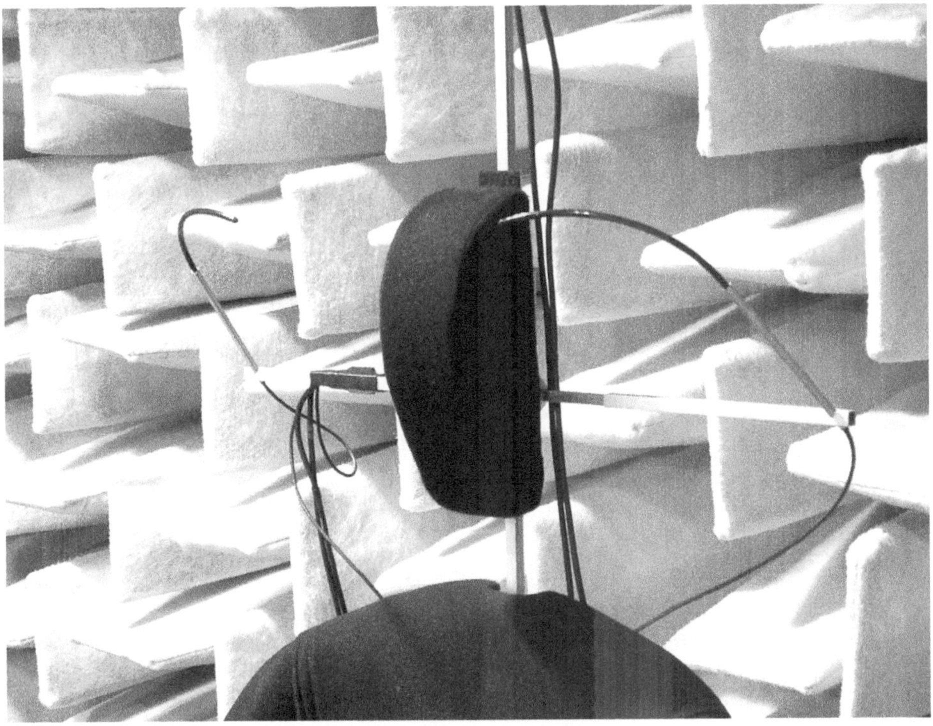

Fig. 2. The scheme of setting the measurement microphones.

The system is operated via portable IBM PC computer to control measurements and data acquisition (Pruchnicki & Plaskota, 2008). The device communicates with the computer through a USB interface. At the same time signals operating the device, measurements signals and camera pictures are transmitted via interface. A special feature of the device is its compact construction and modularity which makes it very easy to assemble or disassemble and convenient to transport.

4.2 Measurement algorithm

The measurement of a single HRTF is accomplished using a transfer method, which is popular in digital measurements systems. A wide spectrum measurement signal is used for stimulation. The system uses the following signals: chirp, MLS, white noise, pink noise, Golay codes. The length of a generated signal can be changed within the range from 128 up to 8192 samples. Sampling frequency is 48 kHz but it is possible to decrease it. The stimulating signal is repeated several times in order to average the answer of the system in the time domain. This operation allows improving the S/N ratio of received responses. There is no need to apply longer measurement signals because, according to other researches, HRTFs may be presented even with such resolution as 100 Hz. On the other hand, responses determined in the system will be used for convolution with real signals and therefore they cannot be too long. Moreover, long measurement signals make the assessment time longer.

The whole measurement procedure is comprised of two parts: the measurement of reference responses and the measurement of regular HRTFs. The measurement of reference responses is made for all measurement spots determined by the system operator. During this procedure microphones, loudspeakers and the whole system work exactly like during any regular measurement. The only difference is that there are no test participants. The HRTF measurement results obtained in the second part are related to reference responses obtained before.

Using a reference response for each measurement point in the space allows limiting many inconvenient effects which decline measurement accuracy (Plaskota & Pruchnicki, 2006). Especially the influence of frequency responses and directivity responses of loudspeakers and microphones is eliminated. The influence of a test room and the reflection from the device elements on measurement results is partly reduced.

The final result of the measurement process are HRIRs (Head Related Impulse Response, that is HRTF's reverse Fourier transform) produced to allow their direct use in convolution with real signals.

4.3 Measurement procedure

The measurement procedure comprises several phases. The first is the system activation and configuration. It involves determining the horizontal and vertical resolutions of measurements. The next step is the selection and fixing of active test loudspeakers position. At this stage the kind of measurement signal and the number of averages should be chosen as well as the calibration of sound level should be carried out.

In the second phase, participant of the test should be properly positioned in the chair, so that the 0° loudspeaker is placed on the ear canal entrance level and the microphones are located at ear canals entrance. The setup of the loudspeakers' arc in relation to the microphones can be monitored using the camera view.

After the test participant measurement is completed, the reference responses are measured. Once the preparation is finished, regular HRTF measurements are carried out according to earlier parameter setups.

In the last phase of the procedure, measurement results are saved in plain text files in the form of the HRIR. Such storing allows access to the test results from any other application at the same time, and is clear to the user.

4.4 System control software

In order to apply the measurement procedure, dedicated software was designed. The modularity of this software, which consists of two basic elements, is its special feature. Figure 3 presents the main window used to control measurements. Via this interface the operator can influence the measurement course and conditions as well as all configuration parameters. Additionally, there is also a test participant communication part.

A separate element of the software is an OCX control which exchanges data between the device and the user interface. Calling certain functions of the control it is possible to steer such parameters as the armchair rotation, the loudspeakers movement or switching.

Applying this solution allows to use the device for purposes not provided by the user interface of the system.

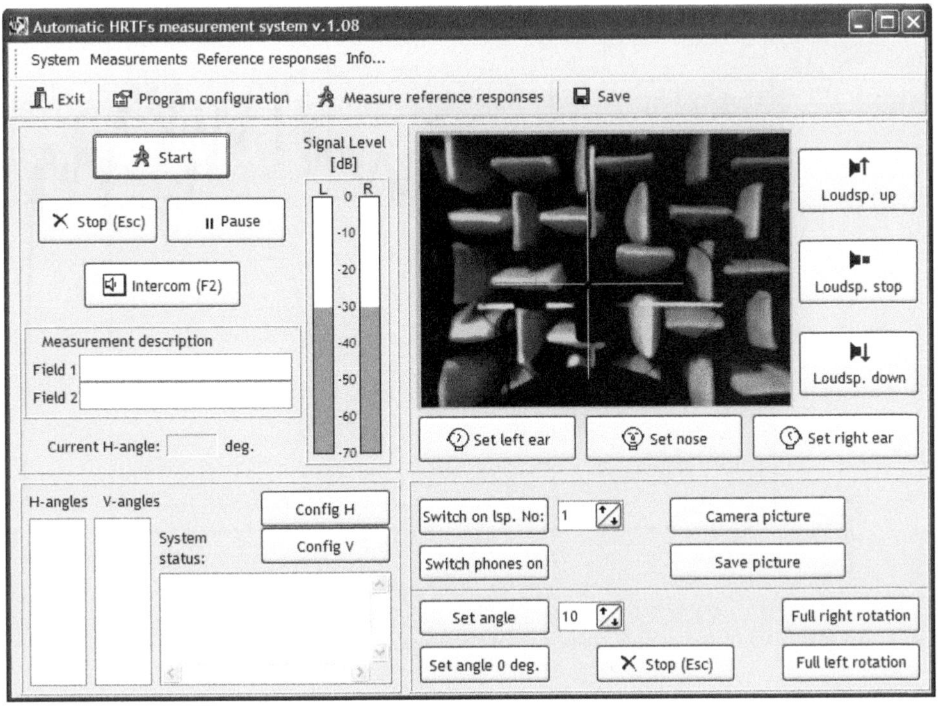

Fig. 3. The main window of the HRTF measurement control software.

4.5 Parameters of the device

The HRTF measuring device has 16 sound sources. The reason for using such number of loudspeakers is the need to conduct tests for many various spots in the listener's surrounding in the shortest time possible. The different positions are found in the following way: the participant in the test turns around his vertical axis while taking a step in defined direction. The distances between the steps define the spatial resolution of the measurement in horizontal dimension. The vertical dimension of spatial resolution is determined by the arrangement of loudspeakers placed on the arc including range of vertical angles between -45° and +90°.

For the precision of the measurement it is important to use a point sound source. The source should produce test signals in the entire operational frequency range of the device. In order to fulfill these conditions two-way car loudspeakers were applied. According to producer data the loudspeakers should operate within a small box. Figure 4 presents an example of amplitude frequency response of the used loudspeakers. The loudspeakers' operational range of frequency is between 200Hz and 20kHz. It should be noted that the frequency responses are not equalized and differ slightly for each loudspeaker less than 4dB. The applied measurement of reference response in the device for each tested spot neutralizes the influence of measuring set on the results of the tests.

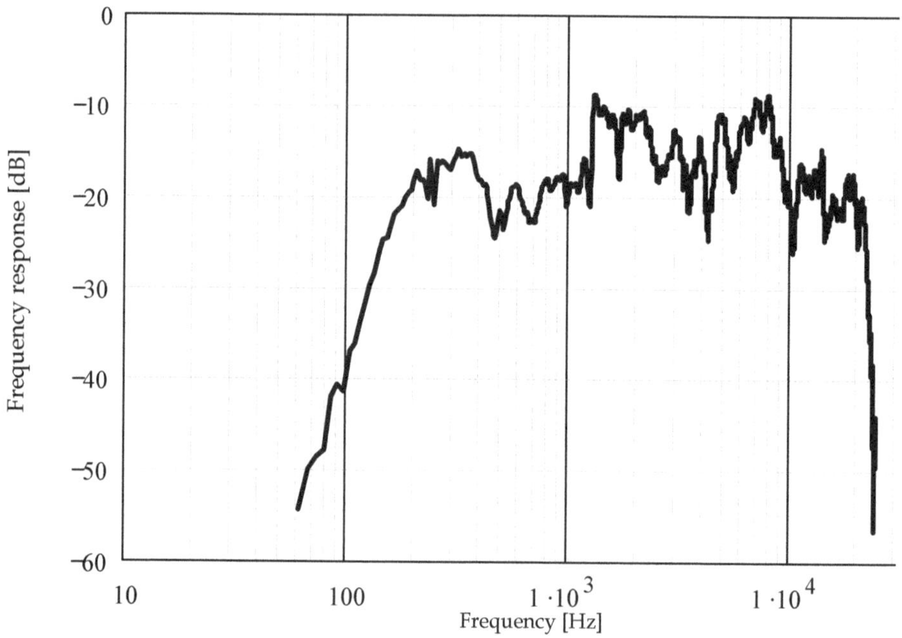

Fig. 4. An example of frequency response of the loudspeaker used in the HRTF measuring device.

The measurement microphones used in the device are the same as those used in hearing aids. It should be underlined that the particular type of microphones has equal frequency response in its entire operational range of ca. 60Hz and 8kHz (Figure 5). It means that these microphones are not commonly used in the hearing problems treatment. The choice of microphones was determined by the importance of the quality of the device and therefore the similarity of frequency responses of each microphone was achieved. The other advantage of this particular type of microphones is their small size. That is indeed a significant feature since it allows reducing the size of the outer cover. This minimizes cover impact on the acoustic field around the head of the test participant.

The operational frequency range of the HRTF measuring device is limited by the lower cut-off frequency of the anechoic chamber in which the tests are conducted. The other factor influencing lower frequency is the operational frequency range of loudspeakers. The lower cut-off frequency within the operational range of the loudspeakers is higher than the value of the cut-off frequency of the anechoic chamber thus the operational frequency range for the entire device starts at around 200Hz.

The upper cut-off frequency limit of the device is determined by the frequency range of the microphones. Hence the upper cut-off frequency is about 8kHz. The other factor carrying impact on operational frequency range of the device is the influence of microphones' covers on the acoustic field around the head of the test participant. The microphones are placed in ca. 5-mm diameter tubes. The wave phenomena for this type of construction elements have a significant impact for 10 kHz frequency and above (Dobrucki, 2006). But that is transversal dimension of applied elements; the length of the microphones cover is more significant dimension in this case and can influence acoustic field within the operational range of the device.

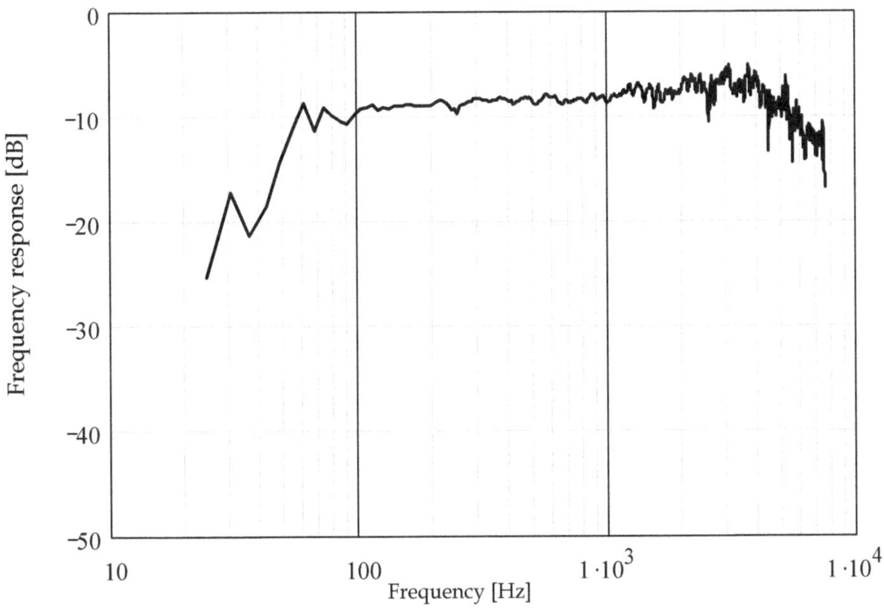

Fig. 5. An example of the measurement microphones frequency response.

One of the methods to eliminate the impact of microphones' cover elements on the acoustic field around the head of the test participant is using microphones placed directly in the matter closing ears' canals (Møller et al.,1995). In this case the usage of cover elements could be avoided and the solution is more advantageous for the precision of the results. On the other hand, the use of plain microphone without a rigid support construction attached to the measurement device gives way to the uncontrolled head motions. The impact of this fact on the tests' results is described in section 5.2. It should be noted that the use of the microphones without rigid support increases the amount of time needed for exact positioning of the participant's head and also makes the measurement of the reference response more difficult.

5. Practical aspects of using the HRTF measuring device

5.1 Verification of the measurements results using dummy head

It is not impossible to verify results of measurements given by presented device directly, but the correctness of measurement results can be verified in indirect process. The first method is a subjective test for a person who had been measured using this device. During the test the signal convolved with a result of HRTF measurement is presented – this operation sets up a virtual sound source in specific point in space around a listener (Dobrucki et al., 2010). The consistence of point determined by convolution process and point indicated by listener is tested. If the consistence is correct, the result of measurement is also correct. Other method for verification of measurement result is a comparison of measurement results with the results of numerical calculation (Dobrucki & Plaskota, 2007).

The correctness of a measurement result was examined by measurement of dummy head (Neumann KU100). The result of measurement was compared to the results of numerical calculation. The dummy head had been placed in measurement device, next the whole measurement process was conducted. The use of a dummy head can eliminate some inconvenient occurring during measurement of a person, i.e. the head movement that provides to large measurement deviation.

The Boundary Elements Method (BEM) has been used to perform the numerical calculation of HRTF (Dobrucki & Plaskota, 2007). The numerical model is a representation of geometrical shape of dummy head, especially with emphasis on accordance of pinna model with geometry of real object. Differences between real object and numerical model are smaller than 0.1 mm (Plaskota, 2007; Plaskota & Dobrucki, 2005). The measurement of acoustical impedance of dummy head has been done (Plaskota, 2006) and the result was used as a boundary condition.

Figure 6 show HRTF measurement and numerical calculation results for azimuth 90°, elevation 0°, for ipsilateral ear (located closer to the sound source). There are three graphs in

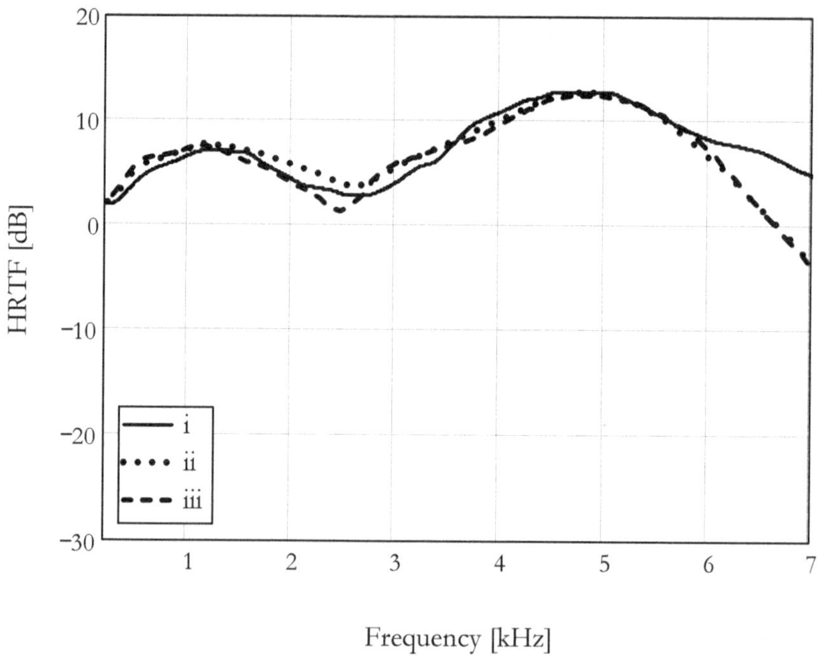

Fig. 6. Measurement and simulation results: azimuth 90°, elevation 0°, ipsilateral ear. Detailed description of symbols in text.

the diagram. The particular letters represent the following cases: i - the measurement result, ii - the result of simulations without impedance boundary conditions (the rigid model), iii - the result of simulations with impedance boundary conditions in whole modeled area except for the pinnas, for which the same boundary condition as for the rigid model was assumed.

Measurements results are in good accordance with calculation results below 6 kHz. On the basis of comparison between the measurement and calculation results, it was found that measurements results are proper. There are some reasons of difference between measurement and calculation results above 6 kHz. At first, the microphone set has been not taken into consideration during the numerical calculation: microphone enclosures probably produce the wave phenomena in frequency range of 8-10 kHz. Secondly, the high cut-off frequency of a numerical model is about 7 kHz.

5.2 Discussion of problems encountered during measuring process

One of the major challenges faced during the tests was positioning of the listeners relatively to the microphones. In the first tryout the microphones were fixed in a way similar to medical stethoscope. Microphones were coupled with flexible wires; these were attached to ears in such a way that the microphones were suspended and their transducers were on the level of ear canals entrances. The head of the human subject was placed on a holder fixed to the extension of the armchair's back. The distance between the head and the head holder was adjusted using cushions of different sizes. By increasing or reducing the amount of cushions the head of the test participant was placed at varied distances from the holder. The position of the head was controlled trough electronic visual system. On the screen the researcher could see the lines matching the position of ear canals entrances and adjust the position of the head accordingly.

This method was verified negatively. The participants during the tests do move their heads slightly. Using a band to fasten the head to the holder did not bring any significant improvement. Those minimal head motions have an impact on the geometry of the measurement arrangement. In the case of high-resolution measurement performance the stability of geometrical configuration: the sounds source - the microphone, is crucial for the accuracy of the measurement.

The other method of attaching measurement microphones was then proposed and tested. The microphones were fixed on a nonflexible construction. The construction had the possibility of adjusting the position of the microphones, though. The microphones were placed on the level of ears' canals entrances like before. Applying the fixed construction resulted in the fact that the test participant felt the microphones support structure limitation. In this case it was easier for the test participant to control head motions: when they appeared, it was a simpler task to put the head back in right position. The other advantage was that the distance between the microphones and the head holder was preset. The researcher avoided long process of positioning the head in relation to the holder. The only thing to be done was locating the listener in a proper elevation according to the sound sources. This solution is presented in Figure 2.

The most important of all the advantages of this particular way of setting the microphones is the possibility of the precise microphones positioning while measuring the reference response and while conducting the tests with human participant, as well. It is very significant for the accuracy of measurements, particularly when the impact of the measuring set and that of the research room should be minimized.

Although conducting the measurement of reference response for each assessment spot excludes the impact of the measurement set, some acoustic phenomena cannot be reduced this way. During the tests it was observed that for the 90° elevation angle and for the angles close to this value, in the reference impulse response the sound reflection from the seat of the armchair was observed. (Figure 7, Time ≈ 7 ms). During the test involving the participant the reflection does not occur because the person is seated in the armchair and therefore covering the seat surface. The phenomena of reflection while measuring the reference response, after the sound reaches the seat of the armchair, could be eliminated by using additional sound diffusion device.

Summing up, in the case of sound reflection from elements covered by the test participant, the use of the reference response is not sufficient. Similar phenomena were observed for different angles but never to such extent as in the case of 90° elevation angle.

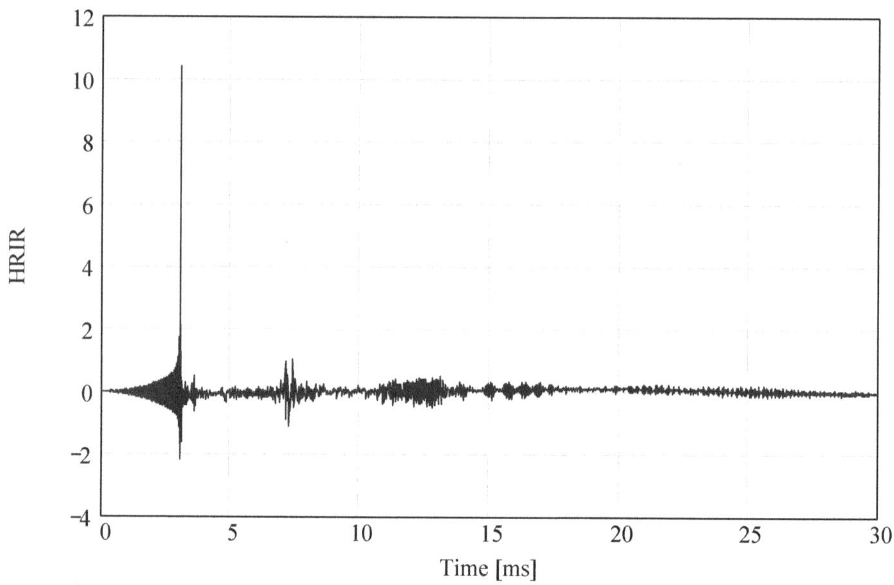

Fig. 7. The impulse response for vertical 90 ° angle.

The impact of the research room is neutralized as much as possible by measuring the reference response. While measuring the reference responses the components with frequencies around 80 Hz were singled out (Figure 8). It could be said that it was the effect of the wave interference inside the room. Although the tests are conducted in anechoic chamber, it is a place designed basically to make measurements involving machines and there is a concrete platform in the middle of the chamber intended for placing machines. This can contribute to forming interference phenomenon. Repositioning the device inside the chamber reduced the presence of the interference occurrence. Nevertheless, the phenomenon was observed only for frequencies outside the operational range of the device.

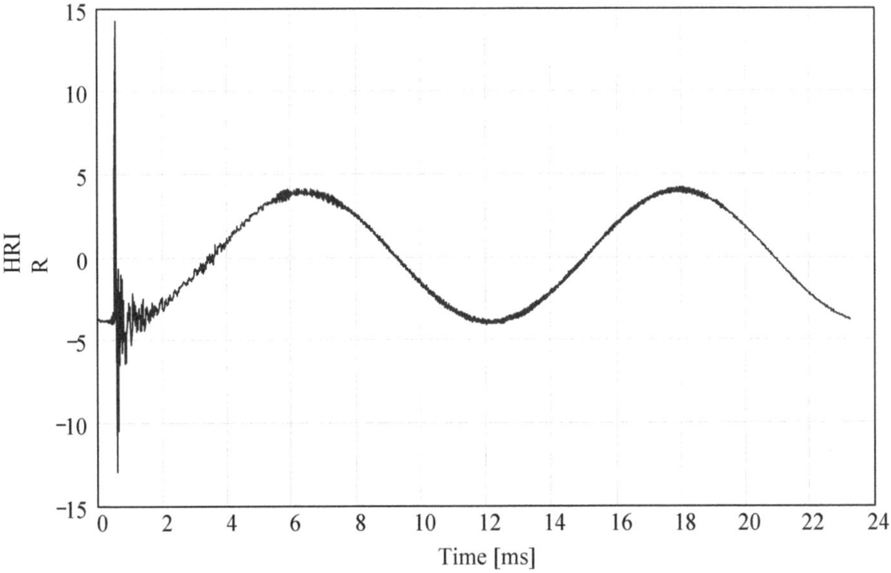

Fig. 8. The impulse response containing a small frequency component.

6. Conclusions

The HRTF measurement system allows a very fast measurement of HRTF with high spatial and frequency resolution. The applied operational algorithms of the system guarantee repeatability of measurements and minimalization of the influence of many disadvantageous factors on measurements results. Compact structure and modularity of construction of the system allows an easy transport of the device. The encountered problems were discussed together with the eventual solutions to them. On the basis of conducted measurements and subjective tests it could be assumed that the device measures the HRTFs accurately enough to recreate the position of the sound source in the space surrounding the listener. The scope for future tests is to verify if the proposed adjustments eliminate the impact of the research room by conducting tests in the reverberation room sufficiently. To eliminate the influence of physical movements of the participant it is recommended that the tests should be conducted using a dummy head.

7. References

Algazi, V.R., Avendano, C. & Thompson, D. (1999). Dependence of subject and measurement position in binaural signal acquisition, *J. Audio Eng. Soc.*, Vol. 47, No. 11, (November 1999), pp. 937–947

Algazi, V.R., Duda, R.O., Thompson, D.M. & Avendano, C. (2001). The CIPIC HRTF database, *Proceedings of IEEE WASPAA 2001*, New Paltz, NY, pp. 99–102

Batteau, D.W. (1967). The role of the pinna in human localization, *Proc. R. Soc. London, Ser. B 168*, 1967, pp. 158–180

Blauert, J. (1997). *Spatial Hearing*, The MIT Press, Massachusetts, 1997

Bovbjerg, B.P., Christensen, F., Minnaar, P., Chen, X. (2000). Measuring the Head-Related Transfer Functions of an Artificial Head with a High Directional Resolution, *Presented at the 109th AES Convention*, 22-25 September 2000, Los Angeles, Preprint 5264

Bujacz, M., Strumiłło, P. (2006). *Stereophonic representation of virtual 3D scenes – a simulated mobility for the blind*, In: New Trends in Audio and Video, Vol. 1, Białystok 2006

Cheng, C.I., Wakefield, G.H. (2001). Introduction to head-related transfer functions (HRTFs): Representations of HRTFs in time, frequency, and space, *J. Audio Eng. Soc.*, Vol. 49, No 4, April 2001, pp. 231-249

Dobrucki, A.B. (2006). *Electroacoustic transducers* (in Polish), Wydawnictwa Naukowo-Techniczne, Warsaw 2006

Dobrucki, A.B., Plaskota, P. (2007). Computational modelling of head-related transfer function, *Archives of Acoustics*, 2007 Vol. 32, No 3, pp. 659-682

Dobrucki, A.B. Plaskota, P., Pruchnicki P., Pec, M., Bujacz, M., Strumiłło, P. (2010). Measurement system for personalized head-related transfer functions and its verification by virtual source localization trials with visually impaired and sighted individuals. *J. Audio Eng. Soc.,* 2010, vol. 58, No 9, pp. 724-738

Gardner, W.G. & Martin, K.D. (1995). HRTF measurements of a KEMAR, *J. Acoust. Soc. Am.* 97, 3907-3908

Grassi, E., Tulsi, J., & Shamma, S.A. (2003). Measurement of headrelated transfer functions based on the empirical transfer function estimate, *Proc. 2003 Intl. Conf. on Auditory Displays _ICAD 2003_*, Boston, MA, pp. 119–121

Hartmann, W.M. (1999). How we localize sound, *Phys. Today 1999*, pp. 24–29

Hartmann, W.M. & Wittenberg, A. (1996). On the externalization of sound images, *J. Acoust. Soc. Am.* 99, pp. 3678–3688

Horbach U., Karamustafaouglu, A., Pellegrini, R., Mackensen, P. & Theile, G. (1999). Design and Applications of a Data-based Auralization System for Surround Sound, *Presented at the 106th AES Convention*, 8-11 May 1999, Munich

Hen, P., Kin, M.J. & Plaskota, P. (2008). Conversion of stereo recording to 5.1 format using head-related transfer functions, *Archives of Acoustics*, 2008, Vol. 33, No 1, pp. 7-10

Minnaar, P., Plogsties, J., Olsen, S.K., Christensen, F. & Møller, H. (2000). The Interaural Time Diference in Binaural Sythesis, *Presented at the 108 AES Convention*, 19-22 February 2000, Paris, Preprint 5133

Møller, H., Sørensen, M.F., Hammershøi, D. & Larsen, K.A. (1992). Head Related Transfer Functions: Measurments on 24 Human Subjects, *Presented at the 92 AES Convention, February 1992, Viena*

Møller, H., Sørensen, M.F., Hammershøi, D. & Jensen, C.B. (1995). Head Related Transfer Functions of Human subjects, *J. Audio Eng. Soc.*, Vol 43, No 5, May 1995, pp. 300-321

Møller, H., Sørensen, M.F., Hammershøi, D. & Jensen, C.B., (1996). Binaural Technique: Do We Need Individual Recordings?, *J. Audio Eng. Soc.*, Vol 44, No 6, June 1996, pp. 451-469

Moore, B.C.J. (1997). *An Introduction to the Psychology of Hearing*, 4th Ed., (1997) Academic, San Diego, 1997

Plaskota, P. (2003). Evaluation of microphone parameters for HRTF measurement (in Polish), *In: X International Symposium the Art of Sound Engineering. ISSET 2003.* Wrocław, Poland, 11-13 September 2003

Plaskota, P. (2006). Measurement of acoustical impedance of human skin, *In: The 6th European Conference on Noise Control EURONOISE 2006, Proceedings, Acoustical Society of Finland.* Tempere, Finland, 30 May - 1 June 2006

Plaskota, P. (2007). Acoustical model of the head for HRTF calculation, *In: LIV Open Seminar on Acoustics.* Rzeszów-Przemyśl, Poland, 10-14.09.2007

Plaskota, P. & Dobrucki, A.B. (2004). Selected problems of HRTF measurement (in Polish), *In: Metrology Congress,* Wrocław, Poland, 6-9 September 2004

Plaskota, P. & Dobrucki, A.B. (2005). Numerical head model for HRTF simulation, *In: The 118th AES Convention, Barcelona,* Spain, May 28-31, 2005, Preprint 6510

Plaskota, P. & Kin, M.J. (2002). The use of HRTF in sound recordings (in Polish*), In: IX New Trends in Audio and Video.* Warsaw, Poland, 27-28 September 2002

Plaskota, P. & Pruchnicki, P. (2006). Practical aspects of using HRTF measuring device, *Archives of Acoustics*, 2006, vol. 31, no 4, suppl., pp. 439-444

Plaskota, P., Wasilewski, M. & Kin, M.J., (2003) The use of Head-Related Transfer Functions in auralization of sound recordings (in Polish), *In: X International Symposium the Art of Sound Engineering. ISSET 2003.* Wrocław, Poland, 11-13 September 2003

Pralong, D. & Carlile, S. (1994). Measuring the human head-related transfer functions: A novel method for the construction and calibration of a miniature in-ear recording system, *J. Acoust. Soc. Am.* 95, pp. 3435–3444. 1994

Pruchnicki, P. & Plaskota, P. (2008). HRTF Automatic Measuring System, *Archives of Acoustics,* 2008, Vol. 33, No 1, pp. 19-25

Weinrich, S.G. (1992). Improved Externalization and Frontal Perception of Headphone Signals, *Presented at the 92 AES Convention, February 1992, Viena*

Zahorik, P. (2000). Limitations in using Golay codes for head-related transfer function measurement, *J. Acoust. Soc. Am.*, Vol. 107, No 3, March 2000, pp. 1793-1796

5

Round Wood Measurement System

Karel Janák
Mendel University Brno,
Czech Republic

1. Introduction

The roundwood volume is one of basic parameters at its sale and purchase being indispensable operational information of every saw mill. The tree stem or its part (log) is an irregular body the form and volume of which can be determined by simple operational procedures only approximately. According to possibilities of measurement (conditions, equipment, time consumption etc.), methods of calculation and approach to problems more procedures have been proposed to determine the volume of roundwood. Many of these methods and procedures are used today. In Central Europe, Huber method became most widespread for manual measurements in the forest. The method compares the tree stem to a cylinder. The main advantage of this method consists in the low demands of measurements (diameter - 2 times perpendicular at each other under bark /u.b./ in the centre of the log length, diameter and length accurate to whole centimetres), easy practicability, no special equipment and sufficient accuracy.

Lines in log yards of present processing plants are ordinarily equipped with optoelectronic sensors and control systems evaluating virtually continuously the roundwood diameter (by at least 10 cm length) accurate to ±1 to ±2 mm and length accurate to ±1 cm. Also at the electronic measurement and calculation of the log volume on the basis of electronic measurements more procedures are used. Due to results in the determination of a mid diameter or volume electronic procedures are not even consistent with manual methods or with each other and their results do not correspond even to the geometric volume of logs. Different results of measurements are often the reason of doubts about the accuracy (rightness) of measurements reflecting also relationships between suppliers and processors of wood.

The aim of the presented paper is the description and analysis of currently used principles of sensing and evaluating the roundwood volume in Central Europe as well as quantification of deviations, which originate due to the use of various method of the determination of roundwood volume at the simultaneous determination of deviations on basic parameters of roundwood (diameter and length). Thus, the paper outlines possibilities to compare results determined according to different procedures.

The paper tries to describe comprehensively problems of sensing and evaluating the roundwood dimensions in order mutual relations to be obvious. In the descriptive part, it deals with methods of sensing the log diameter and length, which are used in Central Europe today. The paper mentions also main rules and regulations, which are related to the measurement and reception of timber. It stems from the author's survey of using sensing and evaluation systems in the Czech Republic and literature data on these problems in surrounding countries (mainly Austria, Germany and Slovakia).

In the analytical and result part, a general algorithm is derived of processing sensed data. Steps are determined leading to different results of particular regulations. The results are compared with results of a method, which tries to evaluate the volume of logs (closest to the geometrical volume) from data taken by systems used at present. In relation to the volume determined in this way properties of existing procedures are also mentioned.

At the end, the paper indicates directions for the further progress of research and agreements as well as legislative regulations with a view to minimize differences in the measurement and determination of the volume of roundwood.

2. Round wood

Dimensions and volume of wood together with its quality are basic data accompanying wood from the stage of a young stand until the end of the wood product service life. In many cases, it is sufficient to know only approximate values of dimensions and volume, e.g. growing stock, the volume of cut, and roundwood supplies on log yards. On the other hand, operations controlling production or where wood is the subject of business - goods, are sensitive to accuracy.

The tree stem or its part, log, is a body, the form of which is mostly compared to a truncated cone, paraboloid or cylinder. However the real form does not correspond to any regular geometric body. Moreover, it is affected by considerable individual diversity given by the tree species, tree age, site, tree position in the stand, care of the stand, type and extent of attack, mechanical damage either natural or caused by man activities and many other effects. From the aspect of dimensions, these effects become evident in various taper, sweep, flattening, root swelling, buttress, burrs and cracks. Production defects of wood are represented by remains of branches and damage to the stem surface at branching and handling, chamfer cuts, hinges (holding wood) after cutting or bucking (cross-cutting), cracks etc. All these properties have to be taken into account at stem or log measurements and at the evaluation of their dimensions and volume. Bark - its thickness and condition, represents a separate problem at the measurement of wood.

3. Ways of dimension scanning

Measurements and methods of their implementation are given by the need of data on wood (in which moment, type of data and satisfactory accuracy) and by the technical and practical feasibility of measurements.

According to the type of manufacture determined for the wood, mass measurements of logs are sufficient (production of agglomerated materials, paper, and cellulose) or it is necessary to measure parameters of particular logs separately (production of sawn wood, veneers). At mass measurements, a basic parameter consists in the total volume of supply; extents of the diameter and length of logs are usually only completive data. Volumes, dimensions and often even the number of particular logs are not recorded at these types of production because they are not decisive.

At the measurement of particular logs we need to obtain following data:

- The log length - according to the length, the following processing the stem is determined. It also serves for classification and calculation of the volume of logs for production and commercial purposes.
- The log diameter in the length centre - it serves for the approximate log volume calculation (at the majority of methods /not exclusively/).
- Log diameter at its top end - it is decisive for the way of log processing, serves for the subsequent classification of logs. For the volume calculation it can serve in cases when the log centre is not available, e.g. at log yards. However, results obtained in this way are less accurate.
- Log diameter at its butt end - it is decisive from the aspect of the line passage (clearance of subsequent machines)
- A diameter continuously along the whole log length - it is necessary for subsequent processing - cross-cutting. The result of measurement consists in the suitable place of cross-cutting (necessary length and top end diameter), removal of defects, and the highest yield.

Dimensions and volumes of particular stems (logs) have to be known already in place of felling, at the latest in places of skidding before the wood haulage from forest (cut records, output of workers and their remuneration). If felling is carried out manually by one-man chain saws then means used for measurements are usually simplest, namely a calliper and tape.

Fig. 1. Electronic calliper with a tape.

Manual measurement and its unpretending equipment is quite satisfactory from the aspect of the accuracy of its results (units - cm) and possibilities to carry out subsequent cross-cutting tree-length logs. Values of diameter and length create, at the same time, basic data for records and calculations of volumes. It is usually determined according to tables. Thus, the accuracy of the measurement of the diameter and length of tree-length logs in the field gives generally the accuracy of values of dimensions and volume of logs given on bills of delivery.

Harvesters are equipped with electronic (electromechanical) scanning systems. Attainable accuracy of the measurement is higher (units - mm). However, results of measurements are substantially dependent mainly on the pressure of particular parts of the scanning equipment (delimbing knives or feeding cylinders). Values of these parameters change continuously according to conditions of felling (tree species, dimensions, and season). Therefore, results of measurements at these systems have to be continuously checked and revised (even several times per shift).

The stem diameter is scanned according to the deviation of delimbing knives or the deviation of arms of feeding cylinders. At scanning by means of delimbing knives (see Fig. 2), the angle of the deviation of two knives which press the stem to the third stable knife is scanned (the third point necessary to define a circle and the subsequent calculation of its diameter). At the measurement by means of feeding cylinders the deviation of arms usually of 3 feeding cylinders is scanned. These cylinders press the stem with each other (3 points of a circle are obtained on principle according to the figure) or a couple of conical cylinders placed against each other (the diameter is scanned directly). The deviation of cylinders is usually scanned by an induction sensor or potentiometer. The stem length is scanned by a pinion (sprocket) pressed to the stem surface during its movement. Turning the pinion is sensed by an impulse generator. Scanning directly by feeding cylinders is not used because at high loading (stem start, delimbing), the slippage of cylinders occurs on the stem surface. Errors resulting from the different length of the stem surface curve and its length are negligible.

Fig. 2. Photo + scheme of scanning the stem diameter and length at harvesters (both figures according to Ponsse).

Log yards of forest enterprises and wood processors are equipped with cross-cutting-sorting carriages (all round cars) or lines. The equipment serves for the preparation of supplies of logs to manufacturing plants or for the preparation of logs according to requirements of subsequent production (With respect to the increasing proportion of harvester logging

/generally/ the importance and number of log conversion depots declines). Scanning equipment for the log diameter and length is always part of the equipment. At scanning the log diameter, the number of taken values of the log diameter in one place of length, direction of scanning, accuracy and density of scanning in the course of length are essential.

One-directional way of scanning - 1D takes the value of the log diameter in one, usually vertical direction. As for possible technical designs, measuring frames are used nearly exclusively. The principle of scanning indicates a scheme on Fig. 3.

One-directional scanning is not able to record the log flattening and only very roughly the log curvature.

Using 1D scanning devices for the purpose of electronic reception is, therefore, unsuitable and results are affected by relatively considerable errors.

Bi-directional measuring (2D) is carried out by two systems perpendicular at each other and placed in one frame. It can be installed in such a way one scanning to proceed in vertical direction and the second scanning in horizontal direction or both measurements perpendicular at each other to proceed at an angle of 45° with respect to a horizontal level.

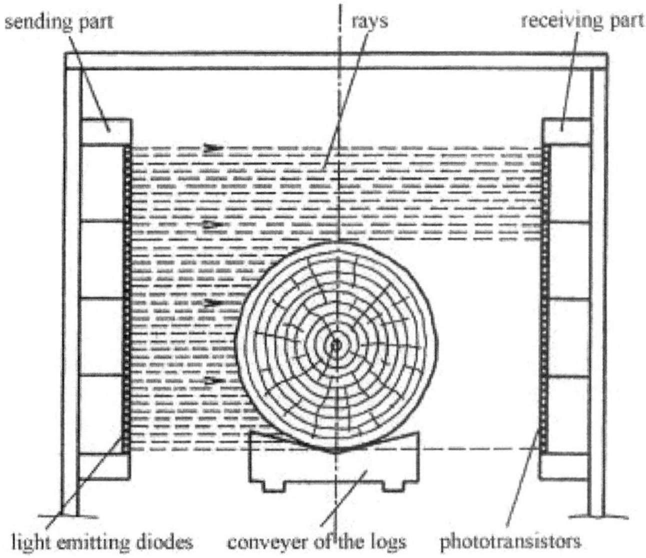

Fig. 3. The principle of scanning the log diameter by a one-directional scanning frame.

The substantial advantage of *2D* scanning consists in the better record of the measured log form and thus a possibility to calculate objective values of the log diameter in the place of measurement. The evaluation of stem curvature (sweep) is markedly more accurate. To calculate the log volume both methods of placing the sensing elements are equivalent. However, vertically and horizontally oriented sensors record flattening the measured logs better - a flattened log shows a tendency "to lie down flat" on the conveyer, i.e. vertically measured diameter shows a smaller value than a diameter measured horizontally. At systems oriented at an angle of 45° this difference disappears and both data approach the average value of log diameter in a given place.

(a) vertical and horizontal position of scanning systems

(b) placing the scanning systems at an angle of 45°

Fig. 4. 2D measurements by a scanning frame. Both systems are placed in one frame. The conveyer can come through the place of measurement (a) or it is interrupted (b).

From the aspect of taking log diameter it is better when the conveyer is interrupted in the place of measurement. Thus, measurements are not affected by "erroneous" data caused by the transport track and passing drive dogs, which have to be subsequently filtrated by special program equipment. However, from the aspect of scanning the log length and mainly evaluation of the stem curvature (generally form) change of the log position on the conveyer is "more dangerous". It will be recorded by the sensing device as a change in the diameter position; however, the sensor is not able to differentiate if it was caused by the log curvature or by the change of its position on conveyers. Unfortunately, the conveyer interruption supports the change of the log position.

Scanning the peripheral curve (3D measurements) makes possible to scan the whole form of the log cross-section in the measured place. There are more principles of scanning and at some equipment, they are also combined. Usually, an intensive narrow light line is projected on the log in the place of measurement perpendicular to its axis. The light "cross section" is subsequently taken by cameras. Based on their signal, the form and position of the cross section is constructed and the area centre is evaluated. Subsequently, distances are evaluated of opposite points (i.e. log diameter) usually at least in an interval of 5° (36 values of the log diameter). Changes of the cross section position in the course of scanning of the whole log length make possible to evaluate the stem curvature and form anomalies.

The accuracy of diameter scanning ranges usually within the limits ±1 to ±2 mm. Values of diameter are evaluated usually at 10 cm intervals of the measured log length.

Systems for scanning log lengths were reduced in the course of development, namely to a laser system using the phase shift between the sent and received ray of laser and a system with a pulse generator and a photocell. The second system is used in the majority of cases in Europe. In the Czech Republic, the system is used exclusively.

The drive of a pulse generator is derived from the conveyer drive. The conveyer wheel diameter, gear ratio, and the number of pulses generated by the generator per one

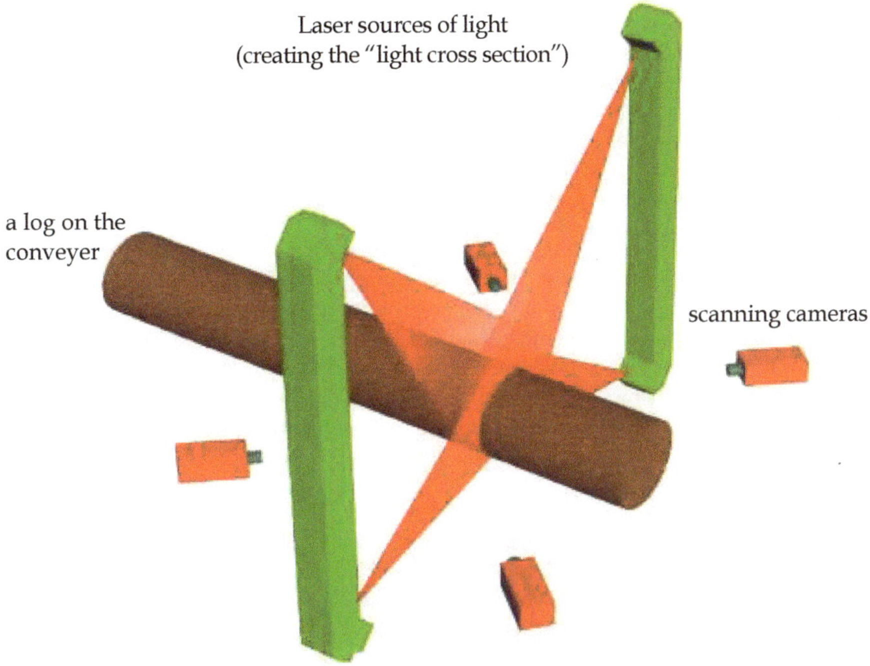

Fig. 5. The lay-out of a configuration for scanning the surface curve (3D measurement) (according to Microtec).

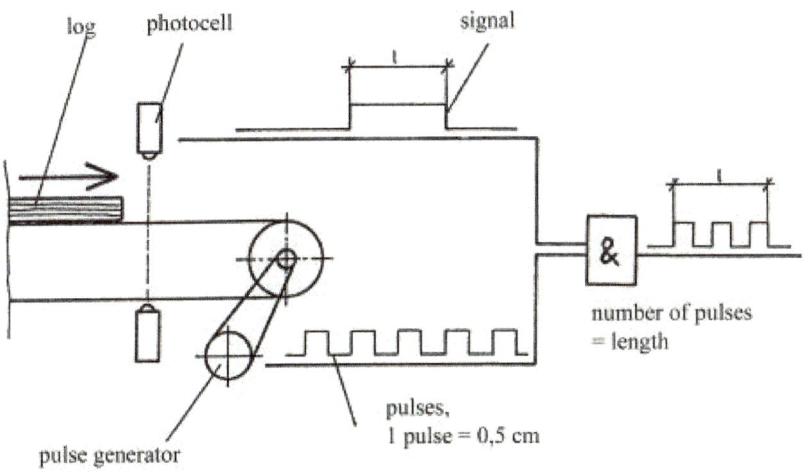

Fig. 6. Principles of measuring the log length by means of a pulse generator and a photocell.

revolution gives the number of pulses per the conveyer line unit. A log moving on a conveyer cuts across the photocell ray and during its shading the photocell sends a signal. The number of pulses sent by the generator during the photocell shading gives the log length. Accuracy reached by this method of measurement ranges from ±1 to ±2 cm. Advantages of this method consist in its simplicity, reliability and non-sensitivity to the conveyer speed and its changes during measurement. Disadvantages consist in the photocell sensitivity to defects on the log end (chamfer cut, torn up fibres). Thanks to this fact, the system has a tendency to give excess values at logs with these defects. Erroneous measurements are also caused by the log shift on a conveyer.

Data on dimensions and shape of logs are also provided by scanners working on the principle of absorption of microwave radiation. Systems equipped with these scanners are primarily determined for scanning quality. Many defects (rot, knots, and growth anomalies) are often not visible from the stem surface being not noted at the visual checking the quality. Evaluation of scanned data at this method of scanning is however, substantially more complicated with respect to mutual relationships of more values (dimensions, density, moisture etc.). Thus, they are not used only for scanning dimensions due to their demands and costs.

4. Processing the scanned data

Dimensions and volumes of logs corresponding "exactly" to their geometrical properties cannot be, naturally, determined in operation. Every determination of the log volume based on the calculation of the volume of geometrical bodies (cylinder, truncated cone, paraboloid etc.) represents only approximation to reality but not its expression. A value achieved by this way represents a nominal "commercial" volume of wood. According to the technical feasibility of measurements under given conditions, availability of means necessary for measuring, requirements and experience of the result users and according to "historical" usage in the given region, many methods of measurements and processing the scanned data are used in Central Europe at present. Any used procedure cannot be considered to be unequivocally bad. Thus, unification of the method of wood measurement, at least in the area of Central Europe, does not appear to be realistic. Therefore, it is necessary at least to recalculate results achieved by particular procedures with each other. Unambiguous recalculation is not possible because at processing results of measurements it is necessary to use statistical methods. Under conditions of Central Europe, the volume of logs is usually given in m^3 under bark (u.b. volume). It is calculated as the volume of a cylinder, the length of which is equal to the nominal length of logs and diameter corresponds to the mid diameter of logs (u.b.).

The log volume calculation as the volume of a cylinder stems from a historical method, which was proposed by a Bavarian forest inspector Franz Xaver Huber in 1825 on the basis of his theoretical analyses and long-time experience. Owing to the simple feasibility of measurements and the method of calculation as well as due to the sufficient accuracy of results, the method was soon used not only in Bavaria but in whole Germany, Austria-Hungary, and in the course of time also in a lot of other European countries.

Simplification of the log form causes that the method of Huber gives generally lower volume compared to reality providing satisfactory results only for the wide average of the large number of logs but not for particular logs. Accuracy of the measurement depends mainly on the stem part, which was used for the log production. The method undervalues the volume of butt logs while top logs are overvalued. F.X. Huber was aware of this fact and, thus, for more accurate measurements, he recommended a section method. This method divides a stem to 1 or 2 m long sections and the volume of each of the sections is calculated separately using the way described above. The volume of a log is then the sum of volumes of particular sections. At the time of its origin, the method was used only for research operations due to its excessive time consumption. Due to similar causes even other procedures requiring more measurements were not used later in practice.

Procedures used at present for wood measurement and determination of its volume (from the aspect of their user in Czech Republic) are as follows:

- Recommended rules for the measurement and classification/grading of wood in the Czech Republic, 2008 (Kolektiv, 2008)
- ÖNorm L 1021 Vermessung von Rundholz, Austria, 2006 (Österreichisches Normungsinstitut, 2006)
- Rahmenvereinbarung für die Werksvermessung von Stammholz, 2005 (Deutscher Forstwirtschaftsrat e.V.& Verband der Deutschen Säge- und Holzindustrie e.V., 2005)

Besides, there is a European standard EN 1309-2 Roundwood and sawn timber - Methods of measuring dimensions - Part 2: Roundwood – Requirements for measurements and rules for the volume calculation, 2006. Although it concerns a relatively new European standard, its use has not been found out in the CR. Its use (anywhere) limits not quite unambiguously determined methods of the determination of a mid diameter. A normative supplement B evokes also certain confusion. The normalized procedure of measurement is presented here as "rules for the measurement and calculation of the log volume valid if there are no state, regional or district rules".

For users in Czech countries, it is suitable to include (from practical aspects) the CSN 48 0050 Standard "Raw timber. Basic and common provisions" into the survey of rules (ČSN 48 0050, 1992). This standard was used in the Czech Republic until the publication of "Recommended rules 2002". At present, it is not legally binding being virtually not used for timber reception. The majority of users are accustomed to results of measurements carried out according to the standard. The users compare often values obtained according to other rules with its results. Moreover, results of measurements carried out according to this standard were in very good agreement with reality.

The analysis of methods of measurements includes all regulations and rules mentioned above talking into account also variants, which are determined or admitted by these regulations. It refers to following variants:

- Recommended rules for the measurement and grading of timber in the Czech Republic 2008 determine separate procedures for manual and electronic measurements, which are not quite identical.
- ÖNorm L 1021 Vermessung von Rundholz makes possible calculations from mid diameter values given in mm or converted to cm in such a way that units in mm are not taken into account. The use of diameter values in cm is preferred.
- Rahmenvereinbarung für die Werksvermessung von Stammholz gives mid diameter and subsequent calculations of the log volume differently in the extent of log diameters up to 20 cm and from 20 cm.

- The CSN 48 0050 Standard "Raw timber. Basic and common provisions" determines different procedures for manual and electronic measurements.

A detailed analysis in the previous determination of procedures and their variants (which does not include all European regulations) has shown that it is possible to analyse them to common elementary steps. Not all steps prescribe all procedures and the method of implementing many steps is different. However, by the exact definition of particular steps, all procedures can be unambiguously and fully characterized (see Fig. 7).

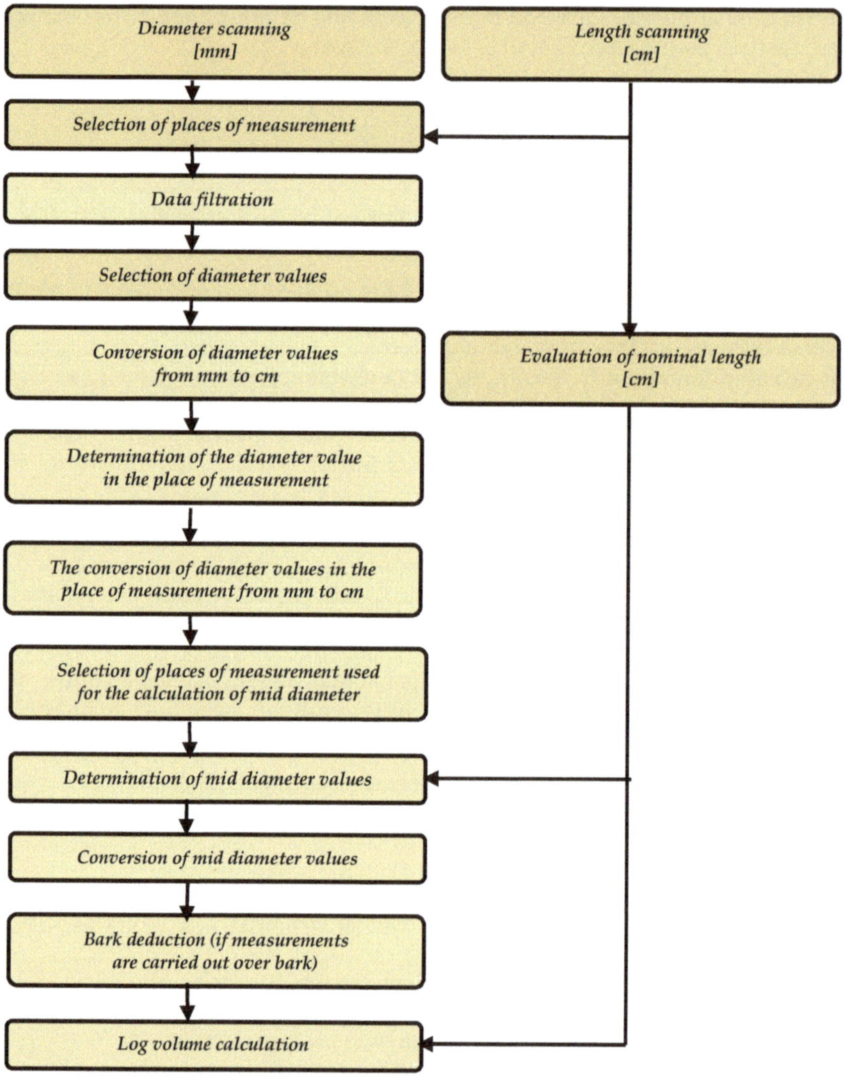

Fig. 7. Elementary steps at scanning and evaluation of dimensions and volume of logs and their sequence.

The way of carrying out particular steps is given in table 1. Variants of the implementation of particular steps (if they are determined or made possible by regulations) are given as separate procedures.

A step, which is consistently ignored by all regulations, is filtration of data. The aim of filtration is to create the stem image, which approaches maximally its actual/real form. It removes error data replacing them by probable data. Extreme values are considered to be "erroneous". These data originate usually by tattered parts of bark, wood, remains of branches etc. These values are replaced (at filtration) by values, which level the form of a log in such a way to correspond (ie with higher probability) its real form.

In addition to defects on the wood surface, filtration is inevitable at systems with a through-way conveyer to filtrate parts of the conveyer (usually driving dogs/carriers and their guide).

At manual measurements, a principle can be considered to be a certain form of filtration, which says that if there is a defect in the place of measurement (e.g. in the log centre), two measurements are carried out at both sides of the defect where the defect already does not appear. An average value from both these measurements is considered to be a value in the original place of measurement.

Methods of filtration at electronic measurements are based on mathematical and statistical procedures, which are usually combined. Basic used procedures usually are as follows:

- *Moving averages* – serve for the general adjustment of the stem surface curve. To each of the places of measurement (x) a value is assigned calculated as an average from values taken in an interval (x - n, ... x, ... x + n). In its centre, there is the place of a given measurement. The number of members serving for the calculation of a moving average (smoothing with) is usually odd. The procedure can be also several times repeated (the depth of smoothing), however, there is a danger of too large idealization of the piece (log) form. By means of this method, it is not possible to calculate values for edge places of measurement. At the determination of the log mid diameter for the calculation of its volume, the shortcoming does not appear but when we need to determine top diameter it is necessary to extrapolate it in the course of moving averages.
- *Moving medians* – serve also for the general adjustment. The procedure is anagogical the previous method, only the average value is replaced by median. Advantages consist in fact that the resulting value is not affected by potential extremes.
- *Top extremes cutting* – serve to eliminate extreme values with positive deviation caused usually by protruding bark, torn up fibres or parts of branches. More types of methods of the elimination of positive extreme values are used. Nevertheless, they do not suffice alone for the filtration, always follows adjustment using some of other methods.
- *Linear regression* – serves for the total adjustment of the stem surface curve to a straight line by the approximation of given values using the polynomial of the first order (straight line) by the method of least squares. If this procedure is applied directly to measured values, considerably distant values can markedly affect (deviate) the whole line. Its use for the set of values, which were already adjusted by mean of another method, was more suitable.
- *Mutual comparison of values of diameters measured in perpendicular directions* (e.g. horizontally - x and vertically - y) in one place with respect to the log length. Differences in values X and Y are compared with a value higher than common flattening. Effects of flattening can be eliminated comparing several successive values of

the X and Y difference. The flattening becomes evident in the whole length of the log or at least on its longer section. The procedure serves to identify "unreliable values". For their replacement, it is necessary to use another method.

- *Mutual comparison of values of diameters measured successively in one direction.* A difference between X_a and X_{a+1} is compared with the maximum admissible size of a difference derived from the possible stem taper. Similarly as the previous method it serves to identify "unreliable values". The method disadvantage is that it does not find an error which consists in the deviation of more successive values. It is not also utilizable for searching defects at the log end.
- *Comparison of the growth coefficient of successive values,* i.e. the relation of X_a and X_{a+1} with a value derived from a possible taper. Members of the proportion may not be successive measured values. Also this method serves to identify defects and also here it is necessary to determine another method or to change the comparison of data at the end of a series.

To record the majority of described erroneous data but, at the same time, for the realistic implementation of filtration there are usually more suitable combined methods. It is possible to combine both various types of basic methods and their succession and in "moving methods" also the number of values taken into account. At the same time, different methods are used for the filtration of the log central part to obtain mid diameter for the volume calculation (or for cross-cutting) and for the filtration of the log end parts to determine end diameter mainly for the purpose of grading. There, the use of "moving" procedures is very limited.

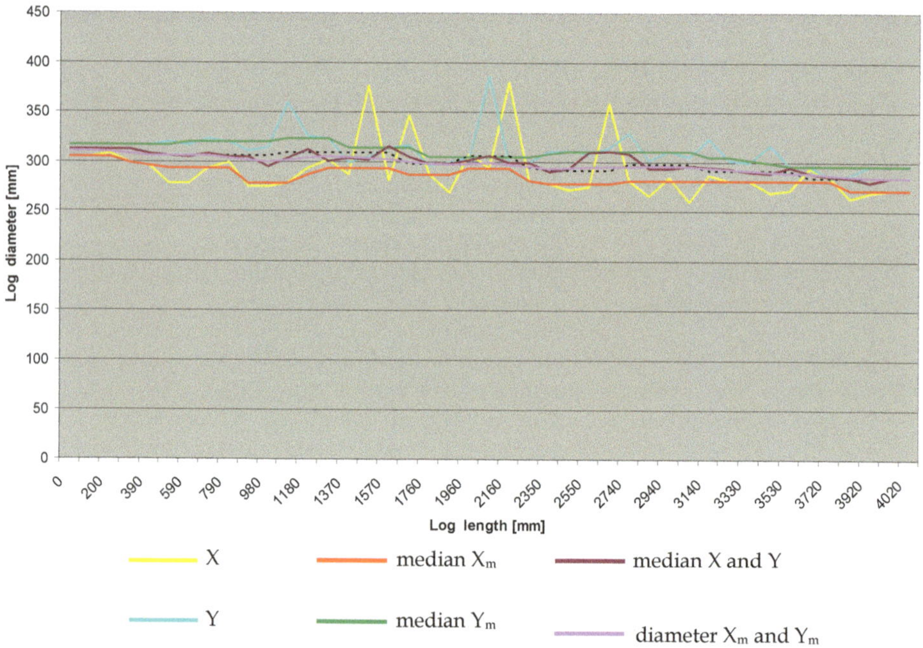

Fig. 8. An example of the effect of various procedures of filtration of scanned data on the resulting image of a log in a control computer (the log shows ragged bark in its central part).

The method of filtration shows substantial effects on values of quantities evaluated in next steps. Different methods of filtration are selected both for various types of sensing devices and assortments of processed raw material. Similarly, it is possible to achieve the required conformity of results of standard measurements with the given reference measurement by the suitable method of filtration. Therefore, manufacturers of sensing (scanning) or control systems use frequently filtration at the final setting the equipment parameters. Thus, it is not possible to define concrete used methods of filtration.

The analysis of particular steps results in the following conclusions:

- methods of the electronic scanning of wood provide very similar data on its geometrical properties,
- methods of the evaluation of electronically scanned data markedly follow the manual method of measurement trying to adapt to its possibilities (number of used measurements, accuracy),
- methods of determination of the stem (log) volume keep the original principle of the log volume calculation as one cylinder although the density of carried out measurements provides the sufficient number of data for the calculation of the log volume according to sections.

Thus, the absence of using possibilities of electronic scanning and evaluation of dimensions, form and volume of wood results in deviations between evaluated (nominal, commercial) volume and real/actual volume.

Differences in the determination of dimensions (above all mid diameter) of logs between particular rules result in differences in the determination of wood volume according to particular rules/regulations. Steps, which cause these differences, are as follows:

- conversion of millimetre values of the log diameter to whole centimetres,
- conversions generally (only exceptional using millimetre values for next calculations),
- the method of conversion (mathematical rounding or removing units in mm),
- the number of conversions (the conversion of particular scanned values and values calculated as an average from values conversed to whole cm already previously),
- determination of the place of measuring the mid diameter of logs (centre of the nominal or actual geometrical length including allowances),
- evaluation of the log mid diameter (average value or a smaller value from diameter values determined in both places of measurement within the measuring area in the centre of the log length),
- selection of diameter values (keeping values taken horizontally and vertically /2D equivalent of scanning/ or searching for a maximum) - only at 3D scanning,
- filtration of scanned data (its effect at present valid regulations cannot be determined because the regulations do not define the way of its implementation).

5. Differences in results

Results of measurements and determination of the volume of logs obtained according to rules given in table 1 are compared with the value of a volume, which approaches most the geometrical volume of logs. Calculation of the "geometrical" volume of logs is based exclusively on values scanned on a regular basis according to present regulations. Thus,

The survey of rules and determined methods of the implementation of particular steps										
No.	Rule	Scanning	Filtration	Places of measurement used for the calculation of the log mid diameter	The conversion of measured values in mm to cm	The method of determining the log diameter in the place of measurement	Conversion of the value of diameter in the place of measuring from mm to cm	The method of determining the log mid diameter	Conversion of the mid diameter value from mm to cm	Volume calculation
1	EN 1309-2 Standard - Roudwood and sawn timber - Methods of dimension measurement Part 2: Roundwood. 2006	2D	not defined	two in the zone of ±10 cm from the centre of a nominal length	mm units are not taken into account (cutting off)	mathematical average from both meaurements	mathematical rounding	mathematical average from measured values	mathematical rounding	as the cylinder volume
2	Recommended rules for the measurement and grading of timber in the Czech Republic 2008.	2D	not defined	two in the zone of ±10 cm from the centre of a nominal length	mm units are not taken into account (cutting off)	not determined	diameter is not determined	mathematical average from measured values	mm units are not taken into account (cutting off)	as the cylinder volume
3	ČSN 48 0050 Standard - Rough timber. Basic and common regulations. 1990. (Manual measurement)	2D	not defined	two in the zone of ±10 cm from the centre of a nominal length	mathematical rounding	not determined	diameter is not determined	mathematical average from measured values	mathematical rounding	as the cylinder volume
4	ČSN 48 0050 Standard - Rough timber. Basic and common regulations. 1990. (Automatic measurement)	2D	not defined	all places by 10 cm in the course of a nominal length	without conversion	not determined	diameter is not determined	mathematical average from measured values	mathematical rounding	as the cylinder volume
5	ÖN L 1021 Vermessung von Rundholz 2006 Log diameter in cm	2D	not defined	two in the zone of ±10 cm from the centre of a geometrical length	mm units are not taken into account (cutting off)	mathematical average from both measurements	mm units are not taken into account (cutting off)	smaller from diameter vallues in both places of measurement	already given in cm	as the cylinder volume
6	ÖN L 1021 Vermessung von Rundholz 2006 Log diameter in mm	2D	not defined	two in the zone of ±10 cm from the centre of a geometrical length	without conversion	mathematical average from both measurements	wihout conversion	smaller from diameter vallues in both places of measurement	without conversion	as the cylinder volume
7	Rahmenvereinbarung für die Werksvermessung von Stammholz. 2005. Roudwood up to a diameter of 20 cm.	2D	not defined	two in the zone of ±10 cm from the centre of a nominal length	without conversion	mathematical average from both measurements	wihout conversion	mathematical average from measured values	mm units are not taken into account (cutting off)	as the cylinder volume
8	Rahmenvereinbarung für die Werksvermessung von Stammholz. 2005. Roudwood diameter of 20 cm and more.	2D	not defined	two in the zone of ±10 cm from the centre of a nominal length	mm units are not taken into account (cutting off)	mathematical average from both measurements	wihout conversion	mathematical average from measured values	mm units are not taken into account (cutting off)	as the cylinder volume
Compa-rative method	Comparative method "the most accurate volume". Volume calculation by 10 cm long sections, diameter in mm	2D	not defined	two at both ends of a section	mathematical rounding	mathematical average from both measurements	wihout conversion	mathematical average from measured values	without conversion	by 10 cm sections

By means of the same colour the same way of step realization at particular rules are indicated

Table 1. The survey of rules and determined methods of the implementation of particular steps.

scanning is realizable at each equipment fulfilling requirements of given regulations. Evaluation of the "geometrical" volume results from a section method: the volume of a log is the sum of volumes of particular sections. Lengths of sections are 10 cm, which corresponds to the distance of particular measurements of the log diameter determined by present rules. The section diameter is equal to the average value from two measurements perpendicular at each other at the beginning and at the end of the section (four values), the section volume is determined as the volume of a cylinder. Thus, the determination of "geometrical volume" is realizable at any existing equipment after the adjustment of its program (software) equipment. Details are given in a table 1, where this method is termed as "comparative".

The comparative measurements were carried out on about 180 000 spruce logs. Dimensional and quality properties of logs correspond to logs for sawmill processing (quality class III, qualities A, B and C, classification according to Recommended rules for the measurement and grading of timber in the Czech Republic 2008, Tab. No. 13, p. 70). Supplies (deliveries) were created in 72% by logs in basic lengths 3 to 6 m with the predominance of lengths 4 and 5 m, 28% supplies were logs in combined lengths 7 to 14 m, however mainly 8 to 12 m (relatively uniform proportion). The average mid diameter of logs ranged about 27 cm. Parameters of each log (values of diameters taken horizontally and vertically at a length interval of 10 cm + measurement location + value of length) obtained by long-term operational measurements at 2 sawmills were stored and subsequently processed according to compared rules. In this way, the consistency of input data was provided. Values of particular comparative coefficients are obtained as medians of values of volumes of particular logs determined according to compared procedures but not as the comparison of total volumes of supplies (deliveries) defined according to compared rules. Values of medians differ from values of averages quite insignificantly, namely at the 4th to the 6th decimal place.

The total comparison of log volume values determined according to particular rules with the comparative method of "geometrical volume" (not distinguishing properties of logs) is given in table 2.

The relationship between the log volume determined according to given rules and a "geometrical" volume determined according to a comparative method (i.e. coefficient) is, in the majority of procedures, substantially dependent on the log diameter and less on the log length (although not negligible). Thus, values given in the table 2 apply only to the considerable number of deliveries of saw logs (thousands logs). Dependencies on other parameters (e.g. taper, flattening) were not examined. Reasons consisted in the rather controversial practical efficiency of these dependences even in cases their effects would be proved. However, particular deliveries of logs (ordinarily 80 - 200 logs) differ in their properties and the mean value of their dimensions is often different from a long-term average. Therefore, for the practical use of given coefficients, it is necessary to specify their average values according to properties of actual deliveries. With respect to the "step character" of deviations at the determination of volume of the same log according to different procedures (mainly due to the conversion of mm to cm) standard statistical processing does not provide too objective image on actual properties of particular procedures. The graphic representation of properties of particular procedures is well-arranged.

	Rule	*Coefficient*	*Lower 95%*	*Upper 95%*
1	EN 1309-2 Standard Roundwood and sawn timber - Measurement of dimensions - Part 2: Roudwood. 2006	0.981072	0.980692	0.981453
2	Recommended rules for the measurement and grading of timber in the Czech Republic 2008.	0.941042	0.940679	0.941406
3	ČSN 48 0050 Standard Rough timber. Basic and common regulations. 1990 Manual measurement	1.000764	1.000394	1.001134
4	ČSN 48 0050 Standard Rough timber. Basic and common regulations. 1990 Automatic measurement	0.995303	0.995149	0.995457
5	ÖN L 1021 Vermessung von Rundholz 2006 (log diameter in cm)	0.936393	0.936037	0.936749
6	ÖN L 1021 Vermessung von Rundholz 2006 (log diameter in mm)	0.984777	0.984447	0.985107
7	Rahmenvereinbarung für die Werksvermessung von Stammholz. 2005 (generally)	0.941534	0.941172	0.941896
8	Rahmenvereinbarung für die Werksvermessung von Stammholz. 2005 (only a method up to a diameter of 20 cm)	0.958114	0.957753	0.958474

Table 2. The total comparison of log volume values determined according to particular rules with the comparative method of "geometrical volume" (not distinguishing log properties). The average mid diameter of logs of the basic population ranged between 29 and 30 cm.

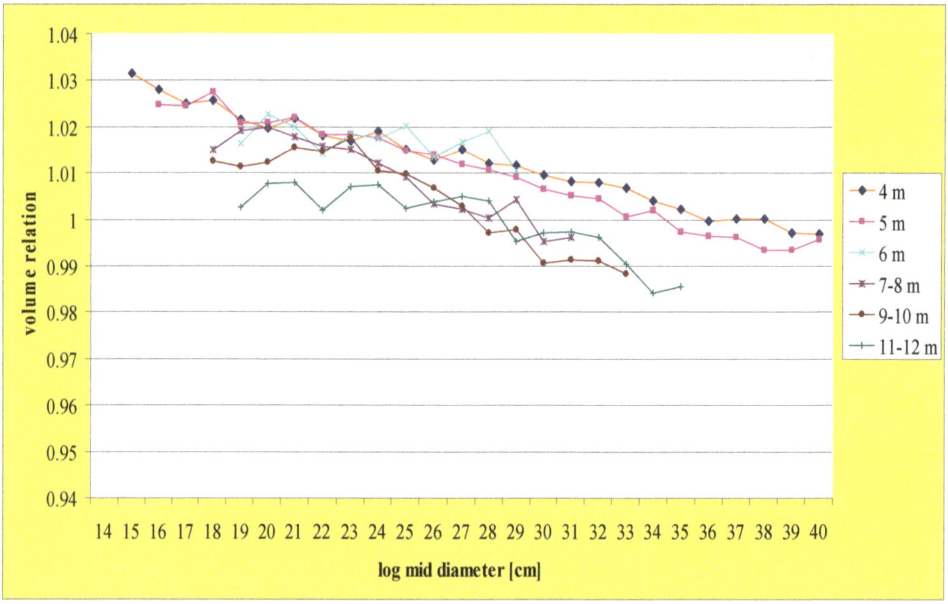

Fig. 9. The relationship between the volume of logs determined according to the ČSN 48 0050 Standard - Rough timber - manual measurement and a comparative method. The dependence of the relationship on the log diameter (x axis) and the log length (particular runs).

The ČSN 48 0050 Standard - Manual measurement is not nearly used in practice. It is chosen as an example because its procedure is consistent with the original Huber method and the dependence of a coefficient on the log diameter and length is (as in Huber method) very marked. (Note: Values of coefficients at sets of logs the number of which did not reach 250 were not plotted - minimum for 0.5% accuracy for 95% results at 0.04 variability determined according to a control sample).

Results of measurements demonstrate a well-known fact that the Huber method generally overvalues small-diameter top logs and undervalues large-diameter but logs (Kolektiv 1959).

The different stem form and subsequently also different evaluation of the log volume becomes also evident in connection with the log length. Regardless of the log diameter the overvaluation of short logs is higher than the overvaluation of longer logs. Particularly visible differences occur between logs in basic and combined lengths. A next diagram, Fig. 10 is well-arranged. It contains only the course of the dependence of logs of length 3 - 6 m (altogether, red course) and 7 - 14 m (also altogether, blue course). In addition to this, the whole group of logs is evaluated regardless of the log length (green course).

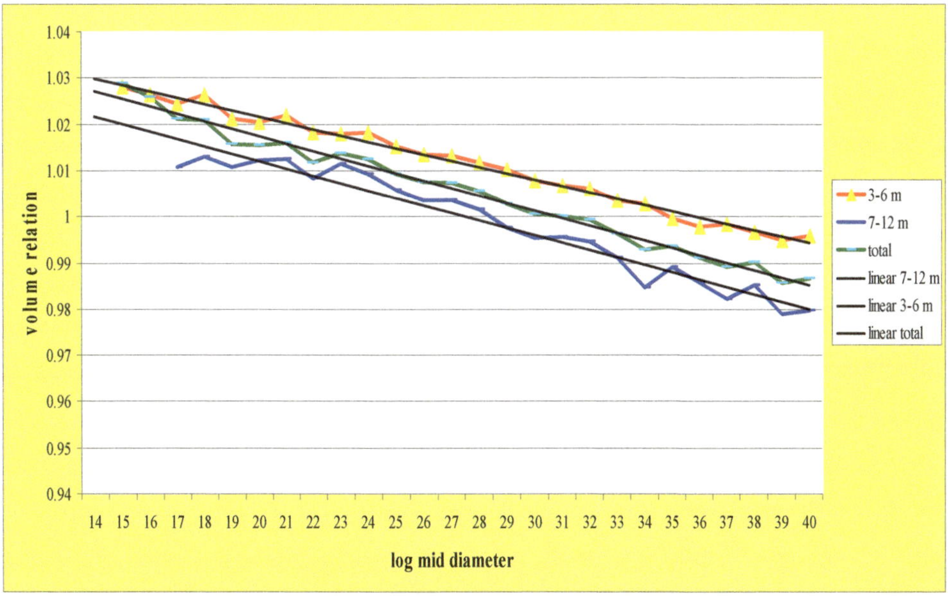

Fig. 10. The relationship between the volume of logs determined according to the ČSN 48 0050 Standard - Rough timber - manual measurement and a comparative method. The dependence of the relationship on the log mid diameter (x axis) and the length of logs (logs 3-6 m and logs 7-12 m). The total dependence (not distinguishing lengths) is expressed as a green line.

On the basis of measurements it is possible to state that on average:

- with the decline of log mid diameter by about 8 cm the value of its volume (determined according to Huber method) increases roughly by 1% as compared with geometrical volume. The dependence very approaches linear dependence.

- together with the geometrical volume, logs of mid diameters about 30 cm are evaluated. Logs in basic lengths are overvaluated already from 35 - 36 cm of a mid diameter while logs in combined lengths from about 27 - 28 cm.

F. X. Huber and also other sources (Šmelko, 2003) explain this fact by the different form of a stem in its butt (large diameter) or top (small diameter) part.

The average value of mid diameters of coniferous logs delivered to sawmills in the Czech Republic ranges between 26 and 28 cm and it is possible to expect its gradual decline. It appears from this that the method mentioned above is inconvenient for consumers of the raw material (particularly logs) and, on the other hand, profitable for suppliers.

Automatic measurement carried out according to the ČSN 48 0050 Standard tries to express the exact actual (geometrical) volume of timber preserving the Huber method. The only possibility is to specify a mid diameter. It is calculated as the average value of all diameter measurements carried out within the nominal log length.

Results are given in Fig. 11. The course is rather balanced within the whole zone of monitored diameters. Thanks to the calculation of the log mid diameter from values taken within the whole (nominal) log length the overvaluation of small-diameter logs (typical of Huber method) is very limited. It becomes evident only at logs with a mid diameter up to about 18 cm, the volume of which is overvalued (on average) by 0.5%. The volume of logs within a mid diameter about 18 - 20 cm is evaluated according to the real volume. The volume of logs of larger diameters is evaluated similarly if they are in basic lengths (3 - 6 m). A lower than real volume (on average 0.5 to 0.7%) is evaluated only at logs of combined lengths and this deviation is rather balanced at all logs of mid diameters over 20 - 22 cm.

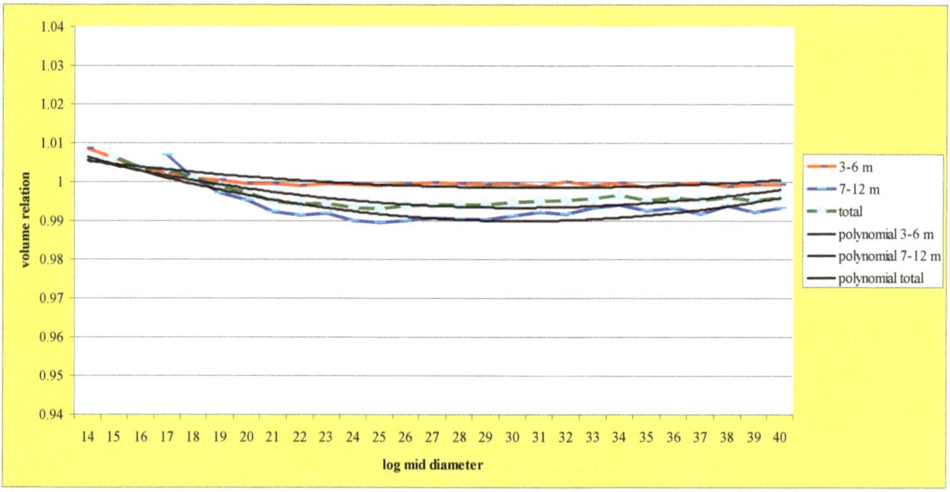

Fig. 11. The relationship between the volume of logs determined according to the ČSN 48 0050 Standard - Rough timber - automatic measurement and a comparable method. The dependence of the relationship on a mid diameter (x axis) and the log length (logs 3-6 m and logs 7-12 m). The total dependence (not distinguishing lengths) is expressed as a green line. The course of measured values is approximated to the polynomial of the 2nd degree.

The EN 1309-2 Standard – Roundwood and sawn timber – "Methods of measuring dimensions" is used only rarely at timber reception (ČSN 49 0018, EN 1309-2, 2006). In the CR, its use has not been noted at all.

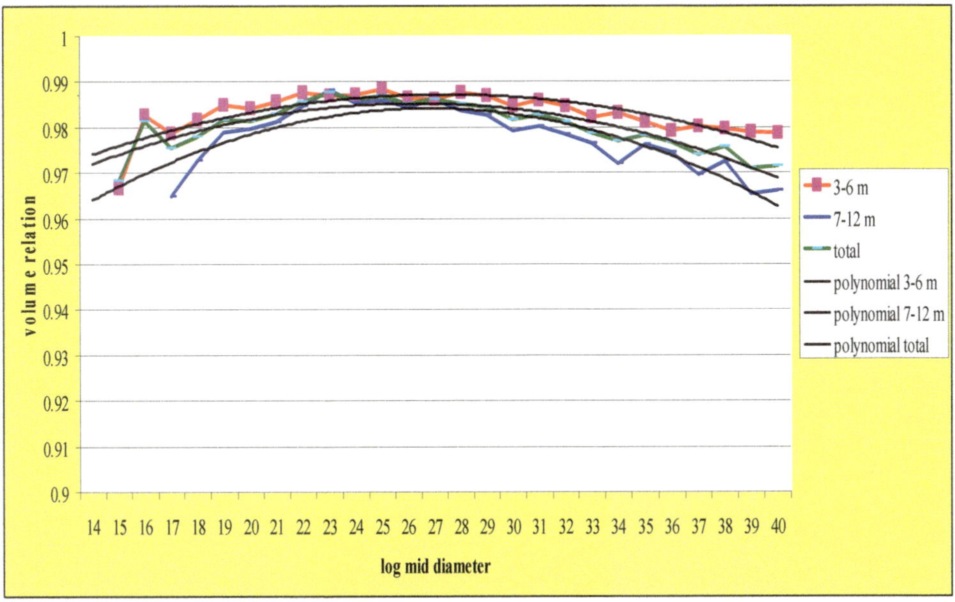

Fig. 12. The relationship between the volumes of logs determined according to the EN 1309-2 Standard - Roundwood and sawn timber - Methods of measuring dimensions and a comparative method). The dependence of the relationship on a mid diameter (x axis) and the log length (logs 3-6 m and 7-12 m). The total dependence (not distinguishing lengths) is expressed as a green line.

Properties of the EN 1309 - 2 Standard are formed by two antagonistic effects - Huber method, which is a basis of the standard and the conversion of mm values of the log mid diameter to cm ("cutting off" mm).

The Huber method tends to undervalue large-diameter logs and to overvalue small-diameter logs. The effect of Huber method prevails in total properties of the EN 1309 - 2 Standard in the field of large-diameter logs where it manifests itself by the fall of characteristics.

At the conversion of millimetre values of the mid diameter to centimetres the standard combines i.e. mm are not taken into account (values of particular measurements) and mathematical rounding (values of the diameter in places of measurements and at the expression of a resulting mid diameter).

The primary "cutting off" of measured diameter values causes the total marked decline of resulting values of log volumes. It becomes evident particularly at small-diameter logs where the decline of a value of a subsequently calculated volume as compared with mathematical rounding can achieve even 5 - 7%.

The size of the fall of total characteristics of the EN 1309 - 2 Standard in the area of small-diameter and large-diameter logs is also dependent on the length of logs. Volumes of logs of basic lengths both the smallest (14 - 15 cm) and large (39 - 40 cm) diameters are evaluated by the method by about 2.5 to 3% lower than it corresponds to the geometrical volume. The volume of logs in combined lengths is (in the same comparison) lower by about 3.5%. Logs of medium diameters (between about 23 and 29 cm) are undervalued equally - by about 1.2 to 1.7%. Differences between evaluations of volumes in basic and combined lengths are small in the middle zone - the volume of logs in combined lengths is usually evaluated by 0.2 to 0.3% lower, than the volume of logs in basic lengths.

Recommended rules for the measurement and grading of timber in the Czech Republic 2008 determine for manual and electronic measurements virtually the same procedure. At manual measurements, the log mid diameter is an average from two values measured perpendicular each other in the centre of its nominal length. Only in case of anomaly in the log centre, two measurements are carried out near the anomaly in the same distance from the log centre, the mid diameter being the average value from four measurements. Electronic measurements determine the log mid diameter from these four values always.

Note: A German general agreement *Rahmenvereinbarung für die Werksvermessung von Stammholz (2005)* defines the mid diameter determination in the same way, however, only for log diameters ≥ 20 cm.

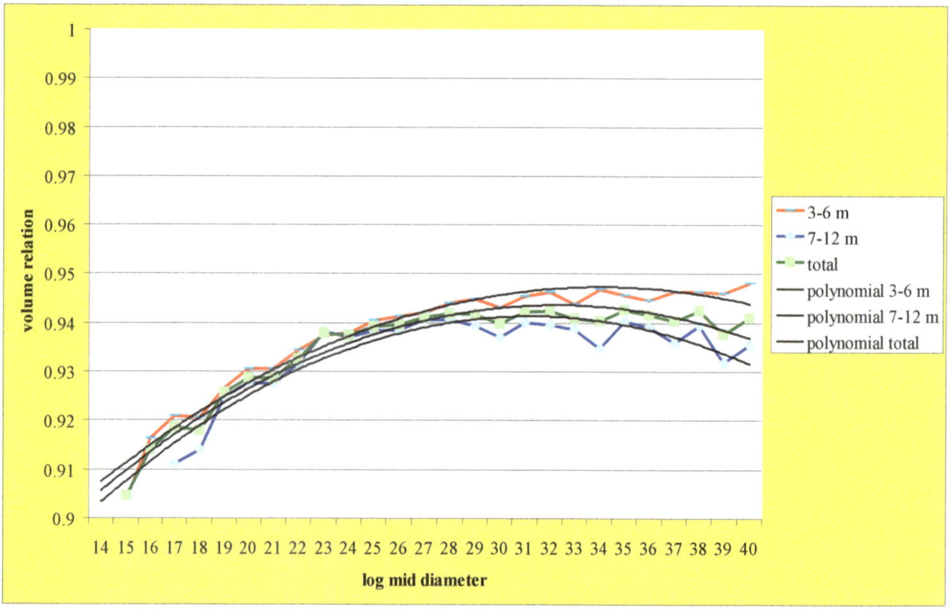

Fig. 13. The relationship between volumes of logs determined according to Recommended rules for the measurement and grading of timber in the Czech Republic 2008 and a comparative method. Dependence of the relationship on the log mid diameter (x axis) and the length of logs (logs 3-6 m and 7-12 m). The total dependence (not distinguishing lengths) is expressed as a green line.

Cutting off (millimetres are not taken into account) at the conversion of values of measurements given in mm to whole cm and subsequently also the conversion of an average value from these data in the same way becomes evident by the marked decline of values of the volume of all logs. This decline is particularly evident at small-diameter logs (up to 26 - 27 cm). Within the range of mid diameters (15 to 26 cm) the fall ranges from about 8.5% (to 91.5%) to about 5.5% (to 94.5%) as compared to the geometrical volume. The effect of the log length is not significant at small-diameter logs. At further increasing the log mid diameter the value of the difference does not increase. However, effects of the log length start to manifest. Logs in basic lengths keep their deviation from a geometric volume on a stable value about 5.3% (94.7% value of the log geometric volume). Logs in combined lengths slightly increase the value of their deviation (at 40 cm diameter, they reach the value of a deviation about 6.2%, i.e. 93.8% geometrical volume.

Comparing the courses with the previous "ČSN EN 1309-2 (49 0018) Standard - Roudwood and sawn timber" we can state that the double "cutting off" mm units at the determination of the log mid diameter becomes evident by more than double increasing the volume deviations (in percent) at logs of all diameters. The trend of decline is identical.

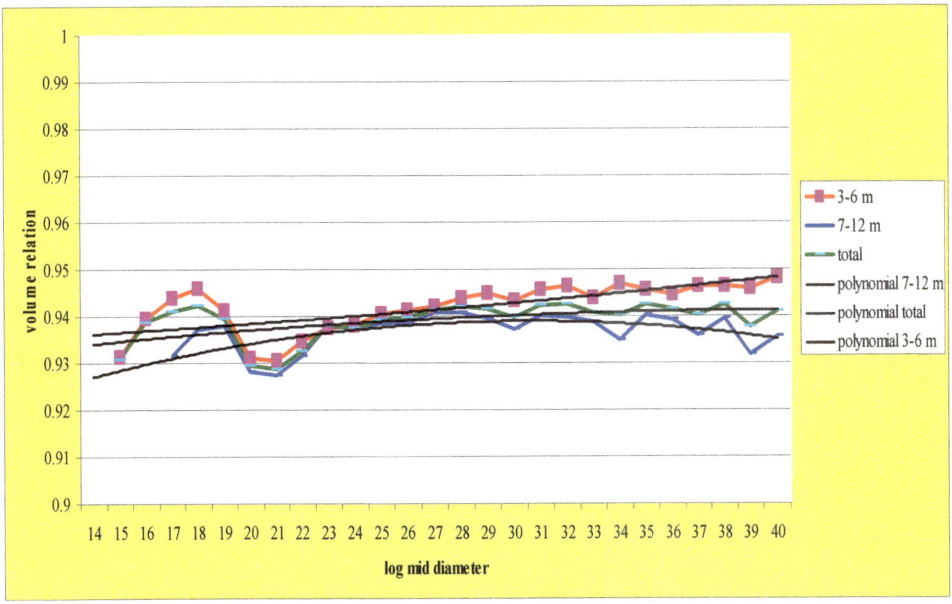

Fig. 14. The relationship between the volumes of logs determined according to Rahmenvereinbarung für die Werksvermessung von Stammholz 2005 and a comparative method. The dependence of the relationship on a mid diameter (x axis) and the length of logs (logs 3-6 m and 7-12 m). The total dependence (not distinguishing lengths) is expressed as a green line.

Rahmenvereinbarung für die Werksvermessung von Stammholz, a general agreement for the measurement of round timber at sawmills used in Germany tries at least partly to balance the decline described below. Therefore, at logs up to 20 cm, the conversion of measured values is not carried out. Only the calculated value of the mid diameter is converted. Thus, "only" one cutting off mm is carried out.

The described adjustment gradually balances the general characteristics of the procedure and the evaluated volume is generally approached to reality (it evaluates on average 94% geometrical volume, average deviation is 6%). However, this value is only of orientation character. In reality, it substantially depends on the diameter structure of logs in the actual delivery.

The Austrian standard *Ö-Norm L 1021 Vermessung von Rundholz* (2006) approaches Recommended rules for the measurement and grading of timber in the Czech Republic 2008 as for the determination of the log mid diameter and calculation of its volume. The log mid diameter is not determined as an average diameter but as the smaller one from diameters in both places of measurement and the position of a mid diameter is derived from a geometric and not nominal length. Thus, the place of measurement is shifted by half the length of an actual allowance towards the top end. The value of mid diameter and subsequently the log volume are thus slightly lower. The higher value of allowances (the volume of these allowances is not included into the log volume) results paradoxically in the lower evaluated diameter and volume of logs.

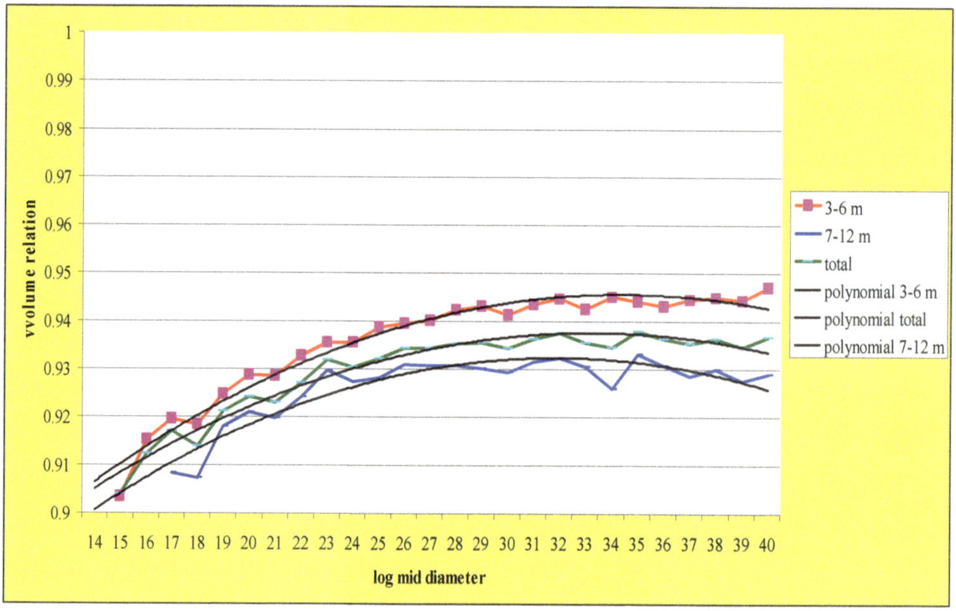

Fig. 15. The relationship between the volumes of logs determined according to the Austrian Ö-Norm L 1021 (2006) Standard and a comparative method. Values of a mid diameter are given in cm. The dependence of the relationship on a mid diameter (x axis) and the length of logs (logs 3-6 m and 7-12 m). The total dependence (not distinguishing lengths) is expressed as a green line.

The determination of a mid diameter as a smaller but not average value from both places of measurements becomes evident (thanks to a very small distance of both measurements /10 cm/) by the total decline of evaluated volumes only very little - by about 0.2%. This value determined by a separate analysis is significant only at the total delivered volume of timber for a certain period (e.g. a month) or at least of whole supplies (deliveries). At particular logs, it virtually coincides with the accuracy of measurement. The practical result of the effect of allowances is different at logs in basic lengths and combined lengths. Logs in basic lengths have allowances rather exact in length (up to 2.5%) and their effect on decreasing the calculated value of the log volume ranges at a level of 0.1 - 0.2%, however, often only in hundredths percent. The effect can be statistically specified, however, it is not possible to prove it virtually by measurements at particular logs (contrary to calculations). At logs in combined lengths, larger (longer) allowances are usually left for the purpose of subsequent cross-cutting. It results in higher values of the decline of evaluated volume by 0.2 to 1.0%.

Note: For the purpose of cross-cutting longer allowances than generally accepted 2 or 1.5% are locally negotiated at logs in combined lengths. In order this step to have no negative impacts on roundwood suppliers additional charges are also usually negotiated for these allowances.

ÖN L 1021 Vermessung von Rundholz makes possible to evaluate mid diameters and subsequently also the volume of logs from values given in mm without conversion to whole cm. Otherwise, the procedure is consistent with the previous procedure.

This alternative markedly approaches evaluated diameters and thus also volumes of all logs to their real geometrical values and, at the same time, avoids the decline of evaluated volumes of logs of small diameters. However, undervaluation of large-diameter logs characteristic of Huber method remains.

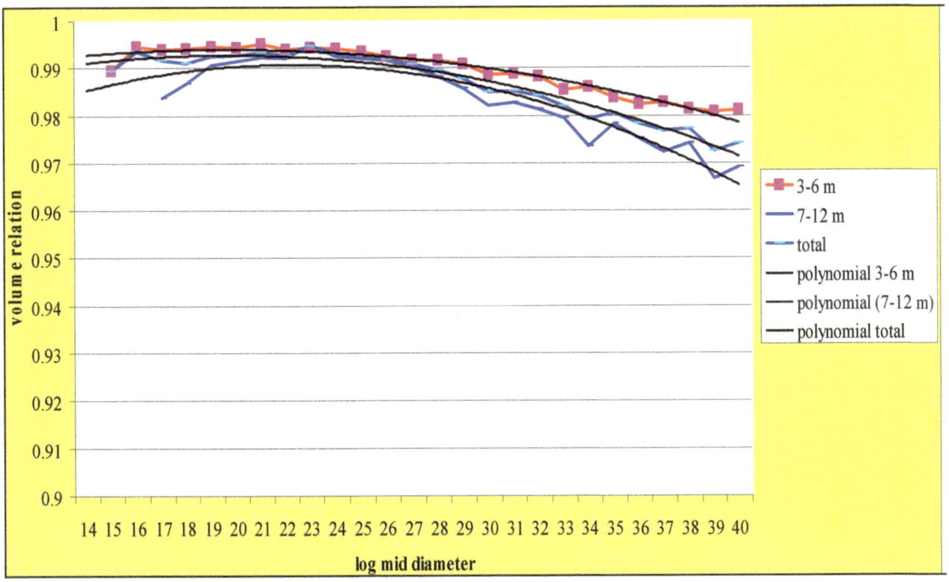

Fig. 16. The relationship between the volume of logs determined according to the Austrian Ö-Norm L 1021 (2006) Standard and a comparative method. Values of a mid diameter are given in mm. The dependence of the relationship on a mid diameter (x axis) and the length of logs (logs 3-6 m and 7-12 m). The total dependence (not distinguishing lengths) is expressed as a green line.

By its results, the procedure is comparable with the EN 1309 – 2 Standard. The higher value of the evaluated volume of logs according to the Ö-Norm L 1021 Standard by about 1% in the area of log diameters over 24 cm is caused mainly by removing the conversion of mm values of the mid diameter to whole cm not taking mm units into account. A fact that a difference obvious in the diagram is lower than a difference mentioned here causes the reduction of the log volume value at ÖN L 1021 placing the mid diameter measurement into the centre of a geometrical length and determination of the mid diameter as a lower but not average value from measurements in both places of the measuring area.

The advantage of the ÖN L 1021 Standard at the evaluation of the mm values of measurements consists in the even course of characteristics in the zone of log diameters below about 24 cm. Methods using "cutting off " mm units (although "only" once, such as EN 1309-2) show an appreciable decline in this area. However, no sawmill was found, which would virtually use the given version of Ö-Norm L 1021.

6. Possibilities of minimizing the differences

At present, the accuracy of sensing and evaluating the log length ranges about ± 2 cm, diameter ±1mm, directions of the diameter measurement are usually 2 perpendicular at each other (at 2D sensing) up to 180 (i.e. at 1° angular rotation at 3D sensing), the diameter measurement density is between 1 and 10 cm log length.

For the purpose of wood processing the accuracy and details of taken and evaluated parameters of logs are quite sufficient. At trading with timber, an error occurring at the determination of dimensions or volume of logs is not a difficulty (problem), but differences in values of these quantities, which originate at using different methods of measurement (sensing and evaluating results obtained).

An ideal procedure to increase the accuracy of evaluating the log volume and to remove differences originating among particular procedures of measurements is to create and accept one technically unambiguous legislative rule (law, standard) obligatory for the electronic measurement of timber for the purpose of trading, and thus also electronic reception. However, expectations to create and mainly to accept such rules and regulations are not real at present both at a national and international level. To minimize effects of different rules it is necessary to start from following facts:

- At present, the accuracy of log dimension measurements is not limited by accuracy or the density of sensing (accuracy ordinarily ± 1 mm, sensing frequency commonly in kHz) or possibilities of evaluating the results of measurement.
- Accuracy of the evaluation of results is given by procedures derived from procedures for the manual measurement (determination of a mid diameter only from measurements carried out in the area of the log centre, determination of the log volume as the volume of one cylinder of a diameter equal to the log mid diameter and a length equal to the log nominal length).
- Regulations do not describe quite exactly all steps of processing the scanned data necessary at the electronic measurement. Thus, there is an area for the individual interpretation of the rules and affecting the results of measurements without the legislative disturbance of rules. It refers mainly to the filtration of scanned data (it is not

defined at all) and directions where the mid diameter is evaluated. In some rules, these are determined only as "perpendicular at each other" and concrete directions are not specified.

- The result of different interpretation consists in using 2D and 3D measurement the application of which and differences in results are not affected by the rules.
- The inspection of measuring devices (setting and accuracy of measurements) is also derived from the inspection of traditional mechanical means of measurement. It is carried out by a tape and circular measurement standards (etalons) in the static condition of a line. By means of such checking it is possible to reveal possible inaccuracies of sensing but not the effect of filtration and subsequent evaluation of the mid diameter and volume.

If we suppose the use of more methods in the future, the value of the log volume (reached at the electronic measurement of timber) will correspond rather to commercial needs than to the geometrical volume. Thus, it is necessary to understand it as a "commercial volume". Following steps are derived to increase the stability of measurements (repeatability with the same or near results) by any method and to reduce deviations.

It is inevitable to determine unambiguously the method of assessment and implementation of particular steps of the algorithm of data processing. Without these conditions it is not possible to determine more exactly properties of the given procedure (regulations, standards). It represents:

- To determine exactly the method of filtration of scanned data. For the central part of a log a combined procedure is recommended removing extreme values in the first part and in the following part slightly balancing the course not fundamentally affecting or eliminating values of local roughness (e.g. stem curvature, burls, root swelling). "Sliding" conditions tend to increase total values (effects of extreme values). For end parts (butt end, top end), it is suitable to start from the regression analysis of the course of diameters in the related log part (Hunková, 2011).
- To define particular directions where the log diameter is to be evaluated. The determination of only two measurements perpendicular at each other not determining their directions (e.g. horizontally and vertically) or without the assessment of the direction equability of the evaluated measurement at a measured log makes possible to find out minima and thus to decline the resulting value. On the contrary, the determination of searching for a minimum value right in the rule disadvantages or even forbids the use of 2D sensors common at present (Janák, 2007).

It is suitable to determine the quality of roudwood where results of measurements are still valid. The worst quality of saw logs is supposed. At the worse quality of logs (sweep, stem curvature, knots, buttress) it is possible to suppose higher misrepresenting effects of data filtration and thus the decline of measurement accuracy.

Properties of used procedures (mainly dependences of the value of the calculated log volume) have to be determined with respect to a reference method. The comparison of two methods is suitable only for particular cases and cannot be used for the determination of actual properties of procedures. The volume of logs determined by a reference method has to approach geometrical volume. Comparison with a geometrical volume determined e.g. by a volumetric method (Hauffe, P. & Müller, L, 2002) is ideal however laborious, time consuming and not practicable everywhere. A procedure of measuring the logs by sections appears to be sufficiently accurate and generally practicable. The length of sections derived from the minimum requirements of present regulations is 10 cm. In addition, the method accuracy is supported by the lower sensitiveness of the section method to the type of filtration. Misrepresenting effects of filtration become evident in the volume deviation of one section but not of the whole log (in a limiting case). Properties of particular procedures given in this paper are expressed in the same way.

To check and calibrate the equipment, procedures are proposed, which affect both accuracy of sensing geometric parameters of timber (logs) and the method of their evaluation. It is achieved by static sensing the etalons/measurement standards/ (the present method of the inspection of accuracy of setting the equipment and sensing the log dimensions) and subsequent operational (dynamic) measurement (checking the filtration and processing results of measurements). For calibration in this way it is necessary to determine (in rules, regulations) allowable tolerances of particular measurements and resulting evaluated values. It has not been determined for dynamic measurements yet.

The inspection of equipment is also related to the determination of authorities qualified to check and calibrate electronic measuring devices (existing in the majority of countries) and determination of sanctions at the infraction of agreed procedures of measurements and determination of geometrical parameters of logs (so far not existing in the Czech Republic). However, it refers to themes occurring out of the field of measuring systems.

7. Conclusion

For the electronic reception of roundwood, 2D and 3D systems of the measurement of roundwood diameter are used virtually exclusively. 1D systems do not provide exact data on geometrical properties of logs being also not permitted by any of standard rules (not mentioned in the survey).

Regulations, which are mostly used for the electronic reception of roundwood are as follows: Recommended rules for the measurement and grading of timber in the Czech Republic 2008 (both manual and electronic measurements), Austrian ÖNorm L 1021 (a version which gives roundwood diameter in cm) and German Rahmenvereinbarung für die Werksvermessung von Stammholz. Measurements according to these standards (particularly ÖNorm L 1021) are also used by many company regulations of the supra-national processors of roundwood.

All these regulations determine the roundwood volume as the volume of one cylinder of a diameter equal to the mid log diameter and length equal to the nominal length of a log, thus according to traditional methods of the manual measurement. According to manual measurements, diameter is also given in whole cm (exceptions are only logs up to a diameter of 20 cm at Rahmenvereinbarung für die Werksvermessung von Stammholz. Thus, for the determination of the log volume not even accuracy or the density of sensing, which is characteristic of present electronic systems (required by present regulations), are used.

The algorithm of the roundwood volume determination of any rule mentioned above does not result in the determination of the log geometric volume or a volume near the value but a volume, which is lower. The decline is mainly caused by the mid diameter value given in whole cm. Units of mm, which are included in electronically taken data, are not taken into account at the conversion. This "cutting off" is carried out even 2 times one after another in the course of calculations, namely at basic measured values and at average values.

The rate of decline of values of the log volume achieved according to particular procedures is compared with the volume of logs determined by a section method (sections as cylinders of a length of 10 cm and diameters given in mm). The method is selected because it evaluates (from given data on the log) a volume, which is closest to the log geometrical volume. The rate of decline varies from -5.5% (German Rahmenvereinbarung für die Werksvermessung von Stammholz, logs of medium diameters) to -9% (Austrian Ö-Norm L 1021, mid diameter value given in whole cm). Procedures, which give lower deviations (EN 1309-2 - 1.5 to 3%, ÖNorm L 1021, mm version - 0.7 to 3% or the ČSN 48 0050 Standard 48 0050 - 1 0.3 to 0.8%) are used minimally in practice (in the CR, their use was not found at all). Results achieved correspond to a comparison carried out in Germany (Sauter, U., Staudenmaier, J. & Verhoff, S. 2010).

The values are affected by the way of filtration of taken data. The filtration is a necessary step, which is carried out always. The method of its implementation is not, however, included in any known Central-European regulation. Through various types of filtration applied at the same taken data on logs even about 2% deviations (from values presented in the paper) were achieved.

At the majority of procedures, the fall (deviation) is considerably dependent on the roundwood diameter. It is less dependent (but not insignificantly) on its length. In the paper, the dependences are presented in diagrams. Values of the roundwood/log volume obtained by one procedure cannot be therefore "converted" (with operational accuracy) to values obtained by another procedure using one universal coefficient. It is always necessary to take into account properties of the concrete delivery of roundwood.

From the operational aspect it is not necessary to know the volume of roundwood exactly. Present technically achievable accuracy is quite sufficient for timber processing. From commercial aspects, the actual deviation is not important but its various values at various procedures.

An ideal solution into the future is to unify procedures. However, experience from negotiations does not make possible the author to believe in the early realization of this solution, because the will of parties concerned is missing. Therefore, it is necessary to aim at:

- Exact defining all steps of sensing and processing the data of algorithms, the regulations to provide minimum space for "individual interpretation" and thus affecting results (at present e.g. 2D and 3D sensing but also filtration).
- Definite determination of a method of the mutual comparison of results. In addition to the higher accuracy of comparison there is also the decline of commercial profitability to use various procedures.
- The inspection and calibration of measuring equipment, which will involve and affect checking the implementation of all steps including the value of a resulting volume (at present, it does not include e.g. data filtration). Calibration should be worked out both technically and legislatively.

8. References

ČSN 49 0018, EN 1309-2 (2006). *Kulatina a řezivo - Metody měření rozměrů. Část 2: Kulatina - požadavky na měření a pravidla pro výpočet objemu.* Český normalizační institut, Praha, Czech Republic

ČSN 48 0050 (1992). *Surové dříví. Základní a společná ustanovení.* Federální úřad pro normalizaci a měření. Praha, Czech Republic

Deutscher Forstwirtschaftsrat e.V.& Verband der Deutschen Säge- und Holzindustrie e.V. (2005). *Rahmenvereinbarung für die Werksvermessung von Stammholz*, Version 2005-01-14.

Hauffe, P.& Müller, L. (2002). Rundholzvermessung in Europa vereinheitlichen. *Holz-Zentralblatt* N. 77, 28. 6. 2002, p. 948

Hunková, V. (2011). *Faktory ovlivňující přejímku kulatiny při elektronickém měření jejích rozměrů.* Mendel University Brno, Czech Republic

Janák, K. (2007). Differences in round wood measurements using electronic 2D and 3D systems and standard manual method. *Drvna Industrija* Vol 58 (3) 2007 pp. (127-133), ISSN 0012-6772

Kolektiv (1959). *Naučný slovník lesnický*, Státní zemědělské nakladatelství Praha, Czech Republic

Kolektiv (2008). *Doporučená pravidla pro měření a třídění dříví v České republice 2008.* ISBN 978-80-87154-01-4 Lesnická práce, Praha, Czech Republic

Ö-Norm L 1021 (2006). *Vermessung von Rundholz.* Österreichisches Normungsinstitut, Vienna, Austria

Sauter, U., Staudenmaier, J. & Verhoff, S. (2010). Mehr Transparenz im Rundholzgeschäft *Holz-Zentralblatt* N. 50, 17. Dezember 2010- pp. 1269-1271

Šmelko, Š. et al. (2003). *Meranie lesa a dreva.* ISBN 80-89100-14-7 Zvolen, Slovakia

6

Fringe Pattern Demodulation Using Evolutionary Algorithms

L. E. Toledo[1], F. J. Cuevas[1], J.F. Jimenez Vielma[2] and J. H. Sossa[2]
[1]Centro de Investigaciones en Optica A.C.,
Dept. of Computer Vision and Artificial Intelligence, Optical Division, Leon,
[2]Center for Computing Research, National Polytehnical Institute,
Artificial Inteligence Laboratory,
Mexico

1. Introduction

Interferometers are used in metrology to measure temperature, displacement, stress and other physical variables. A typical interferometer split a laser beam using a beam divisor. Beam A is called reference, and is projected directly over a film or a CCD camera using mirrors or fiber optic. Beam B interact with the physical phenomenon to be measured. The interaction modifies the optic path of beam B; then it is projected over the same film or CCD camera that beam A. The total irradiance is modelled on eq. 1.

$$I(x,y) = a(x,y) + b(x,y)\cos(\varphi(x,y)) \tag{1}$$

The information about the measure is embodied on an interferogram, that is, a fringe pattern image. In optical metrology, a fringe pattern carries information embedded in its phase, that represents the difference in optical path between beam A and beam B. x,y are integer values representing coordinates of the pixel location in the fringe image, $a(x,y)$ is the background illumination, $b(x,y)$ is the amplitude modulation, and $\varphi(x,y)$ is the phase term related to the physical quantity being measured. Figure 1 shows an interferogram and its associated phase $\varphi(x,y)$.

(a)

(b)

Fig. 1. Fringe pattern(a) and its phase map (b).

The problem is to recover the phase map from the fringe pattern image. The demodulation process can be achieved by different methods, depending on the characteristics of the fringe pattern. If the fringe pattern, or interferogram has open fringes (see fig. 2c) by adding a carrier or tilt onto the phase, phase is obtained using Takeda's Fourier Transform method (Takeda & Kobayashi, 1982) or the Phase Locked Loop (Servin & Rodriguez Vera, 1993). A carrier is a phase that increases or decreases linearly with x and or y.

PLL and Fourier methods can not be used if the interferogram has closed fringes or is not normalized. A normalized fringe pattern means $a(x,y)\approx 0$ and $b(x,y)\approx 1$ (see fig 3). Many methods can be used to normalize a fringe pattern (Quiroga et al, 2001). An interferogram can be normalized due to a tilt in the plane waves of beams, defocus, speckle noise, etc.

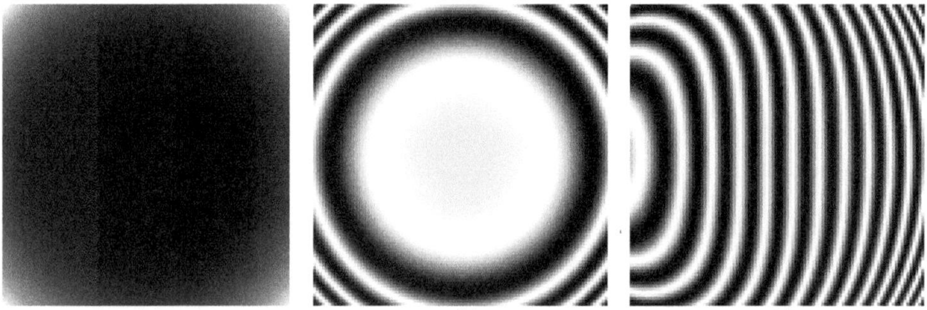

(a) (b) (c)

Fig. 2. Original phase (a), fringe pattern without carrier; (b), fringe pattern by adding a carrier (c).

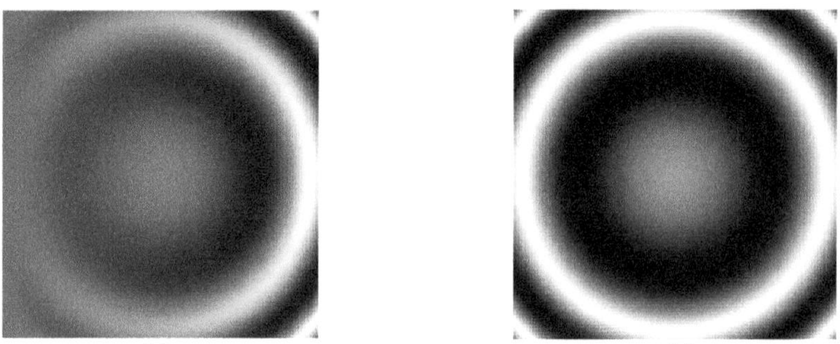

(a) (b)

Fig. 3. (a)Non-normalized fringe pattern. $a(x,y)=0.001y$, $b(x,y)=2x$; (b) Normalized fringe pattern.

If it is possible to take three or more images and add a constant on the phase, a constant shift that is different for each fringe pattern (see fig. 4), phase shifting techniques are suitable (Malacara et al, 1998). It is not necessary to normalize the fringe pattern, but it is assumed that $a_1(x,y)\approx a_2(x,y)\approx a_3(x,y)$ and $b_1(x,y)\approx b_2(x,y)\approx b_3(x,y)$.

A drawback to phase shifted method is the real phase are not obtained, but a mod 2π of the phase (fig. 5a). An unwrapping method (Ghiglia & Romero, 1994) is necessary to obtain the real phase (fig. 4b). The fringe patterns used on phase shifting should be normalized.

(a) (b) (c)

Fig. 4. (a) Original fringe pattern I; (b) Adding a constant phase of 120 degrees, I2; (c) Adding a phase of -120 degrees, I3.

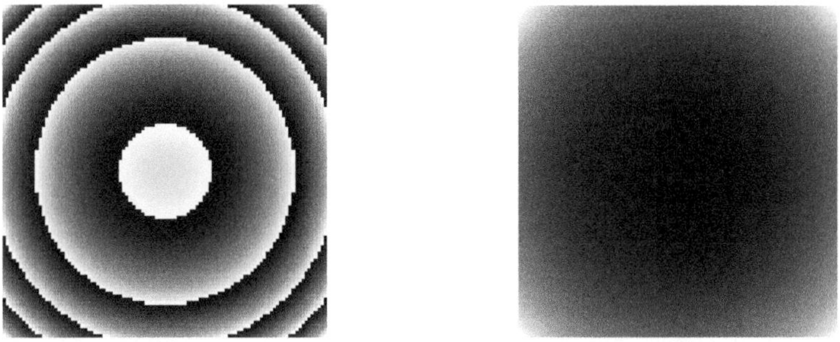

(a) (b)

Fig. 5. (a) Wrapped phase; (b) Unwrapped phase.

Methods like the Phase Tracker (Servin et al, 2001a) and the Two-dimensional Hilbert Transform (Larkin et al, 2001) are used for closed fringes, normalized images. These methods are robust again a large amount of noise, but a subjacent condition is to fulfil Nyquist condition. Phase tracker gives an unwrapped phase so there is not necessary to use an unwrapping method. The phase tracker and Hilbert transform proposed a cost function that depends of some measure of the difference between the real phase and the estimated phase. The real phase is unknown so the original interferogram is used and compared to the fringe pattern of the proposed phase. More terms are added to introduce restrictions.

A problem with minimize a cost function is the danger of fall in a local minimum, far away from the optimal point. It is also possible to use soft computing algorithms, such as neural networks and evolutionary algorithms (EA). In the neural network technique, a multilayer neural network (MLNN) is trained by using fringe patterns, and the phase gradients associated with them, from calibrated objects (Cuevas et al, 2000); after the training, the MLNN can estimate the phase gradient when the fringe pattern is presented in the MLNN input. A genetic algorithm (GA) is a particular type of EAs. GA´s are optimization algorithms that simulate natural evolution (Holland, 1975), and whereas GAs do not search for the best solution to a given problem, they can discover highly precise functional solutions and are very useful for nonlinear optimization problems or in the presence of multiple minimums (Goldberg, 1989), where classic techniques like gradient descent, deterministic hill climbing or random search (with no heredity) fail.

Methods using GA (Cuevas et al, 2002), approximate the phase through the estimation of parametric functions. The chosen functions could be Bessel in the case of having fringes from a vibrating plate experiment, or Zernike polynomials, in the case of an optical testing experiment, and when not much information is known about the experiment, a set of low degree polynomials $p(\mathbf{a},x,y)$ can be used. A complicated pattern is demodulated dividing it into a set of partially overlapping windows fitting a low dimensional polynomial function in each window, so that no further unwrapping is needed (Cuevas et al, 2006).

2. Genetic algorithms

Genetic Algorithms (GAs) (Bäck et al, 2000), are adaptive heuristic search algorithm premised on the evolutionary ideas of natural selection and genetic. The basic concept of GAs is designed to simulate processes in natural system necessary for evolution, specifically those that follow the principles first laid down by Charles Darwin of survival of the fittest. As such they represent an intelligent exploitation of a random search within a defined search space to solve a problem.

First pioneered by John Holland in the 60s, Genetic Algorithms has been widely studied, experimented and applied in many fields in engineering worlds. Not only does GAs provide an alternative method to solving problem, it consistently outperforms other traditional methods in most of the problems link. Many of the real world problems involved finding optimal parameters, which might prove difficult for traditional methods but ideal for GAs. However, because of its outstanding performance in optimization, GAs has been wrongly regarded as a function optimizer. In fact, there are many ways to view genetic algorithms.

In a genetic algorithm, a population of strings (called chromosomes or the genotype of the genome, fig. 6), which encode candidate solutions (called individuals, creatures, or phenotypes) to an optimization problem, evolves toward better solutions. Traditionally, solutions are represented in binary as strings of 0s and 1s, but other encodings are also possible. The evolution usually starts from a population of randomly generated individuals and happens in generations. In each generation, the fitness of every individual in the population is evaluated, multiple individuals are stochastically selected from the current population (based on their fitness), and modified (recombined and possibly randomly mutated) to form a new population. The new population is then used in the next iteration of the algorithm. Commonly, the algorithm terminates when either a maximum number of generations has been produced, or a satisfactory fitness level has been reached for the population.

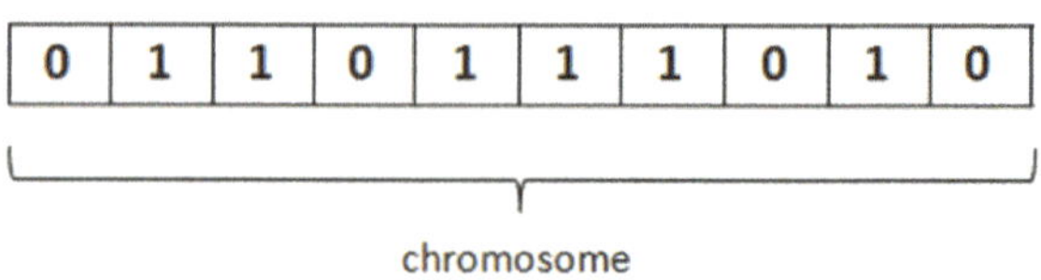

Fig. 6. A chromosome.

Genetic algorithms find application in bioinformatics, computational science, engineering, economics, chemistry, manufacturing, mathematics, physics and other fields.

A typical genetic algorithm requires:

1. A genetic representation of the solution domain, see fig. 7.
2. A fitness function to evaluate the solution domain.

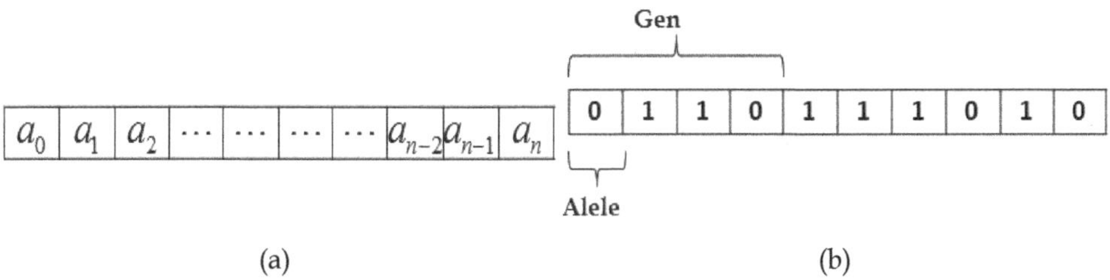

Fig. 7. (a) Representation of the solution domain; (b) Each gene is codified with a bit string.

A standard representation of the solution is as an array of bits. Arrays of other types and structures can be used in essentially the same way. The main property that makes these genetic representations convenient is that their parts are easily aligned due to their fixed size, which facilitates simple crossover operations. Variable length representations may also be used, but crossover implementation is more complex in this case.

The fitness function is defined over the genetic representation and measures the quality of the represented solution. The fitness function is always problem dependent. In some problems, it is hard or even impossible to define the fitness expression; in these cases, interactive genetic algorithms are used.

Once we have the genetic representation and the fitness function defined, GA proceeds to initialize a population of solutions randomly. Improve it through repetitive application of mutation, crossover, inversion and selection operators.

2.1 Initialization

At the first iteration many individual solutions are randomly generated to form the population. The population size depends on the nature of the problem, but typically contains several hundreds or thousands of possible solutions. Traditionally, the population is generated randomly, covering the entire range of possible solutions (the search space).

2.2 Selection

During each successive generation, a proportion of the existing population is selected to breed a new generation. Individual solutions are selected through a fitness-based process, where fitter solutions (as measured by a fitness function) are typically more likely to be selected. Certain selection methods rate the fitness of each solution and preferentially select the best solutions. Other methods rate only a random sample of the population, as this process may be very time-consuming.

A generic selection procedure may be implemented as follows:

1. The fitness function is evaluated for each individual, providing fitness values, which are then normalized. Normalization means dividing the fitness value of each individual by the sum of all fitness values, so that the sum of all resulting fitness values equals 1.
2. The population is sorted by descending fitness values.
3. Accumulated normalized fitness values are computed (the accumulated fitness value of an individual is the sum of its own fitness value plus the fitness values of all the previous individuals). The accumulated fitness of the last individual should be 1 (otherwise something went wrong in the normalization step).
4. A random number R between 0 and 1 is chosen.
5. The selected individual is the first one whose accumulated normalized value is greater than R.

Retaining the best individuals in a generation unchanged in the next generation, is called elitism or elitist selection. It is a successful (slight) variant of the general process of constructing a new population.

2.2.1 Roulette-wheel selection

Fitness proportionate selection, also known as roulette-wheel selection, is a genetic operator used in genetic algorithms for selecting potentially useful solutions for recombination.

In fitness proportionate selection, as in all selection methods, the fitness function assigns a value to possible solutions or chromosomes. This fitness level is used to associate a probability of selection with each individual chromosome. If f_i is the fitness of individual i its probability of being selected is $p_i = \frac{f_i}{\sum_j^N f_j}$, where N is the number of individuals in the population.

This could be imagined similar to a Roulette wheel in a casino. Usually a proportion of the wheel is assigned to each of the possible selection based on their fitness value. This could be achieved by dividing the fitness of a selection by the total fitness of all the selections, thereby normalizing them to 1. Then a random selection is made similar to how the roulette wheel is rotated.

While candidate solutions with a higher fitness will be less likely to be eliminated, there is still a chance that they may be. Contrast this with a less sophisticated selection algorithm, such as truncation selection, which will eliminate a fixed percentage of the weakest candidates. With fitness proportionate selection there is a chance some weaker solutions may survive the selection process; this is an advantage, as though a solution may be weak, it may include some component which could prove useful following the recombination process.

The analogy to a roulette wheel can be envisaged by imagining a roulette wheel in which each candidate solution represents a pocket on the wheel; the size of the pockets are proportionate to the probability of selection of the solution. Selecting N chromosomes from the population is equivalent to playing N games on the roulette wheel, as each candidate is drawn independently.

2.2.2 Stochastic universal sampling

Stochastic universal sampling (SUS) is a technique used in genetic algorithms for selecting potentially useful solutions for recombination. It was introduced by James Baker.

SUS is a development of fitness proportionate selection which exhibits no bias and minimal spread. Where fitness proportionate selection chooses several solutions from the population by repeated random sampling, SUS uses a single random value to sample all of the solutions by choosing them at evenly spaced intervals.

While candidate solutions with a higher fitness will be less likely to be eliminated, there is still a chance that they may be. Contrast this with a less sophisticated selection algorithm, such as truncation selection, which will eliminate a fixed percentage of the weakest candidates. With fitness proportionate selection there is a chance some weaker solutions may survive the selection process; this is an advantage, as though a solution may be weak, it may include some component which could prove useful following the recombination process.

The analogy to a roulette wheel can be envisaged by imagining a roulette wheel in which each candidate solution represents a pocket on the wheel; the size of the pockets are proportionate to the probability of selection of the solution. Selecting N chromosomes from the population is equivalent to playing N games on the roulette wheel, as each candidate is drawn independently.

2.2.3 Tournament selection

It involves running several "tournaments" among a few individuals chosen at random from the population. The winner of each tournament (the one with the best fitness) is selected for crossover. Selection pressure is easily adjusted by changing the tournament size; if it is larger, weak individuals have a smaller chance to be selected.

Deterministic tournament selection selects the best individual (when $p=1$) in any tournament. A *1-way* tournament ($k=1$) selection is equivalent to random selection. The chosen individual can be removed from the population that the selection is made from if it is desired, otherwise individuals can be selected more than once for the next generation.

Tournament selection has several benefits: it is efficient to code, works on parallel architectures and allows the selection pressure to be easily adjusted.

2.3 Crossover

Crossover is a genetic operator used to vary the programming of a chromosome or chromosomes from one generation to the next. It is analogous to reproduction and biological crossover, upon which genetic algorithms are based. Cross over is a process of taking more than one parent solutions and producing a child solution from them.

Crossover techniques :

- One-point crossover (fig. 8a).
- Two-point crossover (fig. 8b).

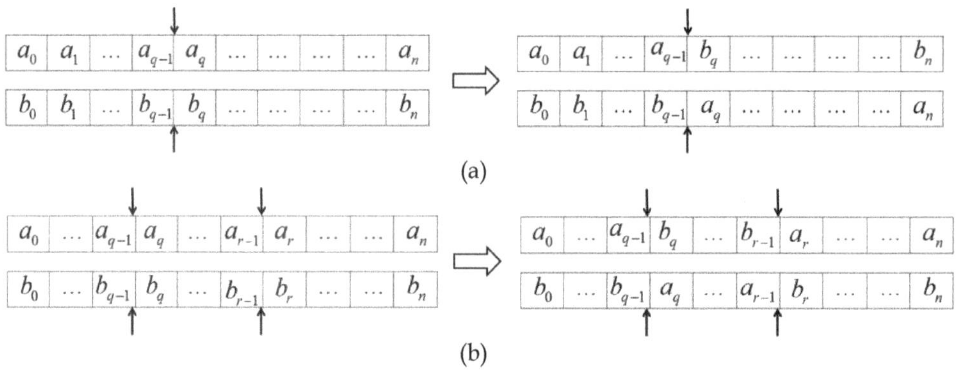

Fig. 8. (a) One point crossover; (b) Two point crossover.

2.4 Mutation

It is a genetic operator used to maintain genetic diversity from one generation of a population of algorithm chromosomes to the next. It is analogous to biological mutation. Mutation alters one or more gene values in a chromosome from its initial state. In mutation, the solution may change entirely from the previous solution. Hence GA can come to better solution by using mutation. Mutation occurs during evolution according to a user-definable mutation probability. This probability should be set low. If it is set to high, the search will turn into a primitive random search.

The classic example of a mutation operator involves a probability that an arbitrary bit in a genetic sequence will be changed from its original state. A common method of implementing the mutation operator involves generating a random variable for each bit in a sequence. This random variable tells whether or not a particular bit will be modified. This mutation procedure, based on the biological point mutation, is called single point mutation. Other types are inversion and floating point mutation. When the gene encoding is restrictive as in permutation problems, mutations are swaps, inversions and scrambles.

The purpose of mutation in GAs is preserving and introducing diversity. Mutation should allow the algorithm to avoid local minima by preventing the population of chromosomes from becoming too similar to each other, thus slowing or even stopping evolution. This reasoning also explains the fact that most GA systems avoid only taking the fittest of the population in generating the next but rather a random (or semi-random) selection with a weighting toward those that are fitter.

3. Genetic algorithm applied to phase recovery

The fringe demodulation problem is difficult to solve since many solutions are possible, even for a single noiseless fringe pattern.

The image on a fringe pattern $I(x,y)$ does not change if $\varphi(x,y)$ in Eq.(1) is replaced by another phase function $\widehat{\varphi}(x,y)$ given by:

$$\hat{\varphi}(x,y) = \pm\varphi(x,y) \pm 2\pi k, \tag{2}$$

A fringe pattern of *R*x*C* pixels dimension is segmented into a window overlapping set of sub-images of *R1*x*C1* pixels dimensions, and with origin coordinates at (*r*,*c*). The GA is used to carry out the optimization process, where a parametric estimation of a non linear function is proposed to fit the phase on the sub-images.

The fitness function is modelled by the next considerations: a) The similarity between the original fringe image and the genetic generated fringe image, and b) the smoothness in the first and second derivatives of the fitting function.

The fitting function is chosen depending on prior knowledge of the demodulation problem, but when no prior information about the shape of $\varphi(x,y)$ is known, a polynomial fitting is adequate. The adequate dimensionality of the polynomial depends on the interferogram complexity, but if we only want to estimate the phase in a small region, the dimensionality of the function can be low.

A *r-degree* approximation is used so that the phase data can be modelled like:

$$\begin{aligned} p(\mathbf{a},x,y) &= a_0 + a_1 x + a_2 y + a_3 x^2 + a_4 y^2 + a_5 xy \\ &+ a_6 x^2 y + a_7 xy^2 + a_8 x^3 + a_9 y^3 + \ldots + a_q y^r \end{aligned} \tag{3}$$

3.1 Decoding chromosomes

As it was said earlier, the GA is used to find the function parameters, in this case, vector $\mathbf{a}$. If we use this function, the chromosome can be represented by the vector:

$$\mathbf{a} = \begin{bmatrix} a_0 & a_1 \ldots a_q \end{bmatrix} \tag{4}$$

A *k-bit* long bit-string is used to codify an allele; then, the chromosome has qxk bits in length. We define the search space for these parameters. The bit-string codifies a range within the limits of each parameter. The decoded value of the *i-th* parameter is:

$$a_i = L_i^B + \frac{\left(L_i^U - L_i^B\right)}{2^k - 1} N_i, \tag{5}$$

where a_i is the *i-th* parameter real value, L_i^B is the *i-th* bottom limit, L_i^U is the *i-th* upper limit, and N_i is the decimal basis value of the bit-string .

L_i^B and L_i^U are redefined for each sub-image. The maximum value for each parameter is calculated taking into consideration the maximal phase value on that window. These maximum values can be expressed as:

$$L_0^B = -\pi \quad \text{and} \quad L_0^U = \pi \tag{6}$$

$$L_i^U = -L_i^B \tag{7}$$

$$L_i^u = \frac{4\pi F}{R1_i^m C1_i^n} \tag{8}$$

where F is twice the maximum fringe number on the window:

$$F = 2 \cdot \max\left(F_x, F_y, \sqrt{F_x^2 + F_y^2}\right) \tag{9}$$

F_x and F_y are the maximum fringe numbers in the x and y directions. m is the relative grade for x of the *i-th* term, and n is the relative grade for y of the *i-th* term.

For the special case a_0 $(i=0)$, the limits are considered to be between $-\pi$ and $+\pi$. a_0 is eliminated from parameter vector $\mathbf{a}$ to redefine a new vector $\mathbf{a}'$:

$$\mathbf{a}' = \begin{bmatrix} a_0 & a_1 \ldots a_q \end{bmatrix} \tag{10}$$

so, $p(\mathbf{a},x,y)$ can be expressed as:

$$p(\mathbf{a},x,y) = p(\mathbf{a}',x,y) + a_0 \tag{11}$$

and replacing Eq. 11 into Eq 1:

$$I(x,y) = a(x,y) + b(x,y)\cos\left(p(\mathbf{a}',x,y) + a_0\right) \tag{12}$$

Additionally, a_0 can be expressed as $a_0 = 2\pi l + a_0'$, l is an integer, and, $a_0' < 2\pi$ so Eq. 12 becomes:

$$I(x,y) = a(x,y) + b(x,y)\cos\left(p(\mathbf{a}',x,y) + a_0' + 2\pi l\right) \tag{13}$$

The cosine function is periodical with period 2π, so:

$$I(x,y) = a(x,y) + b(x,y)\cos\left(p(\mathbf{a}',x,y) + a_0'\right) \tag{14}$$

Eq. 14 demonstrates that limits for a_0 within a range of 2π are enough to represent the phase of the fringe pattern.

In the next section, it will be seen that a_0 can be separated from $\mathbf{a}$ and calculate it independently. Mutation and crossover operators can be applied only over $\mathbf{a}'$, and then add the calculated value for a_0 to $\mathbf{a}'$, and recover vector $\mathbf{a}$.

3.2 Fitness function

The GA, as was described in section 2, is an optimization procedure. Fitness function is always positive, and the optimal point is one minimum value. A negative sign transforms a problem from minimization to maximization. In a given generation, the maximal and minimal values or the fitness values are searched, and they are used to linearly adjust the dynamic range from 0 to A. The values are now positive, and are called aptitude. x and y

are the coordinates in the fringe images. (r,c) are the absolute coordinates of the origin coordinate of the sub-image, f is the function that is adjusted in the current window to approximate the phase term.

The fitness function is applied over each window and swept over the entire image. The path that the process follows can be an arbitrary choice. It is recommended that the window has between 40% and 60% overlapped area with previously demodulated data. This condition is required so the new demodulated phase can be coupled to the previously demodulated phases.

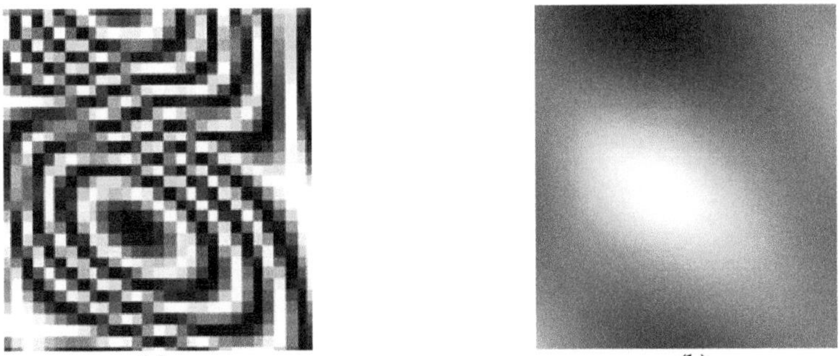

(a) (b)

Fig. 9. (a) Fringe pattern subsampled. (b) Demodulated phase.

A fitness function $U(\mathbf{a}^p)$ for each sub-image is used to obtain the fitness value for the *p-th* chromosome in the population. It form could be diverse, but most of them have a term that compares the RMS error between the original fringe pattern and the fringe pattern obtained from the estimated phase (similarity):

$$U(\mathbf{a}^p) = \sum_{y=r}^{r+R1} \sum_{x=c}^{c+C1} \left[I_N(x,y) - \cos\left(f\left(\mathbf{a}^p, x, y\right)\right)\right]^2 \tag{15}$$

Additional terms are added to the fitness functions to give restrictions on the proposed phase. In (Cuevas et al, 2006) fitness function has three criteria: similarity, smoothness and overlapped phase similarity with a previously estimated phase.

$$R(\mathbf{a}^p) = \sum_{y=r}^{r+R1} \sum_{x=c}^{c+C1} \left\{ \begin{array}{l} \lambda_1 \left[\left(f\left(\mathbf{a}^p, x-r, y-c\right) - f\left(\mathbf{a}^p, x-1-r, y-c\right)\right)^2 \right. \\ \left. + \left(f\left(\mathbf{a}^p, x-r, y-c\right) - f\left(\mathbf{a}^p, x-r, y-1-c\right)\right)^2 \right] m(x,y) \\ + \lambda_2 \left[\left(f\left(\mathbf{a}^p, x-r, y-c\right) - \Phi(x,y)\right)^2 \right] \end{array} \right\} \tag{16}$$

$R(\mathbf{a}^p)$ is the total amount the restrictions add to the fitness function, for a given window whose origin is on (r,c); $m(x,y)$ is a mask that indicates where, inside the image, exist the fringe pattern, $n(x,y)$ indicates the overlapping zone between the phase of a given window, φ, and the total phase, Φ.

The third term, associated with λ_2, allow the GA to extrapolate the trend of the adjacent demodulated regions into the actual window. As a consequence, the GA can demodulate interferograms that are noisy and are subsampled.

In WFPD (Cuevas et al, 2006) the fitness function has three criteria: similarity, smoothness and overlapped phase similarity with a previously estimated phase. It is eliminated third criterion, this simplified fitness function can make robust the phase retrieval in one window from demodulation errors in another window. The phase in different windows can be demodulated in parallel. The resulting phase segments are splicing sequentially. Noise filtering and fringe normalization are solved using alternative low-pass filtering techniques. We suppose a smooth phase with continuity in first and second derivatives.

A fitness function $U(\mathbf{a}^p)$ for each sub-image is used to obtain the fitness value for the *p-th* chromosome in the population, it can be written as:

$$\begin{aligned} U(\mathbf{a}^p) = -\sum_{y=r}^{r+R1}\sum_{x=c}^{c+C1} & \Big\{\left[I_N(x,y) - \cos\left(f\left(\mathbf{a}^p,x,y\right)\right)\right]^2 \\ & \lambda_1\Big[\left(f\left(\mathbf{a}^p,x,y\right) - f\left(\mathbf{a}^p,x-1,y\right)\right)^2 \\ & +\left(f\left(\mathbf{a}^p,x,y\right) - f\left(\mathbf{a}^p,x,y-1\right)\right)^2\Big] \\ & \lambda_2\Big[\left(f\left(\mathbf{a}^p,x+1,y\right) - f\left(\mathbf{a}^p,x-1,y\right)\right)^2 \\ & +\left(f\left(\mathbf{a}^p,x,y+1\right) - f\left(\mathbf{a}^p,x,y-1\right)\right)^2\Big]\Big\} m(x,y). \end{aligned} \tag{17}$$

where $f\left(\mathbf{a}^p x,y\right) = p\left(\mathbf{a}^p x,y\right)$, λ_1 and λ_2 are the regularization terms used to penalize high first and second phase derivatives, and to assure the smoothness of the phase. $m(x,y)$ is a binary mask which defines the valid and invalid data image area. λ_1 and λ_2 values are chosen empirically, and are dependent on window size and fringe pattern regional frequency contents. Typical values are 0.005 to 0.07 for λ_1 and 0.00025 to 0.002 for λ_2 for a $7\text{x}7$, $9\text{x}9$ or $11\text{x}11$ size window.

The following terms were used in the fitness function:

a. Fringe similarity criterion:

$$\left[I_N(x,y) - \cos\left(f\left(\mathbf{a}^p,x,y\right)\right)\right]^2 \tag{18}$$

The fringe pattern is considered to be normalized on the range [-1,1], or a binary version of the original. There are several methods in the literature for the normalizing of binary threshold fringe patterns [16]. The normalized genetic fringe image is calculated using the cosine of the fitting function. The sum of the squared differences between original fringes and the synthesized fringe pattern is calculated, and those chromosomes that generated a more similar fringe pattern to the original, will have a higher probability of being selected.

As it was said earlier, parameter a_0 can be calculated so that mutation and crossover are only applied to $\mathbf{a}'$. a_0 is spliced onto $\mathbf{a}'$ to recover parameter vector $\mathbf{a}$.

The missing parameter, a_0, is obtained from the fringe similarity criterion by:

$$\frac{\partial U_S}{\partial a_0} = A\sin(a_0) + B\cos(a_0) - C\cos(2a_0) - D\sin(2a_0)$$
where
$$A = \sum 2I_N(x,y)\cos(\mathbf{a}'),$$
$$B = \sum 2I_N(x,y)\sin(\mathbf{a}'), \quad (19)$$
$$C = \sum \sin(2\mathbf{a}'),$$
and
$$D = \sum \cos(2\mathbf{a}').$$

A, B, C, and D are constants for any value given to a_0. a_0 is chosen to be the value that makes $\frac{\partial U_S}{\partial a_0} = 0$ in the domain $[-\pi, +\pi]$. If we define the function:

$$f(a_0) = \left(\frac{\partial U_S}{\partial a_0}\right)^2 \quad (20)$$

the problem is finding the minimum. There are several methods to find it, like the Newton method, the Fibonacci search, the steepest descent method, etc. The search domain in Eq. 20 is well defined and small enough, so an exhaustive search can easily found the desired value of a_0

b. Smoothness criterion:

$$\begin{aligned} &\lambda_1\Big[\big(f(\mathbf{a}^p,x,y) - f(\mathbf{a}^p,x-1,y)\big)^2 \\ &+\big(f(\mathbf{a}^p,x,y) - f(\mathbf{a}^p,x,y-1)\big)^2\Big] \\ &\lambda_2\Big[\big(f(\mathbf{a}^p,x+1,y) - f(\mathbf{a}^p,x-1,y)\big)^2 \\ &+\big(f(\mathbf{a}^p,x,y+1) - f(\mathbf{a}^p,x,y-1)\big)^2\Big] \end{aligned} \quad (21)$$

The weighted sum of the discrete approximation (squared differences) of the first and second derivatives is calculated. The goal is to achieve smooth solutions in the first and second derivatives. This term contributes in a negative way to the maximization fitness process, so the chromosome decreases its fitness value.

This term is necessary because of the use of a polynomial function. We use a high degree polynomial interpolation, and oscillations are present on the estimated phase; they can introduce errors into the demodulation process. λ_1 and λ_2 are chosen to minimize these oscillations.

3.3 Selection operator

Chromosomes are evaluated using the fitness function. The result of the evaluation is their fitness value. Fitness value of the entire population is stretched in the range 0 to A. In each generation, the minimum and maximum fitness values are obtained to produce a normalized fitness value:

$$NFv(i)_C = \left(FV(i)_C - \min(i)\right)\frac{A}{\max(i) - \min(i)}, \tag{22}$$

where $NFv(i)_c$ is the normalized fitness value in the *i-th* generation for the *c-th* chromosome, and *min*(*i*) and *max*(*i*) are the minimum and maximum values of $FV(i)_c$, the fitness value for a member in the *i-th* generation.

The normalized fitness value is used to calculate the probability selection P_s of a given chromosome. In order to represent a sexual reproduction, pairs of chromosomes are used to produce a new population. Two chromosomes are randomly selected with a P_s that is proportional to its normalized fitness value.

For the *c-th* chromosome, a random number r_s is generated. If $r_s < P_s$, the chromosome is selected. Two chromosomes are picked this way, and the crossover operator is applied over them to create two new chromosomes for the new population.

P_s can be calculated in many ways. The easiest is the Roulette Wheel method, which P_s is calculated by:

$$P(i)_S = \frac{NFv(i)_S}{\sum_S NFv(i)_S} \tag{23}$$

The Roulette wheel method has the disadvantage of being able to produce a premature convergence, so a Boltzmann selection method was used to avoid this inconvenient:

$$P(i)_S = \frac{\exp\left[\left[NFv(i)_S\right]/T(i)\right]}{\sum_S \exp\left[\left[NFv(i)_S\right]/T(i)\right]}, \tag{24}$$

where $T(i)$ *T*(*i*) is the temperature in the *i-th* generation. *T*(0) is a large value, so P_s is similar for all chromosomes in the initial generation. T is decreased over the generations, so the P_s for the best adapted is progressively higher than for the least adapted. In this way, there is the opportunity to explore all the space of solution, so the probability of falling into a local minimum is lowered. T is varied by:

$$T(i) = T_o \exp(-i / k) \tag{25}$$

where k is a constant that indicates in which generation $T(i) \approx T_0 / 3$. The process of selection and crossover is repeated until an entire new population is obtained.

3.4 Crossover operator

In the GA, a Crossover probability P_c is given to exchange the genetic information between two chromosomes, so that if a randomly generated number is smaller than it, the chromosomes are mixed to produce two new individuals, and if the number is bigger that P_c, the two original chromosomes are added to the new population. In the phase recovery from a fringe pattern, a two point crossover was used, where two crossover points are randomly generated, In this case, it is required to swap the central segments between chromosomes.

3.5 Mutation operator

Mutation is the best known mechanism to produce variations. Alleles of the chromosomes are randomly replaced by others in a random way. Mutation is treated like a background operator to ensure variety in the population.

In GA, a mutation probability P_m is defined. For each position, a random number is generated, and if it is smaller than P_m, the allele is changed for another. In a binary chromosome code, a '0' is changed for '1' and a '1' is changed for '0'.

3.6 GA convergence

GA convergence depends mainly on population size. With a large population, convergence is achieved in a smaller number of iterations, but the processing time is increased . To stop the GA process, different convergence measures can be employed. The maximum number of iterations is chosen at B. The algorithm can be stopped when a relative error ε is smaller than a predefined limit δ:

$$\varepsilon = \left| \frac{\max(i) - \max(i-1)}{\max(i)} \right|. \tag{26}$$

3.7 Splicing process

As it was mentioned before, phase demodulation is achieved through window segmentation of the fringe pattern. The GA demodulates the phase inside each window, independently from others, so this process can be made in parallel. It is supposed that the demodulated phase field in each window is differentiated from the corresponding real phase field only by the concavity and the DC bias. Then, a splicing procedure is required to connect different GA fitted phase windows and determine the whole phase field $\Phi(x,y)$. The splicing process is carried out in a sequential way (e.g., row by row).

It is described as follows:

1. The demodulated phase from the first window is used as the initial reference.

2. From the GA current fitted phase window $\varphi(x,y)=f(\mathbf{a}^p,x,y)$, i a second phase field is calculated $\varphi'(x,y)$ with a negative concavity) as $\varphi'(x,y)=-\varphi(x,y)$, or $\varphi'(x,y)=f(-\mathbf{a}^p,x,y)$.
3. Two DC bias are calculated, one for $\varphi(x,y)$ and one for $\varphi'(x,y)$ using:

$$DC_1=\frac{\sum_{x,y\in N}\left(\Phi(x,y)-\varphi\left(\mathbf{a}^p,x,y\right)\right)}{A} \text{ and } DC_2=\frac{\sum_{x,y\in N}\left(\Phi(x,y)-\varphi'\left(\mathbf{a}^p,x,y\right)\right)}{A}, \quad (27)$$

where N is the overlapped neighbourhood region, and A is the overlapped area (pixel^2) of N.

4. The RMS error for the two alternative phase window fields, $\varphi(x,y)$ and $\varphi'(x,y)$, compared against $\Phi(x,y)$ is calculated as:

$$RMS_1=\frac{\sum_{x,y\in N}\left(\Phi(x,y)-\varphi\left(\mathbf{a}^p,x,y\right)-DC_1\right)^2}{A} \text{ and } RMS_2=\frac{\sum_{x,y\in N}\left(\Phi(x,y)-\varphi'\left(\mathbf{a}^p,x,y\right)-DC_2\right)^2}{A} \quad (28)$$

5. The phase described by the function with the minimum *RMS* error value ($\varphi+DC_1$ or $\varphi'+DC_2$) is spliced onto the demodulated phase field Φ.
6. If there are more windows to splice, the next window in the sequence is labelled as the current window and goes to step 2. Otherwise, the splicing process is finished.

4. Adjustable genetic algorithm

In previous works, the window has a fixed dimension. The *RMS* error for a given window varies due to frequency content of the image, the window size and the values given in the smoothness criterion. High frequency zones are best demodulated using small windows; the demodulating process in low frequency zones is better when using a large window. In fact, using a small window in low frequency zones introduces unnecessary noise due to the splicing process for a certain region. This frequency estimator is used again to determine the best demodulation path; beginning in low frequency regions, it develops across increasing complexity regions. As a result, regions that are easily demodulated can be used to guide the splicing of higher difficulty regions.

Finally, a modification on the GA is presented to determine the smoothness criterion automatically. In a previous work, some values to λ_1 and λ_2 were proposed to be used for certain window values. In this work, this smoothness criterion is calculated by the GA itself, adding two genes to the chromosome; this codifies this criterion in real numbers. As a result, the GA calculates the coefficient vector of the polynomial and the smoothness parameters needed for the correct estimation of the phase at the same time. This modification allows one to have an algorithm that depends less on external characterization.

4.1 Local frequency estimator

In this section, a method to classify regions on the image according to their frequency content is described. This process is necessary to determine the size of the window needed to demodulate a certain region. A given fringe pattern is segmented into two, three or more regions.

1. From the original image, we obtain the image that shows the gradient. This image is obtained by correlating image $I(x,y)$ of RxC dimensions with the Sobel operators.

$$S_x = \begin{bmatrix} +1 & 0 & -1 \\ +2 & 0 & -2 \\ +1 & 0 & -1 \end{bmatrix} \tag{29}$$

$$S_y = \begin{bmatrix} +1 & +2 & +1 \\ 0 & 0 & 0 \\ -1 & -2 & -1 \end{bmatrix} \tag{30}$$

The correlation is defined as:

$$C_k(x,y) = \sum_i \sum_j I(i,j) S_k(x+i, y+j) \tag{31}$$

At each point, the magnitude of the gradient is obtained:

$$|S| = \sqrt{S_x^2 + S_y^2} \tag{32}$$

2. A median filter is used to detect the zones where fringe frequency increases. A median filter of dimensions NxN is defined as:

$$M(x,y) = \frac{\sum_{i=-N/2}^{N/2} \sum_{j=-N/2}^{N/2} I(x+i, y+j)}{NxN} \tag{33}$$

3. The image obtained by means of this process shows brighter zones in the regions with high frequency. This image is segmented into three regions of low, medium and high frequency content.

The brightest point, $max(M)$, and the darkest point, $min(M)$, are located. Their values are used to perform the segmentation of image $M(x,y)$ using:

$$F(x,y) = \frac{[M(x,y) - \min(M)]}{\max(M) - \min(M)} \tag{34}$$

$F(x,y)$ is thresholded in three zones:

$$FT(x,y) = \begin{cases} 0 & 0.3 > \frac{[M(x,y)-\min(M)]}{\max(M)-\min(M)} \geq 0 \\ 1 & 0.7 > \frac{[M(x,y)-\min(M)]}{\max(M)-\min(M)} \geq 0.3 \\ 2 & 1 \geq \frac{[M(x,y)-\min(M)]}{\max(M)-\min(M)} \geq 0.7 \end{cases} \tag{35}$$

Each zone indicates the relative level of frequency content in $I(x,y)$. $F(x,y)$ and $FT(x,y)$ are used to obtain the demodulation map, and to indicate the window size suited for a certain region.

4.2 Quality map and demodulation path

The windows where the polynomial is adjusted are extracted from $I(x,y)$ following the fringes, like in the phase tracker method. To achieve this result, in a previous work a quality map is obtained from $I(x,y)$ by thresholding it on Q levels, and using a filling algorithm to obtain a continuous path along the fringes.

In this work, we propose a method to obtain a different quality map. It is desirable that the demodulation path fulfils more conditions other than following the fringes. The conditions proposed are:

a. Follow the fringes.
b. Follow the frequency contents from low frequencies to high frequencies.
c. Sequentially demodulate regions with increasing levels of difficulty.

We postulate that criterions b and c are identical, so the problem is reduce to incorporating the second criterion into the quality map.

To achieve this goal, $F(x,y)$ is used. $F(x,y)$ is added to $I(x,y)$, and the new image $IF(x,y)$ is used to obtain the quality map, by thresholding it on Q levels.

$$IF(x,y) = I(x,y) + F(x,y) \tag{36}$$

$$IFT(x,y) = Q\frac{\left[IF(x,y) - \min(IF)\right]}{\max(IF) - \min(IF)} \tag{37}$$

IFT is used to generate a new demodulation path that begins on low frequencies and follows the fringes until it reaches high frequencies.

The algorithm used to generate the demodulation path is :

a. An array with al point labelled with some level (quality map $IFT(x,y)$) is needed
b. In the frequency map, $F(x,y)$, the minimum frequency area is searched, and the point that is $min(F)$ is chosen.
c. A matrix $L(x,y)$ of RxR dimensions is defined. All its values are set to '0'.
d. An array of stacks $[S_1, S_2, \ldots, S_Q]$ is defined. All stacks have RxC dimensions. The initial point is placed on the stack corresponding to its value. i.e. if $IFT(x,y) = 1$, then coordinates (x,y) are stored in stack S_1.
e. An array of coordinates $C(n)$ of RxC dimensions is defined. The coordinates of the first point are stored in it. An index n is defined, and its value is set to '0'. $L(x,y)$ is set to '1':

$$\begin{gathered} n \leftarrow 0 \\ C(n) \leftarrow (x,y) \\ L(x,y) \leftarrow 1 \end{gathered} \tag{38}$$

f. The points surrounding (x,y) are scanned while varying index i and index j from -1 to +1, and their coordinates stored in their corresponding stacks:

$$\begin{aligned} &\text{if} \quad L(x+1,y+1)=0 \\ &S_{IFT(x+i,y+j)} \leftarrow (x+i,y+j) \end{aligned} \tag{39}$$

g. Beginning on S_1, a non-empty stack is searched. The point (x,y) on top of that stack S_i is brought out. The next actions are performed:

$$\begin{aligned} n &\leftarrow n+1 \\ C(n) &\leftarrow (x,y) \\ L(x,y) &\leftarrow 1 \end{aligned} \tag{40}$$

h. If $n \leq RxC$ steps 'f' and 'g' are repeated. Points stored in $C(n)$ indicate the demodulation path to follow.

4.3 Genetic algorithms internal parameters automatic adjust

GAs have great possibilities to develop robust algorithms, varying their internal parameters according to the task to be performed. In this paper, we propose using GAs to adjust the internal parameters of this algorithm.

4.3.1 Regularization terms

The original image is shown in fig. 10. A media filter is applied onto the gradient image, yielding high values in regions with high fringe density.

Fig. 10. Original image.

Maximum and minimum values are localized, and the image is discretized into three main regions. A given window size is assigned to each region (Fig 11).

When the demodulation window is positioned in a given region, the size window is varied to obtain a better adjustment. The last stage is to introduce these regions into the demodulation process.

A quality map is used to determine the demodulation path. Low frequency regions are demodulated first by adding the discretized map of frequencies to the original image. A filling algorithm is used to determine the demodulation path using this quality map.

(a) (b) (c)

Fig. 11. Gradient (a) smoothed with a 9x9 window (media filter) (b) Classification on high, medium and low frequencies.

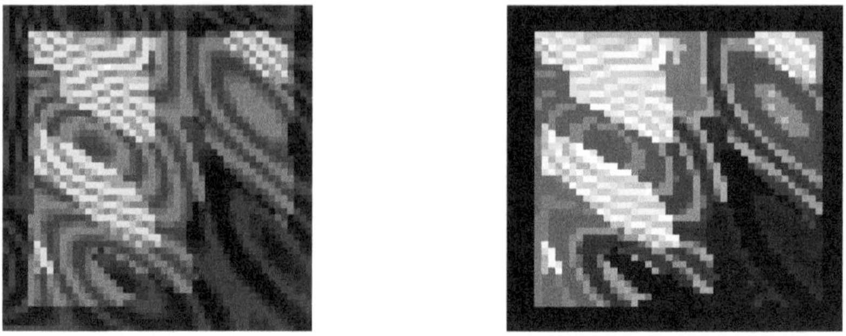

(a) (b)

Fig. 12. (a) Quality map, 15 levels; (b) Demodulation path that follows fringes and increased frequency regions.

5. Results

The proposed algorithm is used to demodulate a shadow moiré image taken from a modeled figure of a dolphin. The noise in the original shadow moiré is filtered through the use of a median filter. This filter is used only on an area determined by a mask that indicates the allowed area on which to perform the demodulation process.

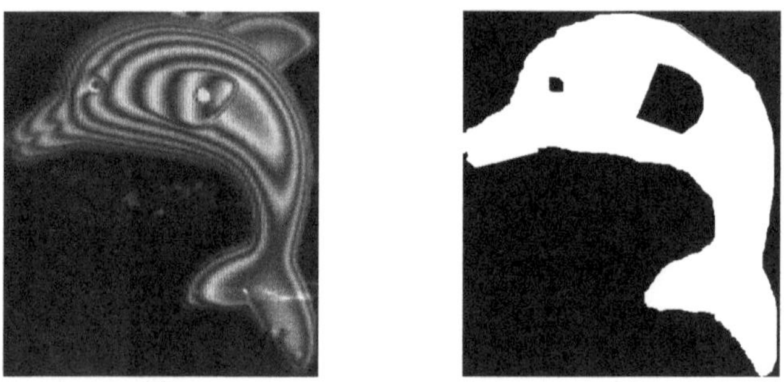

(a) (b)

Fig. 13. (a) Fringe pattern shadow moiré; (b) Mask $m(x,y)$.

The algorithm needs surfaces that are continuous on the first and second derivatives, so, regions in the original dolphin figure that do not fulfill these requirements are cut from the mask.

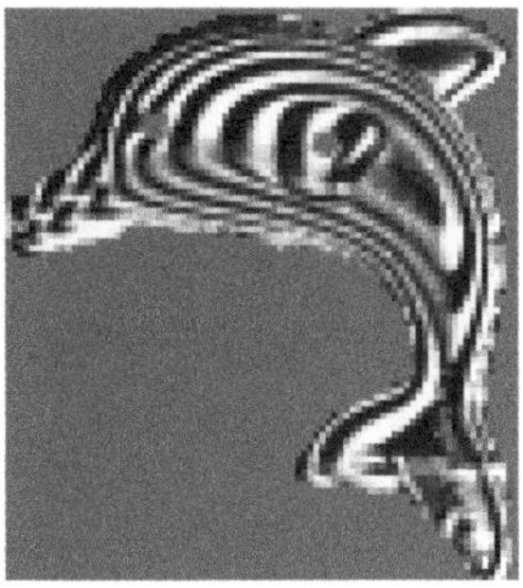

Fig. 14. Image after noise filtering using a median filter.

(a) (b)

Fig. 15. (a) Result of binarize fig 14; (b) Fringe pattern associated with the demodulated phase.

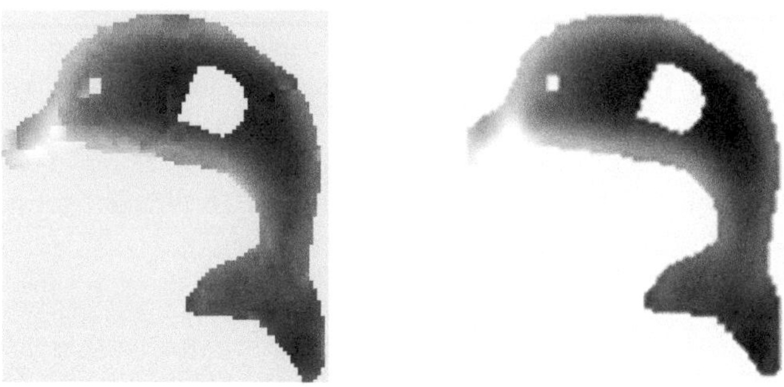

(a) (b)

Fig. 16. (a) Phase demodulated from fig. 15(a); (b) Phase smoothed using a media filter 7x7.

The resulting image is binarized to reduce noise interference in the demodulation process. This binarized image is fed to the algorithm.

The demodulated phase is smoothed using a 7x7 media filter, after using a contrast enhancement and a low pass filter.

6. Conclusions

Computational power has been increased in the last years. This increased processing power allows to develop EA as a practical tool with application in patter recognition, computational intelligence, image processing, automatization, and others. EA's, fuzzy logic and neural networks are called soft computing, because they can deal with problems where there are not a good knowledge of the problem, information is incomplete or inconsistent, or are large amount of noise.

In this chapter is shown only a single application of EA on fringe pattern demodulation. There still a lot of variations that can be explored to improve the performance of actual algorithms.

GA based methods have two advantages over regularized phase tracker: they can work on low resolution images and they can follow changes in concavity. These advantages are the consequence of taking upper grade terms in the interpolated function.

The technique showed in (Toledo, Cuevas 2009), called FPIW, is based in two suppositions: it is not necessary to know the phase on the neighborhood to estimate the phase in a given window, and the estimated phase in a window differ only by its concavity sign and a DC bias, from the real phase in the region framed by the window. As a consequence, the overlapped similarity criterion used in the WFPD (Cuevas et al, 2003) method can be eliminated from the fitness function in the FPIW method. In exchange, FPIW works near Nyquist, but on sub-Nyquist, WFPD is better.

The phase in a given window is estimated without known nothing about the phase in other windows. It is possible to demodulate simultaneously all windows, that is, FPIW method described has implicit parallelism. WFPD demodulate the windows sequentially.

7. Acknowledgments

We acknowledge the support of "Consejo Nacional de Ciencia y Tecnologia" of Mexico, "Consejo de Ciencia y Tecnología del Estado de Guanajuato," and "Centro de Investigaciones en Optica, A.C." We also thanks Mario Ruiz Berganza for his aid proofreading this paper; and thanks to Guillermo Garnica for his invaluable technical support. H. Sossa thanks "Consejo Nacional de Ciencia y Tecnologia" for grant 155014 and "SIP-IPN" for grant 20121311.

8. References

Bäck T, Fogel DB, Michalewicz Z (2000) *Evolutionary computation.* Institute of Physics Publishing, Bristol.

Bone, D.J., *Fourier fringe analysis: the two dimensional phase unwrapping problem,* Appl. Opt, 30, 3627-3632, 1991,

Buckland, J.R., Huntley, J.M., and Turner, S.R.E., *Unwrapping noisy phase maps by use of a minimum-cost-matching algorithm,* Appl. Opt., Vol. 34, No. 23 (August 1995) pp 5100-8.

Cuevas FJ, Servin M, Rodríguez-Vera R (1999) *Depth recovery using radial basis functions.* Opt Communication 163:270–277.

Cuevas FJ, Servin M, Stavroudis ON, Rodríguez-Vera R (2000). *Multi layer neural network applied to phase and depth recovery from fringe patterns.* Opt Commun 181:239–259.

Cuevas F.J., Mendoza F., Servin M., Sossa-Azuela J.H. (2006) *Window fringe pattern demodulation by multi-functional fitting using a genetic algorithm.* Opt. Commun. 261:231-239.

Cuevas FJ, Sossa-Azuela JH, Servin M (2002) *A parametric method applied to phase recovery from a fringe pattern based on a genetic algorithm.* Opt Commun 203:213–223.

Fernández A, Kaufmann GH, Doval AF, Blanco-Garcia J, Fernández JL (1998) *Comparison of carrier renoval methods in the analysis of TV holography fringes by the Fourier transformmethod.* Opt Eng 37:2899–2905.

Ghiglia, D.C., and Romero, L.A., *Robust two dimensional weighted and unweighted phase unwrapping that uses fast transforms and iterative methods,* J. Opt. Soc. Am. A, 11, 107-117, 1994.

Goldberg D (1989) *Genetic algorithms: search and optimization algorithms.* Addison-Wesley, Reading, MA.

Holland JH (1975) *Adaptation in natural and artificial systems.* University of Michigan Press, Michigan.

Ichioka Y, Inuiya M (1972) *Direct phase detecting system.* Appl Opt 11:1507–1514.

Juan Antonio Quiroga, José Antonio Gómez-Pedrero, Ángel García-Botella, *Algorithm for fringe pattern normalization,* Optics Communications, Volume 197, Issues 1-3, 15 September 2001, Pages 43-51

Juan Antonio Quiroga, Manuel Servin, *Isotropic n-dimensional fringe pattern normalization ,* Optics Communications, Volume 224, Issues 4-6, 1 September 2003, Pages 221-227.

L.E. Toledo & F.J. Cuevas, Optical *Metrology by Fringe Processing on Independent Windows Using a Genetic Algorithm,* Springer Verlag Experimental Mechanics 48, pp 559-569.

Larkin KG, Bone DJ, Oldfield MA (2001) *Natural demodulation of two-dimensional fringe patterns in general background of the spiral phase quadrature transform.* J Opt Soc Am A 18:1862–1870.

Malacara D, Servin M, Malacara Z (1998) *Interferogram analysis for optical testing.* Marcel Dekker, New York.

Noé Alcalá Ochoa, A.A. Silva-Moreno, *Normalization and noise-reduction algorithm for fringe patterns,* Optics Communications 270: 161-168.

Quiroga JA, Gómez-Pedrero JA, García-Botella A (2001) Algorithm *for fringe pattern normalization.* Opt Communications 197:43.

Servin M, Rodriguez-Vera R (1993) *Two dimensional phaselocked loop demodulation of interferogram.* J Mod Opt 40:2087–2094.

Servin M, Marroquín JL, Cuevas FJ (2001) *Fringe-follower regularized phase tracker for demodulation of closed-fringe interferograms.* J Opt Soc Am A 18:689–695.

Servin M, Quiroga JA, Cuevas FJ (2001) *Demodulation of carrier fringe pattern by the use of non-recursive digital phase locked loop.* Opt Commun 200:87–97.

Takeda M, Ina H, Kobayashi S (1982) *Fourier-transform method of fringe-pattern analysis for computer based topography and interferometry.* J Opt Soc Am 72:156.

7

Measurement System of Fine Step-Height Discrimination Capability of Human Finger's Tactile Sense

Takuya Kawamura, Kazuo Tani and Hironao Yamada
Department of Human and Information Systems, Gifu University, Japan

1. Introduction

In this study, to measure the human finger's tactile sensation capability of recognizing a fine surface texture using psychophysical experiments, a computer-controlled measurement system that presents fine step-heights of 0 to 1000 μm to human subjects' fingers at various presentation angles were developed. The measurement system can control four parameters of fine step presentation, i.e., the step-height, presentation velocity, presentation angle, and presentation temperature. In psychophysical experiments of this study, the measurement system calculates the amounts of step-heights based on the Parameter Estimation by Sequential Testing (PEST) method (Taylor & Creelman, 1982) and presents the step-heights to subjects' fingers in order to measure difference thresholds and subjective equalities for fine step-heights. Those values are considered to be the fine step-height discrimination capability of finger's tactile sense.

Human finger's tactile sense is a measurement system that can detect subtle surface roughness and smoothness by touching the surface. This finger's tactile sense is much more robust than the tactile sensors developed so far for robot tactile recognition. These sensors for robot still cannot reach the performance of recognizing such fine roughness or smoothness as humans can. Therefore, it is important for engineering, as well as for psychology, to investigate the finger's tactile recognition mechanism.

So far several researchers have examined the finger's tactile sense mechanism in detail using microneurography and psychophysical experimentation. In the microneurography, a tungsten microelectrode was inserted into tactile-related nerve fibers in an arm of humans or monkeys and the reactions of the tactile sense to the stimuli presented to the hand were examined via the signals sensed by the microelectrode. In the psychophysical experiments, on the other hand, several magnitudes of stimuli were presented to human hands and the responses of the tactile sense to the stimuli are analyzed through the replies to questions regarding the stimulus magnitudes.

The microneurography found out that the human tactile organs consist of four types of mechanoreceptive units: Fast adapting type I unit (FA I), Fast adapting type II unit (FA II), Slowly adapting type I unit (SA I), and Slowly adapting type II unit (SA II) (Vallbo & Johansson, 1984; Salentijn, 1992), and it is considered that FA II responds to a subtle mechanical vibration, FA I or FA II to surface unevenness and SA I to a pattern like Braille dots, respectively (Heller, 1989). On the other hand, the psychophysical experiments (Miyaoka et al., 1993, 1996, 1997) determined that the human tactile mechanism is able to detect a mechanical vibration of 0.2 μm in amplitude and a surface unevenness of 3 μm in amplitude. Also, the psychophysical experiment (Kawamura et al., 1996) revealed that FA I plays an important role in discriminating the magnitudes of step-height of around 10 μm. From these experimental results, it is considered that, like the human visual sense, the human tactile sense has several kinds of module mechanisms, and it is supposed that the human tactile modules are classified based on the stimulus magnitudes they can detect and discriminate and their information processing characteristics: the subtle stimulation detection module, fine texture recognition module, two-dimensional pattern recognition module, and three-dimensional shape recognition module. So far the authors have been investigated the tactile sensation capability of recognizing fine step-heights with respect to the fine texture recognition module.

Using a measurement system that presents fine step-heights of about 10 μm to subjects' fingers (Miyaoka et al., 1996; Kawamura et al., 1996), the difference thresholds for a 10 μm step-height were determined when the subjects actively touched the step-height with their fingers moving over the step-height and when they were passively touched the step-height presented to their fingers by the movement of the step presentation device. As a result, the difference thresholds for a 10 μm step-height in the active- and passive-touch experiments agreed approximately. Therefore, it was concluded that the finger's discrimination capability of fine step-heights of about 10 μm does not depend on the touching manners. Also, the paper (Kawamura et al., 1998) suggested that when the subjects discriminated a pair of the 10 μm step-heights presented at the different presentation velocities of 20 and 40 mm/s to their fingers, they perceived the height of the fast moving step-height to be a larger stimulus than that of the slowly moving step-height due to the influence of the stimulus velocity. Furthermore, the authors developed a measurement system that can create fine step-heights of 0 to 1000 μm and present the step-heights to subjects' fingers at various presentation angles (Kawamura et al., 2009).

In this paper, to measure the finger's tactile sensation capability of discriminating fine step-heights, the developed measurement system is used. In the psychophysical tests, the presentation angle of a step is defined as the angle to finger's length and several pairs of fine step-heights of 0 to 100 μm are presented to subjects' fingertips at various presentation angles. This paper first describes the measurement system that controls the amounts of step-heights according to the experiment procedure based on the PEST method in order to determine subjective equalities and difference thresholds for fine step-heights, then examines the effects of the touching manner of human finger, finger's motion direction, and fingertip region on the tactile recognition of fine step-heights. In the psychophysical tests, first, the subjects discriminate step-heights of 10 to 100 μm in active- and passive-touch manners using the center of their fingertips. Next, the subjects discriminate step-heights of around 10 μm using the top and center of their fingertips in various motion directions of their fingers.

2. Psychophysical experiment

2.1 Subjective equality and difference threshold

In the psychophysical experiments of this study, the relationships between the stimulus magnitudes of fine step-heights and the sensitivity of the finger's tactile sensing mechanism are examined. Subjective equalities and difference thresholds for fine step-heights determined using the experiments are important values for investigating the human tactile sensation. The meanings of those values are explained in the following (Gescheider, 1985).

In an experiment, human subjects touch several pairs of stimuli with their fingers and try to distinguish them. One of the stimulus pair is the standard stimulus and the other is the comparison stimulus. The magnitudes of the standard and comparison stimuli are denoted by δ_s and δ_c, respectively. The standard stimulus is designed to be constant and the comparison stimulus is variable. Several pairs of δ_s and δ_c are presented to the subjects and for each pair they are asked to tell which stimulus of δ_s and δ_c they feel stronger. When δ_c is smaller than δ_s, the proportion of the responses that δ_c is chosen as stronger than δ_s is supposed to be low. Conversely, when δ_c is greater than δ_s, the proportion of the responses that δ_c is chosen as stronger than δ_s is supposed to be high. Figure 1 shows a characteristic curve of the proportion that δ_c is chosen as stronger than δ_s. The horizontal axis shows the comparison stimulus while the vertical axis shows the proportion of the subjects selecting the comparison stimulus. The comparison stimulus magnitudes for the proportions equal to 0.25, 0.5, and 0.75 are denoted by $S_{0.25}$, $S_{0.5}$, and $S_{0.75}$, respectively. The value of $S_{0.5}$ is called the subjective equality for δ_s. If the standard and comparison stimuli are presented under the same condition, $S_{0.5}$ should be equal to δ_s.

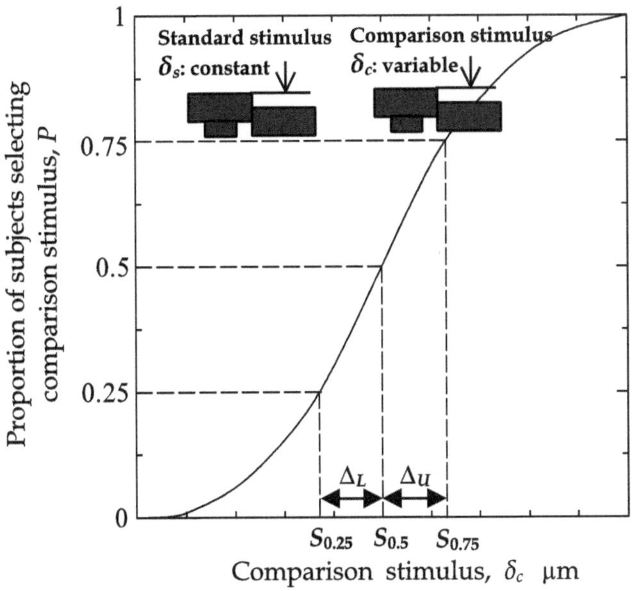

Fig. 1. An example of a discrimination characteristics curve.

The values of $\Delta_U = S_{0.75} - S_{0.5}$ and $\Delta_L = S_{0.5} - S_{0.25}$ are the upper and lower thresholds for δ_s, respectively. Moreover, the average of the thresholds, $\Delta = (\Delta_U + \Delta_L)/2$, is called the difference threshold. In addition, these thresholds usually have very close values because the upper and lower thresholds become almost equal. Also the value of the ratio of Δ to δ_s is called the Weber fraction. The value is known to be constant over the range of stimulus magnitude in tactile sensing mechanisms, as well as in visual and auditory.

2.2 Parameter Estimation by Sequential Testing (PEST) method

Taylor and Creelman developed the PEST method to determine the above-mentioned difference thresholds and subjective equalities through the process of a psychophysical experiment without calculating the characteristics curve (Taylor & Creelman, 1982). The PEST algorithm consists of three groups of rules in the following, and, as shown in Fig. 2, calculates the magnitudes of comparison stimuli to present to a subject based on the subject's responses in the experiment. In this study, the authors have developed the measurement system that calculates the magnitudes by computer based on the PEST algorithm and determines the difference thresholds and subjective equalities.

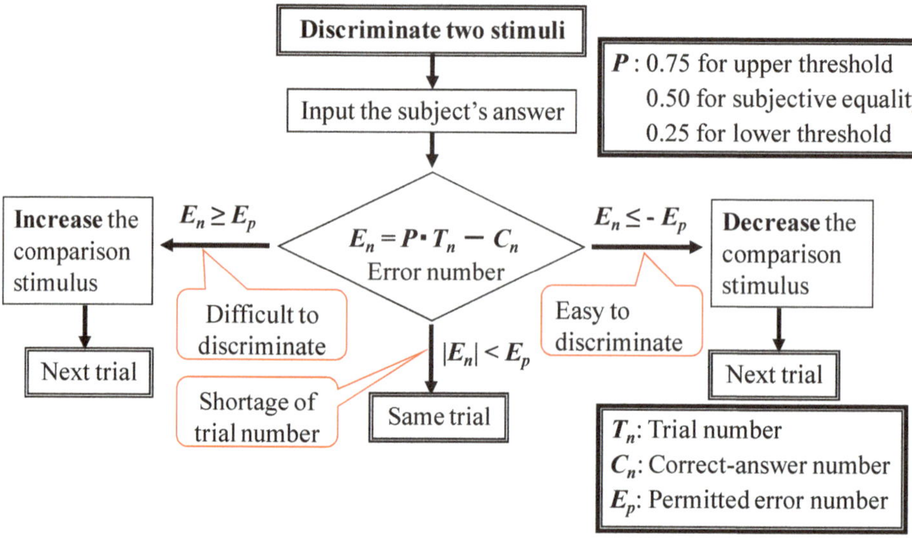

Fig. 2. Flowchart of changing the magnitude of comparison stimulus using the PEST algorithm.

Rule #1: Condition for changing the magnitude of comparison stimulus

A PEST sequence consists of several trial blocks composed of several trials as shown in Fig. 3. Let us consider the *n*-th trial block. The magnitude of comparison stimulus is constant throughout the same block. Let δ_{cn}, T_n and C_n be the comparison stimulus magnitude, the trial number and the number of the human subject's correct answers in the current *n*-th trial block, respectively. For a specified P, the proportion of C_n against T_n, the error number E_n is given as follows:

$$E_n = P \cdot T_n - C_n \,, \tag{1}$$

where the value of P is 0.25, 0.5, or 0.75 to obtain the lower threshold, the subjective equality, or the upper threshold, respectively. Let E_p be the permitted error number. If the condition:

$$| E_n | \; < \; E_p \tag{2}$$

is satisfied, then the experiment continues with the same comparison stimulus. If the condition is not satisfied, then δ_{cn} is varied and the trial block is incremented to the $(n + 1)$-th trial block. δ_{cn+1} is decreased whenever (3) is satisfied and increased whenever (4) is satisfied. Equations (3) and (4) are given as follows:

$$E_n \leq -E_p , \tag{3}$$

$$E_n \geq E_p . \tag{4}$$

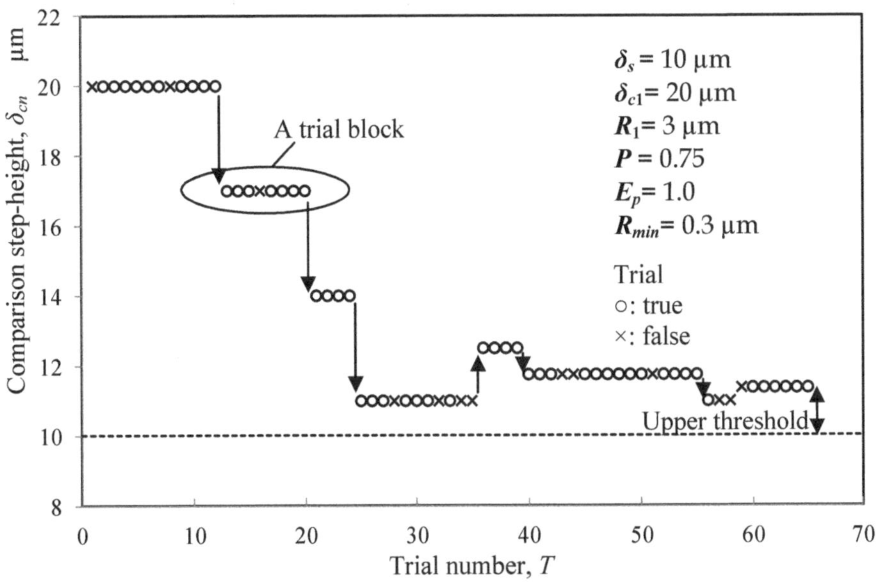

Fig. 3. An example of variation in comparison step-height calculated by the PEST algorithm.

Rule #2: Incremental stimulus magnitude

The incremental range of the comparison stimulus magnitude in the n-th trial block, R_n, should decrease in order for δ_{cn+1} to converge as the number of trials increases. Here δ_{cn+1} is given as follows:

$$\delta_{cn+1} = \delta_{cn} \pm R_n . \tag{5}$$

If δ_{cn} differs considerably from the convergent value, R_n should increase to reach rapidly the convergent value. Taylor and Creelman empirically determined the rules for the adjustment of the incremental range. In their rules, the convergence condition is judged by the variation in fluctuation direction of the stimulus magnitude. The fluctuation direction (increase or decrease) in the n-th trial block is denoted by D_n. R_n is specified as follows:

a. If the direction D_n becomes contrary to the direction D_{n-1} of the $(n - 1)$-th trial block, then R_n is set half R_{n-1}.
b. If D_{n-1} and D_n are the same direction, then R_n is set the same as R_{n-1}.

c. If D_{n-2}, D_{n-1} and D_n are the same direction and R_{n-2} is half R_{n-3}, then R_n is set the same as R_{n-1}. However, if D_{n-2}, D_{n-1}, and D_n are the same direction and R_{n-2} is equal to R_{n-3}, then R_n is set twice R_{n-1}.
d. If D_{n-3}, D_{n-2}, D_{n-1}, D_n, ... continue in the same direction, then R_n, R_{n+1}, R_{n+3}, ... are each twice the previous incremental range.

Rule #3: Condition of termination

R_n becomes small as δ_{cn} approaches the standard stimulus magnitude, δ_s. The minimum incremental range, R_{min}, is specified by the PEST algorithm. If the condition of termination:

$$R_n \leq R_{min} \tag{6}$$

is satisfied, then the processing is terminated. The difference between δ_{cn} and δ_s is the threshold if the value of P is 0.25 or 0.75, and δ_{cn} is the subjective equality if P is 0.5.

Experimental results using the PEST method are exemplified in Fig. 3 to explain the above-mentioned PEST procedure. In the example, P, E_p, and R_{min} are set at 0.75, 1.0, and 0.3 μm, respectively. Also, the standard step-height δ_s and the initial comparison step-height δ_{c1}, the initial increment R_1 are 10 μm, 20 μm, and 3 μm, respectively. While the calculated result of (1) satisfies the condition given by (2), the human subject repeats the comparison of δ_s of 10 μm with δ_{c1} of 20 μm. Since after twelve trials the right side of (1) yields $0.75 \times 12 - 10 = -1$ and the result satisfies the condition given by (3), δ_{c2} is reduced to 17 μm ($\delta_{c2} = \delta_{c1} - R_1$) according to Rule #2 (incremental stimulus magnitude). As is evident from Fig. 3, the comparison step-height decreases as the trial number increases. Thereafter, δ_{c5} is increased to 12.5 μm ($\delta_{c5} = \delta_{c4} + R_4$; $R_4 = R_3 / 2$) when the condition given by (4) is satisfied for a trial block with an 11 μm step-height. In the continuous blocks, the comparison step-height is bounded because the calculated results alternately satisfy the conditions given by (3) and (4). However, the comparison step-height decreases gradually due to Rule #2. Finally the calculated R_8 satisfies the condition of (6). The terminated comparison step-height is 11.375 μm and its upper threshold is obtained from the experiment as $\Delta_U = 1.375$ μm.

In the experiments of the paper, E_p is set a constant value of 1.0 and the other values are determined according to the experiment conditions.

3. Measurement system

To measure the human finger's tactile sensation capability of recognizing fine step-heights using psychophysical experiments, a measurement system that creates step-heights of 0 to 1000 μm and presents several pairs of the step-heights to human subject's fingers according to the PEST algorithm were developed (Fig. 4). In the psychophysical experiments of this paper, the subjects touch fine step-heights in active-touch manner (Fig. 5) and passive-touch manner (Fig. 6). The step-height presentation device has the capability of controlling four parameters of the step-height presentation, i.e., the step-height, presentation velocity, presentation angle, and presentation temperature. The first three parameters are controlled by a computer that drives the wedge-shaped Z stage, the X-table and the rotary table, and the presentation temperature is controlled by the Peltier elements.

A fine step is formed between two fine finished stainless steel plates, and the height of the step is a stimulus magnitude. The stepping motor-driven Z stage slides one of the stainless

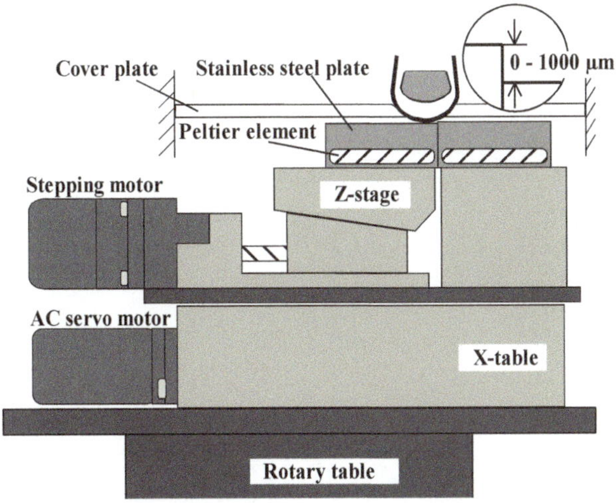

Fig. 4. Step-height presentation device.

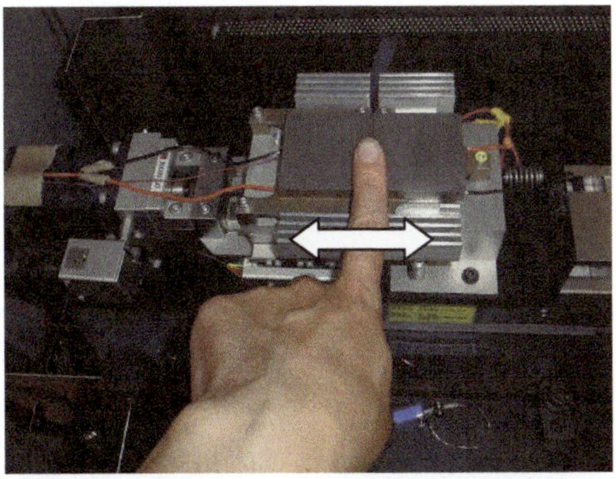

Fig. 5. Scene of psychophysical experiment of active-touch manner.

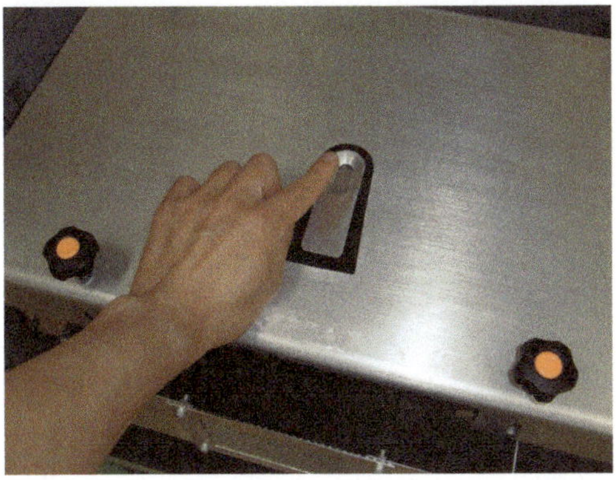

Fig. 6. Scene of psychophysical experiment of passive-touch manner.

plates vertically to control the step-height. The servo motor-driven X-table generates the presentation velocity by its reciprocating movement. The rotary table regulates the presentation angle by rotating the X table placed on it. The Peltier elements maintain, using the Peltier effect to heat or cool, the step plates' temperature by regulating the DC voltage applied to them. Here, the presentation angle of the step plates to a subject's finger is controlled as shown in Fig. 7. The motion direction of the X-table is always perpendicular to the direction of a step edge. Consequently, the presentation device is capable of presenting a fine step-height at the reciprocating velocity of 0 to 60 mm/s and the presentation angle of 0 to 180 degrees. In addition, the step plates' temperature can be controlled within the range of 8 to 50 degrees centigrade.

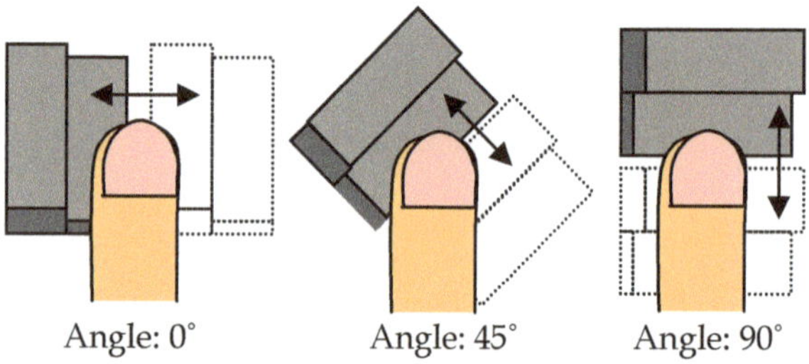

Fig. 7. Presentation angles of step.

In the psychophysical experiments using the measurement system, when the human subjects are required to judge which step-height of the presented step pair is larger, they press each of the right/left computer-mouse buttons to input the answer into the computer. The step-heights of the next trial are calculated by computer based on the PEST algorithm using the subject's answers.

In the passive-touch experiments of this paper, a cover plate with a hole like fingertip profile was installed to cover the step-height presentation device as Fig. 6 showed and the human subjects touched the step plates through the hole using their top and center of their fingertips as shown in Fig. 8. During the experiments, to prevent the sensitivity of human tactile sensation from declining, the step plates' temperature and the room temperature were kept constant approximately 37 and 26 degrees centigrade, respectively. Before the experiments the human subjects washed their hands with soap to keep them clean

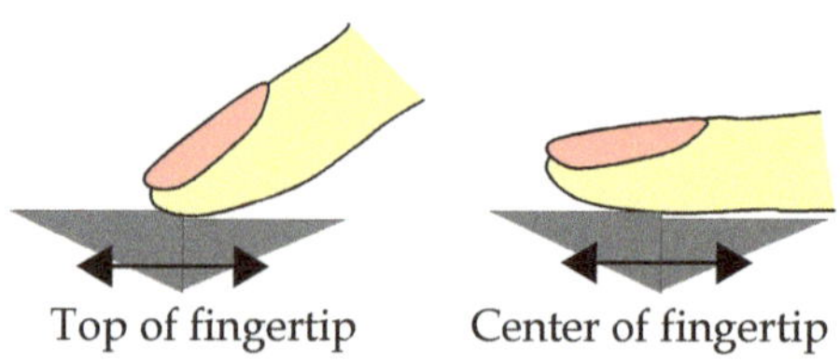

Fig. 8. Fingertip regions.

4. Experimental methods

4.1 Difference thresholds for fine step-heights in active-touch discrimination task

To measure the difference thresholds for fine step-heights in the active-touch experiments, five male subjects in their twenties of age touched and discriminated step pairs with the center of their index fingertips in active-touch manner. The subjects were allowed to touch step pairs in the 0-degree finger-motion direction as long as they wanted as choosing the motion velocity of their fingers arbitrarily. In the active-touch experiments, five step-heights of 10, 40, 70, 100 and 130 μm were used as the standard stimuli.

Table 1 shows the initial values of δ_{c1}, R_1 and R_{min} used in the PEST rules for each standard stimulus of δ_s. Each of the comparison step-heights of δ_{c1} was the value presented in the first trial block of the discrimination tasks. Also the value of P was set at 0.75 to obtain the upper thresholds. During the trials the subjects were required to press the computer-mouse button to input the answers into the computer even if they could not judge the difference between the step pair. The step-heights of the continuous trials were calculated based on the PEST algorithm using the answers and finally the upper thresholds for each standard step-height were determined.

δ_s [μm]	10	40	70	100	130
δ_{c1} [μm]	20	70	110	150	190
R_1 [μm]	3	9	12	15	19
R_{min} [μm]	0.3	0.9	1.2	1.5	1.9

Table 1. Standard step-heights and the initial values used in the PEST rules.

4.2 Difference thresholds for fine step-heights in passive-touch discrimination task

To measure the difference thresholds for fine step-heights in the passive-touch experiments, the six male subjects that had participated in the above-mentioned experiments touched and discriminated step pairs with the center of their index fingertips in passive-touch manner. The steps were moved at the reciprocating velocity of 25 mm/s and the 0-degree presentation angle using the presentation device and the subjects were allowed to touch them through the hole of the cover plate with their fingers as long as they wanted. In the passive-touch experiment, five step-heights of 10, 30, 50, 70 and 100 μm were used as the standard stimuli.

Table 2 shows the initial values used in the PEST rules for each standard stimulus. Also the value of P was set at 0.75 to obtain the upper thresholds. The discrimination tasks in the passive-touch experiment were conducted and as a result, the PEST algorithm determined the upper thresholds.

δ_s [μm]	10	30	50	70	100
δ_{c1} [μm]	20	50	80	110	150
R_1 [μm]	3	6	9	12	15
R_{min} [μm]	0.3	0.6	0.9	1.2	1.5

Table 2. Standard step-heights and the initial values used in the PEST rules.

4.3 Difference thresholds for a 10 μm step presented to the center of a fingertip at various presentation angles

To measure the difference thresholds for a 10 μm step presented to the center of a fingertip at various presentation angles, six male subjects in their twenties of age touched and discriminated step pairs with the center of their index fingertips in passive-touch manner. The magnitude of standard stimulus is 10 μm step-height and the step pairs of the standard and comparison stimuli were presented at the presentation angles of 0, 45 or 90 degrees. The steps were moved at the reciprocating velocity of 30 mm/s by the presentation device and the subjects were allowed to touch the step-heights through the hole as long as they wanted. In the passive-touch experiments, a 0-degree step of 10 μm, a 45-degree step, and a 90-degree step were used as the standard stimuli.

In the experiments, the initial values of δ_{c1}, R_1 and R_{min} used in the PEST rules were 20 μm, 3 μm and 0.3 μm, respectively. The value of P was set at 0.75 to obtain the upper thresholds. The discrimination tasks were conducted and as a result, the PEST algorithm determined the upper thresholds of the center of the subjects' fingertips.

4.4 Difference thresholds for a 10 μm step presented to the top of a fingertip at various presentation angles

To measure the difference thresholds for a 10 μm step presented to the top of a fingertip at various presentation angles, two male subjects in their twenties of age touched and discriminated step pairs with the top of their index fingertips in passive-touch manner. The step pairs of the standard stimulus of a 10 μm step and the comparison stimuli were presented at the presentation angles of 0 or 90 degrees. The steps were moved at the reciprocating velocity of 30 mm/s using the presentation device and the subjects were allowed to touch the step-heights through the hole as long as they wanted. In the passive-touch experiments, a 0-degree step of 10 μm and a 90-degree step were used as the standard stimuli.

In the experiments, the initial values of δ_{c1}, R_1 and R_{min} used in the PEST rules were 20 μm, 3 μm and 0.3 μm, respectively. The value of P was set at 0.75 to obtain the upper thresholds. As a result of the experiments, the PEST algorithm determined the upper thresholds of the top of the subjects' fingertips.

5. Experimental results and discussion

5.1 Effects of touching manner of finger

To evaluate the influence of the touching manner on the finger's fine step-height discrimination capability, the upper thresholds for the step-heights were measured in the active- and passive-touch experiments. Each human subject was tested twice for each standard step-height in the active- and passive-touch experiments and ten upper thresholds in total for each step-height were determined. Tables 3 and 4 show the upper thresholds in the active- and passive-touch experiments, respectively. At the bottoms of the tables the averages of the upper thresholds and the standard deviations are calculated.

Figure 9 describes the relationship among the touching manner, the upper threshold, and the standard step-height. The horizontal axis shows the standard step-height while the vertical axis shows the upper threshold. The threshold magnitudes of the active- and passive-touch discrimination tasks become larger as the magnitude of standard step-height increases in the range of 10 to 100 μm. It is also noticed that the threshold magnitudes for each of the step-heights are almost equal for variations of the standard step-height smaller than approximately 40 μm and that the thresholds of active-touch tasks are smaller than those of passive-touch tasks for variations of the standard step-height greater than 50 μm. The results suggest that the fingertip's tactile sense can increase the sensitivity to the step-heights by touching in active-touch manner. In addition, it could be considered that the tactile recognition module that recognizes fine step-heights of about 10 μm is different from the recognition modules for the step-heights larger than 50 μm.

Human subject	Standard step-height [μm]				
	10	40	70	100	130
A	2.7	9.2	1.8	5.9	9.4
	3.1	6.9	10.8	11.6	2.6
B	1.2	5.8	1.8	9.7	7.1
	1.9	5.8	1.8	4.1	11.6
C	4.6	4.7	13.8	13.4	7.1
	2.7	4.7	7.8	5.9	13.9
D	2.7	3.6	10.8	26.6	20.6
	4.2	10.3	12.3	13.4	16.1
E	2.7	8.1	19.8	30.9	9.4
	3.8	5.8	4.8	11.6	20.6
Ave. [μm]	3.0	6.5	8.6	13.3	11.8
SD	0.97	2.0	5.7	8.4	5.6

SD: standard deviation

Table 3. Upper thresholds for the 0-degree-presented standard step-heights discriminated using the center of an index fingertip of active-touch manner.

Human subject	Standard step-height [μm]				
	10	30	50	70	100
A	2.3	9.1	12.6	30.2	17.2
	1.9	5.4	10.3	16.8	11.6
B	0.8	6.9	8.1	4.8	9.7
	0.1	3.9	5.8	15.3	11.6
C	3.1	7.6	6.9	10.8	9.7
	1.2	2.1	4.7	3.3	9.7
D	3.8	14.4	20.4	9.3	24.7
	1.6	4.6	11.4	16.8	26.6
E	3.8	5.4	21.6	12.3	15.3
	3.8	3.1	12.6	15.3	19.1
Ave. [μm]	2.2	6.3	11.4	13.5	15.5
SD	1.3	3.4	5.4	7.1	6.0

Table 4. Upper thresholds for the 0-degree-presented standard step-heights discriminated using the center of an index fingertip of passive-touch manner.

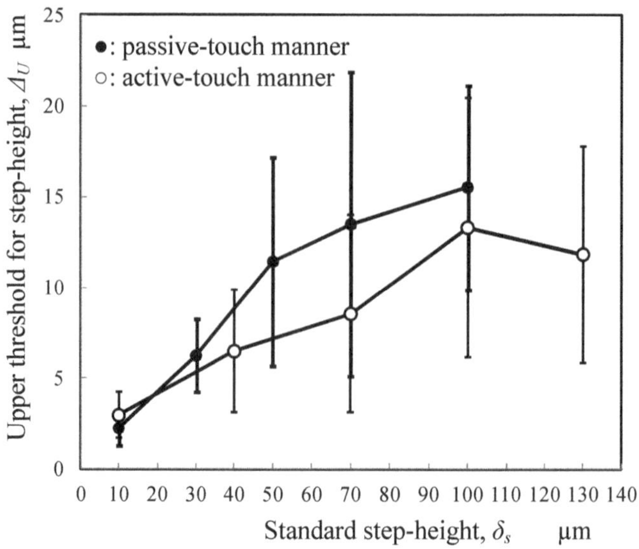

Fig. 9. Relationship among the touching manner, the upper threshold, and the standard step-height.

5.2 Effects of finger motion direction and fingertip region

To evaluate the influences of the finger motion direction and the fingertip region on the fine step-height discrimination capability, the upper thresholds for a 10 μm step-height were determined when the subjects touched the step-height presented at the presentation angles of 0, 45 and 90 degrees using the top and center of their fingertips. Here the finger's motion direction can be defined as the step's presentation angle controlled by the presentation device since it was revealed that the fingertip's discrimination capability of the step-heights of about 10 μm does not depend on the active- and passive-touch manners (Kawamura et al., 1996).

In the discrimination tasks using the center of a fingertip, each subject was tested twice for each of the standard stimuli presented at 0, 45 or 90 degrees and twelve upper thresholds in total for each presentation angle were determined, on the other hand, in the discrimination tasks using the top of a fingertip, each subject was tested four times for each of the standard stimuli presented at 0 or 90 degrees and eight upper thresholds in total for each presentation angle were determined. Tables 5 and 6 show the upper thresholds for a 10 μm step-height presented to the center and top of the subjects' fingertips, respectively. At the bottoms of the tables the averages of the upper thresholds and the standard deviations are calculated.

Figure 10 describes the relationship among the fingertip region, the upper threshold, and the presentation angle. The horizontal axis shows the presentation angle while the vertical axis shows the upper threshold. The magnitude of upper threshold measured at the center of the fingertips almost stays constant or decreases slightly for variations of the presentation angle in the range of 0 to 90 degrees. On the other hand, the magnitude of upper threshold measured at the top of the fingertips becomes smaller as the presentation angle changes from 0 to 90 degrees. It is also noticed that the upper thresholds of the top of the fingertips are smaller than those of the center and that the upper threshold for the 90-degree step presented at the top is the smallest value. Therefore, it is found that the tactile sense of the top of a fingertip is highly sensitive to a 10 μm step-height as compared with that of the center. In addition, the results point out that you can make the most of the fingertip's discrimination ability when you touch a fine step-height with the top of your fingertip moving in the motion direction of 90 degrees.

Human subject	Presentation angle [deg]		
	0	45	90
	Upper threshold [μm]		
F	3.4	2.3	6.1
	6.4	4.2	1.2
G	1.2	1.6	2.7
	4.9	3.4	1.9
H	3.8	5.3	5.7
	3.8	3.4	0.1
I	3.4	3.4	3.4
	4.2	3.8	3.8
J	3.1	2.3	1.6
	3.8	3.4	4.9
K	3.1	3.1	3.4
	2.3	3.4	2.3
Ave. [μm]	3.6	3.3	3.1
SD	1.2	0.75	2.3

Table 5. Upper thresholds for a 10 μm step-height presented to the center of an index fingertip at the presentation angles.

Human subject	Presentation angle [deg]	
	0	90
	Upper threshold [μm]	
L	1.9	0.8
	2.6	1.9
	2.6	1.5
	1.1	1.1
M	2.3	0.0
	1.1	2.3
	1.9	1.1
	3.4	1.5
Ave. [μm]	2.2	1.3
SD	0.72	0.65

Table 6. Upper thresholds for a 10 μm step-height presented to the top of an index fingertip at the presentation angles.

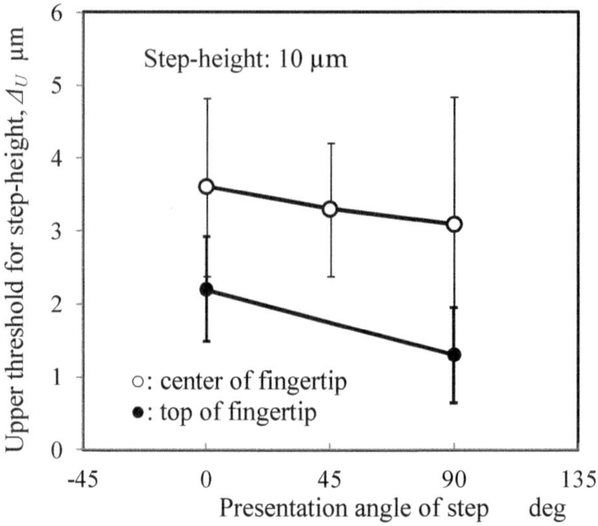

Fig. 10. Relationship among the fingertip region, the upper threshold for a 10 μm step-height, and the presentation angle.

6. Conclusion

In this study, to measure the finger's fine step-height discrimination capability the computer-controlled measurement system that presents fine step-heights of 0 to 1000 μm to subjects' fingers was developed. Using the measurement system the paper examined the effects of the touching manner of human finger, finger's motion direction, and fingertip region on the tactile recognition of fine step-heights. In the psychophysical experiments, to determine the difference thresholds and subjective equalities for fine step-heights the measurement system calculated the amounts of step-height of the step pairs by computer according to the PEST algorithm and presented the step pairs to the subjects.

First, the upper thresholds for the step-heights of 10 to 100 μm were determined in the active- and passive-touch experiments. The resulting thresholds became larger as the magnitude of step-height increased. Also the threshold of active-touch manner for each of the step-heights larger than 50 μm was smaller than that of passive-touch manner and the thresholds of the touching manners for each of the step-heights smaller than about 40 μm were almost equal regardless of the touching manners. Therefore it was found that the fingertip's discrimination ability of the fine step-heights depends on the amounts of step-height and if a step-height is larger than 50 μm, the finger's tactile sense can increase the sensitivity in active-touch manner.

Next, to investigate the effects of the finger's motion direction and fingertip region in recognizing fine step-heights, the upper thresholds for a 10 μm step-height were determined when the human subjects discriminated the pairs of step-heights presented at various presentation angles using the top and center of their fingertips. When the presentation angle of a step-height to a fingertip changed from 0 to 90 degrees, although the thresholds of the center of the fingertips almost stayed constant, the threshold for the step-height presented to the top of the fingertips at 90 degrees became the smallest value. Therefore, it was found that the tactile sense of the top of a fingertip is highly sensitive to the step-height as compared with that of the center.

7. References

Gescheider, G.A. (1985). *Psychophysics: Method, Theory, and Application (Second Edition)*, Lawrence Erlbaum Associates, Hillsdale, New Jersey.

Heller, M.A. (1989). Texture Perception in Sighted and Blind Observers, *Perception & Psychophysics*, Vol. 45, pp. 49-54.

Kawamura, T.; Ohka, M.; Miyaoka T. & Mitsuya, Y. (1998). Human Tactile Sensation Capability to Discriminate Moving Fine Texture, *Proceedings of the 7th IEEE International Workshop on Robot and Human Communication*, Vol. 2 , pp. 555-561.

Kawamura, T.; Ohka, M.; Miyaoka, T. & Mitsuya, Y. (1996). Measurement of Human Tactile Sensation Capability to Discriminate Fine Surface Texture Using a Variable Step-height Presentation System, *Proceedings of the 5th IEEE International Workshop on Robot and Human Communication*, pp. 274-279.

Kawamura, T.; Otobe, Y. & Tani, K. (2009). Effect of Touching Manner and Motion Direction of Human Finger on Human Tactile Recognition, *Proceedings of the 18th IEEE International Symposium on Robot and Human Interactive Communication*, pp. 998-1003.

Miyaoka, T. & Mano T. (1993). A Neural Information Processing Model to Illustrate the Dependency of Vibrotactile Adaptation on the Simulated Site of the Human Hand, *Japanese Psychological Research*, Vol. 35, pp. 41-45.

Miyaoka, T. & Ohka, M. (1997). Tactile Information Processing Mechanisms of Fine Surface Texture Discrimination in Humans: a Study with Ridge Height Discrimination Tasks, *Proceedings of the 13th Triennial Congress of the IEA*, Vol. 7, pp. 264-267.

Miyaoka, T.; Ohka, M.; Kawamura, T. & Mitsuya, Y. (1996). Fine Surface-texture Discrimination Ability Depends on the Number of Mechanoreceptors Participating in the Discrimination Task, *Journal of Acoustical Societies of America*, Vol. 100, No. 4, Pt. 2, p. 2771.

Salentijn, L.M. (1992). The Human Tactile System, In: *Advanced Tactile Sensing for Robotics*, Nicholls, H.R., pp. 123-150, World Scientific, ISBN 981-02-0870-7, Singapore.

Taylor M.M. & Creelman C.D. (1982). PEST: Efficient Estimate on Probability Function, *Journal of Acoustical Societies of America*, Vol. 41, No. 4, pp. 836-855.

Vallbo, Å.B. & Johansson, R.S. (1984). Properties of Cutaneous Mechanoreceptors in the Human Hand Related to Touch Sensation, *Human Neurobiology*, Vol. 3, pp. 3-14.

8

Overview of Novel Post-Processing Techniques to Reduce Uncertainty in Antenna Measurements

Manuel Sierra-Castañer, Alfonso Muñoz-Acevedo, Francisco Cano-Fácila and Sara Burgos
Technical University of Madrid (Universidad Politécnica de Madrid - UPM), Spain

1. Introduction

The error analysis has been investigated since 1975 since recent publications, as shown in references as (Newell et al., 1975) where the classical 18 terms for planar near field systems where established. It is also worth mentioning the more recent studies developed inside the actions under the "Antenna Measurement" activity of the "Antenna Centre of Excellence (ACE)" within the sixth framework research program of the European Union. In particular, that work pretended to establish common error calculation criteria in spherical near-field and far-field antenna measurement systems. The results of that research were summarized in an exhaustive deliverable (Alexandridis et al., 2007), which detailed the observations stated by several research institutions. In that deliverable it was agreed that the causes of uncertainties and errors in a spherical near-field antenna measurement are divided in six categories:

1. **Mechanical uncertainties and errors**: this group includes the axes intersection, the axes orthogonality, the horizontal pointing, the probe vertical position, the probe horizontal and vertical pointing and the measurement distances.
2. **Electrical uncertainties and errors**: this class contains the amplitude and phase drift, the amplitude and phase noise, the leakage and crosstalk, the amplitude non-linearity and the amplitude and phase shift in rotary joints.
3. **Probe-related uncertainties and errors**: this kind of uncertainties takes into account the channel balance amplitude and phase, the polarization amplitude and phase and the pattern knowledge.
4. **Stray signals**: this type of uncertainties consists of the multiple reflections, the room scattering and the AUT support scattering.
5. **Acquisitions errors**: this group involves the scan area truncation and the sampling point offset.
6. **Processing uncertainties and errors**: in this group the spherical mode truncation and the total radiated power are considered.

The second part of the chapter will deal with the description of four different classes of correction techniques for antenna measurements.:

1. Averaging and comparison of measured antenna radiation patterns: pattern comparison technique is a very useful technique to evaluate the quality of the quiet zone in far field systems. This technique can also be used for cancelling source of errors, averaging two measurements including opposite errors. The easiest way is the averaging of two measurements where the distance between the AUT and antenna probe has been modified in a quarter of wavelength: in this case the difference of the reflected rays in both configurations is $\lambda/2$ (180 deg), so the averaging of both acquisitions cancels the effect of this reflection.
2. Application of source reconstruction techniques to reduce the effect of the noise, leakage and scattering: a source reconstruction technique is a method used to obtain the field or the equivalent currents distribution over the antenna under test (AUT) plane from the knowledge of its radiated field (near- or far-field). This information can be employed to identify errors, e.g., electrical errors in arrays or mechanical errors in reflectors. Apart from its classic application, a diagnostic process also gives a complete electromagnetic characterization of the antenna, which is the basis of other applications, like near- to far-field transformations. In this work, the additional information provided by a diagnostic technique is employed to reduce the effect of three common errors in antenna measurements: reflections, noise and leakage.
3. Iterative algorithms to reduce the truncation error in planar, cylindrical and not complete spherical acquisitions: other alternatives to reduce errors in antenna measurements are based on iterative algorithms. This is the case, the method proposed to reduce truncation errors when the measurement is performed over a surface that does not fully enclose the AUT (plane, cylinder or partial sphere). The method is based on the Gerchberg-Papoulis iterative algorithm (Gerchberg, 1974) and (Papoulis, 1975) used to extrapolate band-limited functions and it is able to extend the valid region of the calculated far-field pattern up to the whole forward hemisphere. The extension of the valid region is achieved by applying iteratively a transformation between two different domains.
4. Techniques to improve the quality of the quiet zone in compact ranges: CATR measurement facilities are characterised by a straightforward operation which is able to perform real time antenna measurements along a wide margin of frequencies, whenever reflectors or lenses are used as collimators. In spite of their compactness, CATR facilities operate in far-field conditions, which mean that the test wave received by the AUT should be a plane wave. The operation of a CATR relies on the planarity of this test wave, inside the "quiet zone" volume.

This paper is divided as follows: section 2 explains the main uncertainty sources based on the works developed on the Antenna Centre of Excellence (Alexandridis, 2007). Sections 3 to 6 detail the different aspects mentioned before, including some practical examples.

2. Revision of the uncertainty sources involved in the directivity and gain determinations

The uncertainty constitutes the measure of the quality of a measurement, which allows comparing the measurement results with other outcomes, references, specifications or standards. In order to determine how the uncertainties may influence the measurement results, it is worth reviewing the methodology that allows evaluating the uncertainties in the measured outcomes. (Taylor et al. 1994), (ISO/IEC 98, 1995) and (Gentle et al., 2003) are good basic representative guides, where the general methods to estimate the uncertainty are widely explained. Actually, the above mentioned studies of Hansen in (Hansen, 1988) and Newell in (Newell, 1988), published in 1988 and concerning the uncertainties in antenna measurements, settle the basis to evaluate the inaccuracies that may affect the results achieved from the antenna measurements. However, as the technology has incredibly evolved since then, an update of the uncertainty analysis is considered appropriate and therefore, this section will try to cover the uncertainty study of measurements that nowadays could be carried in indoor facilities, summarizing the work developed in (Alexandridis et al., 2007), under the "Antenna Measurement" activity of the "Antenna Centre of Excellence (ACE)" within the sixth framework research program of the European Union, involving researchers from the National Center of Scientic Research "Demokritos" (NCSRD), the Technical University of Denmark (DTU), the Technical University of Madrid (UPM), SATIMO (Société d'Applications Technologiques de l'Imagerie Micro-Onde S.A.), Saabgroup, FTR&D (France Telecom Recherche & Développement) and IMST GmbH. In particular, that work pretended to establish common error calculation criteria in spherical near-field and far-field antenna measurement systems. Thus, the outcomes of this activity became important instruments to verify the measurement accuracies for each range and to investigate and evaluate possible improvements in measurement set-up and procedures.

It was seen in this research that these errors depend not only on the gain determination technique applied to establish the antenna gain, but also on the AUT that will be measured and on the measurement range employed. Hence, a generic evaluation cannot be carried out and in each case the study has to be particularized and adapted to the AUT and measurement facility considered.

Besides, it is noticeable that the procedure to evaluate the uncertainties has to be different in the near-field systems than in the far-field ranges. This is due to the fact that in the far-field systems the measurement is direct and thus, the uncertainty could also straightly be obtained. On the other hand, in near-field ranges the information and the uncertainties related to this data are processed and so the near-to-far-field transformation has an influence on the both magnitudes. The sources of uncertainty present in the near-field measurements are either systematic or random quantities. The systematic errors can be characterized using the mean. On the other hand, for the random sources of uncertainties, the mean, the variance (σ^2) and the probability distribution are employed to carry out the description of the magnitude studied. The first step to accomplish a complete uncertainty study is to identify all the sources of inaccuracies (uncertainties or errors) affecting the measurements. In the exhaustive deliverable carried out under the "Antenna Measurement" activity of the ACE, the observations, stated by the participant research institutions, were detailed and the causes of uncertainties and errors in a spherical near-field antenna measurement were divided in six categories:

1. Mechanical uncertainties and errors: this group includes the axes intersection, the axes orthogonality, the horizontal pointing, the probe vertical position, the probe pointing and the measurement distances. Fig. 1 illustrates how is defined a non-intersection error in a spherical coordinate system.
2. Electrical uncertainties and errors: this class contains the amplitude and phase drift, the amplitude and phase noise, the leakage and crosstalk, the amplitude non-linearity and the amplitude and phase shift in rotary joints.
3. Probe-related uncertainties and errors: this kind of uncertainties takes into account the channel balance amplitude and phase, the polarization amplitude and phase and the

pattern knowledge. In Fig. 1 "P1" and "P2" are the lineal polarization relations in each port and "Q1" and "Q2" the circular polarization relations in each port.

4. Stray signals: this type of uncertainties consists of the multiple reflections, the room scattering and the AUT support scattering. Fig. 3 shows a schematic representation of these types of reflections in an anechoic chamber.
5. Acquisitions errors: this group involves the scan area truncation and the sampling point offset.
6. Processing uncertainties and errors: in this group the spherical mode truncation and the total radiated power are considered.

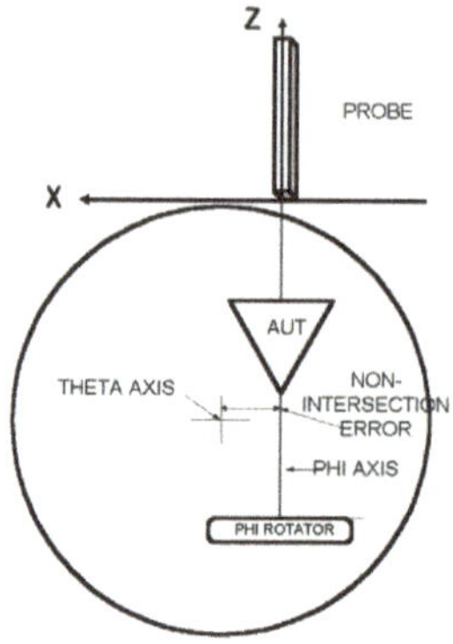

Fig. 1. Non-intersection error.

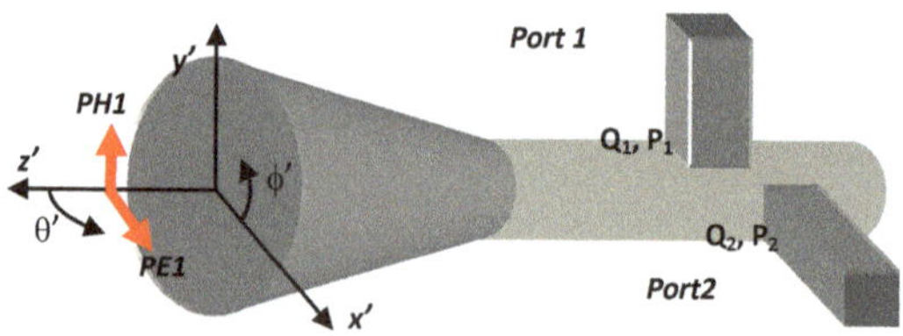

Fig. 2. Ports of a dual-polarized probe.

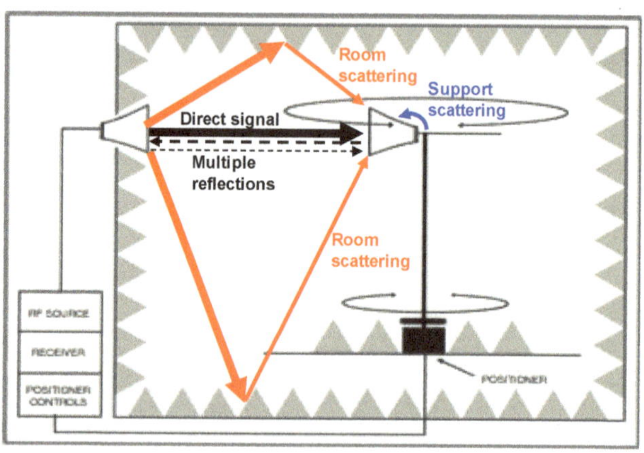

Fig. 3. Different reflections and scattering in an anechoic chamber.

Table 1 summarizes all the uncertainties and errors related to peak directivity in spherical near-field measurement that contribute to the total uncertainty, assuming that the probe is not an array. This table has extended the results from spherical near-field systems to a more general acquisition. Some of the terms affect more spherical systems and other ones are more important for cylindrical or planar systems.

Category	Uncertainty	Description
Mechanical	1. Axes intersection	Lateral displacement between the horizontal and vertical axes
	2. Axes orthogonality	Difference from 90° of the angle between the horizontal and vertical axes
	3. Horizontal pointing	For θ = 0°, horizontal mispointing of the horizontal axis to the probe reference point
	4. Probe vertical position	Vertical displacement of the probe reference point from the horizontal axis
	5. Probe horizontal and vertical pointing	Horizontal or vertical mispointing of the probe z-axis
	6. Measurement distance	Error in the measurement of the distance between the AUT and probe reference point
Electrical	7. Amplitude and phase drift	Systematic amplitude and phase change at still AUT
	8. Amplitude and phase noise	Random amplitude and phase change at still AUT
	9. Leakage and crosstalk	Extraneous signal dependent on angle
	10. Amplitude non-linearity	Difference from linear dependence of measured value versus input signal level
	11. Amplitude and phase shift in rotary joints	Systematic amplitude and phase change with angle of rotary joints
Probe-related	12. Channel balance amplitude and phase	Amplitude and phase difference between the two polarization channels
	13. Polarization amplitude and phase	Amplitude and phase of the probe polarization coefficients
	14. Pattern knowledge	Deviation of the known or assumed probe pattern from the true one
Stray signals	15. Multiple reflections	Contribution to the received signal due to interactions between the AUT and the probe
	16. Room scattering	Contribution to the received signal due to finite reflectivity of the anechoic chamber
	17. AUT support scattering	Contribution to the received signal due to scattering from the AUT support structure
Acquisition	18. Scan area truncation	Error due to partial sphere, cylinder or plane acquisition
	19. Sampling point offset	Error due to continuous positioner rotation
Processing	20. Spherical mode truncation	Changes in the far-field due to different number of modes included in the field calculation
	21. Total radiated power	Error in the calculation of the total power

Table 1. Summary of uncertainty contributions.

3. Averaging and comparison of measured antenna radiation patterns

The comparison among patterns, or pattern comparison technique, is a very well known technique to evaluate the performance on an antenna measurement system. For instance, Fig. 4 shows the patterns measured in far field for the same antenna, but sliding the probe to different positions. The change in the distance between AUT and antenna probe makes completely different the interference between the direct ray, the reflected rays in the walls of the anechoic chamber and the multiple reflections between AUT and antenna probe positioners and absorbers. As it is observed in Fig. 4, the effect is negligible in the co-polar peak, since the level of the direct ray is much higher than the level of the reflected ray. However, the effect is noticeable for the cross-polar radiation and the side lobes. Assuming that the main effect is due to multiple reflections between AUT and antenna probe, if the distance between probe and AUT is moved a quarter of the wavelength, the total distance of the ray due to the first reflection is a half of the wavelength larger. A vectorial addition of both contributions eliminates the effect of this first reflection, while the two direct rays are added in quadrature. Also, if the distance is modified half wavelength, the direct rays are with opposite phase while the first reflections are with the same phase. Therefore, a vectorial subtraction of both magnitudes also cancels the reflected rays.

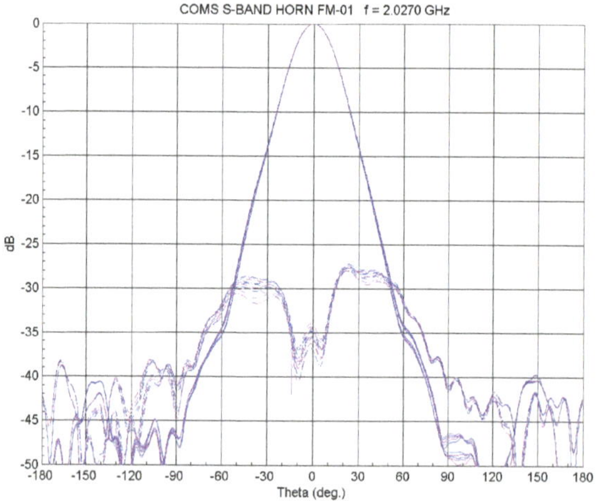

Fig. 4. Measurement of a horn antenna at different probe to AUT distances.

This technique has been applied to the measurements of the VAST12 ("12 GHz Validation Standard Antenna") designed and manufactured at the Technical University of Denmark in 1992 under the contract from the European Space Research and Technology Center and used as Reference antenna for intercomparison purposes. Fig. 6 shows the application of this technique for the measurement of the VAST12 antenna. There, both individual patterns are compared with the averaged pattern, and a slightly improvement in the crosspolar is observed. For the side lobe levels, it is not possible any improvement, since they can be affected by reflections in the walls (not corrected with this averaging). This technique can also be applied to reduce the effect of the scattering in the walls. However, in this case, the sliding has to be different, since half wavelength should be the difference between the direct ray and the reflected ray in the walls. Since the scatterers can be at different positions, a compensation technique is the measurement at different distances and the vectorial averaging of them. In this case, the averaging does not cancel the effect of the scattering, but reduces its effect. Also, in (Burgos, 2009), this technique was used to reduce the effect of the mechanical uncertainties due to the positioners and the gravity effect. Four different full-sphere acquisitions schemes were developed, as it is shown in Fig. 7. The far field was averaged and the results are compared with the individual patterns (Fig. 8). In this case, there is one individual acquisition scheme that clearly shows some problems (mainly due to reflections in the chamber). The averaged far field corrects clearly this problem, as it is observed in the crosspolar and in the side lobe levels.

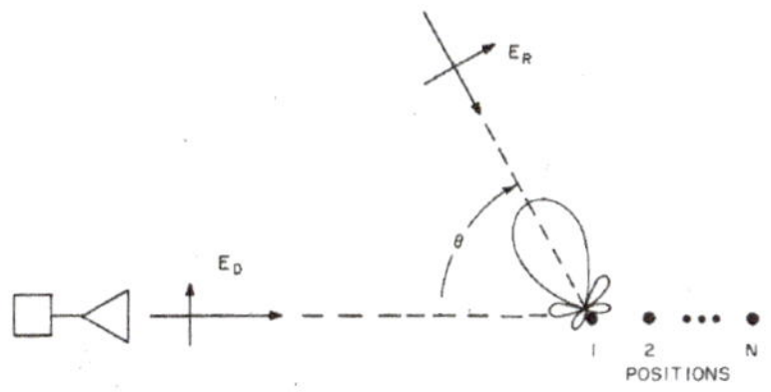

Fig. 5. Sliding for the room scattering compensation.

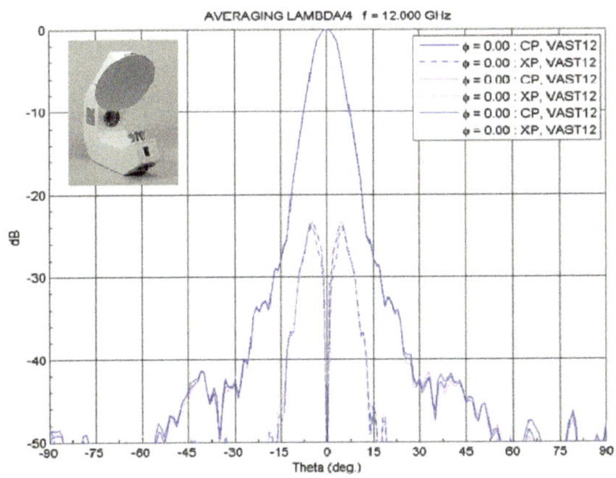

Fig. 6. Measurement of VAST12 antenna at two AUT distances and averaging.

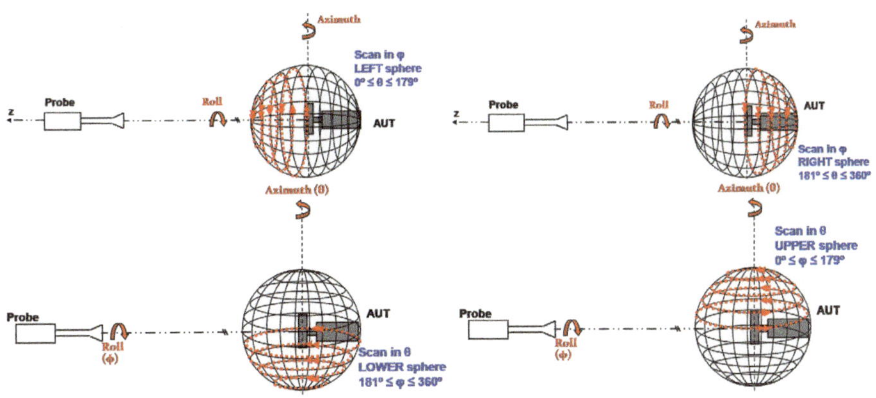

Fig. 7. Four different full sphere acquisitions schemes.

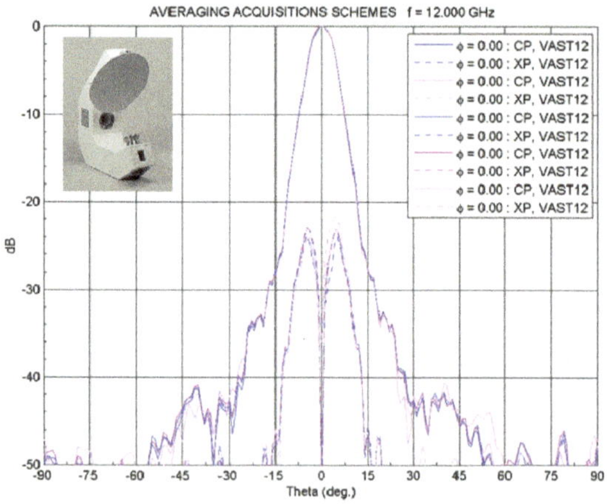

Fig. 8. Measurement of VAST12 antenna for the four acquisition schemes and averaging.

4. Application of source reconstruction techniques to reduce the effect of the noise, leakage and scattering

A sources reconstruction technique is a method to obtain the extreme near-field or the equivalent currents distribution of an antenna from the knowledge of its radiated field (near or far field). This information can be employed to detect errors, and also identify which are the causes of such errors, for example, electrical errors in arrays or mechanical errors in reflectors. Apart from its classic application, a diagnostic process also gives a complete electromagnetic characterization of the antenna, which is the basis of other applications, like near- to far-field transformations, radome applications, radioelectric coverage, etc. In this work, that additional information is employed to reduce the effect of common errors in antenna measurements.

Basically, diagnostic techniques can be divided into two types. The first of them is based on the equivalence principle and the integral equations relating fields and sources (integral equation methods), and the second one on modal expansions (modal expansion methods). Integral equation methods have general characteristics, but from a numerical point of view they are more complex than modal expansion methods because it is necessary to solve a system of integral equations. Several approaches to obtain a solution of such a system (sources from the radiated field) have been studied in (Alvarez, 2007) and (Petre, 1992) for different measurement setup geometries. In these studies, a system of integral equations (1) is solved by using a numerical method like the Finite Difference Time Domain (FDTD) method, the Finite Element Method (FEM), or the Method of Moments (MoM).

$$\vec{E}_{meas}\left(\vec{r}\right) = -\iint_{S_0} \overrightarrow{M}\left(\vec{r}\,'\right) \times \nabla G\left(\vec{r},\vec{r}\,'\right) ds' \tag{1}$$

where $\vec{E}_{meas}(\vec{r})$ is the measured electric field at the observation point $\vec{r}$, $\vec{M}(\vec{r})$ represents the equivalent magnetic current at the source point within the reconstruction surface S_o, and $G(\vec{r},\vec{r}\,')$ symbolizes the three-dimensional Green function.

On the other side, modal expansion methods require lower computational complexity, but due to the fact that the measured field has to be expressed as a superposition of orthogonal functions (vector wave solutions), they can only be used in particular situations, i.e., when the measurement is performed over canonical surfaces; planar, cylindrical and spherical are normally used because complex mechanical scannings are not required. Depending on the coordinate system, one particular kind of expansion (plane wave expansion, cylindrical wave expansion or spherical wave expansion) is carried out (Yaghjian, 1986) and (Johnson, 1973). Then, the plane wave spectrum (PWS) is calculated from the corresponding expansion, and the field over the antenna aperture can be directly obtained once the PWS is known, as shown in (2).

$$\vec{E}_{ap}(x,y)=\frac{1}{2\pi}\int_{-\infty}^{\infty}\int_{-\infty}^{\infty}\vec{P}\left(k_x,k_y\right)\cdot e^{-j\left(k_x x+k_y y\right)}dk_x dk_y \tag{2}$$

where $P(k_x,k_y)$ stands for the electric field PWS in the (k_x,k_y) direction. The last step is to explain how to obtain the PWS from the modal expansions. In planar near-field measurements, the PWS is directly obtained from the modal expansion of the samples. However, this PWS is referenced to the measurement plane, and a back-propagation to the AUT plane (Wang, 1988) has to be used to determine the appropriate PWS. For the spherical near-field case, a transformation from a spherical wave expansion to a plane wave expansion (SWE-to-PWE) is required. This transformation was recently developed in (Cappellin, 2008) and relates spherical and plane coefficients. If a measurement is to be performed in a cylindrical near-field system, up to now there have been no publications explaining the approach to calculate the PWS from the cylindrical coefficients. In this case, the only solution is to apply a cylindrical near-field to far-field transformation (Yaghjian, 1986) and (Johnson, 1973). Next, the PWS components are obtained by solving the system of equations specified in (3).

$$\begin{cases} E_\theta(r,\theta,\phi)=j\dfrac{ke^{-jkr}}{r}\left\{\cos\phi\cdot P_x(k_x,k_y)+sin\phi\cdot P_y(k_x,k_y)\right\} \\ E_\phi(r,\theta,\phi)=j\dfrac{ke^{-jkr}}{r}\cos\theta\left\{\cos\phi\cdot P_y(k_x,k_y)-sin\phi\cdot P_x(k_x,k_y)\right\} \end{cases} \tag{3}$$

This information provided can be used as the basis of methods to suppress certain errors when measuring an antenna. In this work, the next three methods are proposed.

4.1 Reflection suppression method

Normally, the antenna measurements are carried out in anechoic chambers to reduce unwanted contributions, such as reflections or diffractions from the environment. However, there are special cases in which the use of that kind of measurement setup is not possible, and a semi-anechoic chamber or an outdoor measurement system has to be employed. In fact, reflection waves may also appear in anechoic chambers due to imperfections in the radiation absorbing material (RAM). In any case, when the measurement is not performed in a fully anechoic environment, the unwanted contributions could significantly alter the actual antenna properties, producing a ripple in the radiation pattern.

The number of approaches to analyze and cancel the effects of unwanted contributions has increased in recent years. There are methods that are employed not to remove the reflection waves, but rather to characterize the chambers by a figure of merit, usually called the reflectivity level, in frequency domain (Appel-Hansen, 1973) or time domain (Clouston, 1988). Another method, based on measurements at several distances, was presented (Nagatoshi, 2008). Other solutions within the category of so-called compensation methods can be applied to avoid measuring the AUT several times at different distances with respect to the probe (Black, 1995) and (Leatherwood, 2001). Time-gating techniques can also be employed as an option to remove reflected contributions, as it has been presented in (Young, 1973) and (Burrell, 1973). However, the complexity of the pieces of equipment is lower when using frequency-domain measurements as proposed in (Loredo, 2004).

The proposed method is based on a diagnostic technique. The input data are taken over only one arbitrary surface in the frequency-domain. Moreover, compared with some time-gating techniques or frequency decomposition techniques, it is not necessary to measure more than one frequency. Therefore, the measurement time is reduced considerably and, because the frequency-domain is employed, complicated equipment is not needed. If reflected waves are present in antenna measurements, the radiation properties obtained will be perturbed. These same incorrect results are achieved with an equivalent system where reflections can be viewed as direct waves coming from virtual sources, as shown in Fig. 9 (a). Such a replacement is explained in detail via image theory, which was also used in (Lytle, 1972) to study ground reflections as image current distributions. The identification of the virtual sources is not possible with a conventional diagnostic technique, where the field is only reconstructed over the antenna aperture. However, if the field reconstruction is performed over a surface larger than the antenna dimensions, the aforementioned fictitious sources can be found and cancelled with a filtering process.

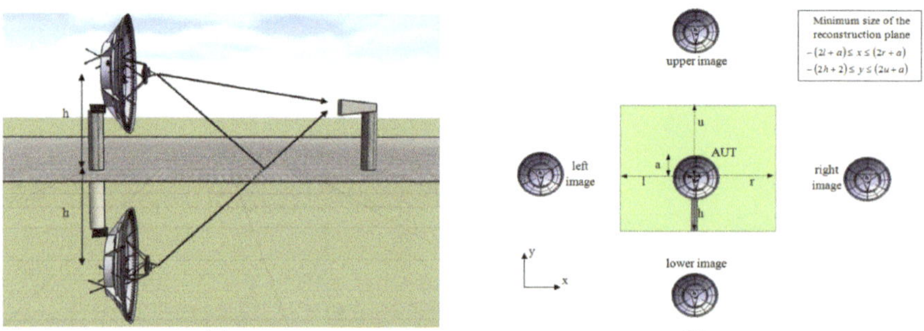

(a) (b)

Fig. 9. (a) Reflections in antenna measurements viewed by means of image theory; (b) Image theory in a general case with reflections on the floor, the ceiling and the side walls.

By applying any of the reconstruction methods explained before, the extended reconstructed field can be found. In the reconstructed domain, real sources and virtual sources are not spatially coincident. Therefore, a spatial filtering can be applied to suppress those virtual sources associated to the reflections. This filtering is defined as follows

$$F_1(x,y) = \begin{cases} 1 & \forall\ (x,y) \notin Z_V \\ 0 & \forall\ (x,y) \in Z_V \end{cases} \tag{4}$$

where Z_V represents the zone where virtual sources are placed.

The steps of the algorithm that implements the proposed method are depicted in detail in Fig. 10. This method is a generic approach because it can be used with any kind of measurement. The only thing to keep in mind is the proper selection of the diagnostic technique. Regardless of the method choice, after the diagnostic stage, the field distribution over the extended AUT plane is known, and virtual sources can be identified. Then, these sources are filtered out by employing the spatial filtering defined in (4). Finally, a new corrected PWS is obtained by taking the inverse Fourier transform of this filtered field distribution.

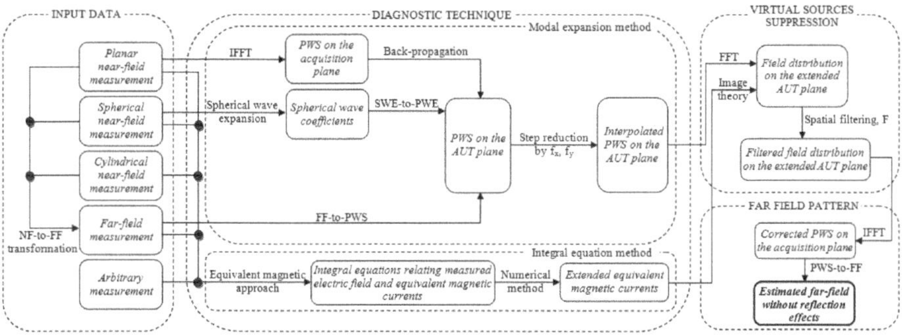

Fig. 10. Diagram of the reflection suppression method.

To verify the accuracy of the presented method, a measurement using the planar near-field range in the antenna test facility of the Technical University of Madrid (UPM) was performed. For the experiment, the probe and the AUT were selected to be a corrugated conical-horn antenna and a pyramidal-horn antenna, respectively, and they were separated from each other by 1.57 m. A reference measurement over a 2.4 m × 2.4 m acquisition plane was recorded in order to have a pattern with which to compare the future results with reflections and reflection suppression. Then, a rectangular metallic plate was placed in the

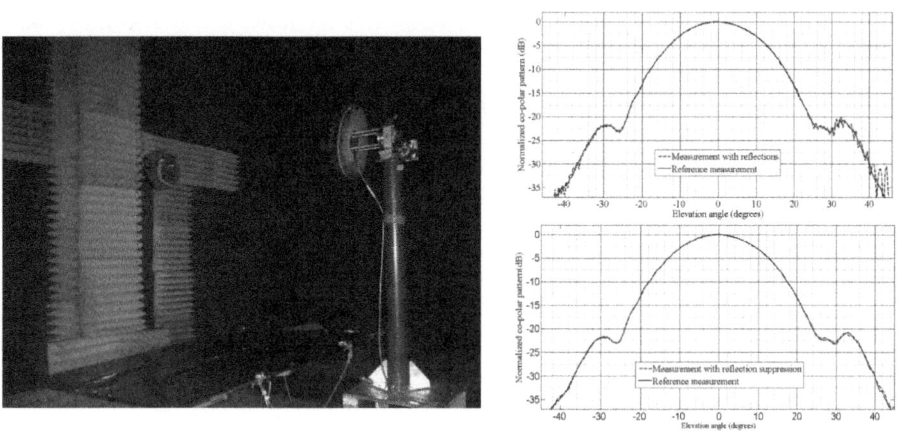

(a) (b)

Fig. 11. (a) Experimental setup with a metallic plate; (b) Comparison between the reference pattern and the pattern with reflections for the φ=90°; (c) Comparison between the reference pattern and the pattern with reflection suppression for the φ=90°.

anechoic chamber, as see in Fig. 11(a), so as to introduce reflections in the measurement with the AUT at a height of 1.3 m above the plate. Fig. 11(b) shows a comparison with the reference far-field, where it is possible to see a large ripple in the upper lateral lobe as a consequence of the disturbance generated by the reflective surface. After the reflection suppression, the estimated radiation pattern was obtained, which shows a very good agreement with the reference pattern, as observed in Fig. 11(c).

4.2 Noise suppression method

Random noise is one of the errors that limit the accuracy of far-field results, particularly, when measuring a low-sidelobe or a high-performance antenna. One possible source for this error is receiver noise that is present in all measurements. Some comprehensive studies for random noise in near-field measurements have already been presented. For the planar system, two independent analyses with similar results have been proposed in (Newell,1988) and (Hoffman, 1988). A similar study for cylindrical near-field measurements was carried out in (Romeu, 1992). These latter publications derived an expression relating the noise power in the near-field and far-field. The second method proposed in this section is also based on diagnostic techniques and tries to increase the signal-to-noise ratio in the far-field pattern obtained from a planar near-field measurement by reducing the noise power. Increasing the signal-to-noise ratio by reducing the noise power was also proposed in (Koivisto, 2004) and (Foged, 2009). The technique described in these studies can be applied to measurements performed in spherical near-field, and it is based on oversampling to obtain a higher number of spherical modes than required (Jensen, 2004), and thus, to be able to apply a modal filtering. In this second method, a spatial filtering instead of a modal filtering is applied.

In the proposed algorithm, once the planar near-field measurement has been performed, the field at the AUT plane (reconstructed field) is computed. Because the desired contribution is theoretically located inside the dimensions of the AUT, filtering can be applied to cancel the outside contribution due to noise. An analysis to assess the noise behavior and to obtain its statistical parameters was carried out. In the analysis, a complex white Gaussian and space-stationary zero mean noise was considered. From this analysis, it was deduced that, for planar near-field noise with the aforementioned statistical characteristics, the noise in the reconstructed field is a complex, stationary, white Gaussian noise, with zero mean and a variance that is equal to the variance of the noise in the scan plane. Because the noise is kept stationary, i.e., the noise power is identically distributed on the reconstructed surface, and the desired contribution is theoretically concentrated in the region where the AUT is located, a filtering can be applied to cancel the noise out of the AUT dimensions. The definition of filtering appears in (5).

$$F_2(x,y) = \begin{cases} 1 & (x,y) \in \omega_{\mathrm{A}} \\ 0 & (x,y) \in \omega_{\mathrm{T}} - \omega_{\mathrm{A}} \end{cases} \tag{5}$$

where w_T and w_A represent the reconstructed region and the AUT region. Because the noise is identically distributed over the reconstructed surface, the signal-to-noise ratio improvement achieved with the proposed spatial filtering method is the ratio between the surfaces of both reconstructed regions (6).

$$\Delta SNR = \frac{S_{\omega_T}}{S_{\omega_A}} \tag{6}$$

All steps of the proposed algorithm are depicted in Fig. 12. First, the measured data are used to obtain the PWS, referenced to the scan plane, by using an inverse Fourier transform (IFT). The next steps are to reference the PWS to the AUT plane and to calculate the reconstructed field. Then, noise filtering can be used. Finally, a new PWS with less noise power is computed using again an IFT. The validation of this method was carried out by employing the data of the reference measurement presented in the previous method. Gaussian noise with 30 dB less power than the maximum of the acquired data was computationally added. The noise power was chosen to be large so as to ensure a negligible measurement noise. Thus, the far-field obtained from the measured data without additive noise can be used as a reference to compare results before and after noise filtering. The results both for the co-polar and cross-polar pattern are shown in Fig. 13. In this particular case, the improvement calculated as indicated in (6) was equal to 28.27 dB.

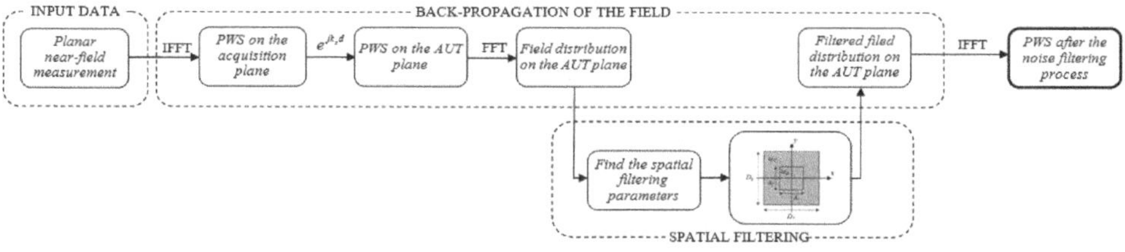

Fig. 12. Diagram of the noise suppression method.

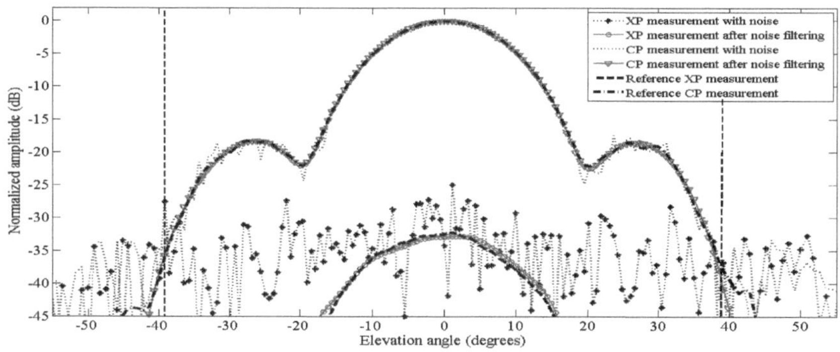

Fig. 13. Comparison between the reference pattern, the pattern with noise and the pattern after the noise filtering in $\phi = 90^{\circ}$ plane. The boundaries of the reliable region are indicated by vertical dotted lines.

4.3 Leakage suppression methods

There are three main sources of leakage signals: the first one is the crosstalk between the reference and the measurement channels in the receiver, although this first source of leakage is normally greatly suppressed by using high quality receivers with a good isolation between channels. The second source is due to connectors, faulty cables or components with poor isolation that act as new emitters distorting the far-field pattern. The last source of leakage is the bias error coming from the receiver's quadrature detector that introduces an imbalance between the two channels. As a consequence, a complex constant is added to every near-field data sample. Several methods have been proposed to detect and cancel leakage signals. Leakage due to loose connectors, faulty transmission cables or components with poor isolation can be detected by terminating the lines connected to the AUT or the probe and measuring the signal picked up by the receiver (Newell, 1988). Other alternatives try to reduce the undesired effects without any additional measurement by means of analytical compensation techniques (Leatherwood, 2001). The leakage from the receiver bias error cannot be reduced with changes in instrumentation. The first option to suppress the leakage bias error is to perform a complete near-field measurement with both the AUT and the probe terminated and the receiver set to its highest level of averaging (Hindman, 2003). If there is no leakage from cables, the signal measured in this way is directly the bias leakage. Other options without requiring additional measurements are based on estimating that constant by averaging all the measured data that are below a given threshold (Rousseau, 1999) or located outside a certain region (Newell, 1999).

In the present section, two methods to cancel leakage from cables and receiver's quadrature detector in planar near-field measurements are proposed. Both methods are based on a diagnostic technique to determine the field distribution over the AUT plane. In the first case, this information is used to filter out a great leakage contribution. The second method estimates the constant associated to the bias error by averaging the field outside the AUT aperture. If there are undesired sources like loosen connectors, faulty rotary joints or any other component with poor isolation that radiate energy, the measured field can be expressed as a sum of the actual field from the AUT and the field from the leakage (7):

$$E_{meas}(x,y,d) = E_{meas,AUT}(x,y,d) + E_{meas,leakage}(x,y,d) \tag{7}$$

After back-propagating the field from the scan plane to the AUT plane, the contribution of the leakage is located outside of the AUT dimensions. Therefore, as in the previous examples, a spatial filtering may be applied to cancel the leakage effects. Consequently, the steps of this first leakage suppression method are the same than in the noise suppression method. The validation of this first method is carried out considering the measurement of the previous sections, but including also another pyramidal-horn antenna of lower gain. Moreover an attenuator was placed before this last horn in order to simulate a leakage source, as observed in Fig. 14 (a). By using the acquired data, the reconstructed field was computed. Then, a spatial filtering was applied to cancel the effect of the leakage source. Finally, a new far-field pattern was calculated by means of an inverse Fourier transform of the filtered field, achieving a great improvement, as observed when comparing the results in Fig. 14 (b) and Fig. 14 (c).

The second method presented in this section tries to reduce leakage bias errors without additional measurements. As in the referenced methods, this one is based on an estimation of the constant added to all measured data. Then, the correction is performed by subtracting that estimated constant from the measured data. The main difference of our proposal is the information used to estimate the constant. Before, the measured data are directly employed in the estimation by averaging the near-field samples that satisfy a certain condition in order to not take into account those samples that are dominated by the AUT. In this method, to use the information on the AUT plane, i.e., the estimation is performed after back-propagating the field from the scan plane to the AUT plane is proposed. Then, the samples located outside of the AUT aperture are employed to calculate the bias constant. Because those samples do not contain any contribution of the AUT, the estimation will be better.

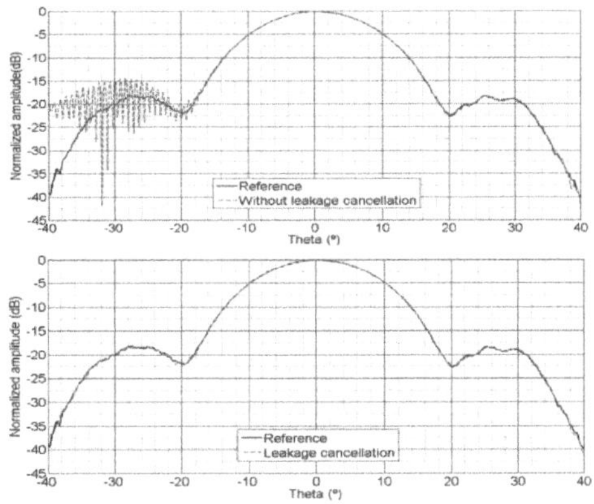

(a) (b)

Fig. 14. (a) Experimental measurement with an additional horn antenna to simulate a leakage source; (b) Comparison between the reference pattern and the pattern with leakage for the φ=90°; (c) Comparison between the reference pattern and the pattern with leakage suppression for the φ =90°.

An example is presented in order to validate this second method. In this case, the reference measurement of the previous section was employed again. Then, a complex constant with 60 dB less amplitude than the maximum of the measured data and phase equal to 45° was computationally added. The estimation of this constant was carried out by using both the method proposed in (Rousseau, 1999) and the alternative proposed here. The results of the estimation are shown in Fig. 15(a) and Fig. 15(b). As observed, a better estimation is obtained by using the method proposed here. Moreover, the range where the constant is

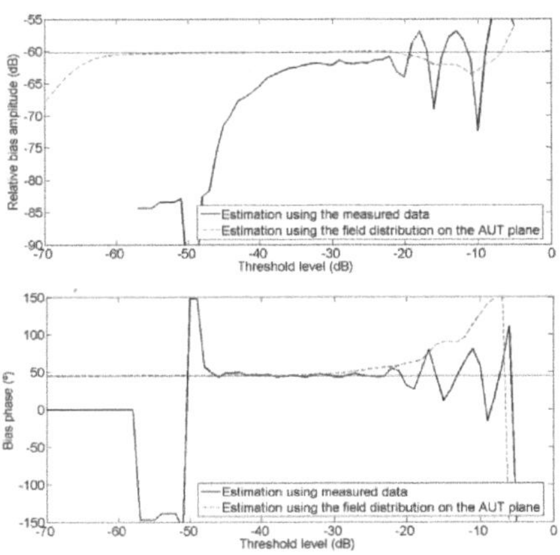

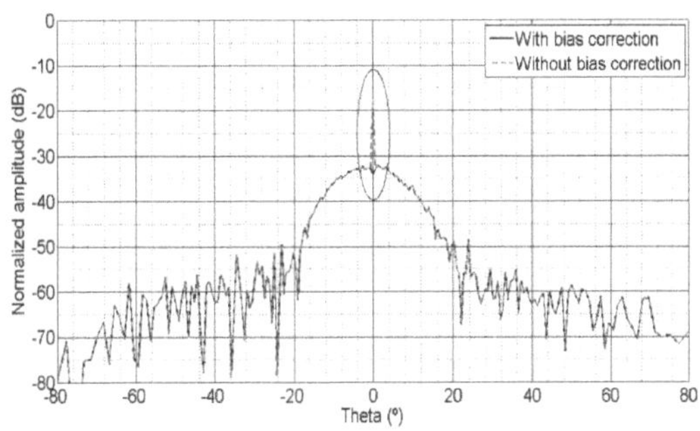

(a) (b)

Fig. 15. (a) Amplitude and (b) phase of the bias error; (c) Comparison between the cross-polar pattern without and with bias correction.

well estimated is much larger. Figure 15(c) demonstrates that after subtracting the estimated constant from the measured data, it is possible to retrieve the reference pattern, removing the peak that appears at the center of k-space.

5. Iterative algorithms to reduce the truncation error in planar, cylindrical and not complete spherical acquisitions

One important requirement to determine exact far-field patterns from near-field acquisitions is the ability to measure the electric or magnetic field that is tangential to an arbitrary surface that encloses the AUT. The planar and cylindrical scanning geometries are mechanically simpler than in the case of the spherical near-field (SNF). Moreover, the spherical near-field to far-field transformation is more complex, requiring more calculations to obtain the far-field pattern from the acquired data. However, the most accurate antenna patterns are obtained using this last type of acquisition because it is the only measurement set-up where the AUT is fully enclosed by the acquisition surface. Therefore, there are no truncation errors in the calculated far-field pattern. In the planar near-field (PNF) and cylindrical near-field (CNF) measurements, because of the finite size of the scan surface, the closed surface condition is never fulfilled, and, consequently, the true far-field pattern is never known in the whole sphere, i.e., the pattern is only valid within the called reliable region. There are some studies that attempt to obtain a good estimation of the true pattern outside of the reliable region. One of these approaches, called the equivalent magnetic current approach (Petre, 1992), presents a method of computing far-fields in the whole forward hemisphere from planar near-field measurements. Another strategy to reduce truncation errors is to rotate the AUT about one or more axes (Gregson, 2005), measuring in different planes and combining them to increase the maximum validity angle. In (Bucci, 2000) and (Bolomey, 2004), the problem of truncation is addressed using a priori information about the AUT. The main idea of this approach is to estimate the near-field data outside of the scanning area by extrapolating the measured data before calculating the far-field pattern. A recent publication (Martini, 2008) also uses a priori information about the AUT in an iterative algorithm to extrapolate the reliable portion of the calculated far-field pattern.

As commented before, truncation errors are always present in PNF and CNF measurements but not in SNF measurements. However, there are special cases, e.g., when measuring electrically large antennas, in which the measurement time may be prohibitively long and an acquisition over the whole sphere is not practical. A solution to reduce the data acquisition time is to measure over a partial sphere. Nevertheless, the new acquisition surface does not fully enclose the AUT, and, as in the PNF and CNF cases, a truncation error appears in the far-field pattern. In (Wittmann, 2002), the truncated spherical near-field data are used to the spherical wave coefficients using forward hemisphere data only, based on a least-squares technique. This method can significantly reduce truncation errors, but it is not efficient for large antennas. This problem is also addressed in (Martini, 2011) where the band-limited property of the spherical wave coefficients is exploited in an iterative algorithm that substitutes the unreliable portion of the measurement sphere with new samples at each iteration. In the present work, a method to reduce truncation errors when measuring the field over truncated surfaces is developed. The method is based on the iterative algorithm that was proposed in (Gerchberg, 1974) and (Papoulis, 1974), and it has already been applied to the PNF case in (Martini, 2008). However, this method can be applied to any scanning geometry by taking certain considerations into account. Therefore, this work can be viewed as a generalization of that method, although in the proposed method it is not necessary to take samples in surfaces external to the actual scanning area. Therefore the reduction of truncation errors is achieved without increasing the measurement time. A bottleneck of the method presented in (Martini, 2011) may be the time required to find the optimum termination point in the iterative procedure. In this work, a faster procedure based on the Gradient Descent algorithm, with which is possible to obtain the iteration number where the error is minimum, is proposed. Before describing all of the steps of the method, the two orthogonal projection operators that play an important role in the method are presented. The first operator is applied in the spectral domain and defines the reliable portion of the PWS. This first operator is given by

$$A\left(k_x,k_y\right)=\begin{cases}1 & \forall\left(k_x,k_y\right)\in\Omega_R\\ 0 & \forall\left(k_x,k_y\right)\notin\Omega_R\end{cases} \tag{8}$$

where A is the spectral-truncation operator and Ω_R is the reliable region. The second operator is called band-limited operator and is applied to the field distribution on the AUT plane.

$$h(x,y)=\begin{cases}1 & \forall(x,y)\in\omega_{AUT}\\ 0 & \forall(x,y)\notin\omega_{AUT}\end{cases} \tag{9}$$

where h stands for the band-limited operator and w_{AUT} is the region where the AUT is located. In the spectral domain, this last operator can be expressed as follows

$$B\,P\left(k_x,k_y\right)=P\left(k_x,k_y\right)*H\left(k_x,k_y\right) \tag{10}$$

being B the band-limited operator defined in the spectral domain, $P(k_x,k_y)$ represents the PWS and H is the inverse Fourier transform of the operator h. The schematic diagram of the method is shown in Fig. 16.

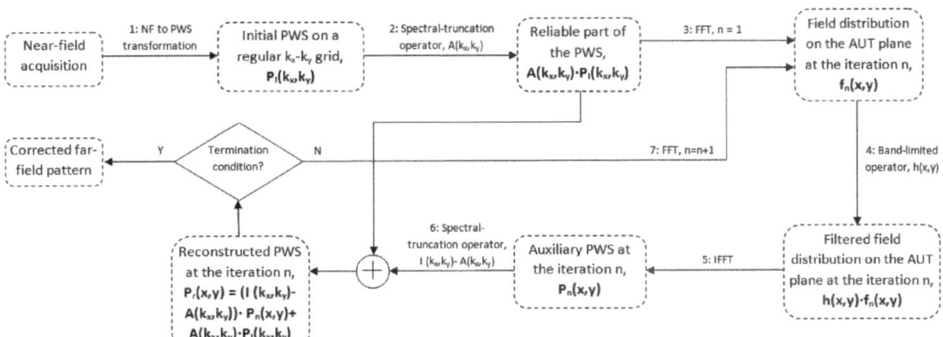

Fig. 16. Diagram of the method to reduce truncation errors.

As mentioned before, the method described in this section can be applied to any scanning geometry. However, the method is only validated here for the CNF case. Other validations for other types of acquisition can be found in (Cano, 2012). For the experiment, a measurement in the CNF range at the UPM was performed. The probe and the AUT consisted of a corrugated conical-horn antenna and a Ku-band reflector with a 40 cm diameter respectively. The data were acquired over a cylinder with a height of 2.7 m and a radius of 2.3 m and with a spatial sampling equal to 0.5λ in the vertical direction and 2.5° in the azimuth. The AUT was also measured in a whole sphere in order to obtain a reference pattern. From inspection of Fig. 17a, it is evident that the truncation error is greatly suppressed, reducing the error outside of the reliable region from 58.2% to 8.9%. A comparison depicted in Fig. 17b shows the reconstructed far-field pattern in the vertical plane versus the truncated and reference far-field pattern.

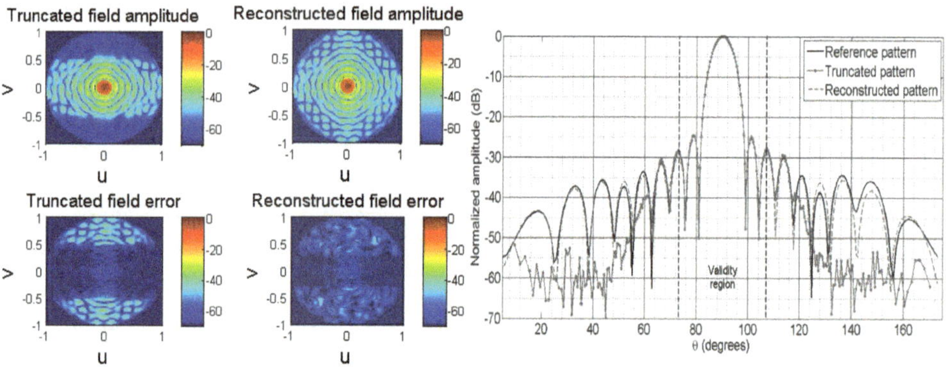

(a) (b)

Fig. 17. (a) Far-field pattern and error in dB before and after applying the iterative method; (b) Comparison among the truncated, reconstructed and reference far-field patterns for the reflector in the φ=90° cut.

6. Techniques to improve the quality of the quiet zone in compact ranges

CATR measurement facilities are characterised by a straightforward operation which is able to perform real time antenna measurements along a wide margin of frequencies, whenever reflectors or lenses are used as collimators. In spite of their compactness, CATR facilities operate in far-field conditions. Therefore, the test wave received by the AUT should be a plane wave. The operation of a CATR relies on the planarity of this test wave, inside the "quiet zone" volume. Electromagnetic theory states that the quiet zone region cannot hold a plane wave field distribution, even not with an arbitrarily small volume, for a finite frequency of operation. This is due to the ability of the range's collimator to concentrate the electromagnetic energy within a reduced set of angular directions, following a distribution which is expected to tend to the delta in the angular domain when moving up in frequency. This theoretical constraint must be seen as the first of a set of contributions that affect the quiet zone planarity, and thus, the performance of the CATR which generates it. Usual implementations of the CATR concept focus on minimizing the following quiet zone degradations:

1. Edge diffraction. The collimators are designed so they are able to generate a plane wave inside the quiet zone, while the edge contributions act in the opposite direction. This is minimized in reflector-type collimators through conformal edge treatment: serrated edge and rolled edge geometries (Muñoz-Acevedo, 2012).
2. Collimation scheme. Varied implementations of the field collimation system have been presented, from the classical reflector scheme towards the hologram configuration, going through the dielectric lens. The surface accuracy requirements of a field collimator are increasingly tough when moving up in frequency, being dramatic quiet zone performance reduction the price to pay if they are not fulfilled (Tuovinen, 1992).
3. Reflection from chamber walls. Through the use of RAM, sometimes following conformal wall shapes and different size RAM pyramids depending on their location on the chamber (Bertino, 1998) and (Aubin, 2011).
4. Time gating techniques. They aim to minimize the contribution of reflected signals from spurious field scatterers inside the chamber: test positioners and various unwanted propagation schemes, such as direct ray from the range's feeder or multiple reflections between the range's elements (feeder antenna and field collimators) (Lemanczyk, 2004) and (Boumans, 1987).
5. Feeding scheme. The collimator acts on an incident wave, collimating its phase and amplitude pattern towards the quiet zone. Conventional feeding alternatives impress an amplitude pattern over the collimator, being influenced both by the feeder's pattern and the unequal spherical wave propagation over the incident directions. The main effect on quiet zone is the appearance of an amplitude tapper which can be minimized if several collimators are used in cascade, feeder pattern are made conformal, or both of them (Jones, 1986) and (Kangas, 2003).

Many of these features are used jointly in recent implementations of CATR setups. For a particular facility, its quiet zone distribution can be characterized in diverse ways. The most common are quiet zone probing (Muñoz-Acevedo, 2011) and RCS measurements of canonical objects (Griendt, 1996). These techniques lead to the knowledge of the quiet zone vector field through the amplitude and phase distributions of the components of interest. The described error depends with the frequency at which the CATR is operated. The collimation capabilities of a CATR increase when the frequency is higher. This is due to the fact that both the field collimators and the edge treatment devices are higher, in terms of wavelengths, and their ability to handle the radiation is, thus, superior. However, besides this fact, the required accuracy in terms of surface distortion and range alignment becomes the limiting factor which degrades the potential field collimation performance of the facility down to the actual one. These notes are summarized in Table 2.

This background makes necessary the use of quiet zone information and correction (Viikari, 2007). Solid knowledge of the quiet zone fields implies extensive acquisition campaigns which deal with diverse feed-probe configurations, so the vector field distribution is fully known. This task is very time consuming to be performed , becoming even more critical when increasing the frequency. It is possible however, to acquire the quiet zone distribution in a more time efficient way, without loss of information. The assessment of a particular CATR facility is able to lead to sampling grids which are intrinsically less time consuming and also imply an increase in the acquired field's signal to noise ratio figure, as proposed in (Muñoz-Acevedo, 2011). Fig. 18 summarizes the data acquisition setup and the involved subsystems. The proposed CATR approach ensures data robustness both in the sense of quiet zone field planarity and distribution stability. The "Field collimation" block follows the notes drawn in (Muñoz-Acevedo, 2012). Accordingly, the direct wave contribution acquired by the probe is isolated from the stray signals through a time gating technique, which the instrumentation "VNA" must be able to perform. The anechoic chamber is properly designed so the measurement environment is isolated from the outer radiation sources. Proper absorber for the frequency range of interest is glued on the inner walls, ceil and floor. These solutions focus on the quiet zone planarity.

Feature	Evolution when increasing frequency
Conformal edge	Increases the scattering performance ($\propto$f).
Collimation scheme	Higher planarity of quiet zone distribution, but stricter surface roughness criteria ($\sigma_{RMS} \propto f^{-1}$).
RAM	Increases the absorbing performance ($\propto$f).
Feeding scheme	Becomes more sensitive to misalignment. ($\propto$f)
Quiet zone acquisitions	Acquisition times for 2D field distributions increase with frequency ($\propto f^2$)

Table 2. CATR features vs frequency.

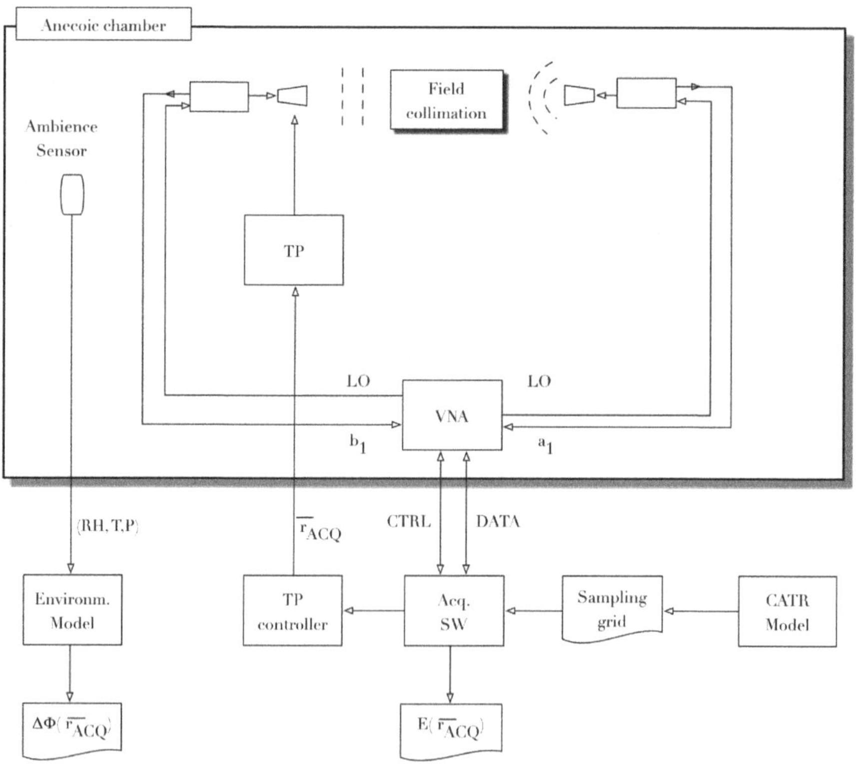

Fig. 18. CATR quiet zone data acquisition setup.

Long term stability of this planarity is achieved through temperature and humidity control inside the chamber and surrounding the instrumentation. The performance of the facility is maximized if these magnitudes are sampled and used as feedback to perform field correction. An ambience sensor samples atmospheric pressure (P), relative humidity (RH) and temperature (T). The environmental model of field propagation uses these magnitudes as inputs and corrects the phase ($\Delta\varphi$) of the probed field taking into account the variation of the air's refractive index *n(T,P,RH)*, as in Eqs. (11, 12). Pressure is measured in *hPa*, T is measured in *K* and Δr states for the path length from the feed towards the test zone, in meters. Coefficients *a*, *b*, *c* can be found in the ITU P.453-9 recommendation.

$$\Delta\phi(T,P,RH) = \frac{2\pi}{\lambda_0} \cdot n(T,P,RH) \cdot \Delta r \tag{11}$$

$$n(T,P,RH) = 1 + \frac{77.6}{T} \cdot \left(P + 4810 \cdot \frac{1}{T} \cdot \frac{RH}{100} \cdot a \cdot \exp\left(\frac{b \cdot (T - T_0)}{(T - T_0) + c} \right) \right) \cdot 10^{-6} \tag{12}$$

The use of well known probe correction techniques (Bolomey, 2004) is able to subtract the effect of using non-isotropic probes in the field acquisition, while it is unnecessary when the quiet zone is known through RCS techniques instead of probing. The data post-processing scheme which corresponds to the proposed facility of Fig. 18, is shown in Fig. 19.

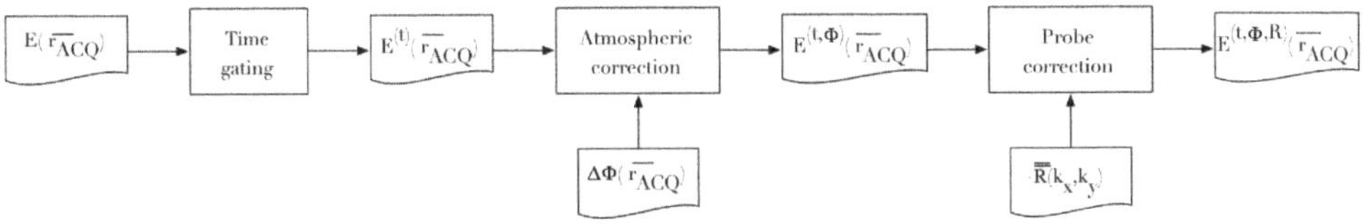

Fig. 19. CATR quiet data post-processing scheme.

7. Conclusions

This chapter has presented different correction techniques to be applied in antenna measurements. In particular, the authors have analyzed the extension of the pattern averaging technique to improve the accuracy of the measurements by minimizing the room scattering or mechanical errors. In the next section, a source reconstruction technique has been used to reduce the measurement errors associated to noise, leakage or scattering. In section 6, the iterative procedure called Gerchberg-Papoulis, applied previously for planar acquisitions, has been extended to cylindrical and partial sphere near field antenna measurements. Finally, section 7 has shown an algorithm to correct some of the classical errors that affect the compact antenna test ranges. In particular, the authors have presented an algorithm to compensate the errors associated to variations in temperature and humidity, important for submillimeter wave frequencies. Therefore, this chapter explains a synthesis of old and new methods for improving the quality of the classical antenna measurement systems, using post-processing tools.

8. Acknowledgments

This work developed in this chapter has been realized under the project CCG10-UPM_TIC-5805, supported by Comunidad de Madrid and Universidad Politécnica de Madrid.

9. References

A. Alexandridis (NCSRD), et al. (December 2007), "Recommendations and Comparative Investigation for Near-Field Antenna Measurement Techniques and Procedures", Deliverable A1.2D2, "Standardization of Antenna Measurement Techniques", Contract FP6-IST 026957, Antenna Centre of Excellence (ACE).

A. C. Newell (June 1988), "Error Analysis Techniques for Planar Near-Field Measurements", IEEE Transactions on Antennas and Propagation, Vol. 36, No. 6, pp. 754-768, June 1988.

A. C. Newell amd C. F. Stubenrauch (June 1988), "Effect of Random Errors in Planar Near-Field Measurement", IEEE Transactions on Antennas and Propagation, Vol. 36, No. 6, pp. 769-773.

A. C. Newell and A. D. Yaghjian (June 1975), "Study of Errors in Planar Near-Field Measurements", Antennas and Propagation Society International Symposium 1975, pp. 470-473.

A. D. Yaghjian (January, 1986), "An overview of near-field antenna measurements," IEEE Trans. Antennas Propagat., vol. AP-34, No. 1, pp. 30-44.

A. Muñoz-Acevedo, L. Rolo, M. Paquay, M. Sierra-Castañer (2011), "Accurate and Time Efficient Quiet Zone Acquisition Technique for the Assessment of ESA's CATR at Millimeter Wavelengths", XXXIII AMTA Symposium, Denver, Colorado.

A. Muñoz-Acevedo, M. Sierra-Castañer (2012) "An Efficient Hybrid GO-PWS Algorithm to Analyze Conformal Serrated- Edge Reflectors for Millimeter-Wave Compact Range", IEEE Transactions on Antennas and Propagation (accepted).

A. Newell (1999), "Methods to estimate and reduce leakage bias errors in planar near-field antenna measurements," presented at Antenna Measurement Techniques Assosiation 1999, Monterey, California, USA.

A. Newell, G. Hindman (November 2008), "Mathematical Absorber Suppression (MARS) for Anechoic Chamber Evaluation & Improvement", Proceedings of the AMTA Symposium, Boston, USA.

A. Papoulis (September 1975), "A new algorithm in spectral analysis and bandlimited extrapolation," IEEE Trans. Circuits Syst., vol. CAS-22, no. 9, pp- 735-742.

B. N. Taylor and C. E. Kuyatt (1994), "Guidelines for Evaluating and Expressing the Uncertainty of NIST Measurement Results", National Institute of Standards and Technology (NIST) Technical Note 1297, 1994 Edition, United States Department of Commerce Technology Administration.

C. Cappellin (September 2007), "Antenna diagnostics for spherical near-field antenna measurements," Ph.D. dissertation, Dept. Elect. Eng., Danmarks Tekniske Universitet, Copenhague, Denmark.

C. Cappellin, A. Frandsen, and O. Breinbjerg (October 2008), "Application of the SWE-to-PWE antenna diagnostics technique to an offset reflector antenna," IEEE Antennas Propagat. Mag., vol. 50, No. 5, pp. 204-213.

D. A. Leatherwood and E. B. Joy (December 2001), "Plane wave, pattern subtraction, range compensation," IEEE Trans. Antennas Propagat., vol. 49, no. 12, pp. 1843-1851.

D. A. Leatherwood, E. B. Joy (December 2001), "Plane wave, pattern subtraction, range compensation," IEEE Trans. Antennas Propagat., vol. 49, No. 12, pp. 1843-1851.

D. G. Gentle, A. Beardmore, J. Achkar, J. Park, K. MacReynolds, J. P. M. De Vreede, (September 2003), "National Physical Laboratory (NPL) Report CETM 46: Measurement Techniques and Results of an Intercomparison of Horn Antenna Gain in IEC-R 320 at Frequencies of 26.5, 33.0 and 40.0 GHz".

D. N. Black and E. B. Joy (April 1995), "Test zone field compensation," IEEE Trans. Antennas Propagat., vol. 43, No. 4, pp. 362-368.

D. W. Hess (October 2002), "An Expanded Approach to Spherical Near-Field Uncertainty Analysis", AMTA 24th Annual Meeting and Symposium, Cleveland, OH.

D. Young, D. E. Svoboda, and W. D. Burnside (July 1973), "A comparison of time- and frequency-domain measurement techniques in antenna theory," IEEE Trans. Antennas Propagat., vol. 21, No. 4, pp. 581-583.

E. Martini, O. Breinbjerg, and S. Maci (November 2008), "Reduction of truncation errors in planar near-field aperture antenna measurements using the Gerchberg -Papoulis algorithm," IEEE Trans. Antennas Propagat., vol. 56, no. 11, pp. 3485-3493.

E. Martini, S. Maci, and L. J. Foged (April 2011), "Spherical near field measurements with truncated scan area," in Proc. European Conf. Antennas Propag. 2011, Rome, pp. 3412-3414.

E. N. Clouston, P. A. Langsford, and S. Evans (April, 1988), "Measurement of anechoic chamber reflections by time-domain techniques," IEE Proc. H, Microwaves Antennas Propagt., vol. 135, Pt. H, No. 2, pp. 93-97.

F. J. Cano-Fácila, S. Burgos, and M. Sierra-Castañer (April 2011), "Novel method to improve the signal-to-noise ratio in the far-field results obtained from planar near-field measurements," IEEE Antennas Propagat. Magazine.

F. J. Cano-Fácila, S. Burgos, F. Martín, and M. Sierra-Castañer (March 2011), "New reflection suppression method in antenna measurement systems based on diagnostic techniques," IEEE Trans. Antennas Propagat., vol. 59, No. 3, pp. 941-949.

F. J. Cano-Fácila, S. Pivnenko, and M. Sierra-Castañer (2012), "Reduction of truncation errors in planar, cylindrical and partial spherical near-field antenna measurements," to be published in Int. J Antennas Propagat.

F. Jensen and A. Frandsen (October 2004), "On the number of modes in spherical wave expansion," in Proc. 2004 Antenna Meaurement. Techniques Assoc., AMTA, Stone Mountain Park, GA, pp. 489-494.

G. A. Burrell and A. R. Jamieson (September 1973), "Antenna radiation pattern measurement using time-to-frequency transformation (TFT) techniques," IEEE Trans. Antennas Propagat., vol. 21, No. 5, pp. 702-704.

G. Hindman, A. C. Newell (February 2007), "Simplified Spherical Near-Field Accuracy Assessment", IEEE Antennas and Propagation Magazine, Vol. 49, No. 1.

G. Hindman, A. Newell, and P. N. Betjes (September 2003), "Error correction techniques for near-field antenna measurements," presented at the International ITG Conference on Antennas 2003, INICA 2003, Berlin, Germany.

I. Bertino, U. Bozzetti, G. Ariano (1998) "A State of the Art Anechoic Chamber for Air vehicle Testing at Alenia Aeronautica", XX AMTA Symposium, Montreal, Canada.

International Organization for Standardization ISO/IEC 98 Publications (1995), "Guide to the Expression of Uncertainty in Measurement (GUM)", International Organization for Standardization, Geneva, Switzerland, 1995, ISBN: 92-67-10188-9.

J. Appel-Hansen (July 1973), "Reflectivity level of radio anechoic chambers," IEEE Trans. Antennas Propagt., vol. AP21, No. 4, pp. 490-498.

J. Aubin, M. Winebrand, V. Vinogradov (July, 2011) "Experimental validation of the "Two - Level GTD" method for design of anechoic chambers," 2011 IEEE International Symposium on Antennas and Propagation, pp.1893-1896.

J. B. Hoffman and K. R. Grimm (June 1988), "Far-Field Uncertainty Due to Random Near-Field Measurement Error", IEEE Transactions on Antennas and Propagation, Vol. 36, No. 6, pp. 774-780.

J. C. Bolomey, et al (February 2004), "Reduction of truncation error in near-field measurement of antennas of base-station mobile communication systems," IEEE Trans. Antennas Propag., vol. AP-52, no. 2, pp. 593–602.

J. E. Hansen (1997), "Definition, design, manufacture, test and use of a 12 GHz Validation Standard Antenna", Executive Summary, ESTEC contract No. 7407/87/NL /PB, Technical Report R672, Tech. Univ. of Denmark.

J. E. Hansen (ed.) (1988), "Spherical Near-Field Antenna Measurements", Peter Peregrinus Ltd., on behalf of IEE, London, UK.

J. J. H. Wang (June 1988), "An examination of the theory and practices of planar near-field measurement," IEEE Trans. Antennas Propagat., vol. 36, No. 6, pp. 746-753.

J. Jones, (1986) "Prime Focus Feeds for the Compact Range", VIII AMTA Symposium, Ottawa, Canada.

J. Lemanczyk, J. Hartmann, D. Fasold (2004) "Evaluation of hard gating in the ESA/ESTEC CPTR", XXVI AMTA Symposium 2004, Atlanta, Georgia.

J. Romeu, L. Jofre and A. Cardama (January 1992), "Far-Field Errors Due to Random Noise in Cylindrical Near-Field Measurements", IEEE Transactions on Antennas and Propagation, Vol. 40, No. 1, pp. 79-84.

J. Tuovinen, A. Vasara, A. Räisänen (1992), "A hologram type of Compact Antenna Test Range", XIV AMTA Symposium, Columbus, Ohio.

L. A. Muth (May 1988), "Displacement Errors in Antenna Near-Field Measurements and Their Effect on the Far-Field", IEEE Transactions on Antennas and Propagation, Vol. 36, No. 5, pp. 581-591.

L. J. Foged and M. Faliero (November 2009), "Random noise in spherical near-field systems," in Proc. 2009 Antenna Meaurement. Techniques Assoc., AMTA, Salt Lake City, UT, pp. 135-138.

M. Boumans, S. Brumley (1987) "Hardware Gating Improves HP 8510 based RCS Measurement Systems", IX AMTA Symposium, Seattle, Washington.

M. Nagatoshi, M. Hirose, H. Tanaka, S. Kurokawa, and H. Morishita (July 2008), "A method of pattern measurement to cancel reflection waves in anechoic chamber," in Antennas Propagt. Soc. Int. Symp. 2008, San Diego, CA, pp. 1-4.

M.A.J. Griendt, V.J. Vokurka, J. Reddy, J. Lemanczyk (1996)"Evaluation of a CPTR using an RCS Flat Plate Method", XVIII AMTA Symposium, Seattle, Washington.

O. M. Bucci, G. D'Elia, and M. D. Migliore (2000), "A new strategy to reduce the truncation error in near-field far-field transformation," Radio Sci., vol. 35, no. 1, pp. 3–17.

P. Koivisto (2004), "Reduction of errors in antenna radiation patterns using optimally truncated spherical wave expansion," Progress In Electromagnetic Research, vol. pier-47, pp. 313-333.

P. Petre and T. K. Sarkar (November 1992), "Planar near-field to far-field transformation using an equivalent magnetic current approach," IEEE Trans. Antennas Propagat., vol. 40, No. 11, pp. 1348-1356.

P. R. Rousseau (1999), "An algorithm to reduce bias errors in planar near-field measurement data," presented at Antenna Measurement Techniques Assosiation 1999, Monterey, California, USA.

R. C. Johnson, H. A. Ecker, and J. S. Hollis (December 1973), "Determination of far-field antenna patterns from near-field measurements," Proc. IEEE, vol. 61, No. 12, pp. 1668-1694.

R. C. Wittmann, C. F. Stubenrauch, and M. H. Francis (2002), "Spherical scanning measurements using truncated data sets," in AMTA Proc. 2002, Cleveland, OH, pp. 279-283.

R. J. Lytle (November 1972), "Ground reflection effects upon radiated and received signals as viewed via image theory," IEEE Trans. Antennas Propagat., vol. AP-20, No. 6, pp. 736-741.

R. W. Gerchberg (May 1974), "Super-resolution through error energy reduction," Opti. Acta, vol. 21, no. 9, pp. 709-720, May 1974.

S. Burgos (September 2009), "Contribution to the uncertainty evaluation in the measurement of the main antenna parameters", , Doctoral Thesis, Technical University of Madrid (UPM), Madrid, Spain.

S. Burgos, M. Sierra Castañer, F. Martín, F. Cano, J. L. Besada, (April 2010), "Error analysis and simulator in cylindrical near-field antenna measurement systems", book entitled "Advances in Measurements Systems", Pages: 289 - 314, Edited by Milind Kr Sharma, ISBN: 978-953-7619-X-X, Vienna, Austria.

S. F. Gregson, C. G. Parini, and J. McCormick (December 2005), "Development of wide-angle pattern measurements using a probe-corrected polyplanar near-field measurement technique," IEE Proc. Microw. Antennas Propag., vol. 152, pp. 563-572, no. 6.

S. Loredo, M. R. Pino, F. Las-Heras, and T. K. Sarkar (February 2004), "Echo identification and cancellation techniques for antenna measurement in non-anechoic test sites," IEEE Antennas Propagat. Mag., vol. 46, No. 1, pp. 100-107.

S. Pivnenko, J. M. Nielsen, O. Breinbjerg (May 2006), "Electrical Uncertainties In Spherical Near-Field Antenna Measurements", Proceedings of the First Antenna

Measurements Techniques Association Europe (AMTA Europe) Symposium, pp.183-186, Munich.

V. Kangas, J. Lemanczyk (2003), "Compact Range Defocused Quiet Zone Characterization", XXV AMTA Symposium, Irvine, California.

V. Viikari (April 2007), "Antenna Pattern Correction Techniques at Submillimeter Wavelengths". Helsinki University of Technology. Dissertation for the degree of Doctor of Science in Technology.

Y. Álvarez, F. Las-Heras, and M. R. Pino (December 2007), "Reconstruction of equivalent currents distribution over arbitrary three-dimensional surfaces based on integral equation algorithms," IEEE Trans. Antennas Propagat., vol. 55, No. 12, pp. 3460-3468.

9

Machine Vision Measurement Technology and Agricultural Applications

Abdullah Beyaz
Ankara University, Faculty of Agriculture, Department of Agricultural Machinery, Turkey

1. Introduction

Many of the techniques of digital image processing, or digital picture processing as it often was called, were developed in the 1960s at the Jet Propulsion Laboratory, Massachusetts Institute of Technology, Bell Laboratories, University of Maryland, and a few other research facilities, with application to satellite imagery, wire-photo standards conversion, medical imaging, videophone, character recognition, and photograph enhancement. The cost of processing was fairly high, however, with the computing equipment of that era. That changed in the 1970s, when digital image processing proliferated as cheaper computers and dedicated hardware became available. Images then could be processed in real time. As general-purpose computers became faster, they started to take over the role of dedicated hardware for all but the most specialized and computer-intensive operations (Anonymous, 2011b). With the fast computers and signal processors available in the 2000s, digital image processing has become the most common form of image processing and generally, is used because it is not only the most versatile method, but also the cheapest. Digital image processing technology for medical applications was inducted into the Space Foundation Space Technology Hall of Fame in 1994.

Today image processing is used in a wide variety of applications for two somewhat different purposes: improving the visual appearance of images to a human viewer, preparing images for the measurement of the features and structures that they reveal. The techniques that are appropriate for each of these tasks are not always the same. This chapter covers methods that are used for measurement tasks.

A machine vision system processes images acquired from an electronic unit, which is like the human vision system where the brain processes images derived from the eyes. Machine vision is a rich and rewarding topic for study and research for agriculture engineers, electronic engineers, computer scientists and many others. Increasingly, it has a commercial future. There are many vision systems in routine industrial use: cameras inspect mechanical parts to check size, food is inspected for quality and images used in biometrics benefit from machine vision techniques (Nixon & Aguado, 2002).

Imaging systems cover all processes involved in the formation of an image from objects and the sensors that convert radiation into electric signals, and further into digital signals that can be processed by a computer. Generally the goal is to attain a signal from an object in such a form that we know where it is (geometry) and what it is or what properties it has (Jähne & Haußecker, 2000).

The human visual system is limited to a very narrow portion of the spectrum of electromagnetic radiation, called visible light. Image processes are sensitive to wavelengths and additional information might be hidden in the spectral distribution of radiation. Using different types of radiation allows taking images from different depths or different object properties (Nixon & Aguado, 2002).

The measurement of images is often a principal method for acquiring scientific data and generally requires that features or structure be well defined, either by edges or unique colour, texture, or some combination of these factors. The types of measurements that can be performed on entire scenes or on individual features are important in determining the appropriate processing steps (Russ, 2006).

2. Acquiring images

2.1 Imaging photometers and colourimeters

Colourimetry is the measurement of the wavelength and the intensity of electromagnetic radiation in the visible region of the spectrum. Colourimetry can help find the concentration of substances, since the amount and colour of the light that is absorbed or transmitted depends on properties of the solution, including the concentration of particles in it. A colourimeter is an instrument that compares the amount of light getting through a solution with the amount that can get through a sample of pure solvent (Fig. 1.). A colourimeter contains a photocell is able to detect the amount of light which passes through the solution under investigation. The more light that hits the photocell, the higher the current it produces, hence showing the absorbance of light. A colourimeter takes 3 wideband (RGB) readings along the visible spectrum to obtain a rough estimate of a colour sample. Pigments absorb light at different wavelengths.

The use of imaging photometers and colourimeters for fast capture of photometric and colourimetric quantities with spatial resolution has attracted increasing interest. Compared with measuring instruments without spatial resolutions, such as spectrometers, this technology offers the following advantages:

- Substantial time-savings with simultaneous capture of a large number of measurements in a single image,
- Image-processing functions integrated in the software permit automated methods of analysis, e.g. calculation of homogeneity or contrast.

However, the absolute measuring precision of imaging photometers and colourimeters is not as high as spectroradiometers. This is because of the operational principle using a CCD Sensor in combination with optical filters, which can only be adapted to the sensitivity of the human eye with limited precision.

Imaging photometers and colourimeters are the instruments of choice for:

- Measurement of luminance and colour distribution of industrial products,
- Measurement of homogeneity, contrast of products,
- Analysis of luminous intensity.

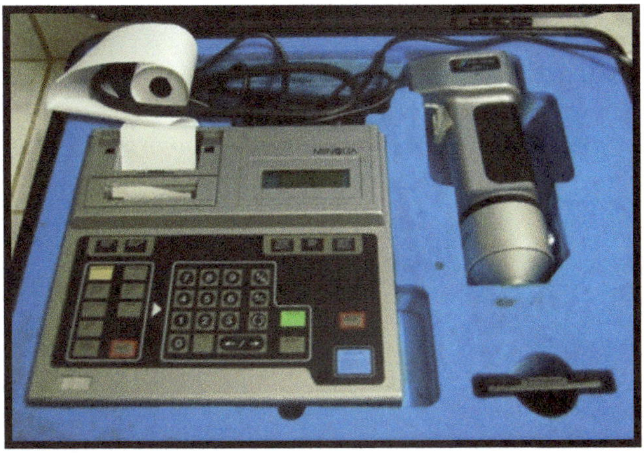

Fig. 1. Minolta Cr 200 colourmeter (Beyaz, 2009a).

2.2 Camera image

In recent years, a massive research and development effort has been witnessed in colour imaging technologies in both industry and ordinary life. Colour is commonly used in television, computer displays, cinema motion pictures, print and photographs. In all these application areas, the perception of colour is paramount for the correct understanding and dissemination of the visual information. Recent technological advances have reduced the complexity and the cost of colour devices, such as monitors, printers, scanners and copiers, thus allowing their use in the office and home environment. However, it is the extreme and still increasing popularity of the consumer, single-sensor digital cameras that today boosts the research activities in the field of digital colour image acquisition, processing and storage. Single-sensor camera image processing methods are becoming more important due to the development and proliferation of emerging digital camera-based applications and commercial devices, such as imaging enabled mobile phones and personal digital assistants, sensor networks, surveillance and automotive apparatus (Lukac & Plataniotis, 2007).

2.2.1 Digital camera architectures

Digital colour cameras capture colour images of real-life scenes electronically using an image sensor, usually a charge-coupled device (CCD), or complementary metal oxide semiconductor (CMOS), sensor, instead of the film used in the conventional analog cameras. Therefore, captured photos can be immediately viewed by the user on the digital camera's display, and immediately stored, processed, or transmitted without any doubt, this is one of the most attractive features of digital imaging (Lukac & Plataniotis, 2007).

2.2.2 Single-sensor device

This architecture reduces cost by placing a colour filter array (CFA), which is a mosaic of colour filters, on top of the conventional single CCD/CMOS image sensor to capture all three primary (RGB) colours at the same time. Each sensor cell has its own spectrally selective filter and thus it stores only a single measurement. Therefore, the CFA image constitutes a mosaic-like grayscale image with only one colour element available in each pixel location. The two missing colours must be determined from the adjacent pixels using a digital processing solution called demosaicking. Such an architecture represents the most cost-effective method currently in use for colour imaging, and for this reason, it is almost universally utilized in consumer-grade digital cameras (Lukac & Plataniotis, 2007).

2.2.3 Three-sensor device

This architecture acquires colour information using a beam splitter to separate incoming light into three optical paths. Each path has its own red, green or blue colour filter having different spectral transmittances and sensors for sampling the filtered light. Because the camera colour image is obtained by registering the signals from three sensors. The mechanical and optical alignment is necessary to maintain correspondence among the images obtained from the different channels. Besides the difficulties of maintaining image registration, the high cost of the sensor and the use of a beam splitter make the three-sensor architecture available only for some professional digital cameras (Lukac & Plataniotis, 2007).

2.3 Real-time imaging systems

Real-time systems are those systems in which there is urgency to the processing involved. This urgency is formally represented by a deadline. Because this definition is very broad, the case can be made that every system is real-time. Therefore, the definition is usually specialized. A "firm" real-time system might involve a video display system, for example, one that superimposes commercial logos. Here, a few missed deadlines might result in some tolerable flickering, but too many missed deadlines would produce an unacceptable broadcast quality. Finally, a "soft" real-time system might involve the digital processing of photographic images. Here, only quality of performance is at issue. One of the most common misunderstandings of real-time systems is that their design simply involves improving the performance of the underlying computer hardware or image processing algorithm. While this is probably the case for the mentioned display orphotographic processing systems, this is not necessarily true for the target tracking system.

Here, guaranteeing that image processing deadlines are never missed is more important than the average time to process and render one frame. The reason that one cannot make performance guarantees or even reliably measure performance in most real-time systems is that the accompanying scheduling analysis problems are almost always computationally complex. Therefore, in order to make performance guarantees, it is imperative that the bounded processing times be known for all functionality. This procedure involves the guarantee of deadline satisfaction through the analysis of various aspects of code execution and operating systems interaction at the time the system is designed, not after the fact when trial-and-error is the only technique available. This process is called a schedulability analysis. The first step in performing any kind of schedulability analysis is to determine, measure or otherwise estimate the execution of specific code units using logic analyzers, the systemclock, instruction counting, simulations or algorithmic analysis. During software development, careful tracking of central processing unit utilization is needed to focus on those code units that are slow or that have response times that are inadequate. Unfortunately, cache and direct memory access, which are intended to improve average real-time performance, destroy determinism and thus make prediction of deadlines troublesome, if not impossible. But, schedulability analysis is usually the subject of traditional texts on real-time systems engineering (Lukac & Plataniotis, 2007).

2.3.1 The issues that must be considered in real-time imaging system

2.3.1.1 Hardware and display issues

An understanding of the hardware support for colour imaging graphics is fundamental to the analysis of real-time performance of the system. Some specialized hardware for realtime imaging applications involves high-performance computers with structural support for complex instruction sets and imaging coprocessors. Inexpensive pixel processors are also available, and scalable structures are increasingly being used for real-time imaging applications. But building systems with highly specialized processors is not always easy, particularly because of poor tool support. Therefore, many commonly deployed colour imaging systems use consumer-grade personal computers. There are many architectural issues relating to real-time performance, such as internal/ external memory bus width, memory access times, speed of secondary storage devices, display hardware issues, and colour representation and storage, to name a few. Collectively, these design issues involve three trade-off problems – schedulability versus functionality, performance versus resolution, and performance versus storage requirements. Real-time design of imaging systems, then, involves making the necessary decisions that trade one quality for another, for example, speed versus resolution. For example, one of the main performance challenges in designing real-time image processing systems is the high computational cost of image manipulation. A common deadline found in many processing systems involves screen processing and update that must be completed at least 30 times per second for the human eye to perceive continuous motion.

Because this processing may involve more than a million pixels, with each colour pixel needing one or two words of storage, the computational load can be staggering. For the purposes of meeting this deadline, then, the real-time systems engineer can choose to forgo display resolution, or algorithmic accuracy. Finding improved algorithms or better hardware will also help meet the deadlines without sacrifice – if the algorithms and hardware behaviorare bounded, which, as mentioned, is not always the case. In this section, however, we confine our discussion to two important hardware issues (Lukac & Plataniotis, 2007).

2.3.1.2 Colour representation and real-time performance

Our interest is in the appropriate representation of colour in the physical hardware, because as previously noted, this has real-time performance implications. The colour buffer may be implemented using one or more of the following storage formats:

- One byte per pixel (indexed or pseudo-colour), which allows 28 = 256 colours,
- Two bytes per pixel (high colour), which, using 16 bits = 65,536 colours,

- Three bytes per pixel (true or RGB colour), which yields approximately 16.8 million colours.

RGB is often considered the standard in many programming languages, and it is used in many important colour image data formats, such as JPEG and TIFF. True colour uses 24 bits of RGB colour, 1 byte per colour channel for 24 bits. A 32-bit representation is often used to enhance performance, because various hardware commands are optimized for groups of 4 bytes or more (Lukac & Plataniotis, 2007).

2.3.1.3 Language issues

Modern languages for real-time imaging must provide an easy interface to hardware devices and provide a framework for maintainability, portability and reliability, among many other features. Many programming languages are commonly used to implement Real-Time Colour Imaging Systems, including C, C++, C#, Java, Visual Basic, Fortran, assembly language, and even BASIC. Poor coding style is frequently the source of performance deterioration in real-time imaging systems. In many cases, the negative effects are due to performance penalties associated with object composition, inheritance, and polymorphism in object-oriented languages. But object-oriented languages are rapidly displacing the lower-level languages like C and assembly language in real-time colour imaging systems, and it is probably a good thing because of the accompanying benefits.

Understanding the performance impact of various language features, particularly as they relate to image storage and manipulation, is essential to using the most appropriate construct for a particular situation. There is no clear answer, and experimentation with the language compiler in conjunction with performance measurement tools can be helpful in obtaining the most efficient implementations.

The following list summarizes key issues when implementing real-time imaging systems in a high-level language:

- Use appropriate coding standards to ensure uniformity and clarity,
- Refactor the code continuously (that is, aggressively improve its structure) with an eye to performance improvement,
- Use performance measurement tools continuously to assess the impact of changes to the hardware and software,
- Carefully document the code to enable future developers to make structural and performance enhancements,
- Adopt an appropriate life cycle testing discipline.

Finally, be wary of code that evolved from non-object-oriented languages such as C into object-oriented version in C++ or Java. Frequently, these conversions are made hastily and incorporate the worst of both the object-oriented and non-object-oriented paradigms simultaneously.

Java is an object-oriented language, with a syntax that is similar to C++ and C#, and to a lesser extent, C. Besides, modern object-oriented languages such as C++ and C# have quite a lot in common with Java (Lukac & Plataniotis, 2007).

3. Commonly used image processing and analysis softwares

3.1 AdOculos

The software has PC-based image processing without the need of extensive programming knowledge. The complete C source code of these DLLs is part of the standard pack. Point, local and global, morphological operations texture, image sequence histograms procedures, colour transformations, automatic counting and interactive measuring, pattern recognition can be done by using this software.

3.2 IMAQ vision

Adds machine vision and image processing functionality to LabVIEW and ActiveX containers (National Instruments).

3.3 Optimas

It is an analytical Imaging - complete commercial image analysis program for windows which is used in biological and industrial measurement environments. Optimas implements hundreds of measurement, image processing, and image management operations, all available from the graphical user interface.

3.4 GLOBAL LAB image

It is an easy to use graphical software application for creating scientific and general purpose imaging application. Use the Point & Click script to develop applications quickly. Simply configure the tools you want to use and add them to your script to create powerful imaging applications without writing any code.

3.5 Matlab ımage processing toolbox

Image Processing Toolbox™ provides a comprehensive set of reference-standard algorithms and graphical tools for image processing, analysis, visualization, and algorithm development. You can perform image enhancement, image deblurring, feature detection, noise reduction, image segmentation, spatial transformations, and image registration. Many functions in the toolbox are multithreaded to take advantage of multicore and multiprocessor computers.

3.6 ImageJ

ImageJ is a public domain Java image processing program inspired by NIH Image. It runs, either as an online applet or as a downloadable application, on any computer with a Java 1.4 or later virtual machine. Downloadable distributions are available for Windows, Mac OS, Mac OS X and Linux.

3.7 Myriad

Myriad image processing software was also used for comparing images and measuring surface area of images. This program has two steps for measuring images. First step was calibration of the program. For this purpose software needed a calibration ruler or a calibration plate. Second step was true selection of images.

4. Measurement types

4.1 Global measurement types

4.1.1 Surface area measurement

Surface areas describe as the boundries between structures. Surface area measurement of the different part of plant such as leaf projection area and fruit projection area is important for process engineering of agricultural products. So scientist use different methods for measuring areas of plant parts and one of them is image processing and analysis techniques. We need to slice products in to the pieces (Sum of these piece area give us the total surface of the agricultural product.) or select the boundries of the agricultural products for surface area measurements (Fig. 2.). They use the surface area measurements for estimating product diameter, weight, fruit volume and pesticide usage estimation.

Fig. 2. Apple area measurement (Selection method) by using Myriad Image Analysis Software.

4.1.2 Length measurement

Length measurement is usually applied in objects that have one more long dimension in comprision of others. For small objects such as seeds, lenght measurement can be done by using ımage analysis and processing techniques. Photographic enlargement of the dimensions help us to measure lenght of agricultural products easly (Fig. 3.).

Fig. 3. Apple lenght measurement by using Myriad Image Analysis Software.

4.1.3 Determining number

In the area of machine vision, blob detection refers to visual modules that are aimed at detecting points and/or regions in the image that are either brighter or darker than the surrounding (Fig. 4.). There are two main classes of blob detectors (i) differential methods based on derivative expressions and (ii) methods based on local extrema in the intensity landscape. With the more recent terminology used in the field, these operators can also be referred to as interest point operators, or alternatively interest region operators.

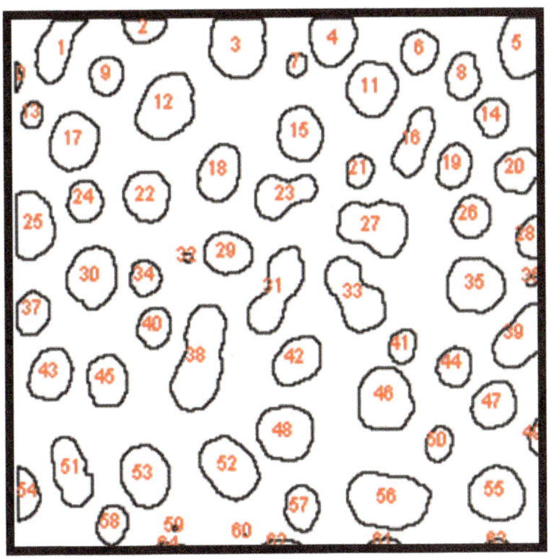

Fig. 4. Counting the number of droplets by using ImageJ Software.

Also determination of the numbers of objects is important for agricaltural applications. We can determine specicif gravity, thousand grain weight, number of droplets in spraying and etc. For example, the numbers of the droplets on a leaf in pesticide application is an important parameter for determining optimum pesticide usage in agricultural products. Because pesticides effects enviroment and using them in high dozes caused to air, soil and water pollution.

4.2 Spesific measurement types

4.2.1 Colour measurements

RGB and CIELab is widely used as a standard spaces for comparing colours. A set of primary colours, such as the sRGB primaries, define a colour triangle; only colours within this triangle can be reproduced by mixing the primary colours. Colours outside the colour triangle are therefore shown here as gray. The primaries and the D65 white point of sRGB are shown in Fig. 5. (Anonymous, 2011c).

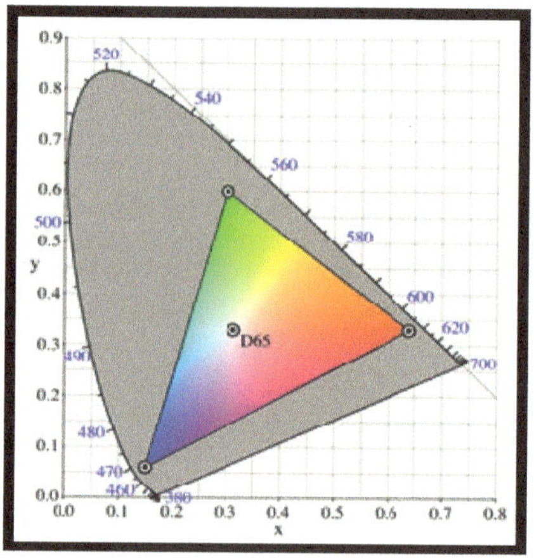

Fig. 5. CIExy1931 sRGB gamut (Anonymous, 2011c).

A Lab colour space is a colour-opponent space with dimension L for lightness and a and b for the colour-opponent dimensions, based on nonlinearly colour space coordinates (Anonymous, 2011c).

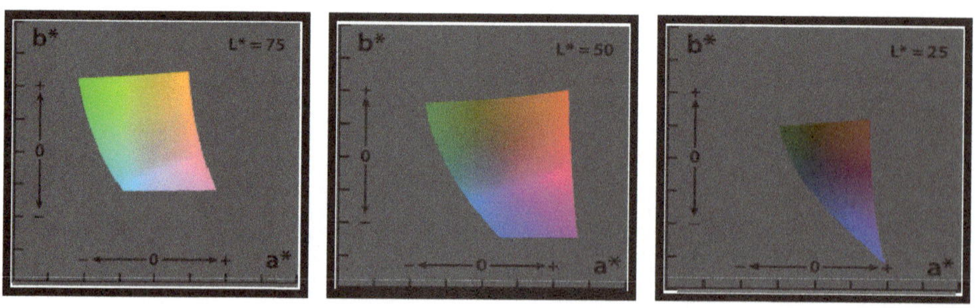

Fig. 6. Lab colour space (Anonymous, 2011c).

The CIE 1976 (L*, a*, b*) colour space, showing only colours that fit within the sRGB gamut (and can therefore be displayed on a typical computer display). Each axis of each square ranges from -128 to 128 (Fig. 6.) (Anonymous, 2011c).

4.2.2 Determining location

The location of features was also needed for identify of the results. There are several definitions of location and pixels in an ordinary image For instance, the x,y coordinates of the midpoint of a feature can be determined simply from minimum and maximum limits of the pixels. Normally, starting from the top left corner of the array, the pixel addresses themselves (Fig. 7.). Global coordinate system gives us information about real pixel dimensions values.

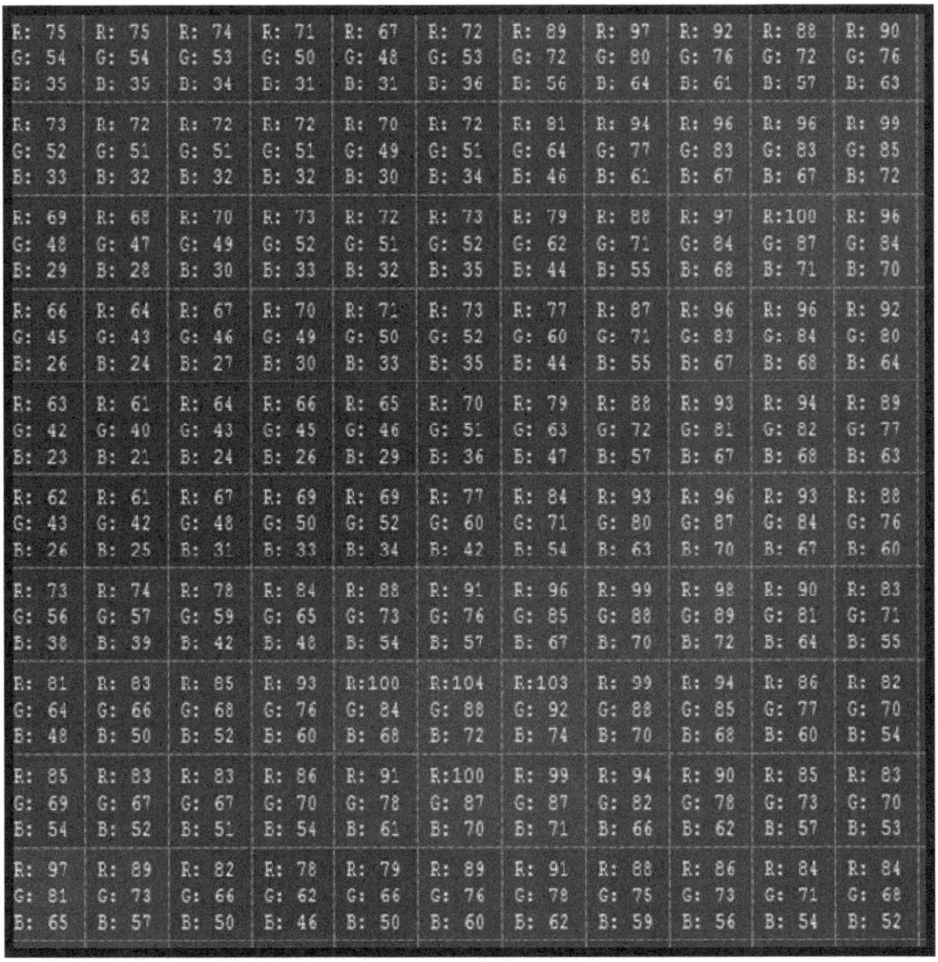

R: 75 G: 54 B: 35	R: 75 G: 54 B: 35	R: 74 G: 53 B: 34	R: 71 G: 50 B: 31	R: 67 G: 48 B: 31	R: 72 G: 53 B: 36	R: 89 G: 72 B: 56	R: 97 G: 80 B: 64	R: 92 G: 76 B: 61	R: 88 G: 72 B: 57	R: 90 G: 76 B: 63
R: 73 G: 52 B: 33	R: 72 G: 51 B: 32	R: 72 G: 51 B: 32	R: 72 G: 51 B: 32	R: 70 G: 49 B: 30	R: 72 G: 51 B: 34	R: 81 G: 64 B: 46	R: 94 G: 77 B: 61	R: 96 G: 83 B: 67	R: 96 G: 83 B: 67	R: 99 G: 85 B: 72
R: 69 G: 48 B: 29	R: 68 G: 47 B: 28	R: 70 G: 49 B: 30	R: 73 G: 52 B: 33	R: 72 G: 51 B: 32	R: 73 G: 52 B: 35	R: 79 G: 62 B: 44	R: 88 G: 71 B: 55	R: 97 G: 84 B: 68	R:100 G: 87 B: 71	R: 96 G: 84 B: 70
R: 66 G: 45 B: 26	R: 64 G: 43 B: 24	R: 67 G: 46 B: 27	R: 70 G: 49 B: 30	R: 71 G: 50 B: 33	R: 73 G: 52 B: 35	R: 77 G: 60 B: 44	R: 87 G: 71 B: 55	R: 96 G: 83 B: 67	R: 96 G: 84 B: 68	R: 92 G: 80 B: 64
R: 63 G: 42 B: 23	R: 61 G: 40 B: 21	R: 64 G: 43 B: 24	R: 66 G: 45 B: 26	R: 65 G: 46 B: 29	R: 70 G: 51 B: 36	R: 79 G: 63 B: 47	R: 88 G: 72 B: 57	R: 93 G: 81 B: 67	R: 94 G: 82 B: 68	R: 89 G: 77 B: 63
R: 62 G: 43 B: 26	R: 61 G: 42 B: 25	R: 67 G: 48 B: 31	R: 69 G: 50 B: 33	R: 69 G: 52 B: 34	R: 77 G: 60 B: 42	R: 84 G: 71 B: 54	R: 93 G: 80 B: 63	R: 96 G: 87 B: 70	R: 93 G: 84 B: 67	R: 88 G: 76 B: 60
R: 73 G: 56 B: 38	R: 74 G: 57 B: 39	R: 78 G: 59 B: 42	R: 84 G: 65 B: 48	R: 88 G: 73 B: 54	R: 91 G: 76 B: 57	R: 96 G: 85 B: 67	R: 99 G: 88 B: 70	R: 98 G: 89 B: 72	R: 90 G: 81 B: 64	R: 83 G: 71 B: 55
R: 81 G: 64 B: 48	R: 83 G: 66 B: 50	R: 85 G: 68 B: 52	R: 93 G: 76 B: 60	R:100 G: 84 B: 68	R:104 G: 88 B: 72	R:103 G: 92 B: 74	R: 99 G: 88 B: 70	R: 94 G: 85 B: 68	R: 86 G: 77 B: 60	R: 82 G: 70 B: 54
R: 85 G: 69 B: 54	R: 83 G: 67 B: 52	R: 83 G: 67 B: 51	R: 86 G: 70 B: 54	R: 91 G: 78 B: 61	R:100 G: 87 B: 70	R: 99 G: 87 B: 71	R: 94 G: 82 B: 66	R: 90 G: 78 B: 62	R: 85 G: 73 B: 57	R: 83 G: 70 B: 53
R: 97 G: 81 B: 65	R: 89 G: 73 B: 57	R: 82 G: 66 B: 50	R: 78 G: 62 B: 46	R: 79 G: 66 B: 50	R: 89 G: 76 B: 60	R: 91 G: 78 B: 62	R: 88 G: 75 B: 59	R: 86 G: 73 B: 56	R: 84 G: 71 B: 54	R: 84 G: 68 B: 52

Fig. 7. Set of RGB coordinate points.

4.2.3 Neighbor relationships

Local coordinates and individuals features of the pixels are important for some applications and neigbour pairs are the easiest way to identify differences between the products. The histogram of the distributions of this features gives the the answers. Fig. 8. shows distribution of features at measurement of rice grains by using Matlab Software.

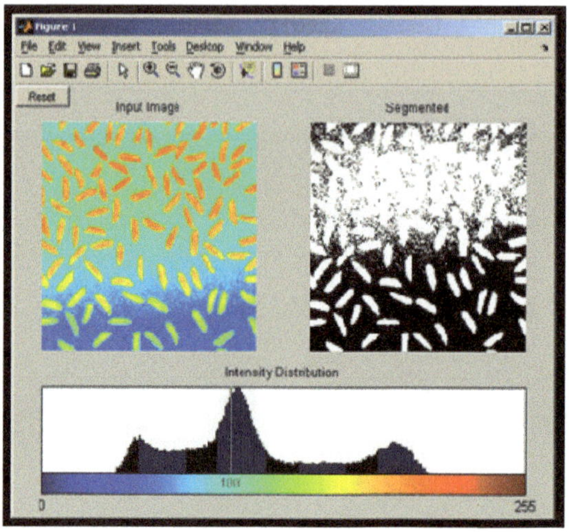

Fig. 8. Measurement of rice grains by using Matlab Software (Anonymous, 2011d).

4.2.4 Perimeter

The perimeter of a feature is a welldefined and familiar geometrical parameter. Measuring a numerical value that really describes the object and these value turns to be easy to determine agricultural products perimeter measurement. Some systems estimate the length of the boundary around the object by counting the pixels but some of them use picture selection method which is based on the selection of the objects manually from images, for investigating perimeter value (Fig. 9.).

4.2.5 Describing shape

Shape and size are inseparable in a physical object, and both are generally necessary if the object is to be satisfasfactoriliy described. Further, in defining the shape some dimensinoal parameters of the object must be measured. Seeds, grains, fruits and vegetables are irregular in shape because of the complexity of their specifications, theorically requires an infinite number of measurements (Fig. 10.). The number of these measurements increases with increase in irregularity of the shape (Mohsenin, 1970).

Fig. 9. Measurement of apple perimeter (Selection method) by using Myriad Image Analysis Software.

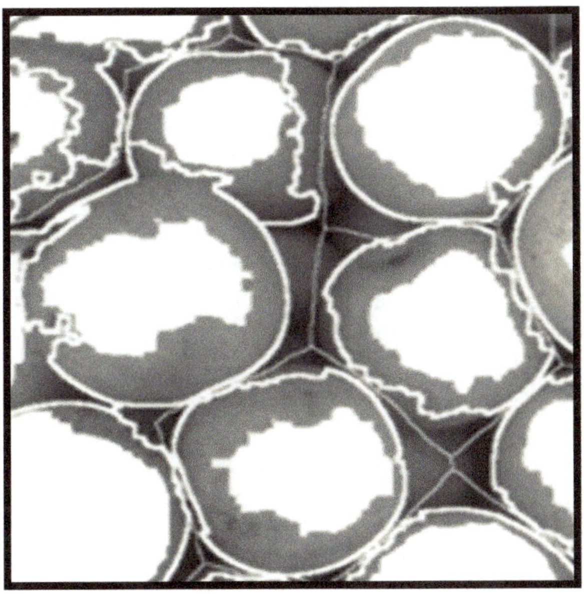

Fig. 10. A group of apple and their shape boundaries.

4.2.6 Three-dimensional measurements

3-D metrics have first been developed for simplification purposes but three-dimensional image assessments causes new challenges (Fig. 11.). There are measurement distortion

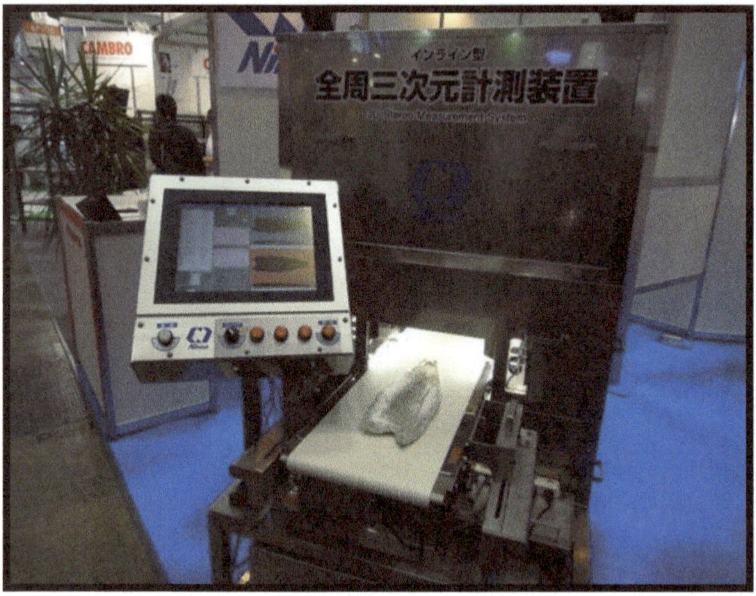

Fig. 11. 3D measurement system that it developed jointly with the National Institute of Advanced Industrial Science and Technology (Anonymous, 2011a.).

between an original 3-D surface and its deformed version. The other important challenge is analysis of 3-D views by using 2-D screens. Image measurements still require identifying the pixels that are connected to each other. In two dimensions, it is necessary to determine differences between touching pixels.

Existing 3D measurement techniques are classified into two major types—active and passive. In general, active measurement employs structure illumination (structure projection, phase shift, moire topography, etc.) or laser scanning, which is not necessarily desirablein many applications. On the other hand, passive 3D measurement techniques based on stereo vision have the advantages of simplicity and applicability, since such techniquesrequire simple instrumentation.

5. Agricultural application gallery

5.1 Image measurement and computation method at applications

A digital camera was used for determining dimensions and digital images were evaluated by Myriad image processing software. Measuring and verification process of digital images can be done with the help of this program safely, also digital images can be compared easly (Fig. 12.). This program has two steps for measuring images. First step was calibration of the program. For this purpose software needed a calibration ruler or a calibration plate. Hence a millimetric paper was used for calibration and testing for software (Fig. 13.). For the calibration, a calibration plate with known dimensions was used while taking digital images. Second step was true selection of images (Beyaz, 2009a).

Fig. 12. Interface of digital image analysis software Myriad v8.0

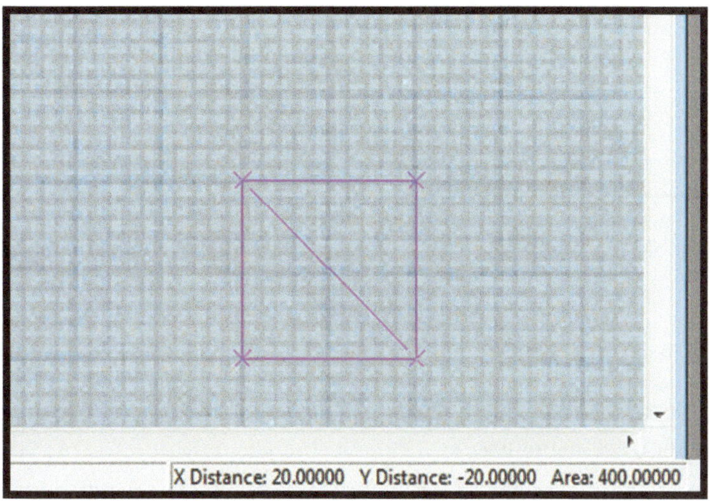

Fig. 13. Milimetric paper test of image processing software.

Colour measurements were determined with Minolta Cr200 model colourmeter which is based on L*a*b* measurement system (Fig. 14.). Avarage of three sample points are used as colour value (Fig. 15.). Minolta Cr200 model colourmeter was calibrated with a white reflective plate.

Fig. 14. Sample colour measurement of an apple.

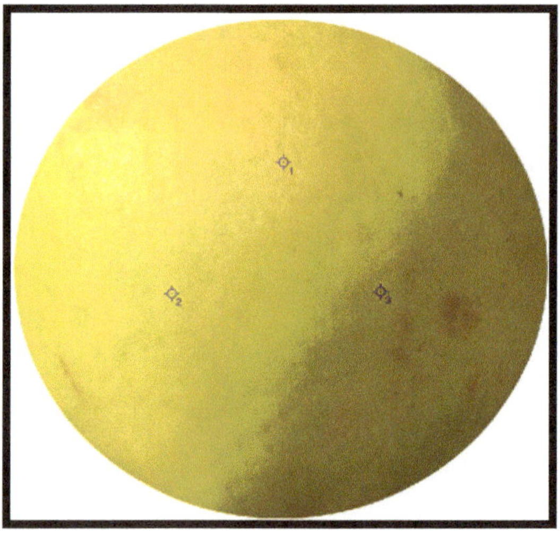

Fig. 15. Selection of L*a*b* sample points at a quice by using colourmeter (Beyaz et al., 2011).

5.1.1 Volume determination of Kahramanmaras red pepper (*Capsicum annuum* L.) by using image analysis technique

The size of an agricultural product is an important parameter to determine fruit growth and quality. It can be used to determine the optimum harvest time as a maturity index. In this study, the image analysis method was tested on Kahramanmaras red pepper (*Capsicum annuum* L.) which may have a non uniform shape. For this purpose; the front, top and left side of each pepper was taken into account for evaluations and projection areas. The effect of each image and image combination has been used to determine the volume of peppers. The regression coefficients between the projection areas and volume values have also been assessed for volume estimation. The relationship between the real volumes of

Kahramanmaras red peppers and the front projection area values (A_F) was pointed as a graph at Fig. 16. Similarly top projection area values were presented at Fig. 17. However left projection areas values and the estimation equation can be seen in Fig. 18 (Beyaz et al., 2009d).

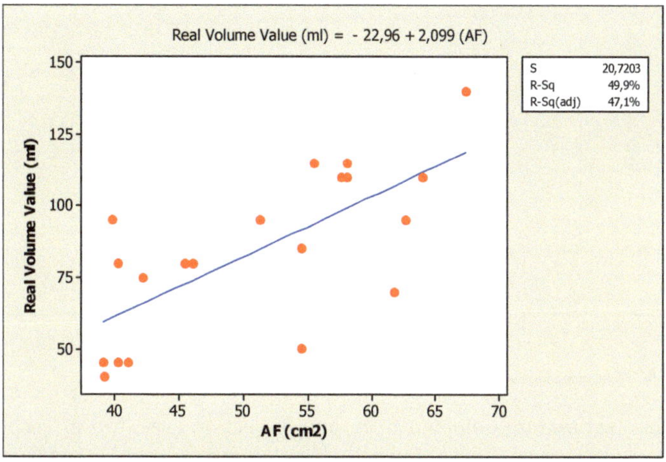

Fig. 16. The regression equation and graph obtained by using A_F projection area (Beyaz et al., 2009d).

According to this assessment, the real volume value which was obtained from the front projection area shows the regression coefficient has reached 49.9%. The regression coefficient obtained from estimation equations by using of the top projection area was 66.6% (Fig. 17.).The regression coefficient obtained by using left projection area was 63.6% (Fig. 18.).

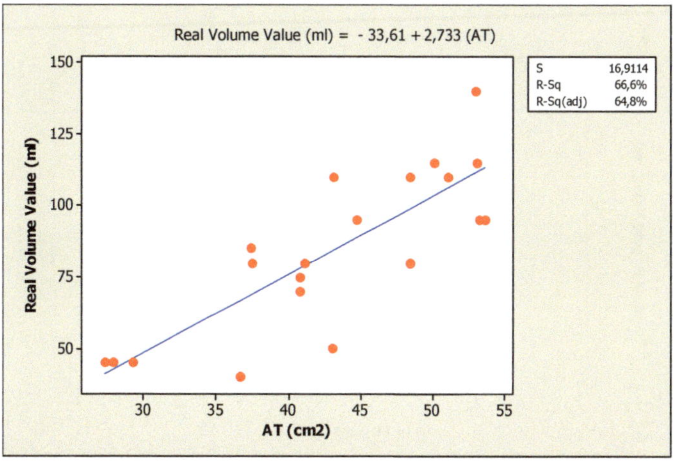

Fig. 17. The regression equation and graph obtained by using A_T projection area (Beyaz et al., 2009d).

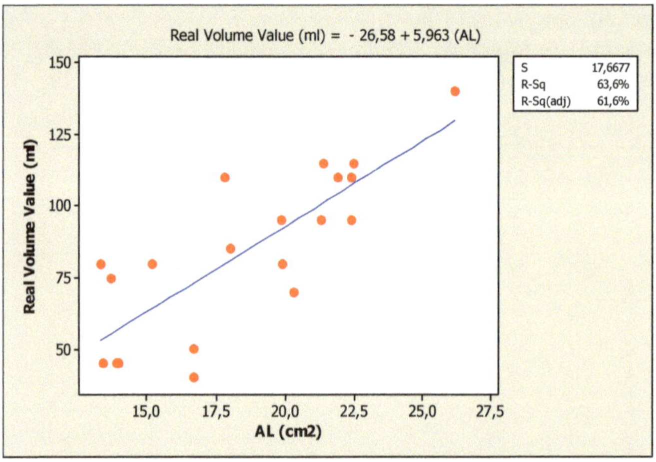

Fig. 18. The regression equation and graph obtained by using A_L projection area (Beyaz et al., 2009d).

When the volume estimation obtained from one projection area, top projection area values has the highest regression coefficient (66.6%). The volume estimation has been examined by making double groups from these three projection areas. Fig. 19. represents the regression graph which is obtained from the sum of top and left projection areas. Estimation equation has reached a regression coefficient 74.7% depending on this assessment.

The most appropriate estimation formula has been calculated from the top and the left projection area. The following equation is the most appropriate equation (Beyaz et al., 2009d):

Real Volume Value (ml) =-47,29+2,132 (A_T+A_L)

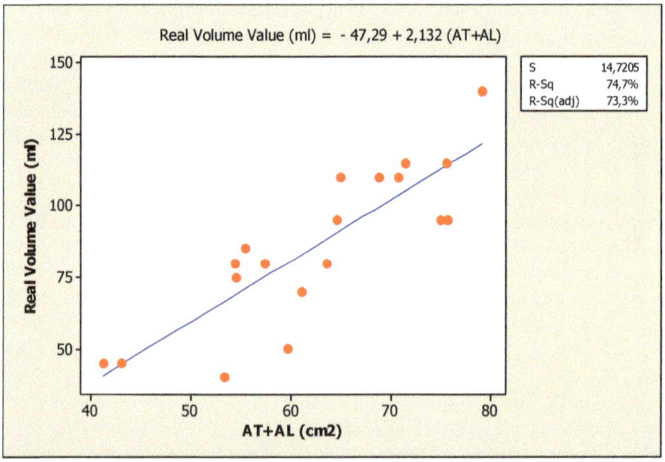

Fig. 19. The regression equation and graph obtained by using A_T + A_L projection areas (Beyaz et al., 2009d).

Fig. 20. shows the regression equation graph which was determined from total value of the front and left projection areas.

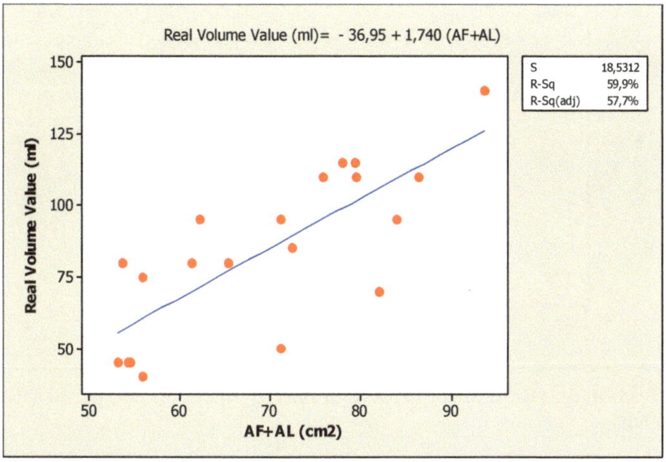

Fig. 20. The regression equation and graph obtained by using $A_F + A_L$ projection areas (Beyaz et al., 2009d).

It can be seen in Fig. 20., the regression coefficient was 59.9%. Fig. 21. shows the sum of the regression graph. Estimation coefficient was obtained as 71.1% depending on the regression equation.

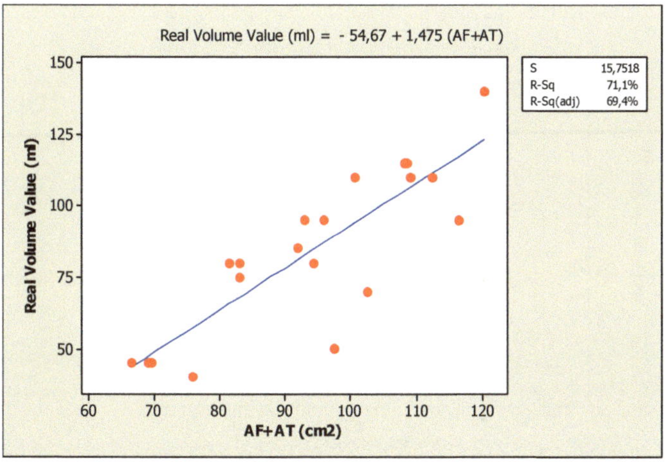

Fig. 21. The regression equation and graph obtained by using $A_F + A_T$ projection areas (Beyaz et al., 2009d).

Fig. 22. shows estimated regression graph obtained from sum of three (front, left and top) projections area values and also describes the regression coefficient 73.9%. This value for the regression gave the second highest value.

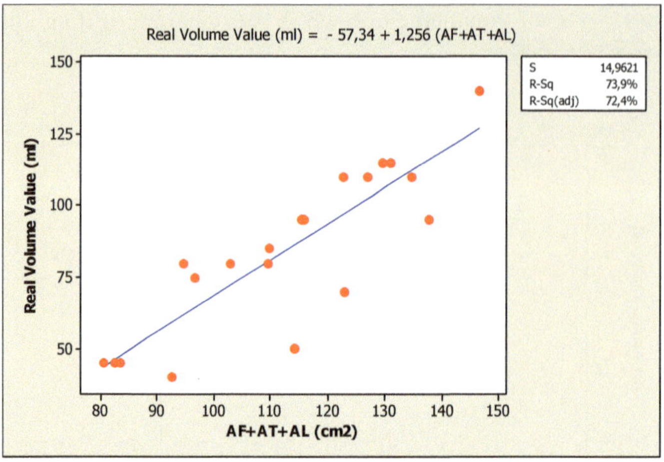

Fig. 22. The regression equation and graph obtained by using $A_F + A_T + A_L$ projection areas (Beyaz et al., 2009d).

The volume estimation with two projection areas, the highest rate of regression estimation value was 74.7% which obtained from the sum of the top and left projection areas. It can be explained by the front view of Kahramanmaras red peppers is more non uniform than top view.Sum of left-top projection area equation which gave the highest regression coefficient was used for estimate volume values. Then relationship between the real volume of peppers and estimated volume values has been compared. The results have been showed in Fig. 23 (Beyaz et al., 2009d).

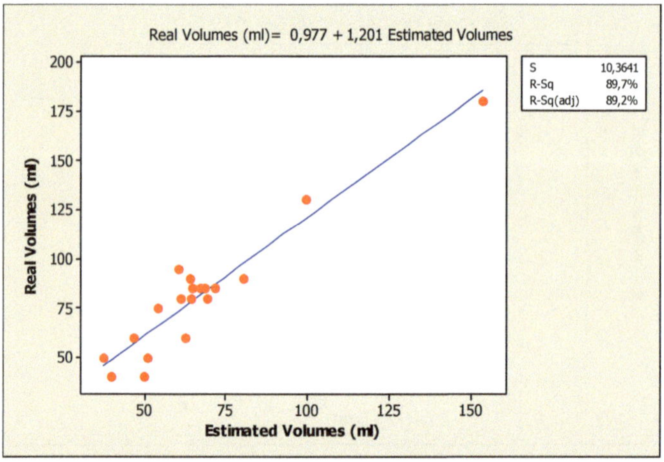

Fig. 23. Comparison of the actual volume and regression equation estimation results by using sum of $A_T + A_L$ projection areas (Beyaz et al., 2009d).

It has been pointed that Fig. 23 the regression coefficient values between the real peppers volume value and the estimated volume values had been found as 89.7%. This volume estimation rate can be used as, before the harvest as a maturity index, after the harvest as a classification and packaging parameter. Overall, at irregular shape products such as pepper is expected to be high regression coefficient relationship between volume and weight. However, classification and packaging system by using image analysis techniques according to the quality of the product is also possible. In this regard research has router features. According to the results of image processing method volume of Kahramanmaras red pepper seems to be appropriate for volume estimation (Beyaz et al., 2009d).

5.1.2 Image analyze technique for measurements of apple and peach tree canopy volume

Image analysis measurements and the real manual measurement method had been compared of apple and peach trees canopy volume in this study. The known of the tree canopy volumes are important from the point of views uniform growing, yield estimation, fertilizers and chemical applications. Also this research gives us some information about tree height, skirt height, parallel diameter. And also we can use these parameters for designing harvesting machines. In this research, totally twenty trees which has different canopy heights and volumes have been used. Apple and peach trees randomly have been selected from apple and peach garden belongs to Ankara University Agriculture Faculty. After the taking photographs of all trees by using digital camera, overall canopy height above the ground, canopy diameter parallel to row near ground level, height to point of maximum canopy diameter, height from ground to canopy skirt data have been obtained from these digital images. Apple and peach canopy volume have been calculated by using a formula that is named Albrigo from the prolate spheroid canopy volume values. The real values and image analysis measurements have been provided dimensions for computing the canopy volume where as image analysis measurements gave information that could be used to compute a image analysis canopy volume index. Tables show manual measurements of the 10 trees for each tree varieties and their corresponding canopy volumes, which were computed using prolate spheroid canopy volume formula (Table 1- 4) (Beyaz et al., 2009c).

Tree ID	H_t (m)	H_c (m)	H_s (m)	D (m)	PS_{CV} (m^3)
1	2,67	1,65	0,67	1,36	2,41
2	2,76	1,48	0,46	2,43	8,68
3	2,65	1,61	0,62	1,73	3,95
4	2,99	1,57	0,70	1,90	5,14
5	2,69	1,58	0,56	2,2	6,68
6	2,66	1,62	0,87	2,49	7,02
7	2,80	1,57	0,64	2,16	6,40
8	2,45	1,58	0,64	2,31	6,36
9	2,45	1,71	0,53	1,6	3,36
10	2,66	1,60	0,56	1,78	4,34
Min	2,45	1,48	0,46	1,36	2,41
Max	2,99	1,71	0,87	2,49	8,68
Mean	2,69	1,60	0,63	2,00	5,43
S.D.	0,53	0,06	0,11	0,38	1,92

Table 1. Two dimensions and canopy volumes calculated using prolate spheroid canopy volume formula from real tree measurements at apple trees (Beyaz et al., 2009c).

Tree ID	H_t (m)	H_c (m)	H_s (m)	D (m)	PS_{CV} (m^3)
1	2,75	1,51	0,67	1,36	2,41
2	2,76	1,54	0,58	2,58	9,26
3	2,34	1,60	0,48	1,66	3,48
4	2,88	1,68	0,66	2,05	6,00
5	2,58	1,40	0,46	2,21	6,62
6	2,53	1,68	0,87	2,49	6,70
7	2,57	1,44	0,64	2,16	5,68
8	2,32	1,41	0,64	2,31	5,76
9	2,22	1,70	0,46	1,60	3,18
10	2,87	1,82	0,48	1,78	5,07
Min	2,22	1,40	0,46	1,36	2,41
Max	2,88	1,82	0,87	2,58	9,26
Mean	2,58	1,58	0,59	2,02	5,42
S.D.	0,53	0,14	0,13	0,40	2,00

Table 2. Two dimensions and canopy volumes calculated using prolate spheroid canopy volume formulae from captured image measurements at apple trees (Beyaz et al., 2009c).

Tree ID	H_t (m)	H_c (m)	H_s (m)	D (m)	PS_{CV} (m^3)
1	2,44	1,76	0,40	2,80	11,16
2	2,66	1,45	0,50	3,00	12,41
3	2,70	1,40	0,58	2,58	8,81
4	1,72	1,20	0,60	1,07	0,85
5	2,59	1,55	0,43	2,60	9,62
6	2,35	1,29	0,60	3,20	11,22
7	1,54	1,04	0,39	1,53	1,80
8	2,36	1,08	0,46	1,19	1,63
9	1,40	1,00	0,22	1,29	1,36
10	1,44	0,92	0,28	1,26	1,22
Min	1,40	0,92	0,22	1,07	0,85
Max	2,70	1,76	0,60	3,20	12,41
Mean	2,12	1,27	0,45	2,05	6,01
S.D.	0,53	0,27	0,13	0,85	4,98

Table 3. Two dimensions and canopy volumes calculated using prolate spheroid canopy volume formulae from real tree measurements at peach trees (Beyaz et al., 2009c).

Tree ID	H_t (m)	H_c (m)	H_s (m)	D (m)	PS_{CV} (m^3)
1	2,54	1,72	0,33	3,15	15,08
2	2,77	1,51	0,38	3,17	15,54
3	2,77	1,52	0,64	2,68	9,66
4	1,86	1,04	0,49	1,25	1,34
5	2,52	1,56	0,38	2,48	8,78
6	2,49	1,25	0,47	3,38	14,40
7	1,56	1,01	0,31	1,57	2,06
8	2,22	1,04	0,44	1,16	1,46
9	1,41	0,8	0,3	2,07	3,04
10	1,39	0,87	0,3	1,19	1,01
Min	1,39	0,8	0,3	1,16	1,01
Max	2,77	1,72	0,64	3,38	15,54
Mean	2,153	1,23	0,4	2,21	7,24
S.D.	0,53	0,32	0,11	0,88	6,15

Table 4. Two dimensions and canopy volumes calculated using prolate spheroid canopy volume formulae from captured image measurements at peach trees (Beyaz et al., 2009c).

Table 1. and Table 3. shows that skirt height for most trees was about 0,63 m at apple trees and 0,43 m at peach trees respectively. Tables also suggest that the difference in volumes for tree each 10 trees. That difference was because the rest of the trees had almost the same parallel (D) diameters. Possibly the captured image measurements provide the easy canopy volume estimation because of fast tree dimensions.

Real tree volumes and tree image volumes regressions have presented in Fig. 24. and 25. Regression coefficient have been calculated as 95,6% for apple tree volume and 96,6% for peach tree volume.

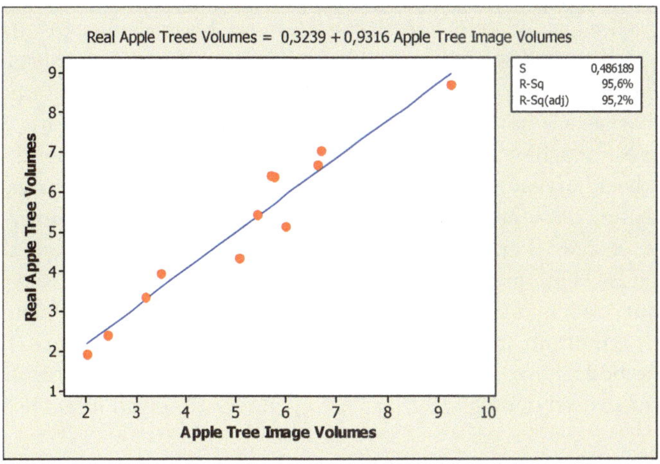

Fig. 24. Real apple tree volumes and apple tree image volumes regression table (Beyaz et al., 2009c).

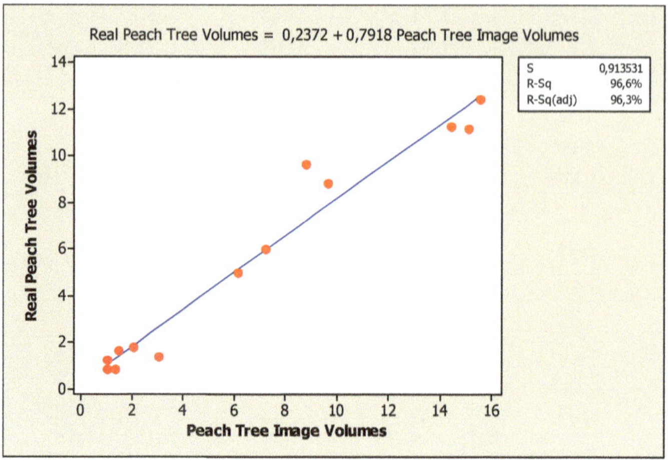

Fig. 25. Real peach tree volumes and peach tree image volumes regression table (Beyaz et al., 2009c).

5.1.3 Determination of sugar beet topping slice thickness by using image analysis technique

Turkey is one of the important sugar beet producers in the world. Turkey has produced 15 488 332 tones sugar beet from 3 219, 806 ha production area and also produced 2 061 000 tones sugar in 2008. Sugar beet harvesting by machine is general in Turkey. During mechanical harvesting several mechanical loads have caused skin and tissue damages of sugar beet. The damages of skin and tissue at sugar beets results in quantitative and qualitative losses. After harvesting, comparisons of the quality of sugar beets as good and bad are important in factory entrance terms of sugar losses. In this study widely used three types of harvesters have been operated. One of these machines is full hydraulic sugar beet harvest machine with arrangeable depth and row, second one is semi-hydraulic sugar beet harvest machine and the last one is full mechanic sugar beet harvest machine. Harvesters have been tried at the same field conditions during September-October months of 2009. Performance values have been obtained from three different evaluation methods. These methods are topping quality determination, the determination of sugar beet injury rate and the soil removal rate of the sugar beet. The determinations of these factors are important to obtain the optimum harvest performance. Image process and analysis methods have been used for evalutions. Descriptive statistics of image processing and measured damaged surface area features are given in Table 5 (Beyaz et al., 2009b).

Correlation coefficients between image processing and measured values at sugar beet head diameter have been determined as 0.98, between length values 0.96 and between surface damage values 0.97. Total yield loss and topping losses differences related to harvest machines are given in Table 6 (Beyaz et al., 2009b).

Variable	Sugar Beet Harvest Machines	Mean	SE Mean
Measured surface area values (mm^2)	Mechanical	685.0	161.0
	Semi-hydraulic	598.0	143.0
	Full-hydraulic	542.2	68.5
Image processing surface area values (mm^2)	Mechanical	540.0	130.0
	Semi-hydraulic	439.0	106.0
	Full-hydraulic	483.3	63.1

Table 5. Descriptive statistics of measured surface area values and image processing surface area values (Beyaz et al., 2009b).

Topping quality class	Sugar beet harvest machines	Percentage	Image processing percentage	Loss factor	Yield loss in each group	Image processing yield loss
1	Mechanical	2.7	0	0.1	0.27	0
	Semi-hyd.	18.6	2.32	0.1	1.86	0.23
	Full-hyd.	5.01	2.32	0.1	0.50	0.17
2	Mechanical	12.1	10.81	0.1	1.21	1.08
	Semi-hyd.	27.56	25.58	0.1	2.75	2.55
	Full-hyd.	26.59	25.58	0.1	2.65	2.83
3	Mechanical	31.64	29.73	0.05	1.58	1.48
	Semi-hyd.	5	30.23	0.05	0	1.51
	Full-hyd.	33.33	30.23	0.05	1.66	0
4	Mechanical	45.46	44.54	0	0	0
	Semi-hyd.	48.84	41.86	0	0	0
	Full-hyd.	35.07	41.86	0	0	1.66
5	Mechanical	8.1	9.51	0.1	0.81	0.95
	Semi-hyd.	0	0	0.1	0	0
	Full-hyd.	0	0	0.1	0	0
6	Mechanical	0	5.40	0.05	0	0.27
	Semi-hyd.	0	0	0.05	0	0
	Full-hyd.	0	0	0.05	0	0

Table 6. Topping yield losses related to harvest machines (Beyaz et al., 2009b).

As shown in Table 6., the total topping yield loss and image processing measurement values have been evaluated. This results respectively, at the mechanical harvesting machine 3.87% and 3.78%, at the semi-hydraulic machine 4.86% and 4.29%, at the fully hydraulic machine 4.81% and 4.66%. According to IIRB ideal topping quality is group is 4. According to the total topping yield loss values, the best machine is mechanical sugar beet harvesting machine (Beyaz et al., 2009b).

5.1.4 Assessment of mechanical damage on apples with image analysis

The aim of the study was to determine oxidation area and skin colour change at the damaged apples by using image analysis technique. Golden Delicious, Granny Smith and Stark Crimson apple varieties were examined with a special pendulum unit which was developed for the impact test. Plastic and wood materials were used as impact surface, and apples were dropped from different heights to the impact surface. Impact to each apple occurred perpendicular to impact surface from predetermined heights. After the impact test, colour change of damaged regions and oxidation areas of each sample were estimated by using a digital camera. Colour measurements were done with a colourmeter based on L*, a*, b* measurement system. The oxidation areas were evaluated clearly after these processes (Beyaz et al., 2010).

Measurements of the apple physical properties which were used for the tests are given in Table 7. The impact tests applied to the three apple varieties changed colour values. Changes of colour values for each apple varieties for five days are presented in Table 8 (Beyaz et al., 2010).

Repeated ANOVA statistics was used for evaluating colour interaction between other variables. When we look at the results of interactions for each of L*, a*, b* values, interaction between the day and apple varieties was significant ($p<0.05$) for all values. The interaction values are presented in Tables 9 - 12. Additionally, interaction between the day and surface variety was found to be significant only for b value presented in Table 12 (Beyaz et al., 2010).

Properties	Golden Delicious	Granny Smith	Stark Crimson
Fruit diameter (mm)	67.08 + 1.48	69.46 + 2.22	70.38+ 2.91
Weight (g)	138.63 + 5.56	201.88 + 5.65	158.63+ 7.68
Geometric mean diameter (mm)	63.74 + 1.33	66.39 + 1.92	66.91 + 2.35
Sphericaly (%)	97.64 + 1.68	96.38 + 2.48	97.36 + 2.94
Hardness of fruit (kg)	3.35 + 0.31	7.52 + 0.44	3.97 + 1.01
L* colour value	75.24 + 2.49	64.53 + 3.69	54.01 + 7.82
a* colour value	-10.7 + 3.45	-16.83 + 8.24	23.39 + 10.47
b* colour value	45.22 + 2.06	45.01 + 2.28	27.03 + 6.19

L*: Brightness of colour, a*: Colour change value from red to green, b*: Colour change value from yellow to blue.

Table 7. Physical properties of apples which were used for the tests (Beyaz et al., 2010).

Golden Delicious		Granny Smith		Stark Crimson	
Days	ΔE	Days	ΔE	Days	ΔE
1. Day	23.48	1. Day	12.81	1. Day	7.44
2. Day	31.52	2. Day	14.11	2. Day	9.32
3. Day	35.59	3. Day	15.61	3. Day	11.45
4. Day	37.60	4. Day	16.85	4. Day	14.02
5. Day	38.30	5. Day	16.96	5. Day	14.65

Table 8. Colour value change for each apple varieties (Beyaz et al., 2010).

Days	Variable		
	Golden Delicious	Granny Smith	Stark Crimson
Day 1	57.40+0.68 Aa	53.60+1.92 Ab	45.63+1.23 Ac
Day 2	51.21+0.94 Ba	53.36+0.63 ABa	44.63+1.22 ABb
Day 3	48.44+1.17 Cb	52.69+0.76 ABa	43.49+1.26 BCc
Day 4	46.96+1.17 Cb	51.37+0.93 Ba	41.70+1.19 CDc
Day 5	46.49+1.30 Cb	51.37+0.94 Ba	41.31+1.17 Dc

Table 9. The interaction between the apple varieties and days for L* value (Beyaz et al., 2010).

Days	Variable		
	Golden Delicious	Granny Smith	Stark Crimson
Day 1	-3.02+0.58 Cb	-12.66+0.81 Bc	20.72+0.95 Aa
Day 2	0.56+0.69 Bb	-11.66+0.75 Bc	19.22+0.92 Ba
Day 3	1.95+0.69 Ab	-10.10+1.15 Ac	17.88+1.01 Ca
Day 4	2.90+0.61 Ab	-10.02+0.98 Ac	16.61+1.09 Da
Day 5	3.00+0.68 Ab	-9.96+0.92 Ac	16.37+1.24 Da

Table 10. The interaction between the apple varieties and days for a* value (Beyaz et al., 2010).

Days	Variable		
	Golden Delicious	Granny Smith	Stark Crimson
Day 1	35.48+0.54 Aa	31.00+0.15 Ab	22.28+1.25 Ac
Day 2	31.77+0.84 Ba	29.71+0.62 ABa	21.53+1.24 ABb
Day 3	28.87+1.12 Ca	29.25+0.76 Ba	20.33+1.28 BCb
Day 4	27.90+1.27 CDa	28.67+0.94 Ba	18.91+1.27 CDb
Day 5	27.27+1.44 Da	28.61+0.90 Ba	18.46+1.39 Db

Table 11. The interaction between the apple varieties and days for b* value (Beyaz et al., 2010).

Days	Surface	
	Wood	Plastic
Day 1	30.24+0.93 Aa	28.93+1.18 Aa
Day 2	28.02+0.88 Ba	27.32+1.12 Ba
Day 3	26.08+1.00 Ca	26.22+1.14 Ca
Day 4	25.29+1.09 CDa	25.03+1.23 Da
Day 5	24.47+1.16 Da	25.08+1.30 Da

Table 12. The interaction between the days and surface varieties for b* value (Beyaz et al., 2010).

When the L*, a*, b* values of Granny Smith varieties were reviewed, L* and a* values increased while b* value decreased (Fig.11). For Stark Crimson variety all L*, a*, b* values decreased. The L*, a*, b* values in Golden Delicious varieties revealed that L* and b* values decreased while a* value increased. The L* value for each Golden Delicious varieties decreased by 65%. The colour value of b* decreased by 60% for Golden Delicious variety among all varieties. Total colour changes in Golden Delicious varieties were significantly different compared to Granny Smith and Stark Crimson indicated in Table 8 (Beyaz et al., 2010).

According to variance analysis technique, two levels of surfaces (wood and plastic surfaces), four levels of drop heights (50, 55, 60 and 65 cm) as height factor and three apple varieties (Golden Delicious, Granny Smith, Stark Crimson) were analysed. The number of irritation had five subgroups. Duncan's multiple range test was used to determine the mean difference between variety, impact surfaces and drop heights. Impact energy was taken as feature covariant (Beyaz et al., 2010).

Statistical analysis and results of variance analysis showed that height and 'Surface x Variety' interaction were significant at $p<0.01$ confidence level. All the variance analysis for 'Surface x Variety' interaction was found to be significant statistically (Beyaz et al., 2010).

The descriptive statistics in the variance analysis are given in Table 13. Duncan's test results of the average of damaged apples were significant at 50, 55 and 60 cm heights. The relations between 60 and 65 cm heights were insignificant (Beyaz et al., 2010).

The descriptive statistics for bruised area and drop height.is presented in Table 13.

Variable	Drop Height (cm)	Average (cm^2)	Std. Error
Impacted Surface Area	50	5.419 A	0.241
	55	4.616 B	0.255
	60	4.225 C	0.311
	65	4.616 C	0.292

Table 13. The descriptive statistics for bruised area and drop height (Beyaz et al., 2010).

The area of damaged surfaces and descriptive statistics for the apple varieties is presented in Table 14. The average plastic surface results of the apple varieties were not observed to be significant statistically. The results of the average impacted surface area of apple varieties for the wood surface are also presented in Table 14. Relations between the Golden Delicious and Stark Crimson varieties had no significant difference, but the Granny Smith variety had a significant difference from the others.

Surface	Varieties	Average of Observations (cm^2)	Std.Error	Std. Deviation
Plastic	Golden 20	3.03 A	0.287	1.282
Impact	Granny 20	4.957 A	0.447	1.998
Surface	Stark 20	4.151 A	0.253	1.131
Wood	Golden 20	5.141 A	0.225	1.006
Impact	Granny 20	5.005 B	0.239	1.07
Surface	Stark 20	6.029 A	0.174	0.777

Table 14. Impacted surface area results for plastic and wood surfaces (Beyaz et al., 2010).

Apple varieties and impact surface interaction were compared and presented in Table 15. The differences between average surface areas were observed significant as indicated in Table 15. for Golden Delicious and Stark Crimson apple varieties.

Varieties	Surface	Average of Observations (cm^2)	Std. Error	Std. Deviation
Golden Delicious Apple	Plastic 20	3.030 B	0.287	1.282
	Wood 20	5.141 A	0.225	1.006
Granny Smith Apple	Plastic 20	4.957 A	0.447	1.998
	Wood 20	5.005 A	0.239	1.07
Stark Crimson Apple	Plastic 20	4.151 B	0.253	1.131
	Wood 20	6.029 A	0.174	0.777

Table 15. Impacted surface area descriptive statistics for Golden Delicious apple varieties, Granny Smith apple varieties and Stark Crimson apple varieties (Beyaz et al., 2010).

Image analysis of the oxidation areas (bruised area) which were created after impact test is important for designing equipments and machines for classification, packaging, transportation and transmission. The results of these impaction tests indicated that the response to impact bruising was dependent on flesh firmness and on the height of the impact (Beyaz et al., 2010).

When impacted apples were analysed according to Duncan's test, differences between 50, 55 and 60 cm heights were significant but difference between 60 and 65 cm was not significant.

We can conclude that differences between average of impacted surface areas at Golden Delicious and Stark Crimson apples were important but this was not the case for the Granny Smith apples. This result shows us that apple varieties, impact surface and tissue of apple varieties affect average of impacted surface area (Beyaz et al., 2010).

For the surface properties, wood surface creates more impacts on apples than plastic surface because of material properties. According to the results, plastic surface affects in a good way apple surfaces which are harvested or processed for postharvest. In this research we have shown that image analysis technique can easily be used for determining impacted surface areas and changes of colour values of the apples and other agricultural products (Beyaz et al., 2010).

6. Future of machine vision measurement technology

The goal of machine vision research is to provide computers with humanlike perception capabilities so that they can sense the environment, understand the sensed data, take appropriate actions, and learn from this experience in order to enhance future performance. The field has evolved from the application of classical pattern recognition and image processing methods to advanced techniques in image understanding like model-based and knowledge-based vision.

In recent years, there has been an increased demand for machine vision systems to address "real-world" problems. The current state-of-the-art in machine vision needs significant advancements to deal with real-world applications, such as agricultural navigation systems, target recognition, manufacturing, photo interpretation, remote sensing, etc. It is widely understood that many of these applications require vision algorithms and systems to work under partial occlusion, possibly under high clutter, low contrast, and changing environmental conditions. This requires that the vision techniques should be robust and flexible to optimize performance in a given scenario (Sebe et al., 2005).

7. Acknowledgement

I wish to personally thank the following people for their contributions to my inspiration and knowledge and other help in creating this chapter:

- Prof. Dr. Ramazan OZTURK
- Prof. Dr. Ali Ihsan ACAR
- Prof. Dr. Mustafa VATANDAS
- Assistant Prof. Dr. Ufuk TURKER
- MSc. Babak TALEBPOUR

8. References

Anonymous. (2011a). FOOMA Japan 3D Measurement System That It Developed Jointly with the National Institute of Advanced Industrial Science and Technology, Date of access: 29.11.2011, Available from: http://www.diginfo.tv/2011/06/20/11-0123-d-en.php

Anonymous. (2011b). History of Digital Image Processing, Date of access: 16.10.2011, Available from: http://en.wikipedia.org/wiki/Digital_image_processing

Anonymous. (2011c). Lab Colour Space, Date of access: 11.10.2011 Available from: http://en.wikipedia.org/wiki/Lab_color_space

Anonymous. (2011d). Matlab Thresholding Tool, Date of access: 16.10.2011, Available from: http://www.mathworks.com/matlabcentral/fileexchange/6770

Beyaz, A. (2009a). Determining of Mechanical Damage on Apples by Using Image Analyse Technique, *Ankara University, Agriculture Faculty, Department of Agricultural Machinery, Master Thesis.*

Beyaz, A., Colak, A., Ozturk, R., Acar, A. I. (2009b). Determination of Sugar Beet Topping Slice Thickness by Using Image Analysis Technique, *Journal of Agricaltural Mechinary Science,* Volume 6, Number 3, pp.185 - 189

Beyaz, A., Ozguven, M. M., Ozturk, R., Acar, A. I. (2009c). Image Analyze Technique for Measurements of Apple and Peach Tree Canopy Volume, *Energy Efficiency and Agricultural Engineering Fourth International Scientific Conferance, Rousse/Bulgaria* - (01-03.10.2009), pp.22 - 30, ISSN 1311 - 9974

Beyaz, A., Ozguven, M. M., Ozturk, R., Acar, A. I. (2009d). Volume Determination of Kahramanmaras Red Pepper (Capsicum Annuum L.) by Using Image Analysis Technique, *Journal of Agricaltural Mechinary Science,* Volume 5, Number 1, pp.103 - 108

Beyaz, A., Ozturk, R., Acar, A. I., Turker, U. (2011). Determination of Enzymatic Browning on Quinces (Cydonia Oblongo) with Image Analysis, *Journal of Agricaltural Machinery Science,* Volume 7, Number, pp. 411 - 414

Beyaz, A., Ozturk, R., Turker, U. (2010). Assessment of Mechanical Damage on Apples with Image Analysis, *Journal: Food, Agriculture & Environment (JFAE),* Vol. 8, Issue 3&4, pp. 476-480 , Online ISSN: 1459-0263, Publisher: WFL.

Jähne, B. and Haußecker, H., (2000). *Computer Vision and Applications- A Guide for Students and Practitioners,* ISBN: 0-12-379777-2, Academic Press, Printed in the United States of America

Lukac, R., Plataniotis, K. N. (2007). *Colour Image Processing Methods and Applications (Image Processing),* ISBN:13: 978-0-8493-9774-5 (Hardcover), CRC Press, Printed in the Canada

Mohsenin, N. N. (1970). *Physical Properties of Plant and Animal Materials,* ISBN: 0677023006, Gordon and Breach Science Publishers, Printed in the United States of America

Nixon, M. S. and Aguado, A. S. (2002). *Feature Extraction and Image Processing,* ISBN: 0 7506 5078 8, An imprint of Butterworth-Heinemann Linacre House a member of the Reed Elsevier plc group, Printed and bound in Great Britain

Russ, J. C., (2006).*The Image Processing Handbook, Fifth Edition (Image Processing Handbook),* ISBN 0-8493-7254-2, CRC Press, Printed in the United States of America

Sebe, N., Cohen, I., Garg, A., Huang, T. S. (2005). *Machine Learning in Computer Vision,* ISBN-13 978-1-4020-3275-2 Springer Dordrecht, Berlin, Heidelberg, New York, Published by Springer, Printed in the Netherlands

10

An Analysis of the Interaction of Electromagnetic and Thermal Fields with Materials Based on Fluctuations and Entropy Production

James Baker-Jarvis
Electromagnetics Division, National Institute of Standards and Technology, Boulder, CO USA

1. Introduction

This paper is an attempt to introduce into measurement science a very unique and unified entropy-production-based method to characterize material parameters, equations of motion, and the related fluctuation-dissipation expressions for electrical and thermal transport properties such as conductivity, noise, and mobility. The approach is general enough to be used to study processes beyond equilibrium and yet it yields the normal transport coefficients near equilibrium. We will emphasize electromagnetic applications and heat transfer.

Transport coefficients are related to fluctuation-dissipation relations (FDRs). These include the Einstein relation, permittivity, resistance, noise, mobility, conductivity, power, and viscosity. In micrometer to nanoscale measurements, FDRs become crucial for modeling and to enhance an understanding of the property being measured.

The method we use in this paper is a projection-operator statistical mechanical approach. The background of this approach has been published and is summarized Baker-Jarvis and Kabos (2001). However, the present paper presents a unified approach that could be applied to a plethora of problems, near or far from equilibrium. The projection-operator approach was pioneered by Mori (1965); Zwanzig (1960). The theoretical approach used here has its roots in the work of Robertson that was based on a generalization and extension of the work of Zwanzig (1960), Rau and Müller (1996); Robertson (1966; 1999). Robertson's theory uses expected values of relevant variables and a nonequilibrium entropy for a dynamically driven system. The results reduce to the relevant thermodynamic potentials, forces, and entropy in the equilibrium limit. The advantage of this approach in studying time evolution of relevant variables is that the equations incorporate both relevant and irrelevant information, are exact, are Hamiltonian-based, have a direct relation to thermodynamics, and are based on reversible microscopic equations.

The system is described by a set of relevant variables, but in order to maintain an exact solution to Liouville's equation, irrelevant information is incorporated by the use of a projection-like operator. This correction for the irrelevant variables manifests and defines relaxation and dissipation Weiss (1999). A common argument about the projection-operator theories is that we do not yet know how to model them in numerical simulators; however,

Nettleton (1999) has made significant progress in this regard, and Eq.(18) of this paper eliminates the explicit projection-like operator in the equations of motion.

The approach in Robertson (1966; 1993) develops the full density operator $\rho(t)$ in terms of the relevant canonical density operator $\sigma(t)$ that is developed from constraints on relevant variables only, plus a relaxation correction term that accounts for irrelevant information. The statistical-density operator $\rho(t)$ satisfies the Liouville equation, whereas the relevant canonical-density operator $\sigma(t)$ does not. By including the irrelevant information by the projection-like operator, this approach is exact and time-symmetric. This theory yields an expression that exhibits all the required properties of a nonequilibrium entropy, and yet is based entirely on time-symmetric equations. In the past, other researchers have developed nonequilibrium statistical mechanical theories by adding a source term to Liouville's equation as in Zubarev et al. (1996), but our approach used here requires no source term. This approach has been used previously to study the microscopic time evolution of electromagnetic properties Baker-Jarvis (2008); Baker-Jarvis and Kabos (2001); Baker-Jarvis et al. (2004); Baker-Jarvis and Surek (2009); Robertson (1967b). The theory can be formulated either quantum-mechanically or classically.

2. Theoretical background of the statistical-mechanical method

In the Robertson projection operator statistical mechanical formulation there are two density operators. The first is the full statistical-density operator $\rho(t)$ that encompasses all information of the system in relation to the Hamiltonian and that satisfies the Liouville equation:

$$d\rho/dt = -i\mathcal{L}(t)\rho(t) = \frac{1}{i\hbar}[\mathcal{H}(t), \rho(t)], \tag{1}$$

where $\mathcal{L}(t)$ is the time-dependent Liouville operator, $\mathcal{H}(t)$ is the time-dependent Hamiltonian, and $[,]$ denote commutator.

In addition to $\rho(t)$, a relevant canonical-density operator $\sigma(t)$ is constructed. Robertson chose to construct $\sigma(t)$ by maximizing the information entropy subject to a limited knowledge on a finite set of constraints on the expected values of operators that are contained in the Hamiltonian, at times t. In the equilibrium limit the expected values of the relevant operators are the thermodynamic potentials and associated generalized forces can be identified as thermodynamic forces. The entropy summarizes our state of uncertainty in the expected values of the relevant variables at time t and is

$$S(t) = -k_B Tr\left(\sigma(t)\ln\sigma(t)\right), \tag{2}$$

where k_B is Boltzmann's constant, Tr denotes trace and in a classical analysis will represent integration over phase variables. We need to note that other forms of the entropy other than Eq.(2) could be used to construct $\sigma(t)$ from the relevant variables contained in the Hamiltonian. The basic generalized thermodynamic quantities that we wish to determine are $< F_n(\mathbf{r}) >\equiv Tr(F_n(\mathbf{r})\rho(t))$. The constraints for the entropy are that the expectations of $F_n(\mathbf{r})$ under both $\sigma(t)$ and $\rho(t)$ are equal for all times

$$Tr(F_n(\mathbf{r})\rho(t)) = Tr(F_n(\mathbf{r})\sigma(t)) =< F_n(\mathbf{r}) > . \tag{3}$$

However note, and this is crucial, that the derivatives $d\sigma(t)/dt$ and $d\rho(t)/dt$ are not equal and also the derivatives of the expectations with respect to $\sigma(t)$ and $\rho(t)$ are not equal and the approach is to find the derivatives in terms of each other and thereby develop an equation of motion. Note that at this stage, neither λ nor $< F_n >$ are known. Throughout the paper unless otherwise noted, the brackets $<>$ indicate expectations with respect to $\sigma(t)$.

Maximization by the common variational procedure leads to the generalized canonical density

$$\sigma(t) = \exp\left(-\lambda(t) * F\right), \tag{4}$$

where we require

$$Tr(\exp\left(-\lambda(t) * F\right)) = 1. \tag{5}$$

We use the $*$ notation for scalar or vector F_n and λ_n

$$\lambda * F = \int d\mathbf{r} \sum_n \lambda_n(\mathbf{r}, t) \cdot F_n(\mathbf{r}). \tag{6}$$

$\lambda_n(\mathbf{r}, t)$, and $< F_n >$ are determined in terms of each other from simultaneous solution of Eqs.(3), and (4), and the equation of motion Eq.(13) that will be developed later in the section. This highlights the difference between this approach and the Jaynesian Maximum Entropy approach where $< F_n >$ would be assumed known and then the λ_n are determined. Even though both methods maximize entropy, the Robertson method does not assume that $< F_n >$ are known but determines them and λ_n by requiring them to in addition satisfy the equations of motion in addition to the constraint equations and thus incorporating irrelevant information and the Liouville, equation into the solution.

As an example involving electromagnetic driving, the relevant variables may be the microscopic internal energy density operator $u(\mathbf{r})$ and polarizations $\mathbf{p}(\mathbf{r})$ and $\mathbf{m}(\mathbf{r})$ with associated intensive quantities with λ's $1/k_B T$, and effective local fields $-\mathbf{E}_p/k_B T$ and $-\mathbf{H}_m/k_B T$. The generalized temperature is $1/\beta = (\hbar\omega/2)\coth(\hbar\omega/2k_B T)$. In a high-temperature approximation this reduces to $1/\beta \rightarrow k_B T$. We need to note that quantities such as internal energy and polarization energies are defined at equilibrium and as a system moves out of thermal equilibrium the interpretation of these quantities change.

The dynamical variables we use are a set of operators, or classically, a set of functions of phase $F_1(\mathbf{r}), F_2(\mathbf{r}), \cdots$. The expectations of these are the observable or measured quantities. For normalization, $F_0 = 1$ is included in the set. We will assume a normalization of the density function σ so that no λ_0 or F_0 is required. The operators $F_n(\mathbf{r})$ are functions of $\mathbf{r}$ and phase variables, but are not explicitly time dependent. The time dependence enters through the driving fields in the Hamiltonian and when the trace operation is performed. These operators are, for example, the microscopic internal-energy density $u(\mathbf{r})$, and the electromagnetic polarizations $\mathbf{m}(\mathbf{r})$ and $\mathbf{p}(\mathbf{r})$ or microscopic electromagnetic induction and electric fields $\mathbf{b}(\mathbf{r})$ and $\mathbf{e}(\mathbf{r})$. Associated with these operators are a set of generalized forces that represent generalized thermodynamic fields that are not operators and do not depend on phase, such as generalized temperature and local electromagnetic fields such as $\mathbf{E}_p(\mathbf{r}, t)$, $\mathbf{H}_m(\mathbf{r}, t)$, and temperature. In any complex system, in addition to the set of $F_n(\mathbf{r})$, there maybe other uncontrolled or unobserved variables that are categorized as irrelevant variables.

This Gibbsian form of the entropy appears to be very reasonable since the condition of maximal information entropy is the most unbiased and maximizes the uncertainty in $\sigma(t)$

consistent with the constraints at each point in time. If we choose not to use the form of Eq.(2) to generate $\sigma(t)$ from a set of constraints, then we must possess additional information beyond the set of constraints. The entropy in Eq.(2) will contain input from Liouville's equation and the Hamiltonian. The λ's and $< F_n(\mathbf{r}) >$ can only be determined by solving Eq.(3) in conjunction with equations of motion developed later in the paper. In Eq.(5), $\lambda(\mathbf{r},t)$ are generalized forces that are not functions of phase, are not operators, and are related to local nonquantized generalized forces, such as temperature and electromagnetic fields and together with $< F_n(\mathbf{r}) >$ are determined by the simultaneous solution of Eqs.(3) and (13). Notice that the information that is assumed to be known is only a subset of the total information that would be required to totally characterize a system. The irrelevant variables are included in the theory exactly by means of a projection-like operator, and these effects are also manifest in the generalized forces.

The dynamical evolution of the relevant variables is the reversible evolution through the Hamiltonian and is denoted by

$$\dot{F}_n(\mathbf{r}) \equiv i\mathcal{L}F_n(\mathbf{r}) = -\frac{1}{i\hbar}[\mathcal{H}(t), F_n(\mathbf{r})]. \tag{7}$$

Note the difference in sign from that of the statistical density operator evolution in Eq.(1). In addition to the dynamical evolution of the relevant variables, the full time derivative of the expected values of the relevant variables is also influenced by the irrelevant variables, and is manifested as dissipation. For an electromagnetic system with internal energy and microscopic fields $\mathbf{d}$ and $\mathbf{b}$ interacting with applied fields $\mathbf{E}$ and $\mathbf{H}$, the Hamiltonian is $\mathcal{H}(t) = \int d\mathbf{r}\{u(\mathbf{r}) - \mathbf{d}(\mathbf{r})\cdot\mathbf{E}(\mathbf{r},t) - \mathbf{b}(\mathbf{r})\cdot\mathbf{H}(\mathbf{r},t)\}$.

Robertson's approach is based on developing an exact integral equation for $\rho(t)$ in terms of $\sigma(t)$. If we use Oppenheim's extended initial condition, Oppenheim and Levine (1979), this relationship is

$$\rho(t) = \sigma(t) + \mathcal{T}(t,t_i)\chi(t_i) - \int_{t_i}^{t} d\tau \mathcal{T}(t,\tau)\{1 - P(\tau)\}i\mathcal{L}(\tau)\sigma(\tau), \tag{8}$$

where the initial condition at time t_i is $\chi(t_i) = \rho(t_i) - \sigma(t_i)$ (note that Oppenheim and Levine (1979) generalized the analysis of Robertson to include this more generalized initial condition). The easiest and generally valid approximation is to set $\mathcal{T}(t,t_i)\chi(t_i) = 0$, for example when $\rho(t_i) = \sigma(t_i)$ or letting $t_i \to -\infty$ and assume the density operators begin the same in the distant past. $\mathcal{T}$ is an evolution operator with $\mathcal{T}(t,t) = 1$ and satisfies

$$\frac{\partial \mathcal{T}(t,\tau)}{\partial \tau} = \mathcal{T}(t,\tau)(1 - P(\tau))i\mathcal{L}(\tau), \tag{9}$$

where $P(t)$ is a nonhermitian projection-like operator defined by the functional derivative

$$\begin{aligned} P(t)A &\equiv \sum_{n=1}^{m}\int d\mathbf{r}\frac{\delta\sigma(t)}{\delta < F_n(\mathbf{r}) >}Tr(F_n(\mathbf{r})A) \\ &= \sum_{m,n=1}^{p}\int d\mathbf{r}'\overline{F}_m(\mathbf{r}')\sigma(t) < F_m(\mathbf{r}')\overline{F}_n(\mathbf{r}) >^{-1} Tr(F_n(\mathbf{r})A)d\mathbf{r}, \end{aligned} \tag{10}$$

for any operator A (Robertson (1966)). As a consequence, $d\sigma/dt = P(t)d\rho/dt$. In Robertson's pioneering work he has shown that Eq.(10) is equivalent to the Kawasaki-Grunton and Grabert's projection operators, and is a generalization of the Mori and Zwanzig projection operators Robertson (1978). Although $P^2 = P$, it is not necessarily hermitian and as a consequence is not a projection operator and we term it a projection-like operator. In an open system, $\rho(t)$ does not evolve unitarily and $\rho(t)$ need not satisfy Eq.(1), but the theory developed in this paper can easily be modified by adding a source term for the interaction with a reservoir (Robertson and Mitchell (1971); Yu (2008)).

An important identity was proven previously in Oppenheim and Levine (1979); Rau and Müller (1996); Robertson (1993):

$$i\mathcal{L}\sigma(t) = -\lambda * \bar{\dot{F}}\sigma, \tag{11}$$

and

$$Tr(i\mathcal{L}\sigma(t)) = -\lambda * < \bar{\dot{F}} >= 0, \tag{12}$$

where the bar is the Kubo transform for any operator $A(\mathbf{r})$, where $\overline{A} = \int_0^1 \sigma^x(t)A\sigma^{-x}(t)dx$ Oppenheim and Levine (1979); Robertson (1967a). In a classical analysis, $\overline{A} = A$.

It has been shown previously that the exact time evolution of the relevant variables can be expressed for a dynamically driven system as Baker-Jarvis (2005; 2008); Oppenheim and Levine (1979); Robertson (1966)

$$\frac{\partial < F_n(\mathbf{r}) >}{\partial t} \equiv -\Delta\{F(\mathbf{r})\overline{F}(\mathbf{r}')\} * \frac{\partial\lambda(\mathbf{r}',t)}{\partial t} =< \dot{F}_n(\mathbf{r}) >$$

$$+Tr(\dot{F}_n(\mathbf{r})\mathcal{T}(t,t_i)\chi(t_i)) - \int_{t_i}^{t} Tr\left(i\mathcal{L}F_n(\mathbf{r})\mathcal{T}(t,\tau)(1-P(\tau))i\mathcal{L}\sigma(\tau)\right)d\tau. \tag{13}$$

$\Delta\{F\overline{F}\}$ can be related to material properties such as heat capacity, susceptibility, etc and the * operator is defined in Eq.(6), and $\Delta\{F\overline{F}\}$ is defined as $\Delta\{F\overline{F}\} =< F\overline{F} > - < F >< \overline{F} >$ as shown in Appendix 10. Note that Eq.(13) is time symmetric (invariant under $t \rightarrow -t$) and yet models dissipation by including the effects of the irrelevant variables. In many problems it is useful to use Eq.(11) in Eq.(13). The first term on the RHS is the reversible contribution, the second term is the initial-condition contribution, and the last term is due to dissipation. Equations (3) and (13) form a closed system of equations, and the procedure for determining the generalized forces in terms of $< F_n >$ is to solve Eqs.(3) and (13) simultaneously. For operators that are odd under time reversal, such as the magnetic moment, the first term on the right hand side of Eq.(13), the reversible term, is nonzero, whereas for functions even under time reversal, such as dielectric polarization and microscopic entropy, this term is zero. However, the third term in Eq.(13) in any dissipative system is nonzero. The relaxation correction term that appears in this formalism is essential and is a source of the time-dependence in the entropy rate. Although these equations are nonlinear, in many cases linear approximations have been successfully made Baker-Jarvis et al. (2007). For open systems, where there is a source that is not in the Hamiltonian, Eq.(13) is modified only by adding a source term Robertson and Mitchell (1971). An exact entropy-evolution equation can be derived from Eq.(13) and this equation will be a key element in this paper.

3. The entropy, entropy rate, and entropy production

Due to the invariance of the trace operation under unitary transformations, the Neumann entropy $-k_B Tr(\rho(t)\ln\rho(t))$, formed from the full statistical-density operator $\rho(t)$ that satisfies the Liouville Eq.(1) in an isolated system, is independent of time and cannot be a nonequilibrium entropy. However, the entropy $-k_B Tr(\sigma(t)\ln\sigma(t))$, is not constant in time and has all the properties of a nonequilibrium entropy, and reduces to the thermodynamic entropy in the appropriate limit. It is important to note that unlike energy, entropy is not a conserved quantity and can be produced in interactions.

We define the entropy from Eqs. (2) and (4)

$$S(t) \equiv k_B \lambda * < F >, \tag{14}$$

and microscopic dynamically-driven entropy rate from Eq.(11) as

$$\dot{s}(t) \equiv k_B \lambda * \dot{F} = k_B \lambda * i\mathcal{L}F. \tag{15}$$

The expected value of the dynamical evolution of the entropy rate vanishes due to Eq.(12) and invariance of the trace under cyclic permutations (for bounded operators) Oppenheim and Levine (1979):

$$< \bar{\dot{s}}(t) >= k_B Tr(\lambda * \bar{\dot{F}}\sigma) = k_B \lambda * Tr(\dot{F}\sigma) =< \dot{s}(t) >= 0. \tag{16}$$

Equation (16) is a result of the microreversibility of the dynamics of the relevant variables and is what would be expected for reversible microscopic equations of motion in a dynamically-driven, but otherwise isolated system with dynamical evolution of the relevant variables. The total entropy evolution equation that contains both relevant and irrelevant effects can be formed from Eq.(13) by multiplying by the $\lambda_n's$, summing, and integrating over space. The entropy evolution is

$$\frac{dS}{dt} - Tr(\chi(t_i)\dot{s}(t)\mathcal{T}(t,t_i)) \equiv \Sigma(t) = \frac{1}{k_B}\int_{t_i}^{t} \langle \dot{s}(t)\mathcal{T}(t,\tau)(1-P(\tau))\bar{\dot{s}}(\tau)\rangle \, d\tau. \tag{17}$$

This equation and Eq.(13) form the foundations of this paper. Note that the entropy rate is antisymmetric about the origin: $dS/dt(t = t_i) - Tr(\chi(t_i)\dot{s}(t_i)) = 0$, and without initial condition, $dS(-t)/dt = -dS(t)/dt$, and $\Sigma(t)$ is the entropy rate. The RHS relates to dissipation and is equivalent to Eq.(49) in Reference Baker-Jarvis and Surek (2009). The macroscopic entropy rate can be expressed in two equivalent forms ($dS(t)/dt \equiv -k_B\lambda * [\Delta\{F\bar{F}\} * \partial\lambda/\partial t]$, see Appendix 10 for definition of Δ), or $dS/dt \equiv k_B \lambda * \partial < F > /\partial t$. Rau and Müller (1996) note that for unbounded operators the cyclic invariance of the trace argument breaks down. The projection operator contribution in Eq.(17) can be re-expressed as (Robertson (1967b))

$$\Sigma(t) = \frac{dS}{dt} - Tr(\chi(t_i)\dot{s}(t)\mathcal{T}(t,t_i)) = \frac{1}{k_B}\int_{t_i}^{t} \langle \dot{s}(t)\mathcal{T}(t,\tau)\bar{\dot{s}}(\tau)\rangle \, d\tau$$
$$-\int_{t_i}^{t} \langle \dot{s}(t)\mathcal{T}(t,\tau)\bar{F}\rangle * < \dot{F} > * < F\bar{F} >^{-1} d\tau, \tag{18}$$

where we have eliminated $P(t)$ and avoided some of complications for simulations using the projection-like operator P. In linear driving the projection operator can be neglected. We see that only the variables that change sign under time reversal, such as magnetic moment, have expected values $< \dot{F}_n > \neq 0$. The even variables satisfy $< \dot{F}_n >= 0$. Equation (18 is related to the energy FDR relation developed by Berne and Harp (1970), but is based on entropy and not energy. These FDR equations reduce to the traditional FDR relations, such as Nyquist's theorem, or conductivity as we approach equilibrium. Equation (17) will form the basis of our applications to various electromagnetic driving and measurement problems. The LHS of Eq.(17) represents the macroscopic dissipation, and the last term on the RHS represents the fluctuations in terms of the microscopic entropy rate $\dot{s}(t)$. For almost all many-body systems, due to incomplete information, there are contributions from the positive semi-definite relaxation terms in Eq.(17). The projection operator, which models the state of knowledge of the system described by the relevant variables, acts to decrease the entropy rate and in the limit of no decoherence, Eq.(17) becomes $dS/dt \to 0$. For an open system Eq.(17) would be modified by adding a term for the entropy source.

To summarize, for a dynamically driven, but otherwise isolated, system, the expected value of the microscopic entropy rate ($< \dot{s}(t) >$) is zero, but the fluctuations in this variable are not zero. This is due to the microscopic reversibility of the underlying equations of motion. However, in a complex system there are other uncontrolled variables in addition to the relevant ones and as a consequence there is dissipation and irreversibility and a net entropy evolution. Whereas $\rho(t)$ satisfies Liouville's equation, $\sigma(t)$ does not, and this is a consequence of irrelevant variables. Equation (17) can model systems away from equilibrium. Transport coefficients and the related FDR relations for conductivity, susceptibility, noise, and other quantities follow naturally from Eq.(13).

The paper is organized as follows. The paper begins with an argument that the author's previously developed entropy-based fluctuation-dissipation equation (EFDR) can be used to obtain classical FDRs and be extended into the realm of nonequilibrium systems. The equations of motion are all time symmetric, in that if one transforms $t \to -t$, the equation remain the same Robertson (1999). We show that if we apply the Euler-Lagrange equations to the difference between the entropy production and dS/dt, or equivalently the net flux through the system, we obtain the statistical mechanical equations of motion. In the last sections of the paper, fluctuation-dissipation equations are derived from the equation for electrical conductivity, noise, thermal conductivity, and the determination of Boltzmann's constant.

4. Entropy rate fluctuation relation and transport coefficients

A generalized entropy rate EFDR relation that is valid arbitrarily far from equilibrium that was developed in recent papers Baker-Jarvis (2005; 2008); Baker-Jarvis and Surek (2009) is the equation of motion for the entropy in Eq.(17)

$$dS/dt = \underbrace{\frac{1}{k_B}\int_{t_i}^{t} \langle \dot{s}(t)\mathcal{T}(t,\tau)(1-P(\tau))\bar{\dot{s}}(\tau)\rangle \, d\tau + < \dot{s}(t) >}_{\text{fluctuation}}$$
$$= k_B \frac{\partial < F >}{\partial t} * \lambda = -k_B\lambda * \left[\Delta\{F\overline{F}\} * \frac{\partial \lambda}{\partial t}\right]. \tag{19}$$

Note that $< \dot{s}(t) >= 0$ for the set of λ satisfying Eqs.(13) and (4). This equation has units of entropy rate and the notation $\Delta\{F\overline{F}\}$ is defined above. The λ_n correspond to applied quantities such as electromagnetic fields, temperature, etc. The relevant variables F_n correspond to generalized thermodynamic potentials such as currents, polarizations, momentum, etc. The LHS relates to fluctuations and the RHS relates to dissipation as entropy production. The RHS was put into two distinct, but equivalent forms. Use of the different forms may be advantageous for different applications. The last term on the RHS is derived using the expression derived in Eq.(70) in Appendix A. The physical interpretation of Eq.(19) is that rate of production of entropy drives fluctuations in the microscopic entropy and fluctuations in the microscopic entropy rate drives the rate of entropy production. Note that temperature does not appear explicitly in the equation. Equation (19) is exact in both classical and quantum-mechanical contexts when driven by non-quantized fields. However, in Eq.(19) we have neglected the effects of the decaying initial condition term. For an open system where there is heat from a reservoir interacting with the system and is not treated as a term in the Hamiltonian, an entropy source term is added to RHS of Eq.(19).

For conserved quantities, Eq.(19) can be cast into another form in terms of generalized microscopic currents $\mathbf{j}_n(\mathbf{r},t)$ that satisfy the conservation conditions $\nabla \cdot \mathbf{j}_n(\mathbf{r},t) = -i\mathcal{L}F_n(\mathbf{r})$. Note $\mathbf{j}_n(\mathbf{r},t)$ have an explicit time dependence since the Hamiltonian is time dependent. Substituting this relationship into Eq.(19), and integrating by parts and discarding surface terms (entropy fluxes), we obtain the following form for the entropy production rate

$$\Sigma(t) = k_B \nabla\lambda(\mathbf{r},t) * \int_{t_i}^{t} \overset{\leftrightarrow}{K}_{\mathbf{j}}(\mathbf{r},t,\mathbf{r}',\tau) * \nabla\lambda(\mathbf{r}',\tau)d\tau \equiv \underbrace{\mathbf{J}(\mathbf{r},t)}_{\text{current}} * \underbrace{k_B \nabla\lambda(\mathbf{r},t)}_{\text{generalized force}}, \tag{20}$$

where

$$\overset{\leftrightarrow}{K}_{\mathbf{j}(mn)}(\mathbf{r},t,\mathbf{r}',\tau) =< \mathbf{j}_m(\mathbf{r},t)\mathcal{T}(t,\tau)(1-P(\tau))\mathbf{j}_n(\mathbf{r}',\tau) >, \tag{21}$$

$$\mathbf{J}(\mathbf{r},t) = \int_{t_i}^{t} \overset{\leftrightarrow}{K}_{\mathbf{j}}(\mathbf{r},t,\mathbf{r}',\tau) * \nabla\lambda(\mathbf{r}',\tau)d\tau \approx -\overset{\leftrightarrow}{L}_{mn}(\mathbf{r}) \cdot \nabla x, \tag{22}$$

where the last expression is in a linear approximation and ∇x is a driving force. Equation (20) is in the form of a flux times a generalized force, which is commonly identified with entropy production. Note we omitted a possible reversible term $< \dot{\mathbf{j}}_n >$.

A very general expression for the transport coefficients in the linear approximation is

$$\overset{\leftrightarrow}{L}_{mn}(\mathbf{r}) = \int_0^\infty \int d\mathbf{r}'\theta(\mathbf{r})\frac{< \mathbf{j}_m(\mathbf{r},0)\mathbf{j}_n(\mathbf{r}',s) >_0}{k_B T(\mathbf{r}')}ds. \tag{23}$$

To obtain the linear approximation, we write $\nabla\lambda = -\theta\nabla x/k_B T$. For example in heat transfer, $\lambda = 1/k_B T$ and $\theta = 1/T$ and the driving force is the temperature gradient $\nabla x = \nabla T$, and then the transport coefficient from Eq.(23) is the thermal conductivity defined in $\mathbf{J} = -\overset{\leftrightarrow}{\kappa} \cdot \nabla T$. As another example, if the electrical potential is ϕ and $\lambda = -\phi/k_B T$, where $\nabla x = \nabla\phi \approx -\mathbf{E}$, and $\theta = 1$. In this case the transport coefficient from Eq.(23) is the electrical conductivity defined in $\mathbf{J} = -\overset{\leftrightarrow}{\sigma}_f \cdot \nabla\phi$.

In the long-wavelength, short-relaxation time limit, $\overleftrightarrow{K}_{\mathbf{j}}(\mathbf{r},t,\mathbf{r}',\tau) = \overleftrightarrow{C}(\mathbf{r},\mathbf{r}',t,\tau)\delta(\mathbf{r}-\mathbf{r}')\delta(t-\tau)$ and then Eq.(20) for the entropy production rate can be written as

$$\Sigma(\mathbf{r},t) = k_B \nabla\lambda(\mathbf{r},t)\cdot \overleftrightarrow{C}(\mathbf{r},t)\cdot\nabla\lambda(\mathbf{r},t). \tag{24}$$

This is a positive semi-definite quantity.

4.1 Steady-state approximation to entropy production

In the case of a linear approximation to Eq.(19) with relevant variables F_n, and $\dot{F}_n = -\nabla\cdot\mathbf{j}_n$, where the system is driven by a constant field, such as a temperature gradient or electric field, we can write the time-independent entropy production as

$$\Sigma(\mathbf{r}) = k_B\lambda(\mathbf{r}) * \int_0^\infty \left\langle \dot{F}(\mathbf{r},0)\dot{F}(\mathbf{r}',\tau)\right\rangle_0 * \lambda(\mathbf{r}')d\tau$$
$$= k_B\nabla\lambda(\mathbf{r}) * \int_0^\infty < \mathbf{j}(\mathbf{r},0)\mathbf{j}(\mathbf{r}',\tau) >_0 d\tau * \nabla\lambda(\mathbf{r}'). \tag{25}$$

For a single variable this reduces to

$$\Sigma(\mathbf{r}) \approx k_B\theta(\mathbf{r})\nabla x(\mathbf{r}) * \int_0^\infty \frac{< \mathbf{j}(\mathbf{r},0)\mathbf{j}(\mathbf{r}',\tau) >_0}{k_B T} d\tau * \theta(\mathbf{r}')\nabla x(\mathbf{r}')$$
$$= \int \frac{\theta(\mathbf{r})}{T(\mathbf{r})}\nabla x(\mathbf{r})\cdot \overleftrightarrow{L}_{mn}(\mathbf{r})\cdot\nabla x(\mathbf{r})d\mathbf{r}. \tag{26}$$

The expectation is under the equilibrium distribution so the correlation function is assumed to be for a stationary system. In the case of steady-state heat transfer where $\theta = 1/T$, the entropy production density is

$$\Sigma_d(\mathbf{r}) = k_B\nabla\lambda\cdot\mathbf{J}_h = \underbrace{\frac{\nabla T(\mathbf{r})}{T^2(\mathbf{r})}}_{-\nabla\lambda}\cdot\underbrace{\overleftrightarrow{\kappa}(\mathbf{r})\cdot\nabla T(\mathbf{r})}_{-\mathbf{J}_h}. \tag{27}$$

For steady-state electrical dissipation where $\theta = 1$ (note the same equation would apply to the chemical potential μ_c instead of ϕ), we have

$$\Sigma_d(\mathbf{r}) = k_B\nabla\lambda\cdot\mathbf{J} = \frac{1}{T(\mathbf{r})}\left(\mathbf{E}(\mathbf{r}) - \frac{\phi(\mathbf{r})\nabla T(\mathbf{r})}{T(\mathbf{r})}\right)\cdot\overleftrightarrow{\sigma}_f(\mathbf{r})\cdot\left(\mathbf{E}(\mathbf{r}) - \frac{\phi(\mathbf{r})\nabla T(\mathbf{r})}{T(\mathbf{r})}\right). \tag{28}$$

5. Exact equations of motion

5.1 Heat transfer

For the first example let us consider heat transfer. For this case $\lambda = 1/k_B T$, $F = u(\mathbf{r})$, where $u(\mathbf{r})$ is the internal energy density. The equation of continuity in this case is $i\mathcal{L}u = \dot{u} = -\nabla\cdot\mathbf{j}_h$. $\lambda * (\Delta\{F\overline{F}\})$ relates to material parameters, in this case the heat capacity. The exact equation of motion, without a source, can be written using Eq.(13) as

$$\frac{\partial < u(\mathbf{r}) >}{\partial t} = \int d\mathbf{r}' \frac{1}{T(\mathbf{r}',t)} \frac{\Delta\{u(\mathbf{r})\overline{u}(\mathbf{r}')\}}{k_B T(\mathbf{r}',t)} \frac{\partial T}{\partial t}(\mathbf{r}',t)$$

$$= \nabla \cdot \left(\int_0^t d\tau \int d\mathbf{r}' \frac{\overleftrightarrow{K_h}(\mathbf{r},t,\mathbf{r}',\tau)}{k_B T^2(\mathbf{r}',\tau)} \cdot \nabla' T(\mathbf{r}',\tau) \right), \tag{29}$$

where $\overleftrightarrow{K_h}(\mathbf{r},t,\mathbf{r}',\tau) =< \mathbf{j}_h(\mathbf{r},t)\mathcal{T}(t,\tau)\mathbf{j}_h(\mathbf{r}',\tau) >$ and $< \mathbf{j}_h >= 0$. Note that Eq.(29) is invariant to the transformation $t \rightarrow -t$. This is not true for the normal heat transfer equation. In addition, due to the time integral, information is not transfered instantaneously, as it is in the normal heat equation where it is transfered without delay.

To see how Eq.(29) reverts to Fourier's equation, consider the case when $\frac{1}{k_B T^2(\mathbf{r}',t)} < u(\mathbf{r})u(\mathbf{r}') >\equiv \rho_d(\mathbf{r}',t)c_p(\mathbf{r}',t)\delta(\mathbf{r}-\mathbf{r}')$ and $\overleftrightarrow{K_h}(\mathbf{r},t,\mathbf{r}',\tau) \equiv k_B T^2(\mathbf{r}',\tau) \overleftrightarrow{\kappa}(\mathbf{r})\delta(\mathbf{r}-\mathbf{r}')\delta(t-\tau)$. This is a long wavelength and short relaxation time approximation. Note from Eq. (23) in the linear approximation we have an expression for the thermal conductivity as

$$\overleftrightarrow{\kappa}(\mathbf{r}) = \frac{\int d\mathbf{r}' \int_0^\infty < \mathbf{j}_h(\mathbf{r},0)\mathbf{j}_h(\mathbf{r}',\tau) >_0 d\tau}{k_B T^2}. \tag{30}$$

If we use this approximation for the heat capacity and thermal conductivity, we obtain the normal heat transport equation

$$\rho_d c_p \frac{\partial T(\mathbf{r},t)}{\partial t} = \nabla \cdot (\overleftrightarrow{\kappa}(\mathbf{r}) \cdot \nabla T(\mathbf{r},t)). \tag{31}$$

Note that the general equation for heat transfer, Eq.(29), is exact and time symmetric, but when the thermal conductivity tensor is defined as above, Eq.(29) loses time symmetry and is no longer invariant under the transformation $t \rightarrow -t$ (Robertson (1999)).

Equation(29) can obtain wave-like properties. To see this take the time derivative of Eq.(29), assuming the expectations are evaluated under an equilibrium distribution function

$$\int d\mathbf{r}' \frac{1}{T(\mathbf{r}',t)} \frac{\Delta\{u(\mathbf{r})\overline{u}(\mathbf{r}')\}_0}{k_B T(\mathbf{r}',t)} \frac{\partial^2 T}{\partial t^2}(\mathbf{r}',t)$$

$$-2\int d\mathbf{r}' \frac{1}{T^2(\mathbf{r}',t)} \frac{\Delta\{u(\mathbf{r})\overline{u}(\mathbf{r}')\}_0}{k_B T(\mathbf{r}',t)} \left(\frac{\partial T}{\partial t}(\mathbf{r}',t)\right)^2$$

$$= \nabla \cdot \left(\int d\mathbf{r}' \frac{\overleftrightarrow{K}_{h(0)}(\mathbf{r},t,\mathbf{r}',t)}{k_B T^2(\mathbf{r}',\tau)} \cdot \nabla' T(\mathbf{r}',t) \right)$$

$$+\nabla \cdot \left(\int_0^t d\tau \int d\mathbf{r}' \frac{\partial \overleftrightarrow{K}_{h(0)}(\mathbf{r},t,\mathbf{r}',\tau)}{\partial t} \cdot \frac{\nabla' T(\mathbf{r}',\tau)}{k_B T^2(\mathbf{r}',\tau)} \right). \tag{32}$$

On the first term on the RHS of Eq.(32) we used Eq.(31) to obtain $\partial T/\partial t$. In the long wavelength, but not short relaxation time special case, when the time dependence is modeled by an exponential decay, we can write $\overleftrightarrow{K_h}= T^2 k_B \overleftrightarrow{\kappa}(\mathbf{r}')\delta(\mathbf{r}-\mathbf{r}')\exp(-(t-\tau)/\tau_h)/\tau_h \overleftrightarrow{I}$.

Here τ_h is the characteristic relaxation time for heat transfer. We also use the approximation for the heat capacity, so that Eq.(32) becomes

$$\tau_h \rho c_p \frac{\partial^2 T}{\partial t^2} - \tau_h \rho c_p \frac{2}{T}\left(\frac{\partial T}{\partial t}\right)^2 + \rho c_p \frac{\partial T}{\partial t} = \nabla \cdot (\overleftrightarrow{\kappa}(\mathbf{r}) \cdot \nabla T(\mathbf{r}, t)). \tag{33}$$

This equation shows that the temperature can obtain wave properties due to the finite time it takes for the information to propagate through the material. This reverts to Fourier's heat equation Eq.(31) in the short relaxation time limit $\tau_h \rightarrow 0$.

5.2 Equation of motion for the entropy production in heat transfer

For heat transfer the entropy production-rate density with a source term using Eq.(19)

$$\int d\mathbf{r}' \frac{\Delta\{\frac{u(\mathbf{r})}{T(\mathbf{r},t)} \frac{\overline{u}(\mathbf{r}')}{T(\mathbf{r}',t)}\}}{k_B T(\mathbf{r}',t)} \frac{\partial T}{\partial t}(\mathbf{r}',t) - \frac{\nabla T(\mathbf{r},t)}{T^2(\mathbf{r},t)} \cdot \int_0^t d\tau \int d\mathbf{r}' \frac{\overleftrightarrow{K}_h(\mathbf{r},t,\mathbf{r}',\tau)}{k_B T(\mathbf{r}',\tau)} \cdot \frac{\nabla' T(\mathbf{r}',\tau)}{T(\mathbf{r}',\tau)} = \frac{g(\mathbf{r},t)}{T(\mathbf{r},t)}. \tag{34}$$

If we define $\mathbf{J}_h(\mathbf{r},t) = -\int_0^t d\tau \int d\mathbf{r}' \frac{\overleftrightarrow{K}_h(\mathbf{r},t,\mathbf{r}',\tau)}{k_B T(\mathbf{r}',\tau)} \cdot \frac{\nabla' T(\mathbf{r}',\tau)}{T(\mathbf{r}',\tau)}$, (which in a linear approximation is equal to $-\overleftrightarrow{\kappa}(\mathbf{r}) \cdot \nabla T/T$), then we can write Eq.(34) as an entropy-density balance equation

$$\frac{\partial S(\mathbf{r},t)}{\partial t} + \nabla \cdot \left(\frac{1}{T}\mathbf{J}_h(\mathbf{r},t)\right) = \frac{\nabla \cdot \mathbf{J}_h(\mathbf{r},t)}{T} + \frac{g(\mathbf{r},t)}{T(\mathbf{r},t)}. \tag{35}$$

5.3 Using the Euler-Lagrange equation applied to entropy production extrema to obtain exact equations of motion

The Euler-Lagrange functional used to search for an extremum is

$$I = \int_{\mathbf{r}_1}^{\mathbf{r}_2} L_e(\lambda(\mathbf{r},t), \nabla\lambda(\mathbf{r},t), \mathbf{r}) d\mathbf{r}. \tag{36}$$

A function that satisfies the Euler-Lagrange condition makes Eq.(36) an extremum when $\delta I = 0$. In order to study the Euler-Lagrange extremum problem in entropy production rate we consider systems where the relevant variables satisfy a microscopic conservation condition $\nabla \cdot \mathbf{j}(\mathbf{r},t) = -\dot{F}(\mathbf{r})$. Examples of this are heat transfer $\nabla \cdot \mathbf{j}_h(\mathbf{r},t) = -\dot{u}(\mathbf{r})$ and charge transfer where $\nabla \cdot \mathbf{j}_c(\mathbf{r},t) = -\dot{\rho}(\mathbf{r})$. We set I to the entropy production minus the total change in entropy density: $\int d\mathbf{r}(\Sigma_d(\mathbf{r},t) - dS_d(\mathbf{r},t)/dt)$, which is equivalent to the entropy flux through the system. Using the RHS of Eq.(19) with λ and $\nabla\lambda$ as variables for the case when the relevant variables satisfy $\dot{F}_n = -\nabla \cdot \mathbf{j}_n$ in the entropy production in Eq.(20) we have

$$I = \underbrace{\int k_B \sum_n \nabla\lambda_{t(n)}(\mathbf{r},t) \cdot \mathbf{J}_n(\mathbf{r},t) d\mathbf{r}}_{\text{entropy production}} + \underbrace{\int \sum_n \lambda_{t(n)}(\mathbf{r},t) \left[\Delta\{F_n(\mathbf{r})\overline{F}(\mathbf{r}')\} * \frac{\partial \lambda_t(\mathbf{r}',t)}{\partial t}\right] d\mathbf{r}}_{\text{-dS/dt}}, \tag{37}$$

where

$$\mathbf{J}(\mathbf{r},t) = \int_{t_i}^{t} \overleftrightarrow{K}_{\mathbf{j}}(\mathbf{r},t,\mathbf{r}',\tau) * \nabla\lambda(\mathbf{r}',\tau) d\tau. \tag{38}$$

If $\lambda_t(\mathbf{r},t)$ is a test function, the Euler-Lagrange equation for extremum is

$$\frac{\partial L_e}{\partial \lambda_t} - \nabla \cdot \frac{\partial L_e}{\partial \nabla \lambda_t(\mathbf{r},t)} = 0. \tag{39}$$

We need the identity for functional derivatives

$$F = \int dr \int B(r,r')n(r')dr'dr, \tag{40}$$

$$\frac{\delta F}{\delta n(r)} = 2\int B(r,r')n(r')dr'. \tag{41}$$

The Euler-Lagrange equations for this problem yield the statistical-mechanical equations of motion in Eq.(13) in analogy to the Lagrange's equations of motion in mechanics. This is similar to Hamilton's Principle and Lagrange's equations of mechanics. Therefore we conclude that an extremum (usually a minimum) of I, defined in Eq.(37) yields the statistical-mechanical equations of motion. Below we will illustrate this by examples in heat transfer and electromagnetics.

a) As an example, if we apply the Euler-Lagrange equation for Eq.(20) for steady-state heat transfer entropy production. In this case dS/dt is a constant and does not contribute. In Eq.(27), with $\lambda = 1/k_BT$, and taking a variation with respect to $\nabla\lambda = -\nabla T/k_BT^2$, we obtain the equation of motion for steady-state heat transfer $\nabla \cdot (\overleftrightarrow{\kappa} \cdot \nabla T) = \nabla \cdot \mathbf{J}_h = 0$. This problem was addressed previously by other researchers by using T and ∇T as the variational variables, but this led to inconsistencies. In our approach we use λ and $\nabla\lambda$ as the variables and this leads to the expected results.

b) As another example, if we apply the Euler-Lagrange equation for steady-state charge transfer entropy production (dS/dt is constant) in Eq.(28) with $\lambda = -\phi/k_BT$, and taking a variation with respect to $\nabla\lambda = -\nabla\phi/k_BT + \phi\nabla T/T^2$, we obtain the equation of motion for steady-state charge transfer $\nabla \cdot \left(\overleftrightarrow{\sigma}_f \cdot \left(\mathbf{E}(\mathbf{r}) - \frac{\phi(\mathbf{r})\nabla T(\mathbf{r})}{T(\mathbf{r})}\right)\right) = \nabla \cdot \mathbf{J}_f = 0$. The time dependent equation of motion is obtained if we include $-dS/dt = \int dr(\phi/T)(\partial < \rho_f > /\partial t)$ and then take the variation to obtain $\partial < \rho_f > /\partial t = -\nabla \cdot \mathbf{J}_f$.

c) As another example, we consider the simple case of one variable for time-dependent heat transfer, $u(\mathbf{r})$ with $\lambda_t = 1/k_BT$ and $\nabla\lambda_t = -\nabla T/k_BT^2$ and $L_e(\lambda, \nabla\lambda, \mathbf{r}) = k_B\nabla\lambda \cdot \mathbf{J}_h - k_B\rho c_p\lambda\partial T/\partial t$, where $\mathbf{J}_h = -\kappa\nabla T$. The Euler-Lagrange equation then yields Eq.(31).

d) If instead we use the exact expression for the current and heat capacity as developed above, in the Euler-Lagrange condition in Eqs.(36) and (37, we obtain the very general heat transfer equation Eq.(29)

$$\int d\mathbf{r}' \frac{1}{T(\mathbf{r}',t)} \frac{\Delta\{u(\mathbf{r})\overline{u}(\mathbf{r}')\}}{k_BT(\mathbf{r}',t)} \frac{\partial T}{\partial t}(\mathbf{r}',t)$$
$$= \nabla \cdot \left(\int_0^t d\tau \int d\mathbf{r}' \frac{\overleftrightarrow{K}_h(\mathbf{r},t,\mathbf{r}',\tau)}{k_BT^2(\mathbf{r}',\tau)} \cdot \nabla' T(\mathbf{r}',\tau)\right). \tag{42}$$

With the definitions of the thermal conductivity and heat capacity in Section 5.1 this becomes the Fourier equation for heat transfer, Eq.(31). Therefore the extrema of Eq.(37) yields the equations of motion. The same approach could be performed for other variables or for a collection of relevant variables and its generalized forces to obtain equations of motion. Therefore we conclude that the test functions that yield an extremum is the set of λ that satisfy the equations of motion for the actual dynamical trajectory. This is similar to Hamilton's Principle, but for statistical-mechanical evolution. Therefore we have a way of deriving equations of motion from the entropy production rate. There could be many more applications of this principle.

5.4 Equation of motion for the charge density and microscopic definition of the electrical conductivity

In our next example, we consider the equation for the free-charge density conservation that satisfies $\dot{\rho}_f = -\nabla \cdot \mathbf{j}_f$. In this example, $F = \rho_f$, $\lambda = -\phi/k_B T$, where ϕ is the electric potential.

$$\frac{\partial < \rho_f(\mathbf{r}) >}{\partial t} = \int d\mathbf{r}' \frac{\Delta\{\rho_f(\mathbf{r})\bar{\rho}_f(\mathbf{r}')\}}{k_B T(\mathbf{r}',t)} \frac{\partial \phi}{\partial t}(\mathbf{r}',t)$$

$$= \nabla \cdot \left(\int_{t_i}^{t} d\tau \int d\mathbf{r}' \frac{\overleftrightarrow{K}_f (\mathbf{r},t,\mathbf{r}',\tau)}{k_B T(\mathbf{r}',\tau)} \cdot \left(\nabla' \phi(\mathbf{r}',\tau) - \frac{\nabla T(\mathbf{r},t)}{T(\mathbf{r},t)} \phi(\mathbf{r},t) \right) \right), \tag{43}$$

where $\overleftrightarrow{K}_f (\mathbf{r},t,\mathbf{r}',\tau) =< \mathbf{j}_f(\mathbf{r},t)\mathcal{T}(t,\tau)\mathbf{j}_f(\mathbf{r}',\tau) >$. The relationship to the electrical conductivity is identified through

$$\frac{< \mathbf{j}_f(\mathbf{r},t)\mathcal{T}(t,\tau)\mathbf{j}_f(\mathbf{r}',\tau) >_0}{k_B T(\mathbf{r}',\tau)} = \overleftrightarrow{\sigma}_f (\mathbf{r},t)\delta(t-\tau)\delta(\mathbf{r}-\mathbf{r}'). \tag{44}$$

In a linear approximation and using $\mathbf{E} \approx -\nabla\phi$, we can write this as the equation of continuity for charge conservation

$$\frac{\partial < \rho_f(\mathbf{r}) >}{\partial t} = C_d(\mathbf{r},t)\frac{\partial \phi(\mathbf{r},t)}{\partial t} = -\nabla \cdot \left(\overleftrightarrow{\sigma}_f (\mathbf{r},t) \cdot \left(\mathbf{E}(\mathbf{r},t) - \frac{\nabla T(\mathbf{r},t)}{T(\mathbf{r},t)} \phi(\mathbf{r},t) \right) \right), \tag{45}$$

C_d is the capacitance density. We see the charge transfer is driven by the electric field and a temperature gradient. The generalized current density is $\mathbf{J}_f = \overleftrightarrow{\sigma}_f (\mathbf{r},t) \cdot \left(\mathbf{E}(\mathbf{r},t) - \frac{\nabla T(\mathbf{r},t)}{T(\mathbf{r},t)} \phi(\mathbf{r},t) \right)$.

In a linear approximation and a stationary system without an impressed temperature gradient, integrating on both sides over time and space for system of volume V reduces Eq.(44) to the fluctuation-dissipation relation for the conductivity

$$\overleftrightarrow{\sigma}_f (\mathbf{r}) = V \int_0^\infty d\tau \frac{< \mathbf{j}_f(\mathbf{r},0)\mathbf{j}_f(\mathbf{r},\tau) >_0}{k_B T}. \tag{46}$$

If we multiply both sides of Eq.(43) by ϕ/T and integrate by parts, then we write Eq.(34) as an entropy-density balance equation

$$\frac{\partial S_e(\mathbf{r},t)}{\partial t} + \nabla \cdot \left(\frac{\phi(\mathbf{r},t)}{T(\mathbf{r},t)} \mathbf{J}_f(\mathbf{r},t) \right) = \frac{\phi(\mathbf{r},t)}{T(\mathbf{r},t)} \nabla \cdot \mathbf{J}_f(\mathbf{r},t). \tag{47}$$

6. Entropy production in electromagnetic driving

6.1 Electromagnetic entropy production

We now consider Eq.(19) for electromagnetic entropy production when the λ_n are the inverse temperature $1/k_B T$, and local fields $-\mathbf{E}_p/k_B T$ and $-\mathbf{H}_m/k_B T$ and the relevant variables are the generalized internal energy density $U =< u >$, the electric displacement, $\mathbf{D} =< \mathbf{d} >$, and the magnetic induction $\mathbf{B} =< \mathbf{b} >$. The macroscopic charge current density $\mathbf{J}_f$ is related to the conductivity $\overleftrightarrow{\sigma}_f (\mathbf{r},t)$ by $\mathbf{J}_f(\mathbf{r},t) = \overleftrightarrow{\sigma}_f (\mathbf{r},t) \cdot \mathbf{E}_p(\mathbf{r},t)$. With these definitions we can write the macroscopic entropy production rate in terms of dissipation and heat flowing through the boundary surfaces ($\overleftrightarrow{Q} (\mathbf{r},t)$)

$$\Sigma(t) = \int d\mathbf{r} \frac{1}{T(\mathbf{r},t)} \left\{ \frac{\partial U(\mathbf{r},t)}{\partial t} - \frac{\partial \mathbf{D}(\mathbf{r},t)}{\partial t} \cdot \mathbf{E}_p(\mathbf{r},t) - \frac{\partial \mathbf{B}(\mathbf{r},t)}{\partial t} \mathbf{H}_m(\mathbf{r},t) \right\}$$
$$= - \int \frac{1}{T} \mathbf{Q}(\mathbf{r},t) \cdot \mathbf{n} dS + \int d\mathbf{r} \frac{1}{T(\mathbf{r},t)} \mathbf{E}_p(\mathbf{r},t) \cdot \underbrace{\overleftrightarrow{\sigma}_f (\mathbf{r},t) \cdot \mathbf{E}_p(\mathbf{r},t)}_{\mathbf{J}_f}. \tag{48}$$

Note that the interpretation of quantities such as internal energy, polarization, and temperature must be generalized when working with nonequilibrium systems, but we use these symbols in order to relate quantities to the thermodynamic limit.

When the frequency dependence of the fields is dominated by a narrow band around ω_0 and there is no external heat source, then we can write Jackson (1999)

$$\Sigma(t) = \int d\mathbf{r} \frac{1}{T(\mathbf{r},t)} \left\{ \frac{\partial U_{eff}(\mathbf{r},t)}{\partial t} - \left\langle \frac{\partial \mathbf{D}(\mathbf{r},t)}{\partial t} \cdot \mathbf{E}_p(\mathbf{r},t) \right\rangle_{\omega_0} - \left\langle \frac{\partial \mathbf{B}(\mathbf{r},t)}{\partial t} \cdot \mu_0 \mathbf{H}_m(\mathbf{r},t) \right\rangle_{\omega_0} \right\}$$
$$= 2\omega_0 \epsilon''(\omega_0) \int d\mathbf{r} \frac{1}{T(\mathbf{r},t)} < \mathbf{E}(\mathbf{r},t)\mathbf{E}(\mathbf{r},t) >_{\omega_0} + 2\omega_0 \mu''(\omega_0) \int d\mathbf{r} \frac{1}{T(\mathbf{r},t)} < \mathbf{H}(\mathbf{r},t)\mathbf{H}(\mathbf{r},t) >_{\omega_0} \tag{49}$$

ϵ'' and μ'' are the loss component of the permittivity and permeability. This shows that the entropy production rate is due to dissipation. We assume that ϵ'' also contains the effects due to dc conductivity. In this equation $<>_{\omega_0}$ indicates time averaging the fields over a period and the effective internal energy is

$$U_{eff} = \Re \left[\frac{d(\omega\epsilon)}{d\omega}(\omega_0) < \mathbf{E}(\mathbf{r},t)\mathbf{E}(\mathbf{r},t) >_{\omega_0} > \right] + \Re \left[\frac{d(\omega\mu)}{d\omega}(\omega_0) < \mathbf{H}(\mathbf{r},t)\mathbf{H}(\mathbf{r},t) >_{\omega_0} > \right]. \tag{50}$$

For time-harmonic fields we have

$$\tilde{\Sigma}(\omega,t) = \frac{\omega}{2}\int d\mathbf{r}\frac{1}{T(\mathbf{r},t)}\left[\epsilon''(\mathbf{r},\omega)|\mathbf{E}(\mathbf{r},\omega)|^2 + \mu''(\mathbf{r},\omega)|\mathbf{H}(\mathbf{r},\omega)|^2\right]. \tag{51}$$

By Eq.(19) the macroscopic entropy production in Eq.(49) must equal

$$\Sigma(t) = \frac{1}{k_B}\int_{t_i}^{t}\left\langle \dot{s}(t)\mathcal{T}(t,\tau)(1-P(\tau))\bar{\dot{s}}(\tau)\right\rangle_\tau d\tau. \tag{52}$$

Using the microscopic Maxwell's equations with a source current $\mathbf{j}_e$ and including a possible external heat transport flux $\mathbf{j}_h$ we have

$$\dot{s}(t) = -\int\frac{1}{T(\mathbf{r},t)}\mathbf{j}_h(\mathbf{r},t)\cdot\mathbf{n}dS + \int d\mathbf{r}\frac{1}{T(\mathbf{r},t)}\mathbf{j}_e(\mathbf{r},t)\cdot\mathbf{E}_p(\mathbf{r},t). \tag{53}$$

where $<\dot{s}(t)>=\int(\{<\dot{u}>-<\dot{\mathbf{d}}>\cdot\mathbf{E}_p-<\dot{\mathbf{b}}>\cdot\mathbf{H}_m\}/T)d\mathbf{r}=0$.

For a thermally insulated system, we can rewrite Eq.(52) after using Eq.(49) for the entropy density as

$$2\omega_0\epsilon''(\omega_0)\frac{1}{T(\mathbf{r},t)}<\mathbf{E}(\mathbf{r},t)\mathbf{E}(\mathbf{r},t)>_{\omega_0} + 2\omega_0\mu''(\omega_0)\frac{1}{T(\mathbf{r},t)}<\mathbf{H}(\mathbf{r},t)\mathbf{H}(\mathbf{r},t)>_{\omega_0}$$

$$= \frac{1}{T(\mathbf{r},t)}\mathbf{E}_p(\mathbf{r},t)\cdot\underbrace{\int_{t_i}^{t}\int d\mathbf{r}'\frac{\left\langle \mathbf{j}_e(\mathbf{r},t)\mathcal{T}(t,\tau)(1-P(\tau))\bar{\mathbf{j}}_e(\mathbf{r}',\tau)\right\rangle_\tau}{k_B T(\mathbf{r}',\tau)}\cdot\mathbf{E}_p(\mathbf{r}',\tau)d\tau}_{\mathbf{J}_f(\mathbf{r},t)} \tag{54}$$

Where the macroscopic current is defined by the underbrace. This is a FDR.

6.2 The permittivity from the entropy production rate equation

If we consider only the internal energy and displacement terms in Eq.(48) and neglect magnetic effects we have

$$\begin{aligned}\Sigma(t) &= \int d\mathbf{r}\frac{1}{T(\mathbf{r},t)}\left[\frac{\partial U(\mathbf{r},t)}{\partial t} - \frac{\partial\mathbf{D}(\mathbf{r},t)}{\partial t}\cdot\mathbf{E}_p(\mathbf{r},t)\right] \\ &= \int d\mathbf{r}\frac{1}{T(\mathbf{r},t)}\frac{\partial\mathbf{D}(\mathbf{r},t)}{\partial t}\cdot\mathbf{E}_{eff}(\mathbf{r},t) \\ &= \frac{\mathbf{E}_{eff}(\mathbf{r},t)}{T(\mathbf{r},t)}\cdot\int_{-\infty}^{t}\int\frac{<\dot{\mathbf{d}}(\mathbf{r},t)\mathcal{T}(t,\tau)(1-P(\tau))\dot{\mathbf{d}}(\mathbf{r}',\tau)>}{k_B T(\mathbf{r}',\tau)}\cdot\mathbf{E}_{eff}(\mathbf{r}',\tau)d\mathbf{r}'d\tau.\end{aligned} \tag{55}$$

Where used an expression for the internal energy without an applied magnetic field, derived from the Hamiltonian in Eq.(58): $\partial U/\partial t = \partial\mathbf{D}/\partial t\cdot\mathbf{E}$. Expanding $<\mathbf{D}>$ through the use of Eq.(3) to first order, we obtain $\mathbf{E}_p \approx \mathbf{D}/\epsilon + \overleftrightarrow{D}_p\cdot\mathbf{E}_p$. Where $\overleftrightarrow{D}_p$ is the depolarization tensor and we used $(\partial\mathbf{D}/\partial t)\cdot\mathbf{D}\approx 0$ and the effective field is then $\mathbf{E}_{eff} = \mathbf{E} - \overleftrightarrow{D}_p\cdot\mathbf{E}_p$.

If we suppress the spatial dependence in a volume V and apply this near equilibrium we have

$$\frac{\partial \mathbf{D}(t)}{\partial t} = V \int_{-\infty}^{t} \underbrace{\frac{< \dot{\mathbf{d}}(t)\mathcal{T}_0(t,\tau)\dot{\mathbf{d}}(\tau) >_0}{k_B T(\tau)}}_{df_d/dt} \cdot \mathbf{E}_{eff}(\tau) d\tau, \tag{56}$$

where f_d is the impulse response function in the limit of a time invariant, linear, isothermal, stationary system. In that approximation the kernel is a function of $t - \tau$ and the Laplace transform can be used. In this limit we can take the Laplace transform ($\mathcal{L}_L$) of this equation and obtain an expression for the permittivity for time-harmonic fields, $\epsilon(\omega)$, where $\tilde{\mathbf{D}} = \overleftrightarrow{\epsilon}(\omega)\, \tilde{\mathbf{E}}_{eff}(\omega)$

$$V\mathcal{L}_L \left[\frac{< \dot{\mathbf{d}}(0)\dot{\mathbf{d}}(\tau - t) >}{k_B T} \right] \to i\omega \overleftrightarrow{\epsilon}(\omega). \tag{57}$$

7. Applications

7.1 Conservation of electromagnetic energy in a nonequilibrium system

Let us write the Hamiltonian under electromagnetic driving in terms of the microscopic electric and magnetic polarization operators $\mathbf{d}(\mathbf{r})$ and $\mathbf{b}(\mathbf{r})$

$$\mathcal{H}(t) = \int d\mathbf{r}\{u(\mathbf{r}) - \mathbf{d}(\mathbf{r}) \cdot \mathbf{E}(\mathbf{r}, t) - \mathbf{b}(\mathbf{r}) \cdot \mathbf{H}(\mathbf{r}, t)\}. \tag{58}$$

The energy dissipated by heat in internal relaxation is

$$\left\langle \frac{\partial \mathcal{H}(\mathbf{r}, t)}{\partial t} \right\rangle = \int d\mathbf{r} \left[\frac{\partial \mathbf{E}(\mathbf{r}, t)}{\partial t} \cdot \mathbf{D}(\mathbf{r}, t) + \frac{\partial \mathbf{H}(\mathbf{r}, t)}{\partial t} \cdot \mathbf{B}(\mathbf{r}, t) \right]. \tag{59}$$

Therefore if we take the time derivative of the expectation of Eq.(58) and subtract the internal relaxation energy that is given by Eq.(59), we have the conservation equation

$$\underbrace{\frac{\partial \langle \mathcal{H}(t) \rangle}{\partial t} - \left\langle \frac{\partial \mathcal{H}(t)}{\partial t} \right\rangle}_{\text{heat added}} = \int d\mathbf{r} \left[\frac{\partial U}{\partial t} - \frac{\partial \mathbf{D}(\mathbf{r}, t)}{\partial t} \cdot \mathbf{E}(\mathbf{r}, t) - \frac{\partial \mathbf{B}(\mathbf{r}, t)}{\partial t} \cdot \mathbf{H}(\mathbf{r}, t) \right]. \tag{60}$$

This is a general energy conservation relation that is valid away from equilibrium. The LHS is the generalized external power delivered to the system beyond that due to the driving fields, such as an open system where heat enters the system. For a system that is isolated except for the dynamically driven fields, the LHS is zero and we obtain the normal energy conservation condition for the fields

$$\frac{\partial U}{\partial t} = \frac{\partial \mathbf{D}(\mathbf{r}, t)}{\partial t} \cdot \mathbf{E}(\mathbf{r}, t) + \frac{\partial \mathbf{B}(\mathbf{r}, t)}{\partial t} \cdot \mathbf{H}(\mathbf{r}, t). \tag{61}$$

7.2 Generalized equation of motion for the polarization and relation to the Debye equation

The electric polarization evolution equation can be obtained using Eq.(13) for the case of $\mathbf{p}(\mathbf{r})$ and $u(\mathbf{r})$ in the Hamiltonian $\mathcal{H}(t) = \int d\mathbf{r}\{u(\mathbf{r}) - \mathbf{p}(\mathbf{r}) \cdot \mathbf{E}(\mathbf{r}, t)\}$, or by taking a variation with respect to $\mathbf{E}_p/k_BT$, to find (Baker-Jarvis et al. (2007))

$$\frac{\partial \mathbf{P}(\mathbf{r},t)}{\partial t} = -\int d^3r' \int_0^t \overleftrightarrow{\mathbf{K}}_e(\mathbf{r},t,\mathbf{r}',\tau) \times \cdot \left(\mathbf{P}(\mathbf{r}',\tau) - \overleftrightarrow{\chi}_0 \cdot \mathbf{E}(\mathbf{r}',\tau)\right) d\tau. \tag{62}$$

Here $\overleftrightarrow{\chi}_0$ is the static susceptibility. The Debye relaxation differential equation is recovered from Eq.(62) when $\overleftrightarrow{\mathbf{K}}_e(\mathbf{r},t,\mathbf{r}',\tau) = \overleftrightarrow{I}\,\delta(t-\tau)\delta(\mathbf{r}-\mathbf{r}')/\tau_e$.

7.3 Generalized equation of motion for the magnetization

The magnetic polarization can be obtained using Eq.(13) for the case of $\mathbf{m}(\mathbf{r})$ and $u(\mathbf{r})$ in the Hamiltonian $\mathcal{H}(t) = \int d\mathbf{r}\{u(\mathbf{r}) - \mu_0\mathbf{m}(\mathbf{r})\cdot\mathbf{H}(\mathbf{r},t)\}$, by taking a variation with respect to $\mathbf{H}_m/k_BT$, to find (Baker-Jarvis (2005; 2008); Baker-Jarvis and Kabos (2001); Baker-Jarvis et al. (2004); Robertson (1967b)):

$$\frac{\partial \mathbf{M}(\mathbf{r},t)}{\partial t} = -|\gamma_g|\mathbf{M}(\mathbf{r},t)\times\mathbf{H}_{eff}(\mathbf{r},t) - \int d^3r' \int_0^t \overleftrightarrow{\mathbf{K}}_m(\mathbf{r},t,\mathbf{r}',\tau)\cdot\chi_0\mathbf{H}_{eff}(\mathbf{r}',\tau)d\tau, \quad , \tag{63}$$

where $\overleftrightarrow{\mathbf{K}}_m$ is a kernel that contains of the microstructural interactions given in Baker-Jarvis and Kabos (2001), γ_g is the gyromagnetic ratio, χ_0 is the static susceptibility, and $\mathbf{H}_{eff}$ is the effective magnetic field. Special cases of Eq.(63) reduce to constitutive relations such as the Landau-Lifshitz, Gilbert, and Bloch equations. The Landau-Lifshitz equation of motion is useful for ferromagnetic and ferrite solid materials (Lax and Button (1962)).

7.4 Maxwell's equations

To derive Maxwell's equations from Eq.(19). For generality, the reversible term $<\dot{s}(t)>$ has been included. Note that $<\dot{s}(t)>=0$ only when the λ's are the functions that satisfy the equations of motion, so we keep it in the variational principle for completeness. The Hamiltonian is $\mathcal{H}(t) = \int d\mathbf{r}\{u(\mathbf{r}) - \mathbf{d}(\mathbf{r},t)\cdot\mathbf{E}(\mathbf{r},t) - \mathbf{b}(\mathbf{r})\cdot\mathbf{H}(\mathbf{r},t)\}$, and the entropy rate satisfies

$$\Sigma(t) = \int d\mathbf{r}\frac{1}{T(\mathbf{r},t)}\{\frac{\partial U_{eff}(\mathbf{r},t)}{\partial t} - \left\langle\frac{\partial \mathbf{D}(\mathbf{r},t)}{\partial t}\cdot\mathbf{E}_p(\mathbf{r},t)\right\rangle_{\omega_0} - \left\langle\frac{\partial \mathbf{B}(\mathbf{r},t)}{\partial t}\cdot\mu_0\mathbf{H}_m(\mathbf{r},t)\right\rangle_{\omega_0}\} =<\dot{s}(t)> + \int d\mathbf{r}\frac{1}{T}\mathbf{J}\cdot\mathbf{E}_p, \tag{64}$$

where $<\dot{s}(t)>=\int(\{<\dot{u}> - <\dot{\mathbf{d}}>\cdot\mathbf{E}_p - <\dot{\mathbf{b}}>\cdot\mathbf{H}_m\}/T)d\mathbf{r}$. To obtain Maxwell's equations we take variations of Eq.(64) with respect to $-\mathbf{E}_p/T$ to obtain the first Maxwell equation, $\partial\mathbf{D}/\partial t = \nabla\times\mathbf{H} - \mathbf{J}$, and the second ($\partial\mathbf{B}/\partial t = -\nabla\times\mathbf{E}$), by a variation with respect to $-\mathbf{H}_m/T$. Here we used the commutation relations in Eq.(7): $<\dot{s}(t)>=<\dot{s}(t),\mathcal{H}(t)>/i\hbar$: $\int d\mathbf{r}' <[\mathbf{d}(\mathbf{r}),\mathbf{b}(\mathbf{r}')\cdot\mathbf{H}(\mathbf{r}',t)]/i\hbar> = -\nabla\times\mathbf{H}(\mathbf{r},t)$ and $\int d\mathbf{r}' <[\mathbf{b}(\mathbf{r}),\mathbf{d}(\mathbf{r}')\cdot\mathbf{E}(\mathbf{r}',t)]/i\hbar> = \nabla\times\mathbf{E}(\mathbf{r},t)$.

7.5 Voltage fluctuations and Nyquist's theorem

We consider the general problem of electrical noise and the related Nyquist problem of dissipation in a resistor. We will begin with a very general analysis that is valid away from thermal equilibrium, and then show how this reduces to Nyquist's result in the equilibrium limit.

The microscopic entropy rate for this electrical system is a function of the electromagnetic energy due to random charge motion. We write the microscopic entropy rate in terms of the microscopic charge current density, $\dot{s}(t) = -\int d\mathbf{r}\dot{\rho}_{f'}\phi/T = -\int d\mathbf{r}\mathbf{D}\cdot\mathbf{E}_p/T$ and since $\dot{\rho}_f = -\nabla\cdot\mathbf{j}_f$ we have $<\dot{s}(t)>= \int d\mathbf{r}\nabla\cdot <\mathbf{j}_f> \phi/T = 0$. We also assume that the macroscopic entropy produced due to a constant bias current I_0 in the resistor is $\Sigma = I_0^2 R/T$. For a system with a resistance R driven by an applied electrical field the RHS of Eq.(19) can be written as

$$\Sigma(t) = -\int d\mathbf{r}\frac{\phi(\mathbf{r},t)}{T}\frac{\partial <\rho_f(\mathbf{r})>}{\partial t} = -\int d\mathbf{r}\frac{1}{T}\nabla\phi(\mathbf{r},t)\cdot\mathbf{J}_e(\mathbf{r}))$$
$$= \int d\mathbf{r}\frac{1}{T}\nabla\phi\cdot \overleftrightarrow{\sigma}_f \cdot\nabla\phi = \int d\mathbf{r}\frac{\mathbf{E}(\mathbf{r},t)}{T(\mathbf{r},t)}\cdot \overleftrightarrow{\sigma}_f \cdot\mathbf{E} \rightarrow \frac{I(t)^2 R}{T}. \tag{65}$$

In this special case, the LHS of Eq.(19) can be written in the following equivalent forms

$$\Sigma(t) = \int_0^t \int d\mathbf{r}d\mathbf{r}'\phi(\mathbf{r},t)\frac{<\dot{\rho}_f(t)\mathcal{T}(t,\tau)(1-P(\tau))\dot{\rho}_f(\tau)>}{k_B T(\mathbf{r},t)}\frac{\phi(\mathbf{r}',\tau)}{T(\mathbf{r}',\tau)}d\tau$$
$$= \int_0^t \int d\mathbf{r}d\mathbf{r}'\left(\mathbf{E}(\mathbf{r},t) - \frac{\nabla T(\mathbf{r},t)}{T(\mathbf{r},t)}\phi(\mathbf{r},t)\right)\cdot\frac{<\mathbf{j}_f(\mathbf{r})\mathcal{T}(t,\tau)(1-P(\tau))\mathbf{j}_f(\mathbf{r}')>}{k_B T(\mathbf{r},t)}$$
$$\times\cdot\frac{\left(\mathbf{E}(\mathbf{r}',\tau) - \frac{\nabla T(\mathbf{r}',\tau)}{T(\mathbf{r}',\tau)}\phi(\mathbf{r}',\tau)\right)}{T(\mathbf{r},\tau)}d\tau$$
$$\approx \frac{1}{k_B}\int_0^t \frac{I(t)}{T(t)} <v_f(t)\mathcal{T}(t,\tau)v_f(\tau)> \frac{I(\tau)}{T(\tau)}d\tau. \tag{66}$$

Here we used $\int d\mathbf{r}\mathbf{j}_f\cdot\mathbf{E} = I_0(V_0 + v_f(t))$, where the constant driving voltage V_0 doesn't contribute and $<v_f(t)>= 0$. This equation is very general and valid away from equilibrium. This approach could be used to generalize the Nyquist result if the temperature was not assumed to be constant in time or space. As we approach a steady state for a constant driving current I_0 and temperature , then the time domain fluctuation-dissipation form of Nyquist's theorem is recovered from the last expression in Eq.(66) when we equate it to $I_0^2 R$:

$$R = \int_0^\infty \frac{<v(0)v(\tau)>_0}{k_B T}d\tau. \tag{67}$$

Equation (67) is an example of Kubo's second fluctuation-dissipation theorem where $<v>_0=$ 0.

For a transmission line with a noisy resistor R that generates a random voltage with zero mean, and load resistor of resistance R over a bandwidth Δf, Eq.(67) with a constant driving current yields $<v^2>_0= 4k_B TR\Delta f$. We can interpret Nyquist's equation in terms of entropy production in the fluctuations. Since $k_B \Delta f$ is the entropy production rate over the bandwidth of the black body due to the voltage fluctuations, this entropy production rate must be equal the entropy produced in equilibrium fluctuations in the resistors, which is the noise power per temperature: $< v^2 >_0 /4RT$. The emissivity for a material with a radiated power P at temperature T can then be defined as the ratio of the entropy rate produced in the body divided that that produced in a pure black body: $e = (P/T)/k_B\Delta f$.

7.6 Estimation of Boltzmann's constant

As noted in Baker-Jarvis (2008), Eq.(19) could be used to obtain values for k_B from measurements of entropy production and it is an exact equation. Note that Eq.(19) does not contain the temperature explicitly. This equation or Eq.(66) can be used to model noise in the limit as we approach equilibrium. Boltzmann's constant can in principle be determined by measurements of any of the transport coefficients , such as κ, $\overset{\leftrightarrow}{\sigma}$, or R, through the equation

$$k_B \overset{\leftrightarrow}{L}_{mn}(\mathbf{r}) = \int_0^\infty \int d\mathbf{r}' \theta(\mathbf{r}) \frac{< \mathbf{j}_m(\mathbf{r},0)\mathbf{j}_n(\mathbf{r}',s) >_0}{T(\mathbf{r}')} ds. \tag{68}$$

8. Conclusion

The goal of this paper was to develop in the context of electromagnetic measurement science, an explanation of how a previously derived exact entropy rate relationship relates to exact equations of motion, fluctuation-dissipation relations, and transport coefficients. Applications in the areas of measureable quantities such as thermal conductivity, electromagnetic response, electrical noise, and Boltzmann's constant were developed. We showed that the concept of entropy production rate can be viewed as a basis for deriving electromagnetic equations of motion, including Maxwell's equations, and extending FDRs. Unlike the classical FDRs, Eq.(19) is valid away from equilibrium. We developed expressions for thermal conductivity, electrical conductivity, and permittivity in terms of correlation functions using an analysis based on entropy-production fluctuations.

Nyquist-noise can also be understood by an analysis based on entropy production instead of an standard argument based on power absorbed and emitted from resistors. Nyquist assumed that for a waveguide in equilibrium terminated by resistors, that in order for the concept of detailed balance to hold, the power absorbed by the resistor at one end of a waveguide must equal the power in the emitted fields that travel down the waveguide and is absorbed by the resistor at the other end. Using the results of this paper we can interpret this as follows. In equilibrium the mean of the microscopic dynamical evolving entropy-production rate is zero, but fluctuations around the mean are nonzero. The principle of detailed balance for equilibrium electrical noise applied to this process requires that any entropy production in the resistor at one end, induced by the microscopic fluctuating voltages, is balanced by the emitted electromagnetic power, with corresponding entropy production, and travels down the waveguide to the other resistor, and, when once absorbed, causes an equivalent production of entropy at that end. Note that the Nyquist result naturally falls out of Eq.(19) without evoking the requirement of an energy of $(1/2)k_B T$ per mode. Using the concept of entropy production rate to study blackbody processes appears to be more fundamental than using the concept of power alone, since it merges the concepts of power and temperature together. Equation (19) gives us an important tool for further study and extension to nonequilibrium analysis. The hope is that this paper is a step in progress toward developing measurement metrology that can be extended to study processes out of equilibrium and relate the measurements to theory.

9. Acknowledgments

We also acknowledge various discussions with members of the NIST Innovative Measurement Science research team for Detection of Corrosion in Steel-Reinforced Concrete by Antiferromagnetic Resonance.

10. Appendix: Alternative expansion of the evolution of the relevant quantities

We can re-express the LHS of Eq.(13) for the various relevant variables in terms of time derivative of the generalized forces. This casts the LHS of Eq.(13) in terms of measurable quantities such as temperature and field rates and thereby produces generalized heat transfer and polarization equations.

$$\begin{aligned} Tr\left(F(\mathbf{r})\frac{\partial \sigma}{\partial t}\right) &= Tr\left((F(\mathbf{r})\frac{\partial}{\partial t}e^{(-\sum_{n=1}\lambda_n(\mathbf{r},t)F_n(\mathbf{r})+\ln Z)}\right) \\ &= -\left(< F(\mathbf{r})\overline{F}(\mathbf{r}) > - < F(\mathbf{r}) >< \overline{F}(\mathbf{r}) >\right)\frac{\partial \lambda_n(\mathbf{r},t)}{\partial t} \end{aligned} \tag{69}$$

where $Z = \sum_n Tr(F_n(\mathbf{r})\sigma(t))$. Therefore

$$\begin{aligned} \frac{\partial < F(\mathbf{r}) >}{\partial t} &= -\int d\mathbf{r}'\{< F(\mathbf{r})\overline{F}(\mathbf{r}') > - < F(\mathbf{r}) >< \overline{F}(\mathbf{r}') >\} * \frac{\partial \lambda(\mathbf{r}',t)}{\partial t} \\ &\equiv -\int d\mathbf{r}\Delta\{F(\mathbf{r})\overline{F}(\mathbf{r}')\} * \frac{\partial \lambda(\mathbf{r}',t)}{\partial t}. \end{aligned} \tag{70}$$

We defined $\Delta\{ab\} \equiv < a\overline{b} > - < a >< \overline{b} >$. Using this equation, the internal energy density can be re-expressed in terms of time derivatives of the Lagrangian multipliers

$$\begin{aligned} \frac{\partial U(\mathbf{r},t)}{\partial t} &= \int d\mathbf{r}'\frac{1}{T}\frac{\Delta[u(\mathbf{r})\overline{u}(\mathbf{r}')]}{k_B T}\frac{\partial T(\mathbf{r}',t)}{\partial t} \\ &- \int d\mathbf{r}'\frac{1}{T}\frac{\Delta[u(\mathbf{r})\overline{\mathbf{p}}(\mathbf{r}')]}{k_B T}\cdot \mathbf{E}_p\frac{\partial T(\mathbf{r}',t)}{\partial t} + \int d\mathbf{r}'\frac{\Delta[u(\mathbf{r})\overline{\mathbf{p}}(\mathbf{r}')]}{k_B T}\cdot\frac{\partial \mathbf{E}_p(\mathbf{r}',t)}{\partial t} \\ &- \int d\mathbf{r}'\frac{1}{T}\frac{\Delta[u(\mathbf{r})\overline{\mathbf{m}}(\mathbf{r}')]}{k_B T}\cdot \mathbf{H}_m\frac{\partial T(\mathbf{r}',t)}{\partial t} + \int d\mathbf{r}'\frac{\Delta[u(\mathbf{r})\overline{\mathbf{m}}(\mathbf{r}')]}{k_B T}\cdot\frac{\partial \mathbf{H}_m(\mathbf{r}',t)}{\partial t} \\ &\equiv (c_{uu} + c_{up(1)} + c_{um(1)})\frac{\partial T(\mathbf{r}',t)}{\partial t} + \mathbf{c}_{up(2)}\cdot\frac{\partial \mathbf{E}_p(\mathbf{r}',t)}{\partial t} + \mathbf{c}_{um(2)}\cdot\frac{\partial \mathbf{H}_m(\mathbf{r}',t)}{\partial t}. \end{aligned} \tag{71}$$

The polarization satisfies

$$\begin{aligned} \frac{\partial \mathbf{P}(\mathbf{r},t)}{\partial t} &= \int d\mathbf{r}'\Delta[\mathbf{p}(\mathbf{r})\overline{\mathbf{p}}(\mathbf{r}')]\cdot\frac{\partial \mathbf{E}_p(\mathbf{r}',t)\beta(\mathbf{r}',t)}{\partial t} \\ &+ \int d\mathbf{r}'\Delta[\mathbf{p}(\mathbf{r})\overline{\mathbf{m}}(\mathbf{r}')]\cdot\frac{\partial \mathbf{H}_m(\mathbf{r}',t)\beta(\mathbf{r}',t)}{\partial t} + \int d\mathbf{r}\frac{1}{T}\frac{\Delta[\mathbf{p}(\mathbf{r})\overline{u}(\mathbf{r}')]}{k_B T}\frac{\partial T(\mathbf{r}',t)}{\partial t} \\ &\equiv (\chi_{pu} + \chi_{pp} + \chi_{pm(1)})\frac{\partial T(\mathbf{r}',t)}{\partial t} + \overleftrightarrow{\chi}_{pp}\cdot\frac{\partial \mathbf{E}_p(\mathbf{r}',t)}{\partial t} + \overleftrightarrow{\chi}_{pm(2)}\cdot\frac{\partial \mathbf{H}_m(\mathbf{r}',t)}{\partial t}. \end{aligned} \tag{72}$$

There is a similar relation for **M**.

An alternative equation for the entropy evolution in terms of time derivatives of the Lagrangian multipliers can also be constructed from Eq.(70)

$$\frac{dS(t)}{dt} \equiv -k_B\lambda * \Delta[F\overline{F}] * \frac{\partial \lambda}{\partial t} \approx \int d\mathbf{r}' \{ \frac{\nabla T \cdot \overset{\leftrightarrow}{\kappa} \cdot \nabla T}{T^2} + \int d\mathbf{r}(k_B\beta^2 \Delta[u(\mathbf{r})\overline{\mathbf{p}}(\mathbf{r}')] \cdot \frac{\partial \mathbf{E}_p}{\partial t}$$
$$+ \frac{k_B}{T}\beta^2 \mathbf{E}_p \cdot \Delta[\mathbf{p}(\mathbf{r})\overline{u}(\mathbf{r}')]\frac{\partial T}{\partial t} + k_B\beta^2 \mathbf{E}_p \cdot \Delta[\mathbf{p}(\mathbf{r})\overline{\mathbf{p}}(\mathbf{r}')] \cdot \frac{\partial \mathbf{E}_p}{\partial t}$$
$$- \frac{k_B}{T}\beta^2 \mathbf{E}_p \cdot \Delta[\mathbf{p}(\mathbf{r})\overline{\mathbf{p}}(\mathbf{r}')] \cdot \mathbf{E}_p \frac{\partial T}{\partial t})\}, \tag{73}$$

where we used the heat equation $c\partial T/\partial t = \nabla \cdot \overset{\leftrightarrow}{\kappa} \cdot \nabla T$.

11. References

Baker-Jarvis, J. (2005). Time-dependent entropy evolution in microscopic and macroscopic electromagnetic relaxation. *Phys. Rev. E* Vol. 72, 066613.

Baker-Jarvis, J. (2008). Electromagnetic nanoscale metrology based on entropy production and fluctuations. *Entropy*, Vol. 10, 411–429.

Baker-Jarvis, J., Janezic, M. D., Riddle, B. (2007). Dielectric polarization equations and relaxation times. *Phys. Rev. E*, Vol. 75, 056612.

Baker-Jarvis, J., Kabos, P., (2001). Dynamic constitutive relations for polarization and magnetization. *Phys. Rev. E*, Vol. 64, 56127.

Baker-Jarvis, J., Kabos, P., Holloway, C. L., (2004). Nonequilibrium electromagnetics: Local and macroscopic fields using statistical mechanics. *Phys. Rev. E*, Vol. 70, 036615.

Baker-Jarvis, J., Surek, J., 2009. Transport of heat and charge in electromagnetic metrology based on nonequilibrium statistical mechanics. *Entropy*, Vol. 11, 748–765.

Berne, J. B., Harp, G. D., (1970). On the calculation of time correlation functions. In: I.Prigogine, Rice, S. A. (Eds.), *Advances in Chemical Physics IVII*. Wiley, NY, p. 63.

Jackson, J. D., (1999). *Classical Electrodynamics (3rd Ed.)*. John Wiley and Sons, New York.

Lax, B., Button, K. J., 1962. *Microwave ferrites and ferromagnetics*. McGraw-Hill, New York.

Mori, H., (1965). Transport, collective motion, and brownian motion. *Prog. Theor. Phys.*, Vol. 33, 423.

Nettleton, R. E., (1999). Perturbation treatment of non-linear transport via the Robertson statistical formalism. *Ann. Phys.* Vol. 8, 425–436.

Oppenheim, I., Levine, R. D., (1979). Nonlinear transport processes: Hydrodynamics. *Physica*, Vol. 99A, 383–402.

Rau, J., Muller, B., (1996). From reversible quantum microdynamics to irreversible quantum transport. *Physics Reports*, Vol. MTT-36, 1–59.

Robertson, B., (1966). Equations of motion in nonequilibrium statistical mechanics. *Phys. Rev.*, Vol. 144, 151–161.

Robertson, B., (1967)a. Equations of motion in nonequilibrium statistical mechanics II, Energy transport. *Phys. Rev.*, Vol. 160, 175–183.

Robertson, B., (1967)b. Equations of motion of nuclear magnetism. *Phys. Rev.*, Vol. 153, 391–403.

Robertson, B., (1978). Applications of maximum entropy to nonequilibrium statistical mechanics. In: *The Maximum Entropy Formalism*. M.I.T. Press, Cambridge, MA, p. 289.

Robertson, B., (1993). Nonequilibrium statistical mechanics. In: Grandy, W. T., Milonni, P. W. (Eds.), *Physics and Probability: Essays in Honor of Edwin T. Jaynes*. Cambridge University Press, New York, p. 251.

Robertson, B., (1999). Comment on remarks on the information theory maximization method and extended thermodynamics. *J. Chem. Phys.*, Vol. 111, 6144–6147.

Robertson, B., Mitchell, W. C., (1971). Equations of motion in nonequilibrium statistical mechanics. III: Open systems. *J. Math. Phys.*, Vol. 12, 563–568.

Weiss, U., (1999). *Quantum dissipative systems*. World Science Publishing Company, Sinapore.

Yu, M. B., (2008). Influence of environment and entropy production of a nonequilibrium open system. *Phys. Lett. A*, Vol. 372, 2572–2577.

Zubarev, D. N., Morozov, V., Ropke, G., (1996). *Statistical Mechanics of Nonequilibrium Processes: Basic Concepts, Kinetic Theory*. John Wiley and Sons, New York.

Zwanzig, R., (1960). Ensemble method in the theory of irreversibility. *J. Chem. Phys.*, Vol. 33 (5), 1338–1341.

11

Risk Performance Index and Measurement System

Seon-Gyoo Kim
Kangwon National University, South Korea

1. Introduction

A mega project can generally be defined as a project that costs more than 1 billion US dollars and includes many risk factors that can cause delays or failures during the project life cycle (Flyvbjerg et al. 2003). Thus, it is important to establish a method and system to manage these risk factors effectively in advance. Moreover, it is necessary to reduce the probability of such risk factors causing failures in the project by measuring the performance of projects from the point of view of risk management. This chapter defines a risk performance index (RPI) that measures the performance of projects by integrating the cost/schedule/risk factors and by adding risk management activities to the EVMS, which is the existing integrated cost/schedule-based performance measurement system for construction projects. We also propose a method to produce and analyze the RPIs to improve the accuracy and efficiency of the general performance measurement for mega projects by extending the conventional cost/schedule-based performance measurement system to include risk management.

2. Survey of existing performance measurement systems

Performance management, which examines and manages whether projects, implemented by either individuals or organizations, are effectively executed, has four components: duty, strategy goal, performance goal, and performance index. A strategy goal is a major policy direction that promotes specific duties including the goal, value, and function of an organization. A performance goal is subordinate to the strategy goal and shows major projects planned in a particular year or a specific goal covering multiple aspects of a business group.

A performance index is a scale to measure the level of achievement of the performance goal. It is important to identify quantitative measures of the goals pursued in the project. The development of a performance index enables the efficiency of the project to be measured by comparing and evaluating quantitatively the achievement and level of the performance goal.

This chapter surveyed three methodologies of performance measurement systems used in existing construction businesses: EVMS, BSC, and KPI.

2.1 EVMS

The Earned Value Management System (EVMS) is the most widely used performance measurement system in construction businesses. The United States Department of Defense

(2008) has described it as "a performance-based management system for measuring actual progress against the criteria configuration for the cost, schedule, and performance goals in projects". Fleming and Koppleman (1996) defined the EVMS as "a continuous measurement for practical works under precisely managed work schedules and a management method that estimates the final cost and schedule in a project through this measurement".

The performance measurement applied by using the EVMS integrates cost and schedule. It helps identify how any difference between the planned budget and the actual cost influences the project, by comparing and managing the performance vs plan and estimating the reduction or delay in the schedule from the earned value to the completion of the project and the excess of the budget. As shown in Table 1, the elements of the EVMS can be classified as plan, performance measurement, measurement for management analysis, and analysis elements.

Terminology		Description
Plan Elements	WBS (Work Breakdown Structure)	A deliverable-oriented grouping of project elements
	CA, (Control Account)	A management control point at which actual cost t may be accumulated and compared to earned value.
	PMB (Performance Measurement Baseline)	The time phased budget against which contract performance is measured
Measurement Elements	BCWS (Budgeted Cost of Work Scheduled)	The sum of the budgets for all planned work scheduled to be accomplished.
	BCWP or EV (Budgeted Cost of Work Performed)	The sum of the budgets for completed work and the completed portions of open work
	ACWP (Actual Cost of Work Performed)	The costs actually incurred in accomplishing the work performed
Analysis Elements	SV (Schedule Variance) SPI (Schedule Performance Index)	BCWP - BCWS, BCWP / BCWS
	CV (Cost Variance) CPI (Cost Performance Index)	BCWP - ACWS BCWP / ACWS
	AV (Accounting Variance) API (Accounting Performance Index)	ACWP - BCWS ACWP / BCWS
	EAC (Estimate At Completion)	ACWP+(BAC-BCWP)/CPI

Table 1. EVMS Terminologies.

2.2 BCS

The Balanced Score Card (BSC) is another representative performance measurement system. The BSC method, proposed by Kaplan and Norton (1993), is a strategic management method that uses traditional financial or accounting measurements to overcome the limits and problems of performance measurement in the short term, and provides a way to establish performance measurement for a general project in the long term. It is widely used to establish performance indexes in construction businesses throughout the world. The BSC is a financial index that represents the results of the project execution and customer satisfaction and that shows operational activities, internal management, and operational indexes for learning and growing.

Although it has the advantage that it performs its management processes strategically by measuring nonfinancial aspects, the BSC's financial and nonfinancial approaches comprehensively differ from traditional measurement methods. During the strategy establishment process, it can be difficult to reach an agreement on should be measured because organizations differ in their strategies, visions, and goals. The balanced performance table is a scale that evaluates the business management. It has certain limitations in the evaluation of the satisfaction level of a project, even though it is useful for evaluating the business management, because it only evaluates the management strategy focused on operational effectiveness.

2.3 KPIs

Key performance indicators (KPIs) are a representative performance measurement system established in Britain based on the construction renovation movement called "Rethinking Construction". The system, which was first promoted in 1998, was intended to improve productivity in construction businesses. It can be used to measure not only construction performance, such as construction cost and duration reduction, but also the performance of a business in terms of profits and productivity (The KPI Working Group, 2000).

The construction renovation movement can be classified into seven major performance indexes: duration, cost, quality, customer satisfaction, design change, project performance, health and safety for the construction culture, recognition, production method, and production system. Performance can be measured based on these classifications and the results are applied to plan the efficiency and productivity of the construction business. It also establishes a partnership between the government, owner, and construction businesses and promotes best practices. It has been shown to improve project performance and cost effectiveness by removing inefficiencies and unproductive factors.

3. Need for risk performance measurement system related to construction processes

In recent years, the main trend in urban regeneration projects and large-scale development projects throughout the world has been the development of three-dimensional mixed-use spaces that include such functions as residential, commercial, business, public, cultural, and leisure and that arrange these in horizontal and vertical spaces.

Although this type of development has the advantage of providing all the facilities required in a specific area, thus simultaneously maximizing the usability of the space, it involves many risks throughout the project, such as complicated interests in the major subjects, mixed development areas dominated by the civil and public spheres, operation and maintenance, and property management. In addition, there have been few studies on performance management for construction businesses because conventional performance management only measures the visible performance in businesses, such as financial and management performance. In particular, few studies have examined the risk factors that affect the performance management of mega projects.

Therefore, it is necessary to create a performance management method related to such risk factors to help estimate these factors' influence on a project in a timely and effective manner by developing a technology that continuously manages performance in relation to the risk factors in the early stages of mega projects and that suggests strategies for responses. Thus, we define an RPI for measuring the performance related to risks in construction businesses and derive calculation techniques and measurement methods. We then propose a new performance measurement method that considers the internal risk factors that affect the success or failure of the project in the context of the conventional cost/schedule-based approach.

4. Risk performance index and measurement systems

4.1 Definition of risk performance index

An Risk Performance Index (RPI) can assess the risk management in three-dimensional mixed-use development projects and can be combined with similar measurement systems such as that for the EVMS. The combined performance measurement index can then be used to measure the performance in the three aspects of cost/schedule/risk.

4.2 Definition of risk performance index

The RPI used in this study recognizes the internal risks in a project from the point of view of risk management and quantizes those risks as schedule and cost risk values based on the estimation of each residual risk. The RPI consists of 18 detailed indexes and variables. Table 2 shows the English terms for these indexes and variables with their descriptions and abbreviations.

No	Terminology	Description	Abb.
1	Cost Risk Performance Index	Performance Index measuring risks related to the project cost	CRPI
2	Schedule Risk Performance Index	Performance Index measuring risks related to the project schedule	SRPI
3	Forecasted Cost Risk Value	Cost Risk Value forecasted at the specified project time	FCRV
4	Forecasted Schedule Risk Value	Schedule Risk Value forecasted at the specified project time	FSRV
5	Residual Cost Risk Value	Cost Risk Value remaining after subtract eliminated cost risk from FCRV	RCRV
6	Residual Schedule Risk Value	Schedule Risk Value remaining after subtract eliminated schedule risk from FSRV	RSRV
7	Forecasted Cost Impact	Cost Impact forecasted at the specified project time	FCI
8	Forecasted Schedule Impact	Schedule Impact forecasted at the specified project time	FSI

No	Terminology	Description	Abb.
9	Actual Cost Impact	Cost Impact actually occurring from cost risk at the specified project time	ACI
10	Actual Schedule Impact	Schedule Impact actually occurring from schedule risk at the specified project time	ASI
11	Cost Impact Variance	Variance between FCI and ACI calculating at the specified project time	CIV
12	Schedule Impact Variance	Variance between FSI and ASI calculating at the specified project time	SIV
13	Actual Response Cost	Cumulative sum of actual costs responding to the forecasted cost risk at the specified project time	ARC
14	Actual Response Days	Cumulative sum of actual days responding to the forecasted schedule risk at the specified project time	ARD
15	Cost Risk Response Variance	Variance between ACI and ARC calculating at the specified project time	CRRV
16	Schedule Risk Response Variance	Variance between ASI and ARD calculating at the specified project time	SRRV
17	Cost Risk Response Effective	Actual cost risk response efficiency calculated from dividing CIV by ARC at the specified project time	CRRE
18	Schedule Risk Response Effective	Actual schedule risk response efficiency calculated from dividing SIV by ARD at the specified project time	SRRE

Table 2. Risk Performance Indexes.

4.2.1 Cost Risk Performance Index (CRPI)

As noted in Equation (1), the cost risk performance index (CRPI) can be calculated by subtracting the residual cost risk variance (RCRV) from the forecast cost risk variance (FCRV) and dividing by the FCRV at a specific point during the business period.

$$\text{CRPI} = (\text{FCRV}-\text{RCRV})/\text{FCRV} \tag{1}$$

where,
CRPI: Cost Risk Performance Index
FCRV: Forecasted Cost Risk Value
RCRV: Residual Cost Risk Value

The analysis of the CRPI can be performed as follows. First, if the CRPI is 1, then the RCRV is 0, showing the perfect elimination of the cost risk. It can also be seen that the residual risk in the project is 0, which is the best condition of the cost risk. Second, if the CRPI is greater than 0 and less than 1, it shows that the RCRV is lower than the FCRV. This means that although there are still some risks in the project, they are at a low level compared with the forecasts and so the cost risk shows a good status. Third, if the CRPI is 0, the FCRV is the same as the RCRV. Because this shows that there has been no reduction in the FCRV, it also shows no reduction in the cost risk. Fourth, if the CRPI is less than 0, it shows that the RCRV exceeds the FCRV, indicating an increase in the cost risk in the project. Table 3 shows the cost risk and its analysis method.

Index	Description
CRPI = 1	Best status, residual cost risk is 0, all cost risks have been eliminated.
0<CRPI<1	Good status, residual cost risks are smaller than forecasted cost risks.
CRPI = 0	Unchanged status, residual cost risks are equal to forecasted cost risks.
CRPI < 0	Bad status, residual cost risks are larger than forecasted cost risks.

Table 3. CRPI Analysis.

4.2.2 Schedule Risk Performance Index (SRPI)

The schedule risk performance index (SRPI) can be computed by subtracting the residual schedule risk variance (RSRV) from the forecast schedule risk variance (FSRV) and dividing by the FSRV at a specific point during the business period. The calculation formula can be expressed as Equation (2).

$$SRPI = (FSRV-RSRV)/FSRV \quad (2)$$

where,
SRPI: Schedule Risk Performance Index
FSRV: Forecasted Schedule Risk Value
RSRV: Residual Schedule Risk Value

The SRPI can be analyzed as follows. First, if the SRPI is 1, it shows that the RSRV is 0, indicating the perfect elimination of the schedule risk. The remaining risk in the project is 0, which shows the best condition of the schedule risk. Second, if the SRPI is greater than 0 and less than 1, it shows that the RSRV is lower than the FSRV. This means that although there are still some risks in the project, they are at a low level compared with the forecasts, indicating that the schedule risk is in an excellent state. Third, if the SRPI is 0, the FSRV is the same as the RSRV. Because this shows there is no reduction in the FSRV, it also shows no reduction in the schedule risk. Fourth, if the SRPI is less than 0, it shows that the RSRV exceeds the FSRV, indicating an increase in the schedule risk in the project. Table 4 shows the schedule risk and its analysis method.

Index	Description
SRPI = 1	Best status, residual schedule risk is 0, all schedule risks have been eliminated
0<SRPI<1	Good status, residual schedule risks are smaller than forecasted schedule risks.
SRPI = 0	Unchanged status, residual schedule risks are equal to forecasted schedule risks
SRPI < 0	Bad status, residual schedule risks are larger than forecasted schedule risks.

Table 4. SRPI Analysis.

4.2.3 Integrated Cost/Schedule Risk Performance Indexes

It is obviously possible to verify the change in the cost/schedule/risk according to the measurement points of the performance index using a method in which the cost/schedule/risk performance can be presented by integrating the CRPI and SRPI in a quadrant, as illustrated in Figure 1.

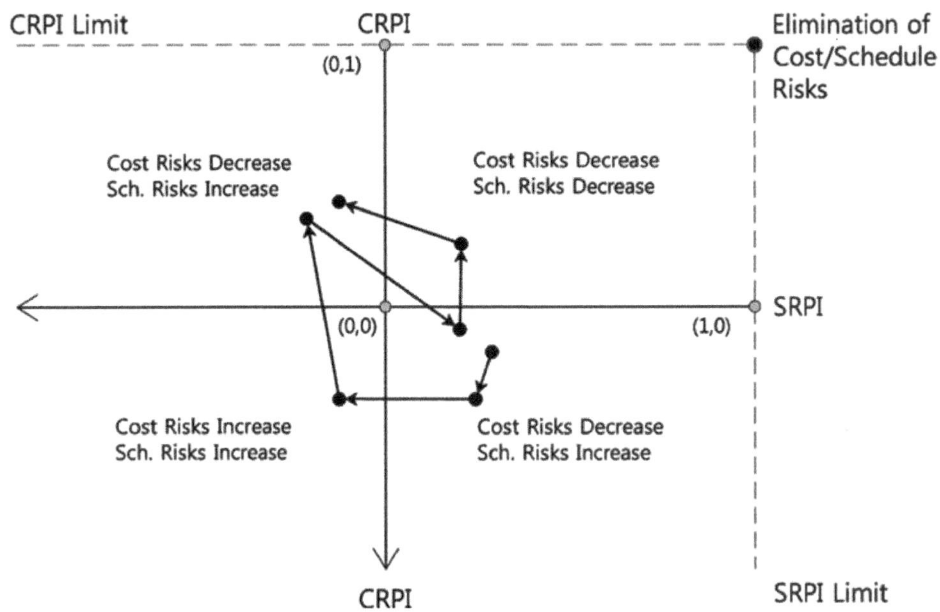

Fig. 1. Integrated Cost/Schedule Risk Performance Indexes.

The analysis of the integrated chart of the cost/schedule RPIs is as follows. First, if the CRPI and SRPI are both 1, it shows that the cost/schedule risks have been totally removed. Second, if the CRPI and SRPI are greater than 0, it shows an excellent condition in which the cost and schedule risks have all been reduced. Third, if the CRPI is greater than 0, but the SRPI is less than 0, the cost risk has decreased, but the schedule risk has increased. Fourth, if the CRPI is less than 0, but the SRPI is greater than 0, the cost risk has increased, but the schedule risk has decreased. Fifth, if the CRPI and SRPI are both less than 0, the cost risk and schedule risk have significantly increased and the project's state has deteriorated.

4.2.4 Cost Impact Variance (CIV), Schedule Impact Variance (SIV)

The cost impact variance (CIV) and schedule impact variance (SIV) verify the effective execution of the response to risks by comparing the cost/schedule impact forecast by the cost and schedule risks at a particular point with the cost/schedule impact that has actually occurred. These can be calculated by using Equations (3) and (4), respectively.

$$CIV = FCI - ACI \tag{3}$$

$$SIV = FSI - ARI \tag{4}$$

where,
CIV : Cost Impact Variance
FCI : Forecasted Cost Impact

ACI : Actual Cost Impact
SIV : Schedule Impact Variance
FSI : Forecasted Schedule Impact
ASI : Actual Schedule Impact

The analysis of the CIV and SIV can be performed as explained in Table 5.

Index	Description
CIV > 0	ACI is less than FCI, risk response has been efficient or cost risk has been decreased.
CIV < 0	ACI is greater than FCI, risk response has been inefficient or cost risk has been increased.
SIV > 0	ASI is less than FSI, risk response has been efficient or schedule risk has been decreased.
SIV < 0	ASI is greater than FSI, risk response has been inefficient or schedule risk has been increased.

Table 5. CIV, SIV Analysis.

4.2.5 Cost Risk Response Variance (CRRV), Schedule Risk Response Variance (SRRV)

The cost risk response variance (CRRV) shows the difference between the actual cost impact and the actual response cost impact investigated at a particular point, and the schedule risk response variance (SRRV) represents the difference between the actual schedule impact and the actual response schedule impact investigated at a particular point. The calculation of these values can be carried out using Equations (5) and (6), respectively.

$$CRRV = ACI - ARC \quad (5)$$

$$SRRV = ASI - ARD \quad (6)$$

where,
CRRV : Cost Risk Response Variance
ACI : Actual Cost Impact
ARC : Actual Response Cost
SRRV : Schedule Risk Response Variance
ASI : Actual Schedule Impact
ARD : Actual Response Days

The analysis of the CRRV and SRRV can be performed as explained in Table 6.

Index	Description
CRRV > 0	Cost risk response strategies are good.
CRRV < 0	Cost risk response strategies are bad
SRRV > 0	Schedule risk response strategies are good
SRRV < 0	Schedule risk response strategies are bad

Table 6. CRRV, SRRV Analysis.

4.2.6 Integrated Cost/Schedule Risk Response Variances

It is possible to examine the total change in the efficiency of the response strategy for the cost/schedule/risk by integrating the CRRV and SRRV in a quadrant as shown in Figure 2. The integrated chart of the cost/schedule risk response variances can be analyzed as follows. First, if the CRRV and SRRV are both greater than 0, it shows that the efficiency of the response strategy is excellent in both cases. Second, if the CRRV is greater than 0, but the SRRV is less than 0, it shows that the efficiency of the strategy of the SRRV is poor. Third, if the CRRV is less than 0, but the SRRV is greater than 0, the efficiency of the CRRV is poor, but the efficiency of the SRRV is good. Fourth, if the CRRV and SRRV are both less than 0, the efficiencies of the CRRV and SRRV are both poor.

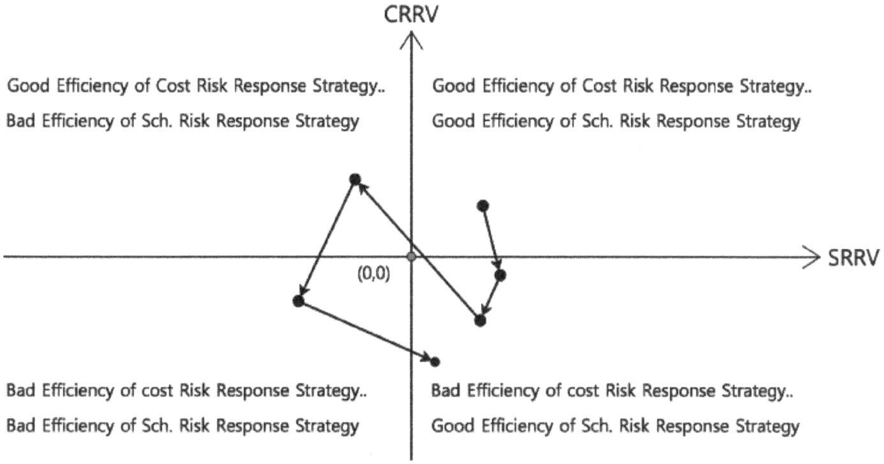

Fig. 2. Integrated Cost/Schedule Risk Response Variance Measurement.

4.2.7 Cost Risk Response Efficiency (CRRE)

The cost risk response efficiency (CRRE) measures the efficiency of the actual cost impact (ACI) vs the forecast cost impact (FCI) at a particular point during the project period. However, the FCI, ACI, and actual response cost (ARC) show different tendencies in their changes. In general, the three curves begin at 0, approach their peaks three-quarters of the way through construction, and return to 0 at the completion of the project. The scale of the changes in the curves is largest for FCI, but the changes in the ACI and ARC are about equal. Figure 3 illustrates the tendency in the change of the forecast vs actual cost impact and response cost. The difference between the FCI and the ACI becomes the CIV, and the difference between the ACI and the ARC becomes the CRRV.

As shown in Figure 3, the CRRE at a particular point during the project period can be obtained by dividing the CIV by the ARC. It can be expressed as Equation (7).

$$CRRE = CIV/ARC \tag{7}$$

where, CRRE: Cost Risk Response Effective
CIV: Cost Impact Variance
ARC: Actual Response Cost

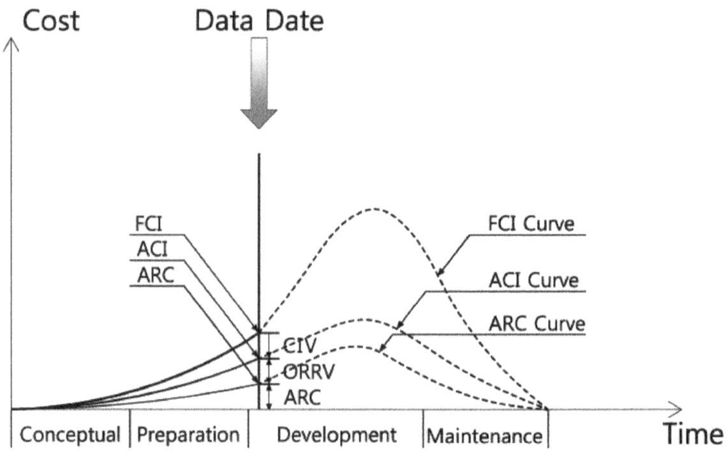

Fig. 3. Relationships between Forecasted/Actual Cost Impact and Response Cost.

The analysis of the CRRE can be performed as follows. First, if the CRRE is greater than 1, it shows an excellent condition for the CRRE because the ARC is greater than the CIV. Second, if the CRRE is 1, there is no CRRE because the CIV is the same as the ARC. Third, if the CRRE is less than 1, the CRRE shows a bad condition because the CIV at that point is greater than the ARC. The analysis of the CRRE is explained in Table 7.

Index	Description
CRRE > 1	Cost Risk Response Efficiency is good.
CRRE =1	Cost Risk Response Efficiency is nothing
CRRE < 1	Cost Risk Response Efficiency is bad

Table 7. CRRE Analysis.

4.2.8 Schedule Risk Response Efficiency (SRRE)

The schedule risk response efficiency (SRRE) measures the efficiency of the actual schedule impact (ASI) vs the forecast schedule impact (FSI) at a particular point during the project period. The difference between the FSI and the ASI becomes the SIV, and the difference between the ASI and the ARD becomes the SRRV. The SRRE at a particular point during the project can be obtained by dividing the CIV by the ARD. It can be expressed as Equation (8).

$$SRRE = SIV/ARD \tag{8}$$

where,
SRRE : Schedule Risk Response Effective
SIV : Schedule Impact Variance
ARD : Actual Response Days

The analysis of the SRRE can be performed as follows. First, if the SRRE is greater than 1, it shows an excellent condition in the SRRE because the ARD is greater than the SIV. Second, if the SRRE is 1, there is no SRRE because the SIV is the same as the ARD. Third, if the SRRE is less than 1, the SRRE shows a bad condition because the SIV at that point is greater than the ARD. The analysis of the SRRE is explained in Table 8.

Index	Description
SRRE > 1	Schedule Risk Response Efficiency is good.
SRRE =1	Schedule Risk Response Efficiency is nothing
SRRE < 1	Schedule Risk Response Efficiency is bad

Table 8. SRRE Analysis.

4.2.9 Relationship between Contingency Reserve (CR) and Actual Risk Cost (ARC)

The relationship between the contingency reserve (CR) and the actual risk cost (ARC) can be generally defined as follows.

As the project proceeds, the contingency reserve at the project start (CR_0) will decrease and the contingency reserve at the project completion (CR_{100}) becomes 0. On the other hand, the actual response cost at the project start (ARC_0) is 0, but as the project proceeds, the actual response cost will increase and the cumulative sum of actual response cost at the project completion (ARC_{100}) matches the contingency reserve at the project start (CR_0). Figure 4 shows the relationship between CR and ARC.

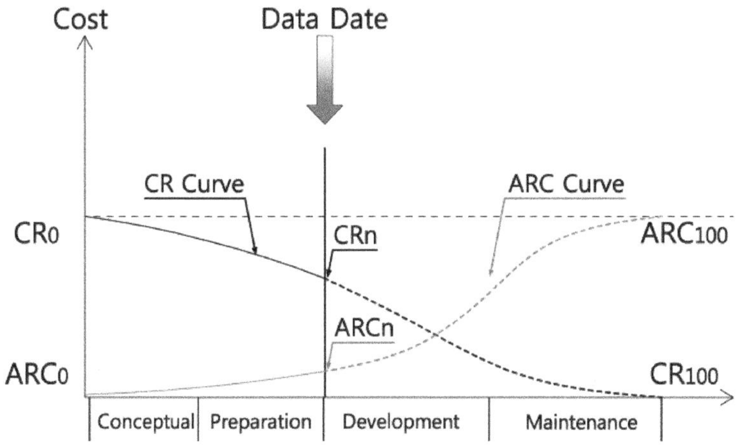

Fig. 4. Relationships between Contingency Reserve (CR) and Actual Risk Cost (ARC).

From Figure 4, the interpretation method of CR_n and ARC_n at a specified project time n is as follows. First, if $CR_0 = CR_n + ARC_n$, that is, if the sum of the contingency reserve and actual response cost is equal to the contingency reserve at the project start (CR_0), we can determine that the contingency reserve at the specified project time is appropriate. Second, if $CR_0 > CR_n + ARC_n$, that is, if the sum of the contingency reserve and actual response cost is less than the contingency reserve at the project start (CR_0), we can determine that project risks are decreasing and the contingency reserve at the specified project time should be reduced because it is too high. Third, if $CR_0 < CR_n + ARC_n$, that is, if the sum of the contingency reserve and actual response cost is greater than the contingency reserve at the project start (CR_0), we can determine that project risks are increasing and that the contingency reserve at the specified project time should be increased because it is too low. The analysis of CR_n and ARC_n at the specified project time n is explained in Table 9.

Index	Description
$CR_0 = CR_n + ARC_n$	Contingency Reserve at the specified project time is proper
$CR_0 > CR_n + ARC_n$	Project risks are decreasing or Contingency Reserve at the specified project time should be reduced because it is too much.
$CR_0 < CR_n + ARC_n$	Project risks are increasing or Contingency Reserve at the specified project time should be increased because it is too low

Table 9. CR_n and ARC_n Analysis.

4.3 Risk performance measurement tables

It is necessary to produce a format that verifies the risk factors existing in a mega project and their influences by analyzing the RPIs and calculation results proposed in this study. Thus, we classified the performance indexes into qualitative aspects that measure the risk performance as indexes, and quantitative aspects that measure risks in monetary amounts. We therefore propose the Qualitative Risk Performance Measurement Table and Quantitative Risk Performance Measurement Table, which can verify each risk factor and the results of the measurement as shown in Figures 5 and 6, respectively.

The Qualitative Risk Performance Measurement Table, in Figure 5, configures the forecast risk value (FRV) and residual risk value (RRV), which can be used as criteria for presenting the RPIs as columns that are calculated on a reference day, and shows the results of the calculation of the CRPI and SRPI based on this table. The Quantitative Risk Performance Measurement Table, in Figure 6, configures the FCI/FSI, ACI/ASI, and ARC/ARD, which can be used as criteria for presenting the risk performance as columns that are calculated on a reference day, and demonstrates the results of the calculation of the CIV/SIV, CRRV/SRRV, and CRRE/SRRE based on this table. It is evident that these risk performance measurement tables help the project manager to judge the scale, influence, and response efficiency of the various risk factors included in the mega project.

4.4 Risk performance measurement example

Figures 5 and 6 show the calculation of risk performance using existing housing redevelopment data. These examples nicely illustrate the theoretical and practical value, as well as the validity, of the risk performance measurement model proposed in this paper. The risk performance measurements in Figures 5 and 6 are evaluated every three months.

A qualitative risk performance measurement for the 'Low rate of apartment sales' on two risk factors is shown in Figure 5. Ratings on the probability scale and cost impact scale for April 1, 2010 were 4 and 5, respectively. Therefore, the forecasted cost risk value (FCRV) was calculated to be 20. Also, the rating on the schedule impact scale was 2, yielding a forecasted schedule risk value (FSRV) of 8. The residual risk values of the 'Low rate of apartment sales' were determined for the base date of July 1, 2010. With this reevaluation, the probability scale and cost impact scale values were lowered to 3 and 2, respectively, making the residual cost risk value (RCRV) 6. On the other hand, because the schedule impact scale value increased to 4, the residual schedule risk value (RSRV) is 12. Using the FCRV, FSRV, RCRV, and RSRV numbers in Equation (1), the cost risk performance index (CRPI) is 0.7. Using Equation (2), the schedule risk performance index (SRPI) is -0.5. A CRPI

			Previous Forecast Date : 2010. 4. 1																		Base Date : 2010. 7. 1																			
Risk ID	Description	Weight	**Forecasted RV**																		**Residual RV**																		**RPI**	
			Probability Scale					Cost Impact Scale					FCRV	Schedule Impac Scale					FSRV	Probability Scale					Cost Impact Scale					RCRV	Schedule Impact Scale					RSRV	CRPI	SRPI
			1	2	3	4	5	1	2	3	4	5		1	2	3	4	5		1	2	3	4	5	1	2	3	4	5		1	2	3	4	5			
II 3B2003	Low rate of apartment sales	0.35				o						o	20		o				8			o				o				6				o		12	0.7	-0.5
II 3B2003	Unreasonable requests from nearby residents	0.24					o			o			15			o			15				o					o		16			o			12	-0.07	0.2

Fig. 5. Qualitative Risk Performance Measurement Table.

(Unit : Thousand Won, Days)

Previous Forecast Date	Base Date
2010. 4. 1	2010. 7. 1

Risk ID	Description	Weight	Cost Risk Impact/Response Effective						Schedule Risk Impact/Response Effective					
			FCI	ACI	CIV	ARC	CRRV	CRRE	FSI	ASI	SIV	ARD	SRRV	SRRE
II 3B2003	Low rate of apartment sales	0.35	200,000	150,000	50,000	30,000	120,000	1.67	65	80	-15	86	-6	-0.17
II 3B2003	Unreasonable requests from nearby residents	0.24	50,000	30,000	20,000	35,000	-5,000	0.57	35	30	5	4	26	1.25

Fig. 6. Quantitative Risk Performance Measurement Table.

between 0 and 1 indicates that the cost risk has been effectively controlled, or the residual cost risks are smaller than the forecasted cost risks, as illustrated in Table 4. However, when the SRPI is less than 0, as it is in this case, the schedule risk has not been effectively controlled, or the residual schedule risks are higher than the forecasted schedule risks (see Table 4). This analysis of the CRPI and SRPI numbers tells the project team that they should focus on controlling the schedule risk of the 'Low rate of apartment sales.'

Figure 6 shows the results of a quantitative risk performance measurement for the same risk item, the 'Low rate of apartment sales.' With respect to cost risk, the forecasted cost risk impact (FCI) based on a previous forecast date was quantitatively determined to be 200,000,000 won, whereas the actual cost impact (ACI) as determined from the base date was 150,000,000 won. Thus, using Equation (3), we can see that the cost impact variance (CIV) is 50,000,000 won. A CIV of 50,000,000 won indicates that the cost risk response was effective, or cost risk has decreased, as shown in Table 5. Also, because the actual response cost (ARC) on the base date was 30,000,000 won, Equation (5) tells us that the cost risk response variance (CRRV) is 120,000,000 won, which means that the cost risk response strategies are good, as is shown in Table 6. Furthermore, using Equation (7), the cost risk response efficiency (CRRE) is calculated to be 1.67, and anything above 1 indicates good CRRE, as shown in Table 7. For schedule risk, the forecasted schedule risk impact (FSI) based on a previous forecast date was quantitatively determined to be 65 days, whereas the actual schedule impact (ASI) based on a base date was 80 days. Thus, using Equation (4), the schedule impact variance (SIV) is -15 days. An SIV less than 0 indicates that the schedule risk response was not effective, or the schedule risk has increased (see Table 5). Also, because the actual response days (ARD) value on the base date was 86 days, Equation (6) yields a schedule risk response variance (SRRV) of -6 days. An SRRV less than 0 means that the schedule risk response strategies are bad, as shown in Table 6. Furthermore, using Equation (8), we can see that the schedule risk response efficiency (SRRE) is -0.17, and anything less than 0 indicates poor SRRE (see Table 8).

4.5 Value and validity of risk performance index and measurement system

Generally, project risk management includes risk identification, analysis, and response at a project-specific time. The traditional EVMS cannot conduct the project performance measurement considering the project uncertainties and risks integrated with the cost and schedule. However, the risk performance indexes and measurement system proposed in this paper account for changing project risks, the evaluation of residual risk values, and the efficiency of risk response strategies by periodically comparing previous forecasted risk performance variables with those at a base date—risk performance indexes are calculated every three months rather than at one project-specific point in time. Furthermore, the measurement system integrates the traditional EVMS and risk management concepts by considering project risks during the project performance measurement.

5. Conclusion

This chapter has proposed risk performance indexes to improve the efficiency of the general performance measurement for mega projects by extending the existing cost/schedule-based performance measurement system. The expected effects of the risk performance index method proposed in this study can be summarized as follows.

First, we constructed our system to be similar to the EVMS, which is the existing cost/schedule integrated performance measurement method. It is therefore possible to conduct three-dimensional integrated performance management using the 18 detailed indexes and variables employed in the risk performance index.

Second, we can perform integrated qualitative performance measurement for cost/schedule/risk by measuring the risk-related cost performance index and schedule performance index.

Third, we can perform integrated quantitative performance measurement for cost/schedule/risk by measuring the cost impact variance, the schedule impact variance, the cost risk response variance, and the schedule risk response variance.

Fourth, we can measure the risk response efficiency by comparing the cost impact variance with the actual response cost, and we have proposed a method to analyze the extra project expenses and actual response cost at a particular point during the project.

Furthermore, using the risk performance measurement of 'Low rate of apartment sales' as an example, the theoretical and practical value and validity of our risk performance indexes and measurement system can be summarized as follows: first, because risk is a dynamic phenomenon, the forecasting and reevaluation of risk factors should be performed periodically; second, our risk performance indexes provide the theoretical foundation for an integrated evaluation of cost and scheduling risks inherent in housing redevelopment projects; and third, by using our risk performance indexes and measurement model, a project team is required to forecast and evaluate project uncertainties and risks continually, thereby generating more proactive and diverse analyses than the traditional EVMS model.

6. Acknowledgment

This study was performed by the 07 high-tech urban development project (Project No: 07 Urban Regeneration B03) implemented by the KICTEP and sponsored by the Ministry of Land, Transportation and Maritime Affairs.

7. References

Department of Energy Guide (2008). *Earned Value Management System*, U.S. Department of Energy

Fleming, Q. W.; Koppleman, J. M. (1996), *Earned Value Project Management*, Project Management Institute

Flyvbjerg, B.; Bruzelius, N.; Rothengatter, W. (2003), *Megaprojects and Risk-An Anatomy of Ambition*, Cambridge University Press

Kaplan, R. S.; Norton, D. P. (1993). *Putting the Balanced Scorecard to Work*, Harvard Business Review

Kim, Seon-Gyoo (2010), Risk Performance Indexes and Measurement Systems for Mega Construction Projects. *Journal of Civil Engineering and Management*, 2010 16(4), pp. 586-594, ISSN 1822-3605

The KPI Working Group (2000). KPI *Report for the Minister for Construction*, Department of the Environment, Transportation and the Regions, UK

12

Measurement and Modeling Techniques for the Fourth Generation Broadband Over Copper

Diogo Acatauassu[1], Igor Almeida[1], Francisco Muller[1], Aldebaro Klautau[1], Chenguang Lu[2], Klas Ericson[2] and Boris Dortschy[2]
[1]*Federal University of Pará*
[2]*Ericsson AB*
[1]*Brazil*
[2]*Sweden*

1. Introduction

Digital subscriber lines (DSL), the broadband data transmission technologies that use the copper cable as channel, are the most used Internet media around the world with more than 300 million users (Oksman et al., 2010), Much of the DSL success is related to the cost-benefit for both operators and served consumers. As the transmission channel is the common copper twisted-pair telephone cable, there is no need for large investments in infrastructure, because the telephone network is largely consolidated and active in almost all the world.

Over the years, DSL systems suffered a step-by-step evolution, being divided in different generations according to the increase of its data rates and technology improvements (Odling et al., 2009). A lot of elements were evolved in this process. Two of them can be highlighted. The first one was the development of modern signal processing techniques to avoid crosstalk effects in DSL transmissions, such as the so-called dynamic spectrum management (DSM) techniques (Cendrillon & Moonen, 2005; Moraes et al., 2010; Oksman et al., 2010; Song et al., 2002). The second one was the reduction of the used copper cable length, resulting in a consequent reduction in the attenuation imposed on the transmission signals. This fact allowed the increase of the used bandwidths, and consequently the data rates.

The standards of the first and the second generations DSL, matching the integrated services digital network (ISDN) (ITU-T, 1991) and asymmetric DSL (ADSL) family (ITU-T, 1999a; 2003; 2005a), were developed to operate over several kilometers copper cables achieving maximum data rates up to 20 Mb/s. The standards of the third generation DSL, matching the very high speed DSL (VDSL) family (ITU-T, 2005b; 2006), were developed to operate over hundred meters copper cables, achieving maximum data rates up to 100 Mb/s. This case, when users do not live near to the service provider's central office, fiber to street cabinets (FTTC) are inevitable to make the copper loop sufficiently short (van den Brink, 2010).

The fourth generation DSL systems will try to explore the copper twisted-pair cable to the maximum, reducing it to few meters and joining it to hybrid optical fiber access architectures, such as fiber to the home (FTTH) (Magesacher et al., 2006; Odling et al., 2009; van den Brink, 2010). The copper cable length reduction process will allow unused frequencies, far above the 30 MHz VDSL2. Recent works described that this frequency values can achieve 100 MHz (Magesacher et al., 2006; Odling et al., 2009; van den Brink, 2010), even reaching 300 MHz, depending on the quality of the used cable (van den Brink, 2010). Results of capacity simulations showed that this transmission channels can achieve near 1 Gb/s, and the use of DSM techniques to avoid crosstalk effects can help ensuring these data rates (Acatauassu et al., 2009; Magesacher et al., 2006; Odling et al., 2009).

It is important to notice that the use of FTTH does not necessary mean that fiber is deployed all the way to a point inside the home. The costs for installation, digging and putting the fiber into every subscriber house are so significant that the required investment is disproportional to the current market (Odling et al., 2009; van den Brink, 2010; Vergara et al., 2010) making it impracticable. So, there will be, in most cases, a lack for the last 50-300m transmission channel, which can be easily used by the existing telephony wiring. To know the behavior of short copper cables, transmitting data in unexplored frequencies, is essential to the development and implementation of the fourth generation DSL systems, and is the focus of this chapter.

The text describes a measurement campaign, that was performed in order to obtain the direct transfer functions and far-end crosstalk transfer functions of 50m, 100m and 200m copper cables. The description of the measurement techniques includes the used equipments, the experimental setup and the reference parameters used during the experiments. Moreover, this chapter introduces a simple procedure for fitting a well-know copper cable model, commonly used for performance evaluation of current DSL standards. The obtained fitted model was compared to the results of the short copper cable measurements, and showed good performance, indicating it can be used in frequency domain-based simulations. In fact, as another contribution, this chapter describes some preliminary simulations results in order to evaluate the fourth generation DSL systems performance, in terms of achievable data rates, optimization of transmission parameters and verification of the rates degradation due to the effects of uncanceled crosstalk.

The text is organized as follows, Section 2 briefly describes the state of art and the standardization effort held by the International Telecommunication Union for the fourth generation DSL systems. Section 3 describes the well-known theory behind the twisted-pair copper cable characterization, and shows some cable reference models used for performance evaluation of the current DSL standards. Section 4 describes a measurement campaign, performed in order to characterize the quality of short copper cables in terms of direct transfer functions (which are related to the loop attenuation) and crosstalk transfer functions (which describe signal leakages of twisted-pairs inside the cables). Continuing, Section 5 describes a modeling technique based on fitting the simplified twisted-pair copper cable model, which is commonly used for performance evaluation of current DSL standards. The data obtained by the developed fitted model is then compared to the results obtained by the short cable measurements. Section 6 shows some preliminary simulations in order to evaluate the performance of the fourth generation DSL systems, including the verification of the achievable data rates and optimization of transmission parameters, and, finishing, Section 7 describes the conclusions of the chapter.

2. State of art and standardization of the fourth generation DSL

Recently, the International Telecommunication Union (ITU) started the fourth generation broadband over copper standardization under the working name G.Fast, where Fast means Fast Access to Subscriber Terminals (van den Brink, 2010). It came due to the desire of Telcos, industries and universities, which observed the potential of this technology, based mainly by the economic point of view. As the costs for installation, digging and putting optical fiber into the ground for every broadband subscriber are so significant, the required investment becomes disproportional to the current market demand (Odling et al., 2009; van den Brink, 2010), so there is a gap that can be bypassed by using the existing telephone wiring.

A comparison between the fourth generation DSL proposal and older DSL generations, in terms of desirable data rates, transmission bandwidth and typical length of the used copper cable is illustrated in Figure 1.

Some initial available works describing the research on the fourth generation broadband over copper can be find in (Acatauassu et al., 2009; Magesacher et al., 2006; Odling et al., 2009; van den Brink, 2010).

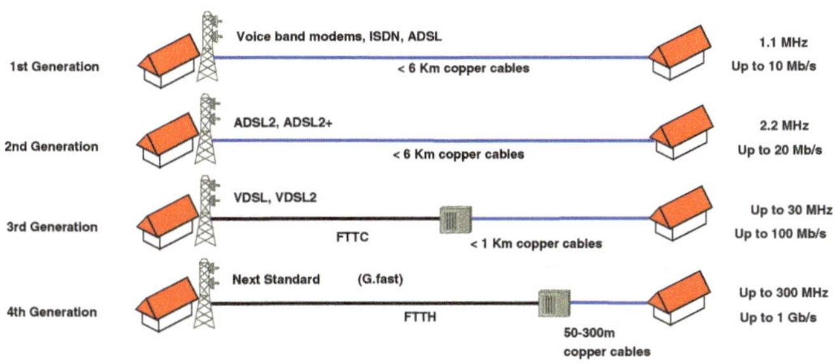

Fig. 1. Comparison of DSL Generations.

3. Twisted-pair copper cables characterization theory

A key point of any communication system, does not matter if wireline or wireless, is to characterize the transmission channel. Copper cables have been used by communication systems for more than a century (Chen, 1998), and play the hole of transmission channel in DSL systems.

Twisted-pair copper cables are usually bundled together in a cable sheath with 25 to 100 twisted-pairs. Each wire is also coated with some type of insulating material such as a paper-based (PULP) or a plastic-based (PIC) material. The rate of twisting is usually in the range of 12 to 40 turns per meter. They are also characterized by the diameter of the copper wire (gauge). The ETSI (European Telecommunications Standards Institute) defines the gauges in millimeters, with diameters ranging from 0.9mm to 0.32mm. The ANSI (American National Standards Institute) defines the gauges using the American Wire Gauge (AWG) designation, with typical values ranging from 19 AWG to 26 AWG (Chen, 1998; Yoho, 2001).

The electrical characteristics of twisted-pair copper cables are defined using the classical transmission line model, which can be described by an equivalent circuit built up with four frequency-dependent parameters: series resistance R, series inductance L, shunt capacitance C and shunt conductance G. The RLGC parameters are known as primary parameters (Chen,

1998). Figure 2 illustrates the transmission line model of a segment of copper twisted-pair, per unit length dz.

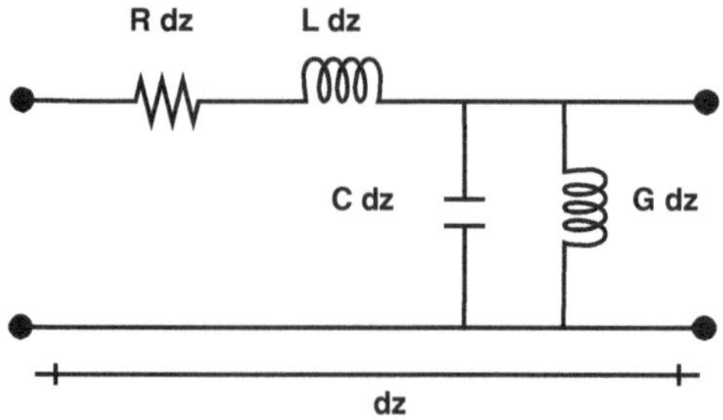

Fig. 2. Transmission line model per unit length dz.

The RLGC primary parameters completely determine a second set of electrical characteristics called secondary parameters. The secondary parameters consist of the propagation constant γ and the characteristic impedance Z_0, which are given in Equations 1 and 2 respectively. Note that both are frequency dependent too.

$$\gamma = \alpha + j\beta = \sqrt{(G + j\omega C)(R + j\omega L)}, \tag{1}$$

$$Z_0 = \sqrt{\frac{R + j\omega L}{G + j\omega C}}, \tag{2}$$

The real part of the propagation constant γ is the attenuation constant α, which represents the loss in the transmission line. The imaginary part, β, represents phase constant. The phase constant relates the wavelength with the phase velocity for each frequency component of the signal (Yoho, 2001).

The secondary parameters are essential to model twisted-pair transmission lines trough the two-port network - 2PN representation. In transmission line theory, a common way to represent a 2PN is to use the transmission matrix, also known as the ABCD matrix (Chen, 1998; Starr et al., 1999). It relates the voltage and current of the input port to the voltage and current of the output port. A cascade connection of two or more two-port networks can be found by simply multiplying the ABCD matrices of each individual two-port network; this aspect of the ABCD parameters allows easy evaluation of copper loops due to the chain-type nature of the topologies (Yoho, 2001). Figure 3 illustrates a two-port network representation, where V_1 and I_1 are the input voltage and input current and V_2 and I_2 are the output voltage and output current. The way they are related are shown in Equations 3 and 4,

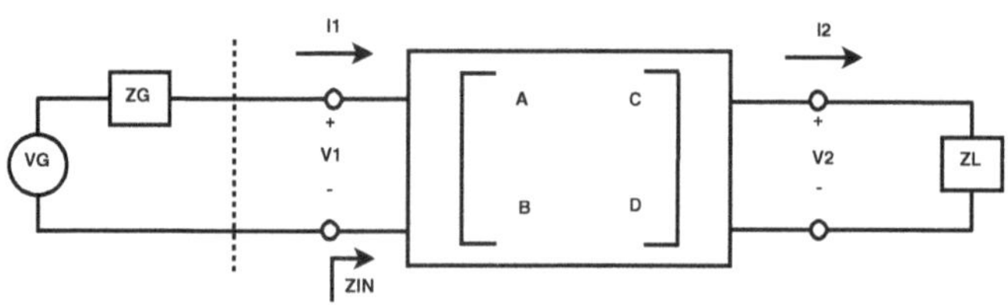

Fig. 3. Two-port network model.

$$V_1 = AV_2 + BI_2, \tag{3}$$

$$I_1 = CV_2 + DI_2, \tag{4}$$

Or as a matrix, as follows,

$$\begin{bmatrix} V_1 \\ I_1 \end{bmatrix} = \begin{bmatrix} A & B \\ C & D \end{bmatrix} \begin{bmatrix} V_2 \\ I_2 \end{bmatrix}, \tag{5}$$

The secondary line parameters γ and Z_0 are related to the ABCD parameters as follows, where l is the cable length,

$$\begin{bmatrix} A & B \\ C & D \end{bmatrix} = \begin{bmatrix} \cosh(\gamma l) & Z_0 \sinh(\gamma l) \\ \frac{1}{Z_0} \sinh(\gamma l) & \cosh(\gamma l) \end{bmatrix}, \tag{6}$$

Figure 3 also shows the source voltage V_G and source impedance Z_G (which are commonly the voltage and impedance of the data transmitter), the load impedance Z_L and the input impedance Z_{IN}, which can be found as follows,

$$Z_{IN} = \frac{V_1}{I_1} = \frac{AZ_L + B}{CZ_L + D}, \tag{7}$$

The direct transfer function of the copper twisted-pair is defined as the ratio of the voltage across the output port of the loop with the voltage placed on the input port. It can be very useful in order to indicates how attenuated a signal becomes after it passed trough the channel. Equation 8 describes it when the input reference is the source voltage.

$$H = \frac{V_2}{V_G} = \frac{Z_L}{Z_G[CZ_L + D] + AZ_L + B}, \tag{8}$$

The measurements of direct transfer functions of short twisted-pair copper cables at unexplored DSL frequencies will be detailed in Section 4.

3.1 Crosstalk

Crosstalk is the leakage into one channel of signal power from another channel. For DSL, this means coupling between twisted-pairs in the same copper cable. The coupling mechanism is a consequence of the cable's construction. It increases both with cable length and frequency (Starr et al., 1999). Crosstalk is very random in nature and its transfer functions differ from wire-pair combination to wire-pair combination. These differences can be significantly: up to tens of dB between the worst and best wire-pair combination (ETSI, 1997-02).

Crosstalk is one of the main impairments for DSL systems, it can reduce the full potential of broadband access over twisted-pair copper cables (Starr et al., 1999). It can present itself in two ways: far-end crosstalk (FEXT), which describes the coupled signals that originate from the end opposite of the affected twisted-pair, and near-end crosstalk (NEXT), which describes the coupled signals that originate from the same end as the affected twisted-pair (Oksman et al., 2010). Figures 4 and 5 illustrate FEXT and NEXT.

Crosstalk is an additional quantity that characterizes twisted-pair copper cables. The measurements of far-end crosstalk transfer functions of short copper cables at unexplored DSL frequencies will be detailed in Section 4.

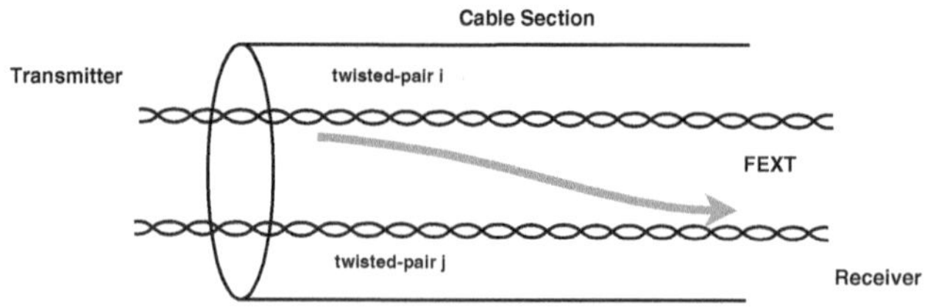

Fig. 4. Far-end crosstalk.

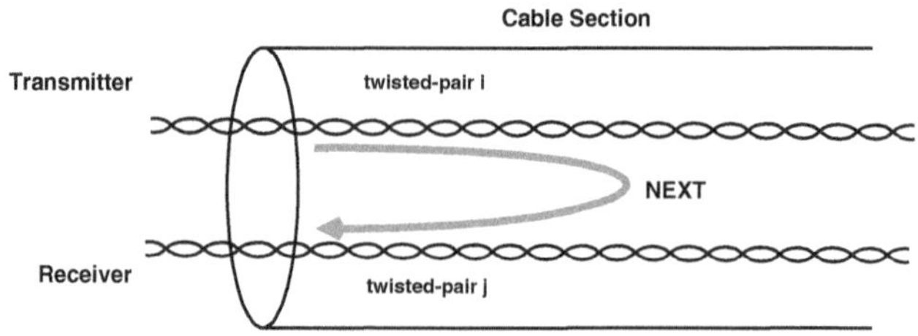

Fig. 5. Near-end crosstalk.

3.2 Cable models

In order to predict the transmissions behavior over twisted-pair copper cables in different topologies and scenarios (without real measurements data), DSL engineers have used the so-called cable models. These mathematical models can be very useful when applied in frequency domain-based computer simulations for performance evaluation of twisted-pair networks. The current cable models have proven to be especially useful to describe the behavior of twisted-pair copper cables over a range of frequencies from DC to tens of MHz with good precision (Chen, 1998; Starr et al., 1999). Two well-known cable models are described in this Subsection. It is important to notice that both models are not suitable for time domain evaluations due their non-causality.

3.2.1 British Telecom models - BT0 and BT1

The British Telecom (BT) models are focused in modeling twisted-pair copper cables by their primary parameters, RLGC (ETSI, 1997-02). Based on the primary parameters, their direct transfer functions can be easily found by the two-port network modeling approach using the ABCD matrix. The RLCG functions were fit to data produced from measurements performed by the BT laboratories. Two models are defined, BT0 and BT1 (ETSI, 1997-02). The RLGC functions of BT0 model are shown in Equations 9, 10, 11 and 12.

$$R(f) = \sqrt[4]{r_{oc}^4 + ac.f^2}, \tag{9}$$

$$L(f) = \frac{l_0 + l\infty\left(\frac{f}{f_m}\right)^b}{1 + \left(\frac{f}{f_m}\right)^b}, \tag{10}$$

$$G(f) = g_0.f^{g_e}, \tag{11}$$

$$C(f) = c_\infty + c_0.f^{-c_e}, \tag{12}$$

The values of the RLGC sub-parameters, e.g, r_{oc}, l_{∞}, ac, c_e, are defined according to the cable's gauge and material. ETSI and ANSI defined reference documents containing the correct values for each RLGC sub-parameter for different types of twisted-pair copper cables (ETSI, 1997-02).

BT1 model uses the same equations for LGC parameters. The only different function is related to R, which is given as follows,

$$R(f) = \frac{1}{\frac{1}{\sqrt[4]{r_{oc}^4 + ac.f^2}} + \frac{1}{\sqrt[4]{r_{os}^4 + as.f^2}}}, \tag{13}$$

3.2.2 Simplified twisted-pair cable model

The simplified twisted-pair cable model is described in (Chen, 1998) and (Starr et al., 1999). The mathematical approach of this model is focused on a function that takes into account the cable gauge, the cable length and its propagation constant for a given frequency range. For a perfectly terminated loop, that is, for a loop which is terminated with its characteristic impedance, the direct transfer function is given as follows, where d is the loop length and f is the frequency,

$$H(d,f) = \epsilon^{-d\gamma(f)} = \epsilon^{-d\alpha(f)}\epsilon^{-jd\beta(f)}, \tag{14}$$

For frequencies higher than 250 kHz, the transfer function can be simplified as follows,

$$H(d,f) = \epsilon^{-d(k_1\sqrt{f}+k_2 f)}\epsilon^{-jdk_3 f}, \tag{15}$$

where k_1, k_2 and k_3 are fitting parameters related to the cable gauge. For frequencies up to 30 MHz, their values are fixed as shown in Table 1.

Gauge	k_1x(10^{-3})	k_2x(10^{-8})	k_3x(10^{-5})
22	3.0	0.35	4.865
24	3.8	-0.541	4.883
26	4.8	-1.709	4.907

Table 1. Parameters for the simplified twisted-pair cable model. This values are defined for frequencies up to 30 MHz.

It can be noticed that the simplified twisted-pair cable model does not takes into account the two-port network modeling approach using the ABCD matrix. It derives the direct transfer-function of a given twisted-pair directly by one equation.

This cable model will be used as basis to derive a fitted twisted-pair copper cable model for the fourth generation DSL systems, that will be described in Section 5.

4. Measurement techniques for the fourth generation DSL

Despite the usefulness of cable models, real measured data is the best way to characterize twisted-pair copper cables.

This section describes a measurement campaign, performed in order to evaluate the direct transfer functions and far-end crosstalk transfer functions of short twisted-pair copper cables.

It was used during the measurements 50m, 100m and 200m cables, including two different gauges, 0.4mm and 0.5mm (26 AWG and 24 AWG respectively). The results of these measurements can serve as a baseline for the observation of short copper cable's behavior when submitted to transmit high frequency signals, such as the fourth generation DSL proposes to implement. Recent works described that this frequency values can achieve 100 MHz, 200 MHz, even 300 MHz, depending on the cable quality (Magesacher et al., 2006; Odling et al., 2009; van den Brink, 2010).

The direct transfer functions and far-end crosstalk transfer functions were measured in frequency domain using a gain/phase-network analyzer. The measurement parameters were sent to the network analyzer via GPIB (General Purpose Interface Bus) interface, through a MATLAB script, and defined, among other things: the measured bandwidth, the number of sub-bands, the number of measurements on each twisted-pair and the calibration standards.

As current DSL systems apply FDD (Frequency Division Duplexing), which can mitigate the effects of near-end crosstalk, no NEXT was measured.

4.1 Equipments, setup and measurement procedures

The equipments used during the measurements were:

- Agilent Network Analyzer 4395A;
- 2 Baluns North Hills 0301BB (10kHz - 600MHz, 50 UNB, 100 BAL);
- Agilent 87512A (Signal Splitter - measurement B/R);
- 100 Ω impedance matching resistors;
- Trim trio conectors;
- 50m 0.4mm 16 pair copper cable;
- 100m 0.5mm 24 pair copper cable;
- 200m 0.5mm 24 pair copper cable;
- 200m 0.4mm 16 pair copper cable;

The used cables were chosen randomly in a set of short copper cables. Trim trio connectors were then soldered in the beginning and in the end of each individual twisted-pair inside this cables. During each measurement, the selected twisted-pair was connected to the network analyzer through the baluns, that was responsible for the balanced and unbalanced signal conversion [1]. The baluns also had the function to convert the system impedance from 50 Ω of the network analyzer to 100 Ω of the copper loop (Agilent, 2000). Figure 6 illustrates the measurements experimental setup, through the connection of the used equipments.

It is important to describe that the temperature conditions were kept constant during the measurements. Figure 7 illustrates the real equipments used during the experiments such as the network analyzer, the twisted-pair copper cables, the baluns and the trim trio connectors.

The measurements management was done by a MATLAB script that was coded to operate this way:

- Calibrates the network analyzer;

[1] The balanced transmission method is used by the copper loop, however almost all measurement instruments have unbalanced input and output ports

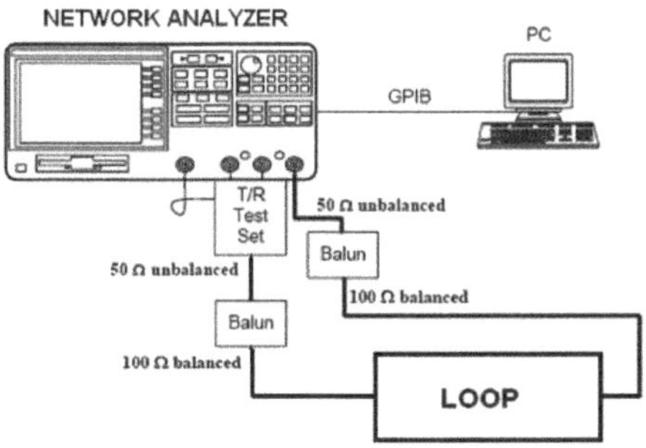

Fig. 6. Experimental setup used during the measurements campaign.

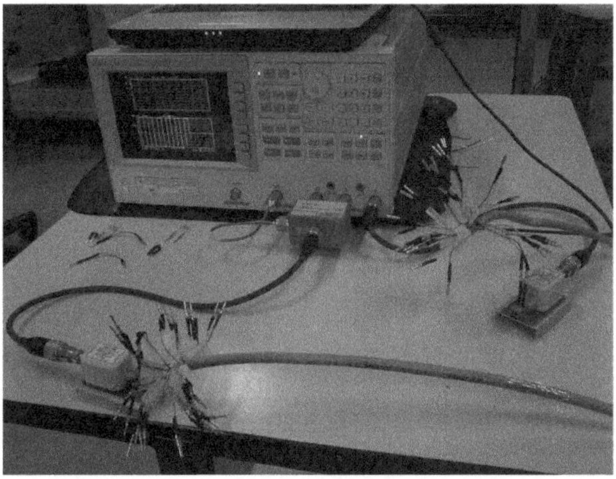

Fig. 7. Overview of the used equipments: network analyzer, twisted-pair cables, baluns and trim trio conectors.

- Sets the defined measurements parameters, such as the measured bandwidth and the number of measurements on each twisted-pair;
- Measures the twisted-pair;
- Checks if there was any outliers by using the Dixon's test;
- Saves the results in ASCII files;

It was defined that three measurements on each pair, without outliers, ensured the results reliability. The main measurements parameters were:

- Start frequency: 0 Hz;
- Stop frequency: 100 MHz;
- Number of sub-bands: 6;
- Number of measurements on each twisted-pair: 3;

- Calibration standards: open, short, through;
- Display of channel 1: Magnitude;
- Display of channel 2: Phase;

The measurements were performed only for positive frequencies, from start frequency to stop frequency, in low-pass mode. It was assumed a uniform sampling of the frequency axis, which means that neighboring points were separated by Δf Hz, which is called frequency resolution. For the measured 50m 0.4mm and 200m 0.4mm cables, Δf was set to 43.125 kHz, while for the measured 100m 0.5mm and 200m 0.5mm, Δf was set to 4.3125 kHz.

4.2 Measurements results

As described in Subsection 4.1, three different lengths of short copper cables were measured. The used 50m cable was a 16 pair 0.4 mm (26 AWG). This cable measurements resulted in 256 channel transfer functions: 16 direct transfer functions and 240 far-end crosstalk (FEXT) transfer functions. The used 100m cable was a 24 pair 0.5mm (24 AWG). This cable measurements resulted in 576 channel transfer functions: 24 direct transfer functions and 552 FEXT transfer functions. For the 200m length, two different gauges were used. The first cable was a 16 pair 0.4mm. This cable measurements resulted in 256 channel transfer functions: 16 direct transfer functions and 240 FEXT transfer functions. The second cable was a 24 pair 0.5 mm. This cable measurements resulted in 576 channel transfer functions: 24 direct transfer functions and 552 FEXT transfer functions. The measured bandwidth for all fours cables was 100 MHz.

Figures 8, 9, 10 and 11 show some recorded data from the 50m, 100m and 200m copper cables measurements in terms of direct transfer functions and FEXT transfer functions. It can be observed the direct channel attenuation increase due to the copper cable length increase. For example, at 100 MHz using the measured 50m 0.4mm cable the magnitude is near -15 dB while at the same 100 MHz using the measured 200m 0.5mm cable the magnitude is near -40 dB.

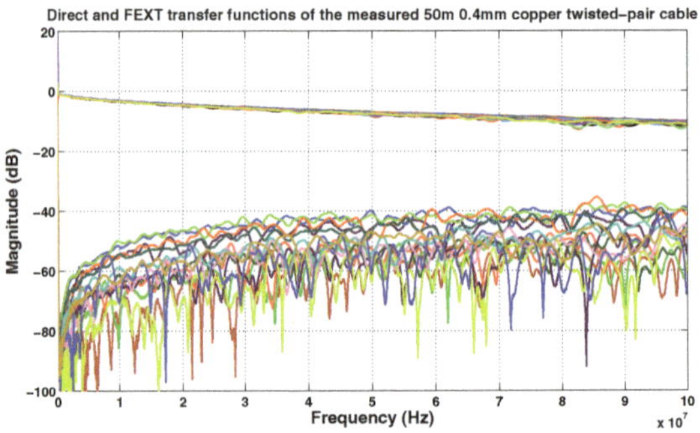

Fig. 8. Direct transfer functions (upper curves) and FEXT transfer functions (lower curves) of the measured 50m 0.4mm copper cable.

5. Modeling techniques for the fourth generation DSL

The current cable models have proven to be especially useful to describe the behavior of twisted-pair copper cables over a range of frequencies from DC to tens of MHz (e.g, 30 MHz), that is, the frequency range of the current DSL standards (Chen, 1998; Golden et al., 2006).

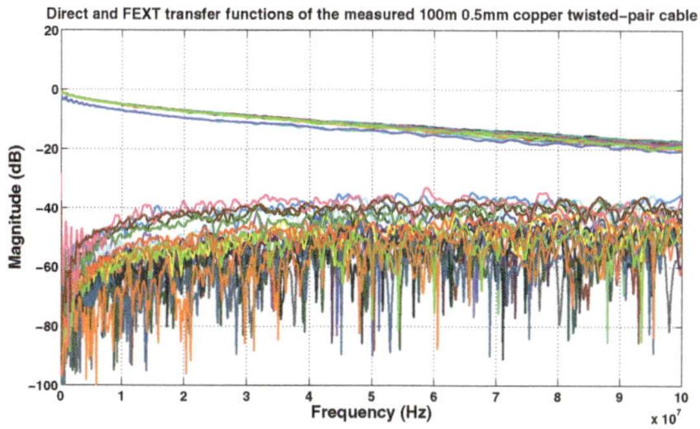

Fig. 9. Direct transfer functions (upper curves) and FEXT transfer functions (lower curves) of the measured 100m 0.5mm copper cable.

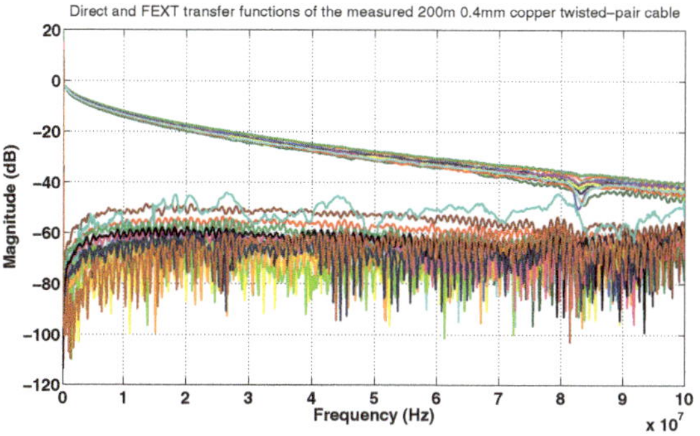

Fig. 10. Direct transfer functions (upper curves) and FEXT transfer functions (lower curves) of the measured 200m 0.4mm copper cable.

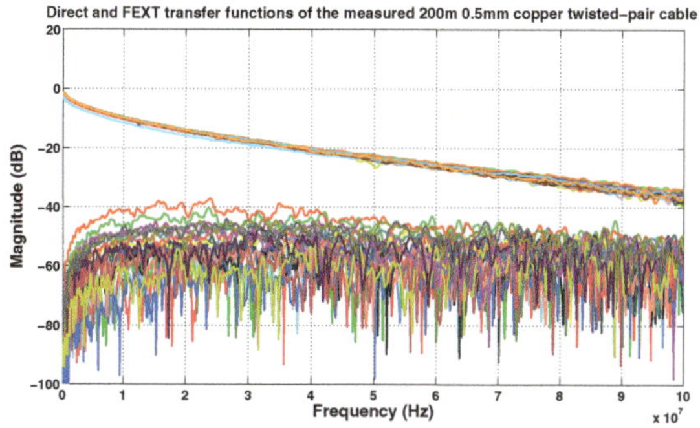

Fig. 11. Direct transfer functions (upper curves) and FEXT transfer functions (lower curves) of the measured 200m 0.5mm copper cable.

In order to help researchers and students all over the world to do their own simulations for performance evaluation of the fourth generation DSL systems, this Section describes a modeling technique for fitting the simplified twisted-pair cable model, described in

Sub-subsection 3.2.2. The objective is to make it suitable to describes the direct transfer functions behavior of short twisted-pair copper cables up to 100 MHz.

The fitting process is performed for one cable gauge (0.4mm), but the same procedure can be done to derive fitted models for other cable gauges. It was focused in find the best values for the three parameters of the simplified twisted-pair cable model, k_1, k_2 and k_3, in order to make the generated transfer functions as near as possible to the obtained by the average of the measurements described in Subsection 4.2.

The fitting procedure was done this way:

- It was chosen the 0.4mm gauge (26 AWG);
- The RLGC parameters were calculated using BT0 equations, described in Equations 9, 10, 11, 12. The values of each RLGC sub-parameter were taken from (ITU-T, 1999b). It is important to notice that these values are well-defined for frequencies up to 30 MHz;
- Based on the RLGC parameters, it was calculated the propagation constant γ, described in Equation 1;
- Based on the propagation constant γ, it was calculated the attenuation constant α;
- k_2 was found by the resolution of the system shown in Equation 16;
- k_1 was found by brute force;
- k_1 and k_2 was set in order to minimize the mean square error between them and α, as shown in Equation 17. k_3 was found by observing β;

$$\begin{aligned} k_1 \sum_{i=1}^{n} f_i + k_2 \sum_{i=1}^{n} f_i^{1.5} &= \sum_{i=1}^{n} f_i^{0.5} \alpha_i(f_i) \\ k_1 \sum_{i=1}^{n} f_i^{1.5} + k_2 \sum_{i=1}^{n} f_i^{2} &= \sum_{i=1}^{n} f_i \alpha_i(f_i), \end{aligned} \tag{16}$$

$$\epsilon_i^2 = (k_1 \sqrt{f_i} + k_2 f_i - \alpha_i(f_i))^2, \tag{17}$$

The obtained parameters for the 0.4mm gauge cable were:

- k_1=3.9e^{-3};
- k_2=0.16750^{-8};
- k_3=3.0e^{-5};

Figures 12 and 13 illustrate the comparison between the direct transfer functions obtained by the fitted simplified twisted-pair cable model (proposed by this work), the average of the real cable measurements (performed by this work), the original simplified twisted-pair cable model (described in (Chen, 1998) and (Starr et al., 1999)) and BT1 model (described in (ETSI, 1997-02)) for 50m 0.4mm and 200m 0.4mm copper cables respectively.

It is clear the the direct transfer functions obtained by the fitted simplified model proposed by this work are closer to the obtained by the real measurements. It indicates that the fitted model can be suitable for frequency domain-based simulations in order to evaluate the performance of the fourth generation DSL systems. It is important to notice that good behavior in frequency

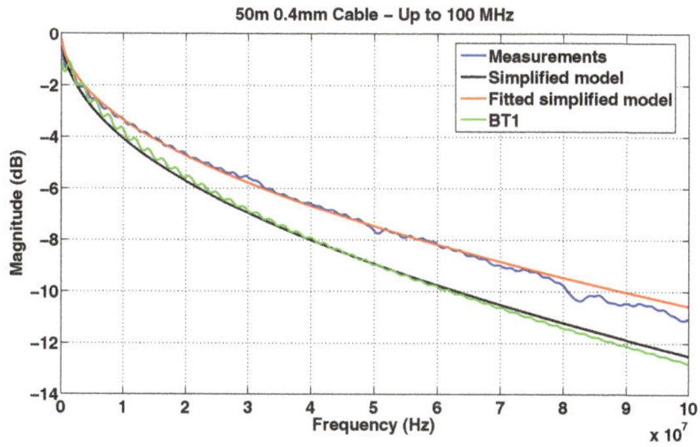

Fig. 12. Direct transfer functions obtained by the fitted simplified model, the average of measurements, the original simplified model and BT1 model. 50m 0.4mm cable.

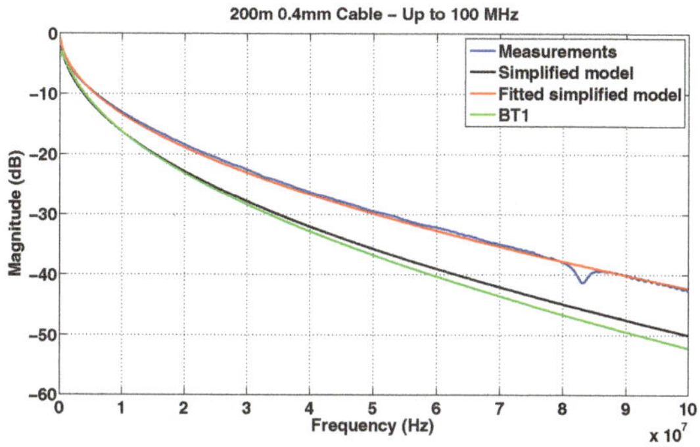

Fig. 13. Direct transfer functions obtained by the fitted simplified model, the average of measurements, the original simplified model and BT1 model. 200m 0.4mm cable.

domain dos not ensures good behavior in time domain, so it can be said that the proposed model is not suitable for time domain simulations.

Figures 14 and 15 illustrate the mean square error between the proposed fitted simplified twisted-pair cable model, the original simplified twisted-pair cable model and BT1 model when compared to the real measurements.

6. Frequency domain simulations for evaluation of the fourth generation DSL

The research on the fourth generation DSL systems is currently based on computer simulations, highlighting frequency domain (or PSD-based) simulations. An example of key results that can be achieved by frequency domain simulations we can cite the achievable data rates obtained by these systems. Initial studies showed that the bandwidth limit beyond where there are no significant data rate increases is 100 MHz (Magesacher et al., 2006), however, recent ITU contributions consider that the transmission bandwidth can be as high as 300 MHz depending on the used cable quality (van den Brink, 2010). This section describes some frequency domain simulations for performance evaluation of the next generation DSL systems. The used channel data was based on the results obtained by the short copper cable

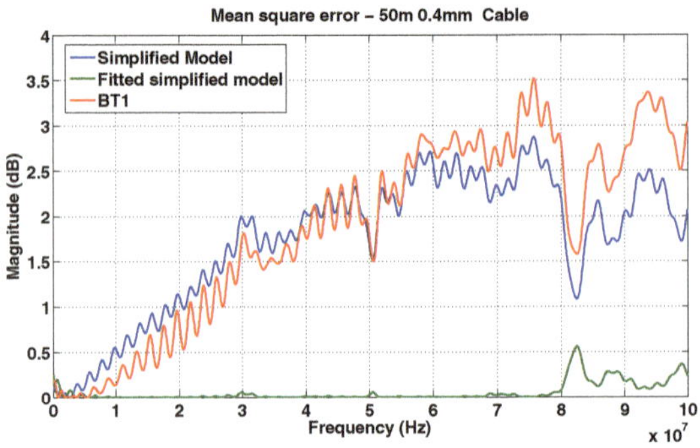

Fig. 14. Mean square error between the proposed fitted simplified model, the original simplified model and BT1 model when compared to the real measurements. 50m 0.4mm cable.

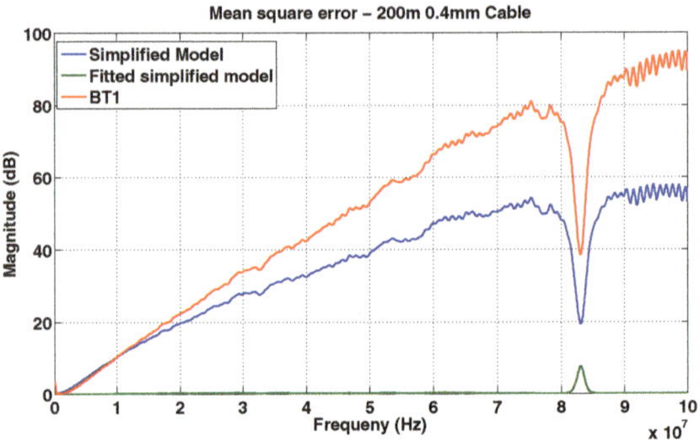

Fig. 15. Mean square error between the proposed fitted simplified model, the original simplified model and BT1 model when compared to the real measurements. 200m 0.4mm cable.

measurements described in Section 4 and cable reference models, including the fitted model proposed in Section 5.

Two approaches for frequency domain simulations are described. The first one focus the maximum data rates achievable by a hypothetical DMT-based fourth generation DSL system. The second one focus the maximum number of bits-per-tone that can be used by theses systems without loss of performance.

6.1 Achievable data rates of fourth generation DSL systems

For the evaluation of the achievable data rates of fourth generation DSL systems, two scenarios are defined. The first one observes an ideal transmission case, that is, without crosstalk impairment. It can be possible if all lines inside the copper cable are connected to a Vectored fourth generation DSL access node. In fact, the next generation DSL systems may offer support for vectoring in order to enjoy the full benefits of high frequency transmissions (Odling et al., 2009; Oksman et al., 2010). The second one observes the achievable data rates degradation resulting when one FEXT disturber is affecting a given user. This case the NEXT effects were mitigated assuming a FDD transmission.

A key point in both simulations is to define the transmission PSD (Power Spectrum Density) masks. For now, there are no definitions done by ITU on these recommendations. So, the first transmission PDS mask used here is defined according to the one proposed by (Magesacher et al., 2006). This mask is derived from the ingress and egress limits described in the CISPR 22 norm. It follows the standard VDSL2 up to 30 MHz, which is -60 dBm/Hz. For frequencies above 30 MHz, the PSD mask decays linearly from -60 dBm/Hz to -80 dBm/Hz at 100 MHz. Figure 16 shows this PSD mask, divided in upstream and downstream bands. Note that the spectrum was divided in a symmetrical way between upstream and downstream transmission, but with higher priority for downstream, as this one uses, for most of its tones, lower (less attenuated) frequencies. Furthermore, four other PSD masks are defined. The objective is to compare the impact of each one to the obtained data rates. Their power limits followed this assumptions:

- -60 dBm/Hz up to 30 MHz, then -80 dBm/Hz up to 100 MHz, flat shape;
- -60 dBm/Hz up to 30 MHz, then -76 dBm/Hz up to 100 MHz, flat shape;
- -60 dBm/Hz up to 30 MHz, then -72 dBm/Hz up to 100 MHz, flat shape;
- -60 dBm/Hz up to 30 MHz, then -68 dBm/Hz up to 100 MHz, flat shape;

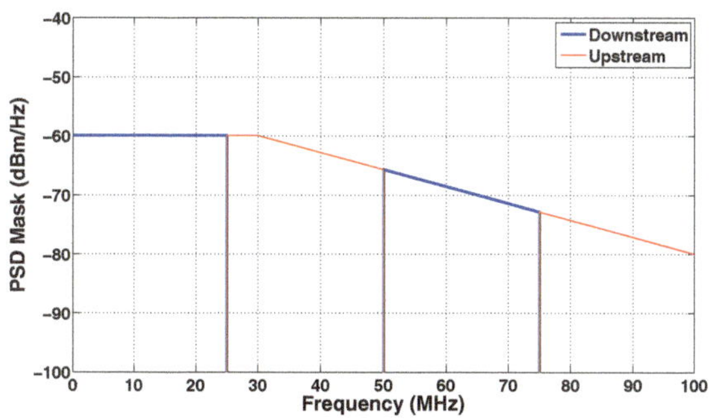

Fig. 16. One of the PSD masks used by the 4th generation DSL systems adopted in this work.

The bits allocation per tone k and for user n is expressed as follows (Starr et al., 1999)

$$b_n^k = \log_2\left(1 + \frac{1}{\Gamma\gamma_n}\frac{\left|h_{n,n}^k\right|^2 p_n^k}{\sum_{n\neq m}\left|h_{n,m}^k\right|^2 p_m^k + \sigma_n^k}\right), \tag{18}$$

where

- $|h_{n,n}^k|^2$ is the the square-magnitude of the direct transfer function gain for user n at tone k;
- $|h_{n,m}^k|^2$ denotes the square-magnitude of the far-end crosstalk transfer function from transmitter m to receiver n at tone k;
- p_m^k denotes the power transmitted by user m at tone k;
- σ_n^k represents the background noise power on tone k at receiver n;

- Γ is the signal-to-noise (SNR) ratio gap, which is a function of the desired bit error rate;
- γ_n is the noise margin of user n;

The total data rate R_n of each user n is calculated as

$$R_n = f_s \sum_{k=1}^{K} b_n^k, \tag{19}$$

where f_s is the DMT symbol rate (Starr et al., 1999).

The main transmission parameters used during the simulations of the two scenarios (with and without crosstalk) are described as follow:

- Background Noise: AWGN, -140 dBm/Hz;
- Tone spacing Δf : 25 kHz;
- Symbol rate: 21.562 kHz;
- DS max power: 14.5 dBm;
- US max power: 12.2 dBm;
- Target SNR margin: 5 dB;
- SNR Gap for uncoded QAM: 9.8 dB;
- Maximum number of bits per tone: 15;

As an initial approach, the first simulated scenario (which assumed no crosstalk impairment), uses as channel data the ANSI TP1 cable model extended up to 100 MHz. This cable model uses BT equations for calculates RLGC parameters. The values of each RLGC sub-parameter are defined in (ITU-T, 1999b).

The obtained results, using the five different PSD masks previously defined, are illustrated in Figure 17

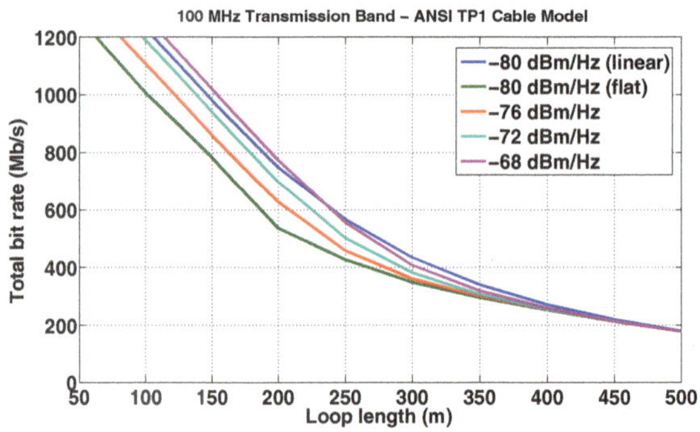

Fig. 17. Obtained bit rates up to 100 MHz using the extended ANSI TP1 as channel data. No crosstalk is taken into account.

It can be observed that, even using anyone of the five PSD masks, the maximum achievable data rates are bigger then 1 Gb/s when the copper loop is shorter than 150m. Using a 300m cable, the data rates becomes near 400 Mb/s, while using a 500m cable it is reduced to near 200 Mb/s.

In order to compare the mismatch between the results of frequency domain-based simulation using real measured data and current cable models, another simulation of the first scenario is performed. This case the channel data are the ones obtained by the short copper cable measurements described in Section 4. Figure 18 illustrates the mismatch for all the defined PSD mask limits.

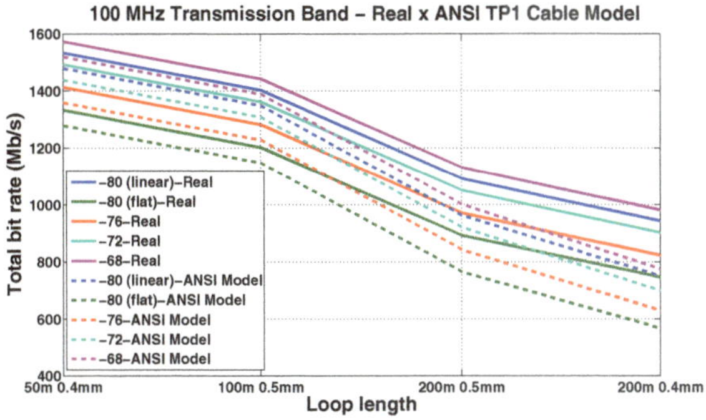

Fig. 18. Mismatch on the obtained bit rates using real channel measured data versus extended cable model.

It is clear the mismatch on the results. The difference can be as high as 200 Mb/s for some PSD masks using the 200m 0.4mm cable. So, in order to prove the quality of the fitted model performed in Section 5, another simulation is shown, this case using the fitted model of the 50m and 200m 0.4mm cables as channel data. Figure 19 illustrates the results.

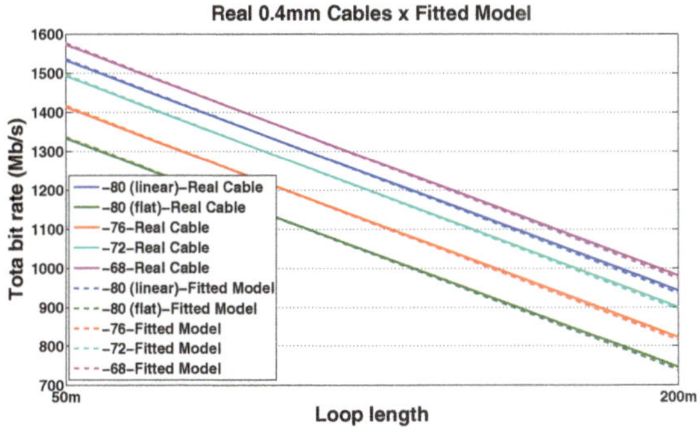

Fig. 19. Obtained bit rates using as channel data the fitted model and the average of the copper cable measurements. 50m and 200m 0.4mm cables.

The results shown in Figure 19 present good match, indicating the fitted model is well-characterized for frequency domain-based simulations in order to evaluate the performance of the fourth generation DSL systems.

For the frequency range being considered for use by the fourth generation DSL systems (100, 200, even 300 MHz), crosstalk can be an even worse impairment than it is for current DSL deployments. Hence, crosstalk mitigation techniques are highly recommended for these systems, allowing them to achieve the best possible performance (Odling et al., 2009; van den Brink, 2010). In order to evaluate the rates degradation of the fourth generation DSL systems transmissions due to uncanceled crosstalk effects, a second frequency domain-based simulation scenario for evaluation of achievable data rates is defined. This time two users share the same binder and are affected by uncanceled far-end crosstalk. Figure 20 illustrates it.

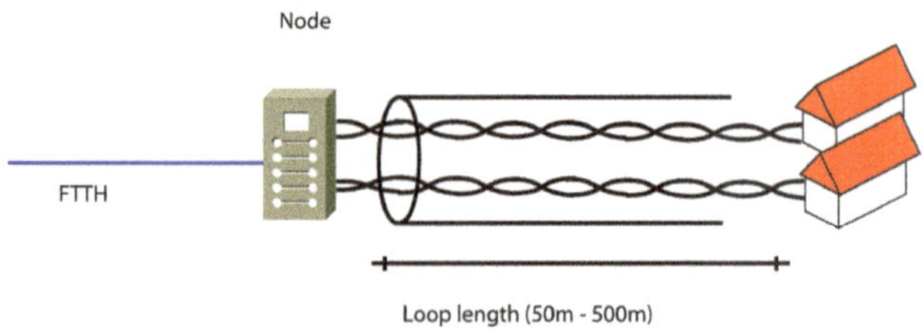

Fig. 20. Second scenario for evaluation of the achievable data rates of 4th generation DSL systems. Crosstalk is taken into account.

The adopted five PSD mask limits and transmission parameters are the same previously described. Figure 21 illustrates the results.

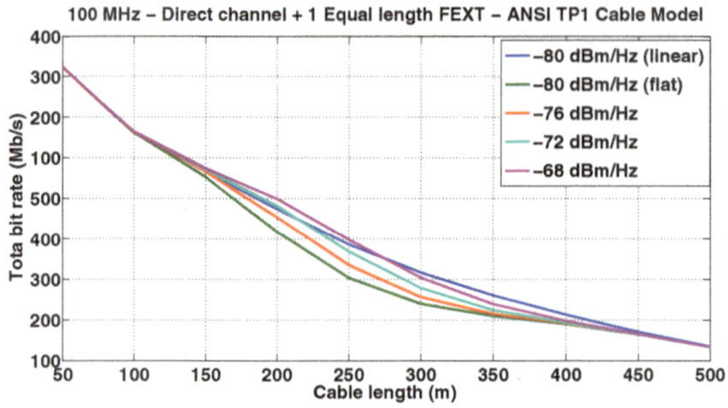

Fig. 21. Obtained bit rates up to 100 MHz using the extended ANSI TP1 as channel data. Uncanceled far-end crosstalk is taken into account.

It is clear that the data rates degrade significantly if no attempt is done to control crosstalk effects. It can be illustrated by the achieved data rate of the -80 dBm/Hz (linear) PSD mask. Using a 200m cable, it was near 800 Mb/s when no crosstalk was taken account, as shown in Figure 17, while it was near 500 Mb/s when crosstalk was considered, a decrease of almost 300 Mb/s.

6.2 Maximum number of bits/tone that should be used by fourth generation DSL systems

After evaluate the possible data rates obtained by a hypothetical DMT-based fourth generation DSL system, a new set of frequency domain simulations are described here. This time it is observed the maximum number of bits per DMT tone, or constellation size, that should be used by these systems, up to 100MHz. The maximum number of bits/tone is related to the constellation size: constellation size = $2^{bits/tone}$. An optimal value of number of bits/tone is a key element to limit complexity of these systems.

The first scenario is defined as a direct loop with no crosstalk effect. Again, as an initial approach, it is used a cable model extrapolated up to 100 MHz. The used cable model is the

ETSI ADSL PEO4 (26 AWG), which uses BT equations for calculates RLGC parameters. The values of each RLGC sub-parameter are defined in (ITU-T, 1999b). During this evaluation, only the PSD mask limit illustrated in Figure 16 is used. Figure 22 illustrates the obtained data rates for each value defined as maximum number of bits/tone.

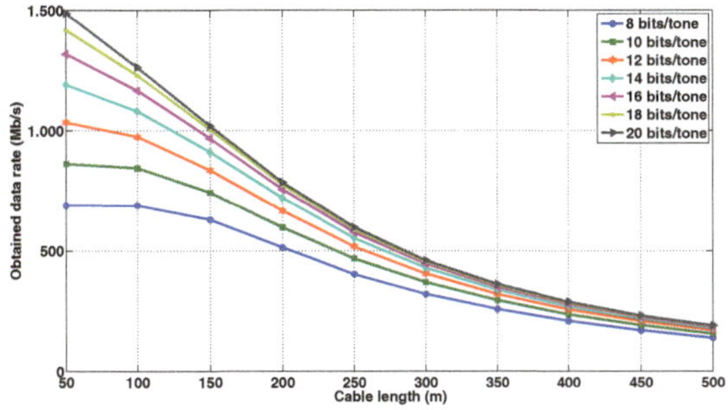

Fig. 22. Obtained data rates using different values of maximum number bits/tone. No crosstalk is taken into account. 0.4mm (26 AWG) cable model.

It can be concluded that an optimal value to be used as maximum number of bits/tone can be found between 12 and 14 bits. Below 12 bits the performance is strongly reduced and above 14 bits there is no much difference in terms of obtained data rates, especially for loop lengths longer than 200m, this way, the cost benefit (computational complexity versus improvements) may not be favorable.

Knowing that current cable models are not well characterized for high frequency DSL simulations, a new round of simulation to observe the achievable data rates using different maximum number of bits/tone is shown. This time the channel is based on the developed fitted model, described in Section 5.

Figure 23 illustrates the obtained data rates for each value defined as maximum number of bits/tone.

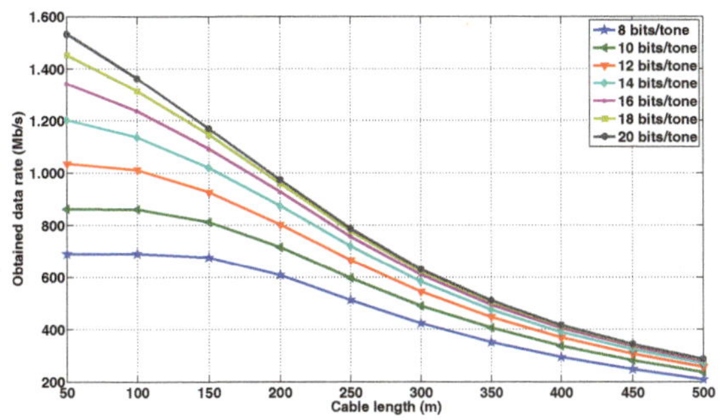

Fig. 23. Obtained data rates using different values of maximum number bits/tone. No crosstalk is taken into account. 0.4mm fitted cable model.

Figure 24 illustrates the expected mismatch between the results using ETSI PE04 cable model and the fitted 0.4mm cable model proposed by this work.

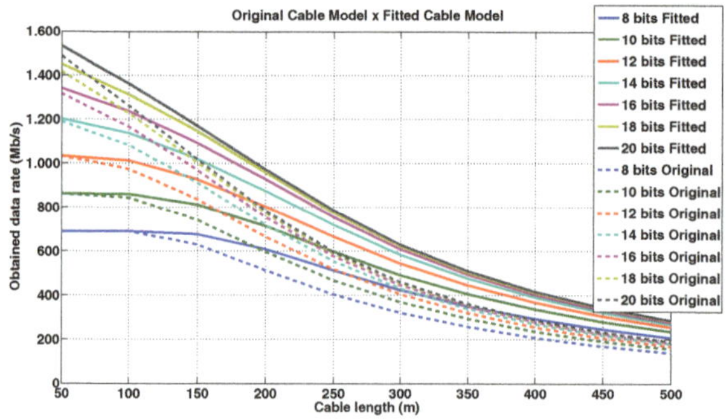

Fig. 24. Difference between the obtained results using ETSI PE04 cable model and the fitted 0.4mm cable model.

Finishing the frequency domain simulations described here, it is shown the impact of uncanceled far-end crosstalk in the achievable data rates of the fourth generation DSL systems even if they use different maximum number of bits/tone. The scenario overview is like the one illustrated in Figure 20. It was used the ESTI PE04 cable model. It was done because no crosstalk models were fitted. Figure 25 illustrates the obtained results, where it becomes clear that the data rate degrades significantly if no attempt is done to control crosstalk effects. The results reinforces that crosstalk mitigation techniques are highly recommended for the fourth generation DSL systems, allowing them to achieve the best possible performance.

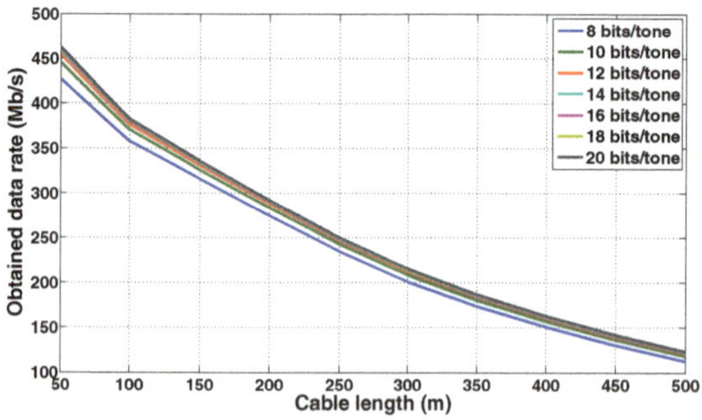

Fig. 25. Performance of 4th generation DSL systems using different values of maximum number bites/tone impaired by uncanceled FEXT.

7. Conclusion

During the last years, broadband technologies over copper increased their end-user bit rates due to a step-by-step evolution.

The fourth generation DSL systems, wish to improve the provided services achieving data rates near 1 Gb/s. As the direct channel attenuation is higher if the copper loop is long, the use of short cables will be inevitable to ensure improved rates. The copper cable reduction process will allow the use of unexplored transmission frequencies, much higher than the current 30

MHz. Unfortunately, the current cable reference models are not able to predict the behavior of short twisted-pair copper cables at unexplored DSL frequencies.

This work described measurement and modeling techniques for characterize short twisted-par copper cables at these new transmission environments. The measurements techniques were exemplified by a measurement campaign that evaluated the direct transfer functions and far-end crosstalk transfer functions of 50m, 100m and 200m copper cables. It was described the used equipments, the measurements parameters and measurements procedures. The measurements results of direct transfer functions showed how attenuated the direct channel of short cables can be. These informations are very important for the determination of channel capacities, which can show the maximum bit rates supported by these transmission channels. The far-end crosstalk measurements results showed how impaired the transmissions over short copper cables can be if no technique to avoid this interference is implemented.

The modeling techniques were exemplified by a fitting process in order to adjust the simplified twisted-pair cable model to ensure a well characterized response up to 100 MHz. The direct transfer functions generated by this fitted model proved to be much more similar to the ones obtained by the measurements of real copper cables. It is important to notice that although it presents good behavior in frequency domain, the proposed model is not suitable for time domain-based simulations.

This chapter also presented some frequency domain simulations in order to evaluate the potential of the fourth generation DSL systems. It was verified that the achievable data rates up to 100 MHz can be as high as 1 Gb/s. Moreover it was verified that an optimal value to be used as maximum number of bits/tone is near 12 bits. As a last contribution this work showed the effects of uncanceled crosstalk in the reduction of the performance of these systems.

8. Acknowledgments

This work was supported by the Innovation Center, Ericsson Telecomunicações S.A., Brazil, and CNPq

9. References

Acatauassu, D., Muller, F. & Klautau, A. (2009). Capacity of MIMO DSL systems using 100 MHz measured channel data, *International Conference on Telecommunications, 2009. ICT '09.*, pp. 266–269.

Agilent (2000). ADSL copper loop measurements - Agilent Technologies Product Note.

Cendrillon, R. & Moonen, M. (2005). Iterative spectrum balancing for digital subscriber lines, *2005 IEEE International Conference on Communications, 2005. ICC 2005.*, Vol. 3, pp. 1937 – 1941 Vol. 3.

Chen, W. Y. (1998). *DSL : Simulation Techniques and Standards Development for Digital Subscriber Lines*, Macmillan Technology Series, Macmillan Technical Publishing, Indianapolis, IN.

ETSI (1997-02). STD. ETSI STC TM6 - Cable Reference Models for Simulating Metallic Access Networks.

Golden, P., Dedieu, H. & Jacobsen, K. (2006). *Fundamentals of DSL Technology*, Auerbach Publications, Taylor & Francis Group.

ITU-T (1991). Integrated services digital network (ISDN)-Basic access interface for use on metallic loops for application on the network side of the NT (layer 1 specification).

ITU-T (1999a). Asymmetric Digital Subscriber Line (ADSL) transceivers.

ITU-T (1999b). ITU-T G996.1 - Test Procedures for Digital Subscriber Line (DSL) Transceivers.

ITU-T (2003). Asymmetric Digital Subscriber Line Transceivers 2 (ADSL2).

ITU-T (2005a). Asymmetric Digital Subscriber Line (ADSL) transceivers - Extended bandwidth ADSL2 (ADSL2+).

ITU-T (2005b). Very high speed digital subscriber line transceivers (VDSL).

ITU-T (2006). Very high speed digital subscriber line transceivers 2 (VDSL2), *ITU-T Recommendation G.993.2* .

Magesacher, T., Rius i Riu, J., Ödling, P., Börjesson, P., Tilocca, M. & Valentini, M. (2006). Limits of ultra-wideband communication over copper, *International Conference on Communication Technology, 2006. ICCT '06.*, pp. 1 –4.

Moraes, R. B., Dortschy, B., Klautau, A. & Riu, J. R. i. (2010). Semiblind spectrum balancing for DSL, *IEEE Transactions on Signal Processing* 58(7): 3717–3727.

Odling, P., Magesacher, T., Host, S., Borjesson, P., Berg, M. & Areizaga, E. (2009). The fourth generation broadband concept, *IEEE Communications Magazine* 47(1): 62 –69.

Oksman, V., Schenk, H., Clausen, A., Cioffi, J., Mohseni, M., Ginis, G., Nuzman, C., Maes, J., Peeters, M., Fisher, K. & Eriksson, P.-E. (2010). The ITU-T's new G.vector standard proliferates 100 MB/s DSL, *IEEE Communications Magazine* 48(10): 140 –148.

Song, K. B., Chung, S. T., Ginis, G. & Cioffi, J. (2002). Dynamic spectrum management for next-generation DSL systems, *IEEE Communications Magazine* 40(10): 101 – 109.

Starr, T., Cioffi, J. M. & Silverman, P. J. (1999). *Understanding Digital Subscriber Line Technology*, Prentice-Hall.

van den Brink, R. F. (2010). Enabling 4GBB via hybrid-FttH, the missing link in FttH scenarios, BBF 2010.1395.

Vergara, A., Moral, A. & PeÌĄ andrez, J. (2010). Costa: A model to analyze next generation broadband access platform competition, *14th International Telecommunications Network Strategy and Planning Symposium (NETWORKS), 2010*, pp. 1 –6.

Yoho, J. J. (2001). *Physically-Based Realizable Modeling and Network Synthesis of Subscriber Loops Utilized in DSL Technology*, PhD thesis, Virginia Polytechnic Institute and State University.

13

Protocol Measurements in BitTorrent Environments

Răzvan Deaconescu
University POLITEHNICA of Bucharest
Romania

1. Introduction

Based on BitTorrent's success story (it has managed to become the number one protocol of the internet in a matter of years), the scientific community has delved heavily in analysing, understanding and improving its performance. Research focus has ranged from measurements (Pouwelse et al. (2004)) to protocol improvements (Tian et al. (2007)), from social networking (Pouwelse et al. (2008)) to moderation techniques (Pouwelse et al. (2005)), from content distribution enhancements (Vlavianos et al. (2006)) to network infrastructure impact (Das & Kangasharju (2006)).

The BitTorrent protocol currently (October, 2011[1]) accounts for one of the largest percentages of the Internet traffic. Designed to provoke peers to give more in order to get more, the BitTorrent protocol is a prime choice for large data distribution. BitTorrent's design relies on several key elements:

- the use of a particular form of *tit-for-tat* that prevents free riding[2] and stimulates peers' selflessness;
- *optimistic unchoke* lets a peer periodically allow a connection from another peer in order to exchange content; (*choking* is the operation by which a peer closes its connection to another peer);
- *rarest piece first* aims to distribute all pieces among peers and increase availability.

In this chapter we present a novel approach involving client-side information collection regarding client and protocol implementation. We have instrumented a libtorrent-rasterbar client[3] and a Tribler[4] client to provide verbose information regarding BitTorrent protocol implementation. These results are collected and subsequently processed and analysed through a rendering interface.

The aim is to measure and analyze protocol messages while in real-world environments. To achieve this aim, a virtualized infrastructure had been used for realistic environments; apar

[1] `http://www.sandvine.com/news/global_broadband_trends.asp`
[2] though this can be circumvented as shown in BitThief (Locher et al. (2006))
[3] http://www.rasterbar.com/products/libtorrent/
[4] http://www.tribler.org/trac/

from it, clients and trackers running in a real-world swarm have been used and instrumented to provide valuable protocol information and parameters. No simulators (see Naicken et al. (2007)) have been used for collecting, measuring and analyzing protocol parameters, rather a "keep it real as much as possible" approach. Information, messages and parameters are collected directly from peers and trackers that are part of a real-world Peer-to-Peer swarm.

An approach to providing a unified model for collecting information would be a standard for developing a logging implementation for various clients, such that if using a common easy to parse output format, all information about messages exchanged between the active participants can be centralized and followed by new improvements. Such an approach would ensure maximum flexibility, albeit at the cost of having to update all clients in use, which is why we have focused on the above on an approach of collecting and analysing logging information directly provided by BitTorrent clients.

The action chronology for measuring parameters had been collecting data, parsing and storing it and then subjecting protocol parameters to processing and analysis. The rest of this chapter presents the measured parameters, approaches to collecting, parsing and storing information into an "easy to be used" format and then putting it to analysis and interpretation.

2. BitTorrent messages and parameters

Analysis of BitTorrent client-centric behavior and, to some extent, swarm behavior, is based on BitTorrent protocol messages[5]. Messages are used for handshaking, closing the connection, requesting and receiving data.

The BitTorrent client will generate at startup a unique identifier of itself known as *peer id*. This is client dependent, each client encoding a peer id based on its own implementation.

2.1 Protocol messages

Each torrent is exclusively identified by a 20-byte SHA1 hash of the value of the info key from the torrent file dictionary which is defined as *info hash*. The peer id and info hash values are important in the TCP connection establishment and are typically logged by trackers.

The handshake is the first sent message. It uses the format:

```
<length><protocol><reserved><info_hash><peer_id>
```

The protocol parameter represents the protocol identifier string and the length parameter represents the protocol name length. Reserved represents eight reserved bytes whose bits can be used to modify the behavior of the protocol. Standard implementations use this as zero-filled. `info_hash` represents the identifier of the shared resource that is requested by the initiator of the connection. `peer_id` represents the initiator's unique identifier.

The receiver of the handshake must verify the `info_hash` in order to decide if it can serve it. If it is not currently serving, it will drop the connection. Otherwise, the receiver will send its own handshake message to the initiator of the connection. If the initiator receives

[5] http://www.bittorrent.org/beps/bep_0003.html

a handshake whose peer_id does not match with the expected one – it must keep a list with peers addresses and ports and their corresponding peer_id's – then it also must drop the connection.

Remaining protocol messages use the format:

```
<length><message ID><payload>
```

The length prefix is a four byte big-endian value representing the sum of message ID and payload sizes. The message ID is a single decimal byte. The payload is message dependent.

- **keep-alive** (<len=0000>)
  The Keep-alive message is the only message without any message ID and payload. It is sent to maintain the connection alive if no other message has been sent for a given amount of time. The amount of time is about two minutes.
- **choke** (<len=0001><id=0>)
  The Choke message is sent when the client wants to choke a remote peer.
- **unchoke** (<len=0001><id=1>)
  The Unchoke message is sent when the client wants to unchoke a remote peer.
- **interested** (<len=0001><id=2>)
  The Interested message is sent when the client is interested in something that the remote peer has to offer.
- **not interested** (<len=0001><id=2>)
  The Not interested message is sent when the client is not interested in anything that the remote peer has to offer.
- **have** (<len=0005><id=4><piece index>)
  The piece index is a 4 bytes value representing the zero-based index of a piece that has just been successfully downloaded and verified via its hash value present in the torrent file.
- **bitfield** (<len=0001+X><id=5><bitfield>)
  The bitfield message may only be sent immediately after the handshake sequence has occurred and before any other message is sent. It is optional and need not be sent if a client has no pieces. The Bitfield payload has the length X and its bits represent the pieces that have been successfully downloaded. The high bit in the first byte corresponds to piece index 0. A set bit indicates a valid and available piece, and a cleared bit indicates a missing piece. Any spare bits are set to zero.
- **request** (<len=0013><id=6><index><begin><length>)
  The Request message is sent when requesting a block. Index is the zero-based index of the piece containing the requested block, begin is the block offset inside the piece and length represents the block size.
- **piece** (<len=0009+X><id=7><index><begin><block>)
  The Piece message is sent when delivering a block to an interested peer. Index is the zero-based index of the piece containing the delivered block, begin is the block offset inside the piece and block represents the X-sized block data.

- **cancel** (<len=0013><id=8><index><begin><length>)
  The Cancel message is sent when canceling a block request sent before. Index is the zero-based index of the piece containing the requested block, begin is the block offset inside the piece and length represents the block size.
- **port** (<len=0003><id=9><listen port>)
  The Port message is sent by clients that implement a DHT tracker. The listen port is the port of the client's DHT node listening on.

Swarm measured data is usually collected from trackers. While this offers a global view of the swarm it has little information about client-centric properties such as protocol implementation, neighbour set, number of connected peers, etc. A more thorough approach (Iosup et al. (2006)) uses network probes to interrogate various clients.

Our approach, while not as scalable as the above mentioned one, aims to collect client-centric data, store and analyse it in order to provide information on the impact of network topology, protocol implementation and peer characteristics. Our infrastructure provides micro-analysis, rather than macro-analysis of a given swarm. We focus on detailed peer-centric properties, rather than less-detailed global, tracker-centric information. The data provided by controlled instrumented peers in a given swarm is retrieved, parsed and stored for subsequent analysis.

We differentiate between two kinds of BitTorrent messages: *status messages*, which clients provide periodically to report the current session's download state, and *verbose messages* that contain protocol messages exchanged between peers (chokes, unchokes, peer connections, pieces transfer etc.).

Another type of messages are those provided by tracker logging. Tracker-based messages provide an overall view of the entire swarm, albeit at the cost of less-detailed information. Tracker logging typically consists of periodic messages sent by clients as announce messages. However, these messages' period is quite large (usually 30 minutes – 1800 seconds) resulting in less detailed information. Their overall swarm vision is an important addition to status and verbose client messages.

2.2 Measured data and parameters

Data and parameters measured are those particular to BitTorrent clients and swarms, that provide support for evaluation and improvements at protocol level. The measured parameters are described in the Table 1, Table 2 and Table 3, depending on their source (either status messages, verbose messages or tracker messages).

2.3 Approaches to collecting and extracting protocol parameters

Peer-to-Peer clients and applications may be instrumented to provide various internal information that is available for analysis. This information may also be provided by client logging enabled for the client. Such data features parameters describing client behavior, protocol messages, topology updates and even details on internal algorithms and decisions.

We "aggregate" this information as messages and focus on protocol messages, that is messages regarding the status of the communication (such as download speed, upload speed) and those with insight on protocol internals (requests, acknowledgements, connects, disconnects).

Parameter	Explanation
Download speed	Current peer download speed – number of bytes received
Upload speed	Current peer upload speed – number of bytes sent
ETA	How long before the complete file is received
Number of connections	Number of remote peers currently connected to this client
Download size	Bytes download so far
Upload size	Bytes uploaded so far
Remote peers ID	IP address and TCP port of remote peers
Per-remote peer download speed	Download speed of each remote connected peer
Per-peer upload speed	Upload speed of each remote connected peer

Table 1. Parameters from Status Messages

Parameter	Explanation
`CHOKE`	Disallow remote peer to request pieces
`UNCHOKE`	Allow remote peer to request pieces
`INTERESTED`	Mark interest in a certain piece
`NOT_INTERESTED`	Unmark interest in a certain piece
`HAVE`	Remote peer possesses current piece
`BITFIELD`	Bitmap of the file
`REQUEST`	Ask for a given piece
`PIECE`	Send piece
`CANCEL`	Cancel request of a piece
`DHT_PORT`	Present DHT port to DHT-enabled peers

Table 2. Parameters from Verbose Messages

Parameter	Explanation
Swarm size	The number of peers in the swarm
Client IP/port	Remote peer identification (IP address and TCP port in used)
Client type	BitTorrent implementation of each client
Per-client download size	Download size for each client
Per-client upload size	Upload size for each client

Table 3. Parameters from Tracker Messages

As such, there is a separation between periodic, status reporting messages and internal protocol messages that mostly related to non-periodic events in the way the protocol works. These have been "dubbed" *status messages* and *verbose messages*.

Status messages are periodic messages reporting session state. Messages are usually output by clients at every second with updated information regarding number of connected peers, current download speed, upload speed, estimated time of arrival, download percentage, etc. Status messages are to be used for real time analysis of peer behaviour as they are lightweight and periodically output (usually every second).

Status messages may also be used for monitoring, due to their periodic arrival. When using logging, status messages are typically provided as one line in a log file and parsed to provide valued information. Graphical evolution and comparison of various parameters result easily from processing status messages log files.

Verbose messages or *log messages* provide a thorough inspection of a client's implementation. The output is usually of large quantity (hundreds of MB per client for a one-day session). Verbose information is usually stored in client side log files and is subsequently parsed and stored.

Verbose information may not be easily monitored due to its event-based creation. When considering the BitTorrent protocol, verbose messages are closely related to BitTorrent specification messages such as `CHOKE`, `UNCHOKE`, `REQUEST`, `HAVE` or internal events in the implementation. Verbose information may be logged through instrumentation of client implementation or activation of certain variables. It may also be determined through investigation of network traffic.

Apart from protocol information provided in status and verbose messages, one may also collect information regarding application behavior such as the piece picking algorithm, size of buffers used, overhead information. This data may be used to fulfill the image of the overall behavior and provide insight on possible enhancements and improvements.

There are various approaches to collecting information from running clients, depending on the level of intrusiveness. Some approaches may provide high detail information, while requiring access to the client source code, while others provide general information but limited intrusiveness.

The most intrusive approach requires placing hook points into the application code for providing information. This information may be sent to a monitoring service, logged, or sent to a logging library. Within the P2P-Next project[6], for example, the NextShare core provides an internal API for providing information. This information is then collected either through a logging service that collects all information or through the use of a monitoring service with an HTTP interface and MRTG graphics rendering tools.

Another approach makes use of logging information directly provided by BitTorrent clients. There are two disadvantages to this approach. The first one is that each client provides information in its own way and a dedicated message parser must be enabled for each application. The second one is related to receiving verbose messages. In order to be able to receive verbose messages, one has to turn on the verbose logging. This may be accomplished through a startup option, an environment variable or a compile option. It may be the case that non-open source applications possess none of these options and cannot provide requested information. This is the approach that we will focus on for the rest of the chapter.

Finally, a network-oriented approach requires a thorough analysis of network packets similar to deep packet inspection. It allows an in depth view of all packets crossing a given point. Its main advantage is ubiquity: it may be applied to all clients and implementation regardless of access to the source code. The disadvantage is the difficulty in parsing all packets and extracting required information (specific to the BitTorrent protocol) and, perhaps more pressing, the significant processing overhead introduced.

[6] `http://www.p2p-next.org`

Messages and information collected are concerned with client behavior. As such, the applications in place work at the edge of the P2P network on each client. No information is gathered from the core of the network, inner routers or the Internet. In order to provide an overall profile of the swarm or P2P network information collected from all peers must be aggregated and unified. While having only edge-based information means some data may be lacking it provides a good perspective of the protocol internals and client implementation. We dub this approach client-centric investigation.

Collected data may be either monitored, with values rendered in real time or it may also be archived and compressed for subsequent use. The first approach requires engaging parsers while data is being generated, while the other allows use of parsers subsequently. When using parsers with no monitoring, data is usually stored in a "database". "database" is a generic term which may refer to an actual database engine, file system entries, or even memory information. A rendering or interpretation engine are typically employed to analyze information in the database and provide it in a valuable form to the user.

3. Log collecting for BitTorrent peers

The log collection approach implying a less intrusive activity but providing a great deal of protocol parameters is the use of logging information from clients. Each client typically presents status information (the dubbed *status message*) consisting of periodic information such as download speed, upload speed, number of connection and, if enabled, a set of enhanced pieces of information (the dubbed *verbose message*). Types of messages and their content have been thoroughly described in Section 2. All or most of BitTorrent clients provide status messages but some sort of activation or instrumentation is required to provide verbose messages.

3.1 Using and updating BitTorrent clients for logging

Throughout experiments we have used multiple open-source clients. All of them provided basic status information, while some were updated or altered to provide verbose information as well. Transmission, Aria2, Vuze, Tribler, libtorrent-rasterbar and the mainline client had been used to provide status parameters, while Tribler and libtorrent-rasterbar had also been instrumented to provide verbose parameters.

As building the setup, deploying peers and collecting information and subjecting it to dissemination is a lengthy process, this has to be automated Deaconescu et al. (2009).

Several approaches had been put to use to collect status information, depending on the client implementation:

- The main issue with **Azureus** was the lack of a proper CLI that would enable automation. Though limited, a "Console UI" module enabled automating the tasks of running Azureus and gathering download status and logging information.
- Although a GUI oriented client, **Tribler** does offer a command line interface for automation.

- **Transmission** has a fully featured CLI and was one of the clients that were very easy to automate. Detailed debugging information regarding connections and chunk transfers can be enabled by setting the `TR_DEBUG_FD` evironment variable.
- **aria2** natively provides a CLI and it was easy to automate. Logging is also enabled through CLI arguments.
- **hrktorrent** is a lightweight implementation over **libtorrent-rasterbar** and provides the necessary interface for automating a BitTorrent transfer, albeit some minor modifications have been necessary.
- **BitTorrent Mainline** provides a CLI and logging can be enabled through minor modifications of the source code.

In order to examine BitTorrent transfer parameters at a protocol implementation level, we propose a system for storing and analysing logging data output by BitTorrent clients. It currently offers support for hrktorrent/libtorrent[7] and Tribler[8].

Our study of logging data takes into consideration two open-source BitTorrent applications: Tribler and hrktorrent[9] (based on libtorrent-rasterbar). While the latter needed minimal changes in order to provide the necessary verbose and status data, Tribler had to be modified significantly.

The process of configuring Tribler for logging output is completely automated using shell scripts and may be reversed. The source code alterations are focused on providing both status and verbose messages as client output information.

Status message information provided by Tribler includes transfer completion percentage, download and upload rates. In the modified version, it also outputs current date and time, transfer size, estimated time of arrival (ETA), number of peers, and the name and path of the transferred file.

In order to enable *verbose message* output, we took advantage of the fact that Tribler uses flags that can trigger printing to standard output for various implementation details, among which are the actions related to receiving and sending BitTorrent messages. The files we identified to be responsible for protocol data are changed using scripts in order to print the necessary information and to associate it to a timestamp and date. Since most of the protocol exchange data was passed through several levels in Tribler's class hierarchy, attention had to be paid to avoid duplicate output and to reduce file size. In contrast to libtorrent-rasterbar, which, at each transfer, creates a separate session log file for each peer, Tribler stores verbose messages in a single file. This file is passed to the verbose parser, which extracts relevant parts of the messages and writes them into the database.

Unlike Tribler, hrktorrent's instrumentation did not imply modifying its source code but defining `TORRENT_LOGGING` and `TORRENT_VERBOSE_LOGGING` macros before building (recompiling) libtorrent-rasterbar. Minor updates had to be delivered to the compile options of hrktorrent in order to enable logging output.

---

[7] http://www.rasterbar.com/products/libtorrent/
[8] http://www.tribler.org/trac
[9] http://50hz.ws/hrktorrent/

Although our system processes and stores all protocol message types, the most important messages for our swarm analysis are those related to changing a peer's state (choke/unchoke) and requesting/receiving data. Correlations between these messages are the heart of provisioning information about the peers' behaviour and BitTorrent clients' performance.

3.2 Storage

Logging information is typically stored in log files. In libtorrent-rasterbar's case, logging is using a whole folder and logging information for each remote peer is using a single file in that folder. Usually information is redirected from standard output and error towards the output file.

As in a given experiment logging information occupies a large portion of disk space, especially verbose messages, files and folders are compressed in archive files. There would generally be a log archive for each client session. When information is to be processed, logging archives are going to be provided to the data processing component. A log archive contains both status messages and verbose messages.

Logging information may be stored in archive files for subsequent use or it may be processed live – that is parsing and interpreting parameters as log files are being generated. When running a live/real-time processing component, compressing logging information may not be required. However, in order to still preserve the original files, some experimenters may choose to retain access to the log archives.

The usefulness of a live processing component is based primarily on relieving the burden of space consumption, in case archiving is disabled. Most of the logged information is not useful, due to the fact that some peers may not be connected to other peers and status information, though provided, consists of parameters that are equal to zero – no connections means `0 KB/s` download speed, `0 KB/s` upload speed and others. On a certain occasion, a log file that had been used for more than 3 weeks, occupied more than 1GB of data but resulted in just 27KB valuable information.

Either when using live parsers or subsequent analysis, parameters are parsed for rapid use. The post-parsing storage is typically a relational database. The advantage of such a storage facility is its rapid access for post processing. When inquiring about given swarm parameters, the user would query the database and rapidly obtain necessary information. If that wouldn't be enabled, each inquiry would require a new parsing activity, resulting in large overhead and CPU consumption. Database storage is the final step of the logging and parsing stage. Parameter analysis, interpretation and advising activities would not be concerned of logging information, but only query the database.

3.3 Experiments

In order to collect information specific to a swarm, one must have access to all clients and logging information from those clients. As such, either all clients are accessible to the experimenter, or users would subsequently provide logging information to the experimenter.

Some remote information may be replaced by that provided by a tracker log file. A tracker logs information regarding the overall swarm view, albeit its periodicity is quite large (typically 30 minutes – 1800 seconds).

An intermediate approach to collecting logging information is a form of aggregation of information on the client side. This information may be either sent to a logging service or stored to be subsequently provided to the user. The former approach it taken by the Logging Service withing the P2P-Next project.

Typical experiments are those that allow full control to the user and provide all information rendered by clients. Deployment, log activation, log collection/archiving and even parsing are accomplished in full automation. One would create a configuration file and run experiments. Log archive files typically result from the experiment and, after gathering all, may be subjected to analysis.

The inclusion of tracker information had been enabled in the use of UPB Linux Distro Experiments[10] as described by Bardac et al. (2009). Tracker log files are parsed live and provide overall swarm parameters. Various information from tracker log files are provided as graphic images that show the evolution of swarm parameters.

Tracker logs have also been enabled in certain experiments. These experiments rely on extensive logging information (verbose messages) provided by seeders and tracker information. The lack of complete access to the all clients in the swarm is balanced out by the usage of verbose logging on the seeders' side. However, remote peers' intercommunication is not logged in anyway. Such that a form of aggregation and collection of remote peers' intercommunication messages is still required.

3.4 Monitoring and post-processing

Log processing, as described in Section 4 refers to parsing and interpreting BitTorrent protocol parameters. Data is parsed into an easy to be accessed database that is provided to the user.

As described above, one may choose to store logging information and then enable analysis. We dub this approach *post-processing*. The other approach is for live analysis of the provided parameters, resulting in client and swarm monitoring. The two approaches may, of course, be combined: while doing parsing of information, it is also stored in a database while various parameters are also monitored.

An overview of a typical architecture for data processing is presented in Figure 4. Separate parsers are used for live parsing and classical parsing. Classical parsing results in a database "output", while live parsing results in both a database "output" and the possibility of deploying live client and swarm monitoring.

4. Protocol data processing engine

As client instrumentation provides in-depth information on client implementation, it generates extensive input for data analysis. Coupled with carefully crafted experiments and message filtering, this will allow the detection of weak spots and of improvement

[10] http://torrent.cs.pub.ro/

possibilities in current implementations. Thus it will provide feedback to client and protocol implementations and swarm "tuning" suggestions, which in turn will enable high performance swarms and rapid content delivery in peer-to-peer systems.

Due to various types of modules employed (such as parser implementations, storage types, rendering engines) a data processing framework may provide different architectures. A sample view of the infrastructure consists of the following modules:

- **Parsers** – receive log files provided by BitTorrent clients during file transfers. Due to differences between log file formats, there are separate pairs of parsers for each client. Each pair analyses status and verbose messages.
- **Database Access** – a thin layer between the database system and other modules. Provides support for storing messages, updating and reading them.
- **SQLite Database** – contains a database schema with tables designed for storing protocol messages content and peer information.
- **Rendering Engine** – consists of a GUI application that processes the information stored in the database and renders it using plots and other graphical tools.

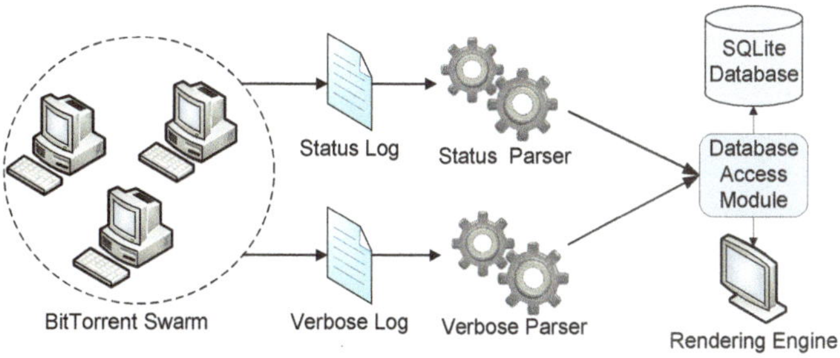

Fig. 1. Logging System Overview

As shown in Figure 1, using parsers specific to each type of logging file, messages are sent as input to the *Database Access* module that stores them into an SQLite database. In order to analyse peer behaviour, the *Rendering Engine* reads stored logging data using the *Database Access* module and outputs it to a graphical user interface.

Once all logging and verbose data from a given experiment is collected, the next step is the analysis phase. The testing infrastructure provides a GUI (*Graphical User Interface*) statistics engine for inspecting peer behaviour.

The GUI is implemented in Python using two libraries: *matplotlib* – for generating graphs and *TraitsUi* – for handling widgets. It offers several important plotting options for describing peer behaviour and peer interaction during the experiment:

- *download/upload speed* – displays the evolution of download/upload speed for the peer;
- *acceleration* – shows how fast the download/upload speed of the peer increases/decreases;
- *statistics* – displays the types and amount of verbose messages the peer exchanged with other peers.

The last two options are important as they provide valuable information about the performance of the BitTorrent client and how this performance is influenced by protocol messages exchanged by the client.

A sample GUI screenshot may be observed in Figure 2 and Figure 3:

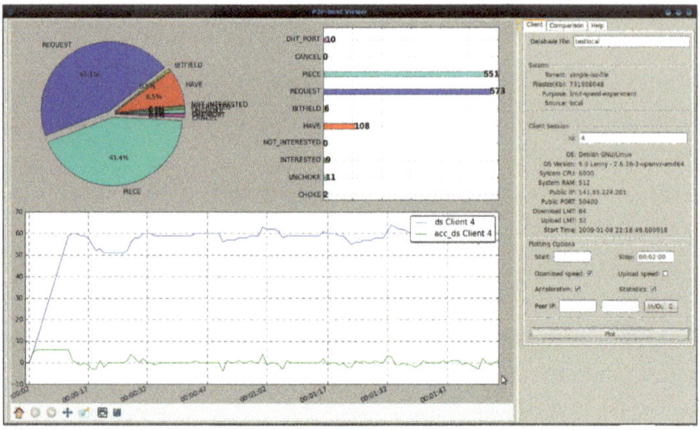

Fig. 2. Rendering Engine for BitTorrent Parameters: Client Analysis

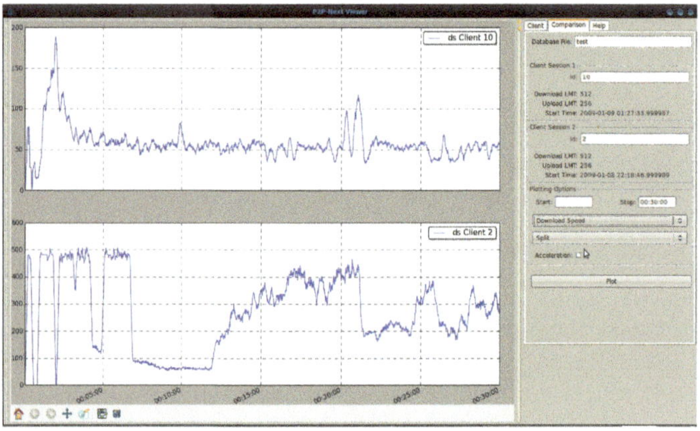

Fig. 3. Rendering Engine for BitTorrent Parameters: Client Comparison

The *acceleration* option measures how fast a BitTorrent client is able to download data. High acceleration forms a basic requirement in live streaming, as it means starting playback of a torrent file with little delay.

The *statistics* option displays the flow of protocol messages. We are interested in the choke/unchoke messages.

The GUI also offers two modes of operation: *Single Client Mode*, in which the user can follow the behaviour of a single peer during a given experiment, and *Client Comparison Mode*, allowing for comparisons between two peers.

4.1 Post processing framework for real-time log analysis

The dubbed post processing framework is used for storing logging information provided by various BitTorrent clients into a storage area (commonly a database). An architectural view of the framework is described in Figure 4.

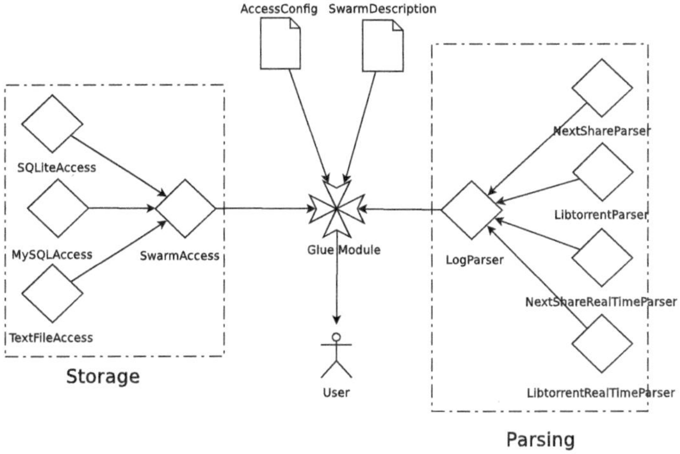

Fig. 4. Post-Processing Framework Architecture

The two main components of the framework are the parser and the storage components. Parsers process log information and extract measured protocol parameters to be subject to analysis; storers provide an interface for database or file storing – both for writing and reading. Storers thus provide an easy to access, rapid to retrieve and extensible interface to parameters. Storers are invoked when parsing messages – for storing parameters, and when analyzing parameters – for retrieving/reading/accessing parameters.

Within the parser component, a `LogParser` module provides the interface to actual parser implementations. There are two kinds of parsers: log parsers and real time log parsers. The former are used for data already collected and subsequently provided by the experimenter. Another approach involves using parsers at the same time as the client generates logging information. This real time parsing approach possesses three important advantages: monitoring may be enabled for status messages, less space is wasted as messages are parsed in real time, and processing time is reduced due to the overlapping of the parsing time and the storing time. The disadvantage of a real time parser is a more complex implementation as it has to consider the current position in the log file and continue from that point when data is available. At the same time, all clients must be able to access the same database, probably located on a single remote system.

The storage component is interfaced by the `SwarmAccess` module. This module is backed by database-specific implementations. This may be RDMBS systems such as MySQL or SQLite or file-based storage. Parameters are stored according to the schema described in Figure 5.

Configuration of the log files and clients to be parsed is found in the *SwarmDescription* file. All data regarding the current running swarm is stored in this file. Client types in the description file also determine the parser to be used. Selection of the storage module is based on the configuration directives in the *AccessConfig* file. For an SQLite storage, this contains the path to the database file; for an MySQL file, it contains the username, password, database name specific to database connection inquiries.

The user/developer is interfaced by a *Glue Module* that provides all methods deemed necessary. The user would call methods in the Glue Module for such actions as parsing a session archive, a swarm archive, for updating a configuration file, for retrieving data that fills a given role.

4.1.1 Parameter storage database

The database schema as shown in Figure 5 is used for relational database engines such as MySQL or SQLite.

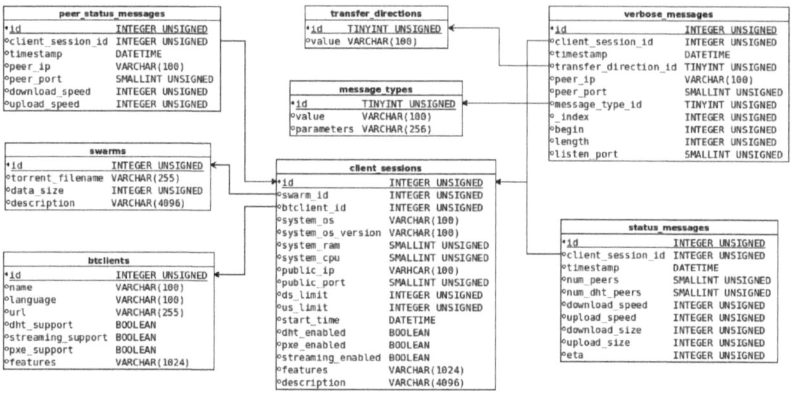

Fig. 5. Database Schema

The database schema provides the means to efficiently store and rapidly retrieve BitTorrent protocol parameters from log messages. The database is designed to store parameters about multiple swarms in the `swarms` table; each swarm is identified by the `.torrent` file its clients are using.

Information about peers/clients that are part of the swarm are stored in the `client_sessions` table. Each client is identified by its IP address and port number. Multiple pieces of information such as BitTorrent clients in use, enabled features and hardware specifics are also stored.

Three classes of messages result in three tables: `status_messages`, `peer_status_messages` and `verbose_messages`. The `peer_status_messages` table stores parameters particular to remote peers connected to the current client, while the `status_messages` stores parameters specific to the current client (such as download speed, upload speed and others). Each line in the `*_messages` tables points to an entry in the `client_sessions` table, identifying the peer it belongs to – the one that generated the log message.

4.1.2 Logfile-ID mapping

When parsing log files, one has to know the ID of the client session that has generated the log file. In order to automate the process, there needs to be a mapping between the log file (or log archive) and the client session ID.

At the same time, the client session ID needs to exist in the `client_sessions` table in the database, together with information such as BitTorrent client type, download speed limitation, operating system, hardware specification etc. This information needs to be supplied by the experimenter in a form that is both easy to create (by the experimenter) and parse.

A swarm description file is to be supplied by the experimenter. This file consists of all required swarm and peer information including the name/location of the log file/archive.

As we consider the INI format to be best suited for this, as it is fairly easy to create, edit and update, it was chosen to populate initial information. The experimenter may easily create an INI swarm description file and provide it to the parser together with the (compressed) log files.

The swarm description file is to be parsed by the experimenter and SQL queries will populate the database. One entry would go into the `swarms` table and a number of entries equal to the number of peers in the swarm description file would go into the `client_sessions` table. As a result of these queries, swarm IDs and client sessions IDs are going to be created when running SQL insert queries (due to the `AUTO_INCREMENT` options). This IDs are essential for the message parsing process and are going to be written down in the Logfile-ID-Mapping-File.

The swarm description file parser is going to parse that file and also generate a logfile-id mapping file. The parser is responsible for three actions:

- parsing the swarm description file
- creating and running SQL insert queries in the `swarms` and `client_sessions` tables
- create a logfile-id mapping file consisting of mappings between client session IDs and log/file

A logfile-id mapping file is to be generated by the swarm description parser and will subsequently be used by the message parser (be it status messages or verbose messages). The mapping file simply maps a client session ID to a log file or a compressed set of log files. A sample file is stored in the repository. The message parser doesn't need to know client session information; it would just use the mapping file and populate entries in the `*_messages` tables.

The message parser is going to use the logfile-id mapping file and the log file (or compressed set of log files) to populate the `*_messages` tables in the database (`status_messages`, `peer_status_messages`, `verbose_messages`.

The workflow of the entire process is highlighted in Figure 6.

There is a separation between the *experimenter* – the one running trials and collecting information and the *parser* – the one interpreting the information.

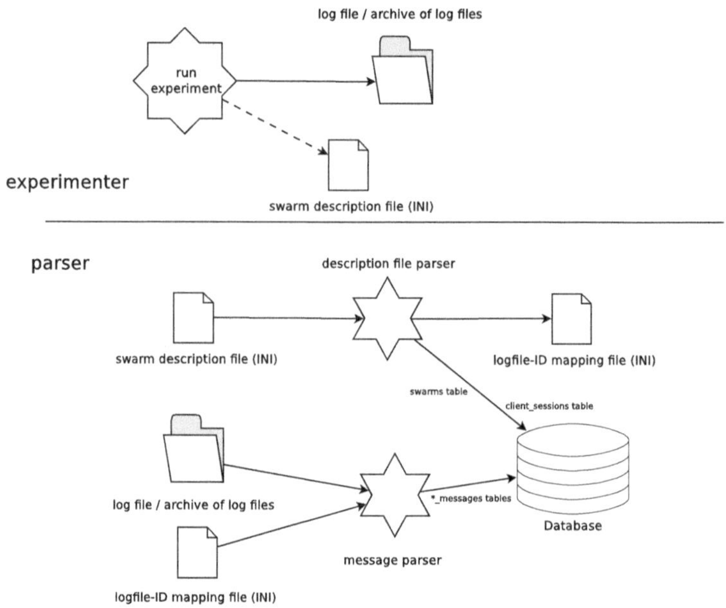

Fig. 6. Workflow of Log Parsing Considering ID Mapping

Trials are run and the experimenter provides a log file or set of log files or archive of log files (the data) and a swarm description file (INI format) consisting of characteristics of clients in the swarm, the file used and the swarm itself (the metadata).

The swarm description file is used to provide an intermediary logfile-id mapping file, as described above. This file may be provided as a file system entry (typically INI), as an in-memory information or it may augment the existing swarm description file (only the client session ID needs to be added).

The logfile-ID mapping, the swarm description file and the log file(s) are then used by the message parser and the description parser to provide actual BitTorrent parameters to be stored in the database. The parsers would instantiate a specific storage class as required by the users and store the information there.

5. Conclusion

In order to provide thorough analysis of Peer-to-Peer protocols and applications, realistic trials and careful measurements were presented to be required. Clients and applications provide necessary parameters (such as download speed, upload speed, number of connections, protocol message types) that give an insight to the inner workings of clients and swarms.

Protocol analysis, centered around BitTorrent protocol, relies on collecting protocol parameters such as download speed, upload speed, number of connections, number of messages of a certain type, timestamp, remote peer speed, client types, remote peer IDs.

We consider two kinds of messages, dubbed *status messages* and *verbose messages* that may be extracted from clients and parsed, resulting in the required parameters.

Various approaches to collecting messages are presented, with differences in the method intrusiveness and quantity and quality of data: certain methods may require important updates to existing clients and, as such, access to the source code, while others may only need access to information provided as log files.

Collection, parsing, storage and analysis of logging information is the primary approach employed for protocol parameter measurements. A processing framework has been designed, implemented and deployed to collect and process status and verbose messages. Multiple parsers and multiple storage solutions are employed. Two types of processing may be used: post-processing, taking into account a previous collection of logging information into a log archive, and real-time processing when data may be monitored as it is parsed in real time.

Protocol parameters are presented to the user through the use of a rendering engine that provides graphical representation of parameter evolution (such as the evolution of download speed or upload speed). The rendering engine makes use of the database results from the processing framework and provides a user friendly interface to the experimenter.

6. References

Bardac, M., Milescu, G. & Deaconescu, R. (2009). Monitoring a BitTorrent Tracker for Peer-to-Peer System Analysis, *Intelligent Distributed Computing* pp. 203–208. URL: *http://www.springerlink.com/index/r528521241850jnl.pdf*

Das, S. & Kangasharju, J. (2006). Evaluation of Network Impact of Content Distribution Mechanisms, *Proceedings of the 1st International Conference on Scalable Information Systems* pp. 35–es.

Deaconescu, R., Milescu, G., Aurelian, B., Rughinis, R. & Tapus, N. (2009). A Virtualized Infrastructure for Automated BitTorrent Performance Testing and Evaluation, *International Journal on Advances in Systems and Measurements* 2(2&3): 236–247.

Iosup, A., Garbacki, P., Pouwelse, J. & Epema, D. (2006). Correlating Topology and Path Characteristics of Overlay Networks and the Internet, *Proceedings of the Sixth IEEE International Symposium on Cluster Computing and the Grid*, CCGRID '06, IEEE Computer Society, Washington, DC, USA, pp. 10–. URL: *http://portal.acm.org/citation.cfm?id=1134822.1134925*

Locher, T., Moor, P., Schmid, S. & Wattenhofer, R. (2006). Free Riding in BitTorrent is Cheap, *Fifth Workshop on Hot Topics in Networks (HotNets-V)* . URL: *http://www.sigcomm.org/HotNets-V/program.html*

Naicken, S., Livingston, B., Basu, A., Rodhetbhai, S., Wakeman, I. & Chalmers, D. (2007). The state of peer-to-peer simulators and simulations, *SIGCOMM Comput. Commun. Rev.* 37(2): 95–98.

Pouwelse, J. A., Garbacki, P., Epema, D. H. J. & Sips, H. J. (2005). The Bittorrent P2P File-Sharing System: Measurements And Analysis, *4th International Workshop on Peer-to-Peer Systems (IPTPS)*. URL: *http://citeseerx.ist.psu.edu/viewdoc/summary?doi=10.1.1.59.3191*

Pouwelse, J. A., Garbacki, P., Wang, J., Bakker, A., Yang, J., Iosup, A., Epema, D. H. J., Reinders, M., van Steen, M. R. & Sips, H. J. (2008). TRIBLER: A Social-based Peer-to-Peer

System: Research Articles, *Concurr. Comput. : Pract. Exper.* 20: 127–138.
URL: *http://portal.acm.org/citation.cfm?id=1331115.1331119*

Pouwelse, J., Garbacki, P., Epema, D. & Sips, H. (2004). A Measurement Study of the BitTorrent Peer-to-Peer File-Sharing System.
URL: *http://citeseerx.ist.psu.edu/viewdoc/summary?doi=10.1.1.3.4761*

Tian, Y., Wu, D. & Ng, K.-W. (2007). Performance Analysis and Improvement for BitTorrent-like File Sharing Systems, *Concurrency: Practice and Experience* 19: 1811–1835.

Vlavianos, A., Iliofotou, M. & Faloutsos, M. (2006). BiToS: Enhancing BitTorrent for Supporting Streaming Applications, *INFOCOM 2006. 25th IEEE International Conference on Computer Communications. Proceedings* pp. 1–6.
URL: *http://dx.doi.org/10.1109/INFOCOM.2006.43*

14

Shape Optimization of Mechanical Components for Measurement Systems

Alexander Janushevskis, Janis Auzins,
Anatoly Melnikovs and Anita Gerina-Ancane
Riga Technical University,
Latvia

1. Introduction

For the topology and shape optimization of structures, the different realizations of homogenization method are widely used (Arora, 2004; Bendsoe & Sigmund, 2003). This method is highly effective for shell constructions and allows implementing topometry and topography, sizing and shape as well as freeform optimization (Vanderplaats, 2004). However, it is a very time consuming procedure because the number of design parameters can reach a million and more. There is the possibility of taking into account some technological factors, nevertheless, in the case of bulky bodies it frequently produces shapes that are difficult to manufacture. As shown in work (Mullerschon et al., 2010), the Hybrid Cellular Automata method does not allow parallelization of computations and PBS queuing system has been used. At the same time the following resource saving approach (Janushevskis et al., 2010; Janushevskis & Melnikovs, 2010) can be used for shape optimization which includes the following steps: 1) Planning the position of control points of NURBS (see, for example, Saxena & Sahay, 2005) for obtaining a smooth shape. 2) Building geometrical models using CAD software in conformity with design of experiment. 3) Calculation of responses for a complete FEM model using CAE software. 4) Building metamodels (surrogate models) for responses on the basis of computer experiment. 5) Using metamodels for shape optimization. 6) Validating the optimal design using CAE software for the complete FEM model.

2. Basics of approach

Latin Hypercube type experimental designs first proposed by Vilnis Eglajs in his work (Audze & Eglajs, 1977), then by McKay (McKay et al., 1979) and used by many other investigators, as well as its improvements (Auzins, 2004) are a very essential aspect utilized in the proposed method. The significance of approximations for the solution of optimization problems proposed by Lucien Schmit in his early works (Schmit & Farcshi, 1971) and nowadays generally recognized as the Response Surface Method is also the foundation of current approach. The use of planned computer experiments and the metamodeling (surrogate model) approach (Forrester et al., 2008; Sacks et al., 1989) ensures great economy of computing time, especially for finite element (FEM) calculations. First of all let us demonstrate our approach on the simple test problems.

2.1 Test problem of plate bending

A clamped square plate is considered under a concentrated load of 500 N applied at centre in a direction orthogonal to its main surface. The isotropic material properties are: the Young's modulus E = 200 GPa, the Poisson's ratio ν = 0.3 and dimensions are 400x400x4.2118 mm. The shape optimization of the plate with constant thickness is carried out to minimize its volume in the case of a single displacement constraint δ = 0.5 mm. The cutout shape of the plate is defined by subsequent techniques shown in Figure 1: 1) with the points that are connected with straight lines; 2) with the NURBS knot points; 3) with the control points of NURBS polygon. Due to symmetry only ⅛ of the plate is considered for cutout definition and ¼ of the plate for problem solution by FEM.

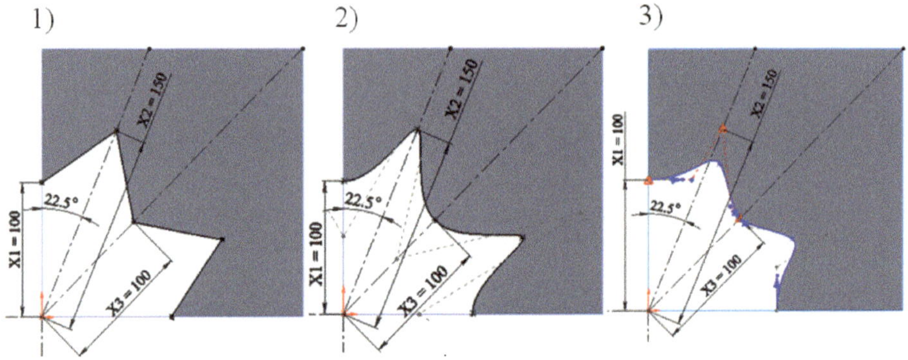

Fig. 1. Techniques for definition of cutout.

Three parameters are stated to define location of points. Parameters are varied in the following ranges: 100≤ X1≤170; 100≤X2≤210; 100≤X3≤230 mm for the first two variants and 100≤X1≤180; 100≤X2≤235; 100≤X3≤230 mm for third variant of definition. In the last case at both end points two continuity vectors are defined additionally with direction normal to the side and to symmetry axis of the corresponding plate and with fixed length of 19 and 3 mm. The design of experiment for 3 factors and 40 trial points is calculated with mean-square error (MSE) criterion (Auzins & Janushevskis, 2002; Auzins, et al., 2006) value 0.4262 using EDAOpt (Auzins & Janushevskis, 2007) - software for design of experiments, approximation and optimization developed in the Riga Technical University. The geometrical models are developed using SolidWorks (SW) for all variants. The shapes for the third variant of definition are shown in Figure 2. In the next step responses of these models are calculated by SW Simulation (Lombard, 2009), using elements with a global size 4 mm and total number of DoF ~100000. Then these responses are used for approximation by EDAOpt. For example, for approximation of response y by quadratic polynomial the following expression (see, for example, Auzins & Janushevskis, 2007) is used:

$$\hat{y} = \beta_0 + \sum_{i=1}^{d} \beta_i x_i + \sum_{i=1}^{d-1} \sum_{j=i+1}^{d} \beta_{ij} x_i x_j + \sum_{i=1}^{d} \beta_{ij} x_i^2 + \varepsilon \qquad (1)$$

where there are d variables $x_1,\ldots, x_d$, $L = (d+1)(d+2)/2$ unknown coefficients β and the errors ε are assumed independent with zero mean and constant variance σ^2.

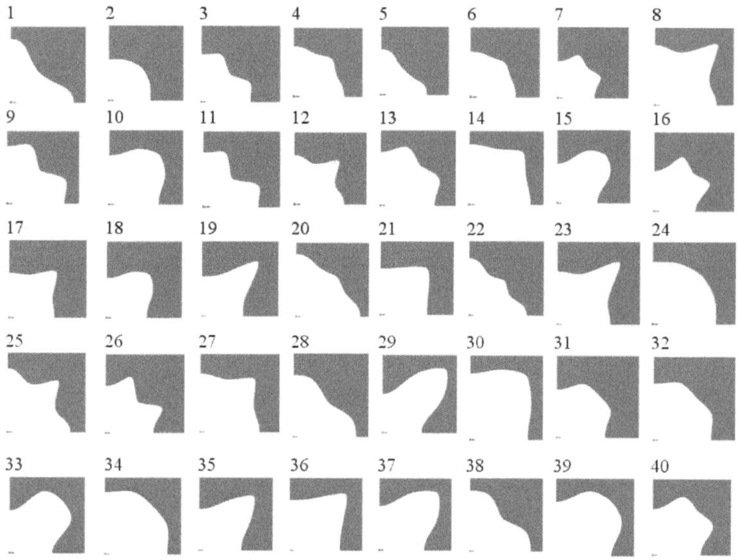

Fig. 2. Shapes of cutout in compliance with design of experiment.

In the case of locally weighted polynomial approximation, coefficients $\boldsymbol{\beta}=(\beta_1, \beta_2, \beta_3,..., \beta_L)$ depend on point x_0 where prediction is calculated and are obtained using the weighted least squares method:

$$\beta = \arg\min_{\beta} \sum_{j \in N_X} w(x_0 - x_j) \times (y_j - \hat{y}(x_j))^2 \tag{2}$$

where the significance of neighboring points in the set N_X is taken into account by Gaussian kernel (weight function):

$$w(u) = \exp\left(-\alpha u^2\right) \tag{3}$$

where u is Euclidian distance from x_0 to current point and α is a coefficient that characterizes significance.

Quality of approximation is estimated by relative crossvalidation error using leave-one-out crossvalidation:

$$\sigma_{err} = 100\% \frac{\sqrt{\frac{1}{n}\sum_{i=1}^{n}(\hat{y}_{-i}(x_i) - y_i)^2}}{\sqrt{\frac{1}{n-1}\sum_{i=1}^{n}(y_i - \bar{y})^2}} \tag{4}$$

where root mean squared prediction error stands in numerator and mean square deviation of response from its average value stands in denominator, n is the total number of experimental trials and $\hat{y}_{-i}(x_i)$ denotes approximated value for response in i-th point, calculated without taking into account the i-th experimental point.

Using the obtained locally weighted polynomial approximations by global search procedure (Janushevskis et. al, 2004), implemented in EDAOpt, the optimal cutout shape is obtained (see Figure 3) for the different aforementioned techniques. In table 1 the results are summarized and compared with volume obtained in work (Liang et al., 2001) using the homogenization method. Variants correspond to shapes shown in figure 3. The value of Gaussian kernel parameter α of the local quadratic polynomial approximation is chosen to minimize relative leave-one-out crossvalidation error σ_{err} of approximations of appropriate responses, i.e. deflection δ and volume v of plate. v_p is the predicted volume calculated using approximations and v_a is the actual volume calculated using a geometrical model. v_p and v_a in % show comparison of appropriate volume in respect to the volume obtained in (Liang et al., 2001). Best results are achieved with the technique using the control points of NURBS polygon. This allows reducing the volume of the plate by 1.38 % (Fig. 3 e) in comparison with the homogenization method.

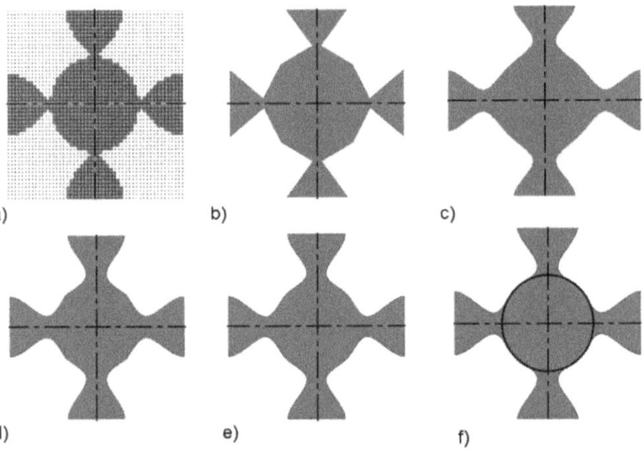

Fig. 3. Shapes of plate obtained by (a) homogenization method (Liang et al., 2001); and by the current approach using different techniques: (b) the points that are connected with straight lines; (c) with the NURBS knot points; (d) with the control points of NURBS polygon; (e) same as "d" but with additionally optimized tangent weighting at the spline endpoints; (f) same as "e" but with circle added.

Variant	α	$\sigma_{err\ \delta}$ %	$\sigma_{err\ v}$ %	v_p mm^3	v_a mm^3	v_p %	v_a %
a	-	-	-	-	68750.00	-	-
b	17	9.81	0.03	69414.78	69331.58	1	0.84
c	17	9.81	0.03	68988.67	68815.88	0.4	0.096
d	15.6	9.81	0.03	68862.32	68721.98	0.16	-0.04
e	3.2	0.79	0.16	67797.524	67800.975	-1.385	-1.38

Table 1. Quantitative indices of the shape optimization of cutout for the plate bending problem.

It should be mentioned that the predicted volume for variant e is in good agreement with the actual value. At the same time the total number of FEM problem calculations (i.e. number of trials of computer experiment) is less on 10 than by using the homogenization method. The distribution of resultant displacement for plate of optimal shape (variant e) is shown in fig. 4.

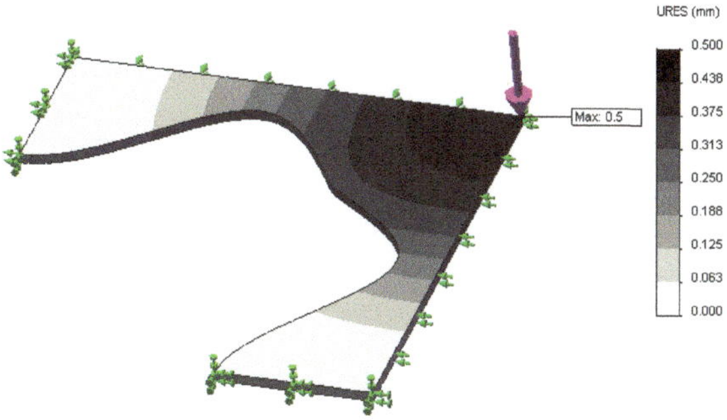

Fig. 4. Displacement of plate for variant e.

2.2 Stretched plate test problem

A square plate with dimensions 1300x1300x0.0001 mm is considered under two axial stretching loads of p = 0.65 N/mm² and p/2 accordingly (see Figure 6). The isotropic material properties are: the Young's modulus E = 210 GPa, the Poisson's ratio ν = 0.3. The shape optimization of the plate with constant thickness is carried out to minimize its volume in case of a single constraint on maximal level of equivalent stress σ_{max} = 4.38 MPa. Due to symmetry only ¼ of the plate is considered for cutout definition and for problem solution by FEM. The cutout shape of the plate is defined by five coordinates of points 2, 3, 4, 5 and 6 situated on straight lines that make angles of 0; 22.5; 45 and 90 degrees with horizontal axis as shown in Figure 5. The initial cutout shape is a circle with radius 250 mm and the volume of plate v is 373.4 mm³.

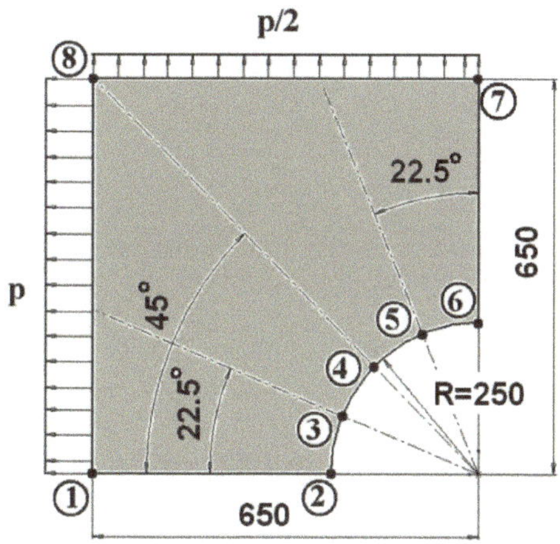

Fig. 5. Scheme of ¼ of plate with initial cutout.

The aforementioned five parameters are varied in the range from 250 to 640 mm. The design of experiment for 5 factors and 112 trial points is calculated with MSE criterion value 0.5394. In the next step responses of these models are calculated by SW Simulation using shell elements with global size 7.5 mm and total number of DoF ~108000. Then these responses are used for approximation by EDAOpt. Using obtained locally weighted polynomial approximations by global search procedure, the optimal cutout shape is obtained (see Figure 6) for the different aforementioned techniques. In table 2 the results are summarized and compared with the volume obtained in work (Papadrakakis et al., 1998). Variants correspond to shapes shown in figure 6.

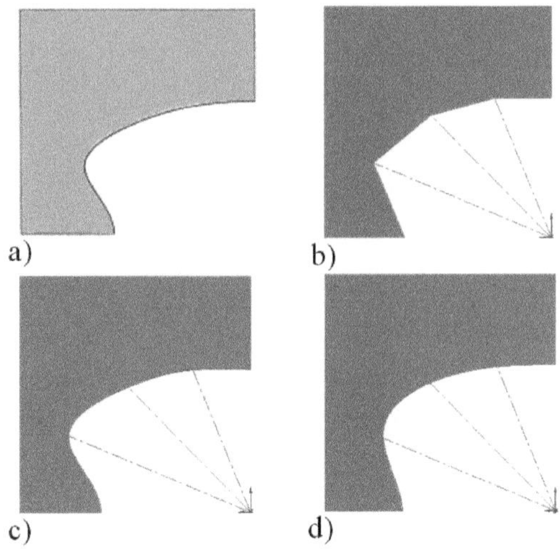

Fig. 6. The obtained plate shapes. Plate shapes obtained (a) in (Papadrakakis et al., 1998); and by current approach using different techniques: (b) the points that are connected with straight lines; (c) with the NURBS knot points; (d) with the control points of NURBS polygon.

Variant	α	$\sigma_{err\ \sigma}$ %	$\sigma_{err\ v}$ %	v_p mm^3	v_a mm^3	$\sigma_{max\ p}$ *MPa*	$\sigma_{max\ a}$ *MPa*
a	-		-	-	280	-	4.38
b	8	48.73	0.00	255.93	255.84	4.38	5.22
c	12.6	47.48	2.27	251.76	251.38	4.38	4.35
d	7.3	31.94	1.17	251.00	250.69	4.38	4.34

Table 2. Quantitative indices of the shape optimization of cutout for the plate stretching problem.

The value of Gaussian kernel parameter α of the local quadratic polynomial approximation is chosen to minimize relative leave-one-out crossvalidation error σ_{err} of approximations of appropriate responses, i.e. maximal equivalent stress σ_{max} and volume v of plate. v_p is predicted volume calculated using approximations and v_a is actual volume calculated using full FEM and geometrical models. Again the best results are achieved with the technique using the control points of NURBS polygon. This allows reducing volume of the plate by

10.5 % (Fig. 6 d) in comparison with the volume obtained in (Papadrakakis et al., 1998). Distribution of equivalent von Mises stress for case d is shown in Figure 7.

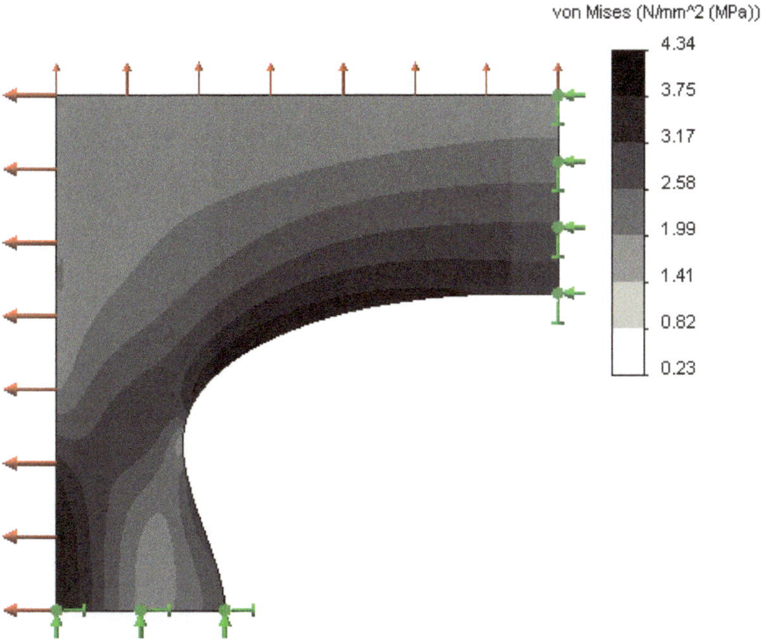

Fig. 7. Equivalent von Mises stress levels in the plate for variant d.

3. Tensometric wheel pairs

At the present time special tensometric wheel pairs are used for the wheel - rail system monitoring. For each type of rolling stock these wheel pairs must fit the vehicle's wheel wearing condition, diameter and bearing box connection type. Using and delivering the tensometric wheel pairs is expensive and takes a lot of time for preparing strength - dynamics tests. In this work removable equipment for monitoring is proposed for mounting on the ordinary wheel pairs. The monitoring wireless system (Grigorov, 2004; Hart, 1986) for 80 tons wagon (freight car) is taken for prototype. The movable part of the equipment (Fig. 8) consists of a removable disk, two transmitters and a transmitting antenna as well as strain gauges bonded to the wheel at defined places. The removable disk is fixedly attached to the wheel pair's axis. A circular transmitting antenna and two transmitters are mounted on the outside of the disk.

Removable equipment must be lightweight to minimize distortions of measurements and at the same time it must be with appropriate durability. During testing dynamical loads caused by rail joints, railroad switches and other irregularities as well as due to defects of wheel geometry are transmitted from wheel pairs to the removable disk which is rigidly mounted on the wheelset axis. Therefore shape optimization of the mounting disk that is the main heavy-weight part of the equipment is very important to reduce its total weight.

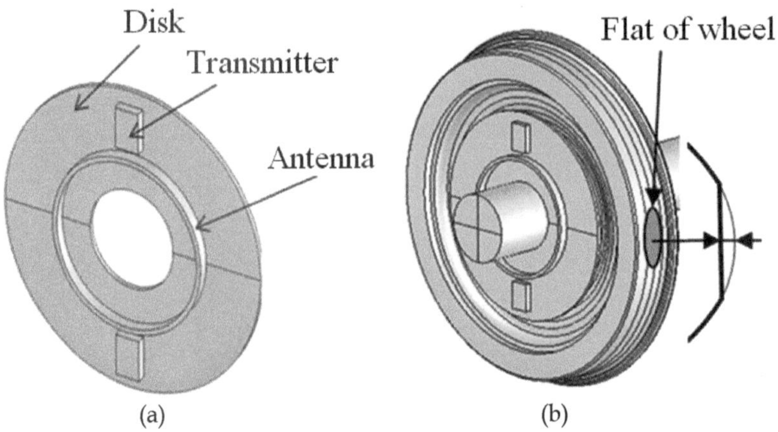

Fig. 8. (a) Removable disc with elements of measurement system and (b) its mounting place.

3.1 Loads acting on wheel pairs

The main loads are acting in vertical direction and are caused by railroad irregularities and wheel defects in the wheel - railroad contact. The removable disk sustains all loads from the wheel pair because it is rigidly fastened. Strength of the removable disk is calculated for maximal possible loading. For example, in the case when the wheel pair has 2 mm flat of wheel (Fig.8 b), the loaded and empty wagon wheelsets undergo different loads in vertical direction at different velocities (see Table. 3) (Sladkowsky & Pogorelov, 2008). For an empty wagon the maximal load is at velocity 5 m/s, but for the loaded wagon at 10 m/s. Strength of the removable disk will be analyzed with maximal vertical load - 620.6 kN for two cases of orientation of the transmitters - horizontal and vertical when the wheel pair acceleration can reach 12 g (the gravitational acceleration g = 9.81 m/s^2 is taken into account).

Velocity of wagon m/s	Maximal load in moment of shock, kN	
	Empty wagon	Loaded wagon
Static load	22.8	104.5
1	136.1	251.3
2	170.2	316.4
5	297.8	367.2
10	271.3	620.6
20	276.6	604.9

Table 3. Load versus velocity of wagon (Sladkowsky & Pogorelov, 2008).

Strength of the disk under centrifugal load will also be analyzed at maximum vehicle velocity 200 km/h (wheel angular velocity = 116.98 rad/s). In this case the disk is considered as new without wear on riding circle.

Besides, frequency analysis was made to find natural frequencies of the wheel pair and evaluate possible resonance in the case of flat of wheel. Obtained results show that excitation frequencies at velocities of operating conditions are significantly smaller than fundamental frequency.

3.2 Disk model for strength calculation

The geometrical model of the disk is created using SW. It takes into account the shape, size and material of the disk and transmitters (dimensions 20x55x80 mm, mass 0.1 kg). The transmitting antenna (Fig. 8) is removed from the calculation model because it has small dimensions and lightweight material. The calculation model also doesn't consider fastening holes and fastening elements. The transmitting antenna works stable if its displacement in axial direction is less than 2 mm (Hart, 1986). This constraint is taken into account in the next optimization.

The strength calculations are performed using SW Simulation. First, the shape of the removable disk (radius R = 300 mm and thickness b = 10 mm) is changed to an ellipse with semi-major axis length 300 mm and semi-minor axis E1 (Fig. 9 a) by simple size optimization of E1. This shape is convenient for the equipment mounting purposes and is taken for initial design.

FE mesh (Fig. 9 b) is generated with second order tetrahedral solid elements and is generated to get high accuracy results. It consists of about 51000 elements with 86000 nodes (258000 DOF).

Removable disk's displacement is restrained on its cylindrical face (Fig. 9 c). The material of the disk is aluminum alloy (1060 H12) with elastic modulus E = 69000 MPa, Poisson's ratio $\nu = 0.33$, mass density $\rho = 2700\ \mathrm{kg/m^3}$ and yield strength $\sigma_y = 27.5742$ MPa.

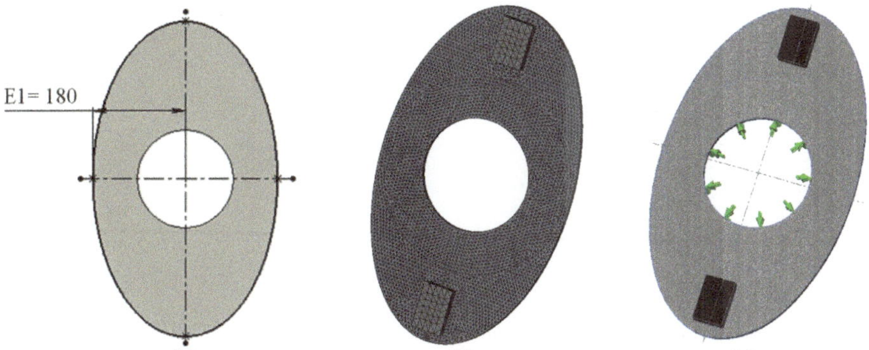

(a) (b) (c)

Fig. 9. (a) Ellipsoidal disk; (b) Computational finite element mesh of the model; (c) Scheme of disk fastening.

The material's ultimate fatigue resistance is calculated as (State Railway Research Institute [SRRI], 1998):

$$\sigma_{-1} = 0.4 \cdot \sigma_y \tag{5}$$

So stresses must be less than $\sigma_{-1} = 11.0297$ MPa. Additionally the value of the factor of safety FOS = 2.75 is assumed to be sure that the disk will be durable in any worst-case situation (SRRI, 1998). The acceptable stress in the disk material is reduced to $\sigma_{max} = 4$ MPa. Von Mises yield stress criterion is used for all strength calculations:

$$\sigma_{vonMises} = \sqrt{\frac{(\sigma_1 - \sigma_2)^2 + (\sigma_2 - \sigma_3)^2 + (\sigma_1 - \sigma_3)^2}{2}}, \tag{6}$$

where $\sigma_1, \sigma_2, \sigma_3$ are principal stresses.

Thereby the von Mises stress at any point of the disk should be less than acceptable stress:

$$\sigma_{vonMises} < \sigma_{\max} \tag{7}$$

3.3 Stresses in initial design disk

Three variants of stressed state of the disk are analyzed, i.e., from loads due to flat of wheel in two cases of the disk orientation: when the major axis of ellipse is vertical and horizontal as well as from centrifugal loads.

We consider a loaded wagon with maximal loading in moment of shock that occurs at velocity 10 m/s (Table 3). Maximal stresses in moment of shock (acceleration a = 119.3 m/s^2) are shown on Fig. 10. As we can see, values of maximal stress levels for both orientations of the disk are very similar.

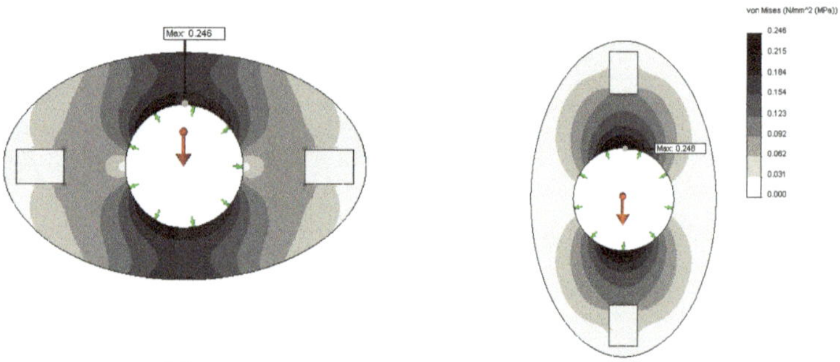

(a) (b)

Fig. 10. Von Mises stresses distribution in initial design disk for (a) horizontal and (b) vertical orientation.

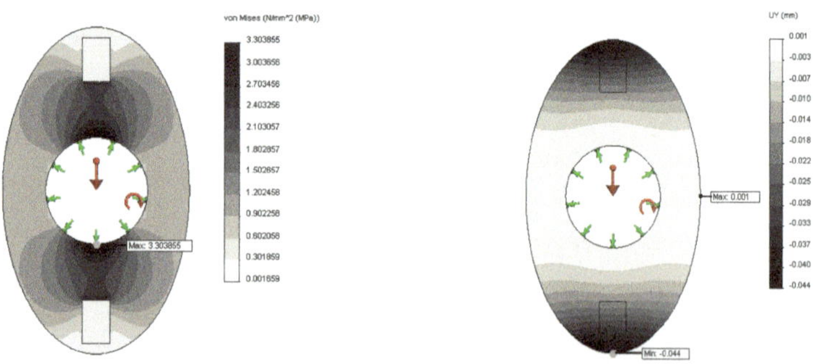

(a) (b)

Fig. 11. (a) Von Mises stresses distribution in initial design disk from centrifugal loads; (b) Disk's displacements in axial direction.

The results for the stressed state of initial design of the disk from centrifugal load are shown in Fig. 11. Maximum von Mises stress in the disk from centrifugal load is at least 12 times greater than in the case of loading from flat of wheel.

3.4 Shape optimization of cross section of the ellipsoidal disk

The shape of cross section of the disk is optimized, taking into account only centrifugal load. The constructive restrictions allow changing the disk cross section shape only at one side and in radial direction at range 150 mm to 300 mm from ellipse center. The section of the disk at radial distance 0-150 mm has constant thickness b = 10 mm. Three methods are used to define the cross section shape (Fig. 12): a) with NURBS knot points, b) with NURBS polygon points and c) with points that are connected with straight lines. Four parameters are stated to define the shape. Parameters are varied in the following ranges: $4 \leq X1 \leq 10$; $4 \leq X2 \leq 10$; $5 \leq X3 \leq 12$; $3 \leq X4 \leq 5$ for variants "a", "c" and $3 \leq X1 \leq 10$; $0.5 \leq X2 \leq 20$; $5 \leq X3 \leq 25$; $2 \leq X4 \leq 5$ for "b". The design of experiments is calculated with MSE criterion for 4 factors and 70 trial points. This design of experiment is also available on the web: http://www.mmd.rtu.lv.

So the 70 strength studies are calculated for each considered method. SW Simulation results (volume, maximal von Mises stress, axial displacement of the disk etc.) are entered into EDAOpt for approximation and subsequent global search.

Some optimization and approximation characteristics are shown in Table 4. Results of variants "a" and "b" are obtained with second order local polynomial approximation. Third order local polynomial approximation is used for variant "c". Gaussian kernel coefficient α was varied for least value of crossvalidation error (4).

Variant	α	Approximation's σ_{err} [%]		Volume v [mm³]			Maximal von Mises stress $\sigma_{vonMises}$ [MPa]		
		$\sigma_{vonMises}$	v	Predicted	Real	Error [%]	Predicted	Real	Error [%]
a	6	20.56	0.06	1003944	1003891	0.005	3.9999	3.833816	4.33
b	3	40.28	2.03	923421	921740	0.018	3.9999	4.200354	4.77
c	4	10.93	0.00	946180	946173	0.001	3.9998	4.125750	3.05
d	-	-	-	-	1394900	-	-	3.3	-

Table 4. Quantitative data of approximation and shape optimization of the ellipsoidal disk cross section.

The obtained metamodels are used for optimization of factors. The ellipsoidal disk volume is minimized by taking into account the specified constraints on displacement and stress level. The obtained shapes are presented in Fig. 12. As shown in table 4, the best results are obtained for variant "b" (Fig. 13), where the volume is lower by 8.2 % in comparison to variant "a" and on 2.6 % - to "c". All 3 variants give significant advantage in volume (28.1 - 33.9 %), comparing to variant "d"- the initial shape design with constant 10 mm thickness.

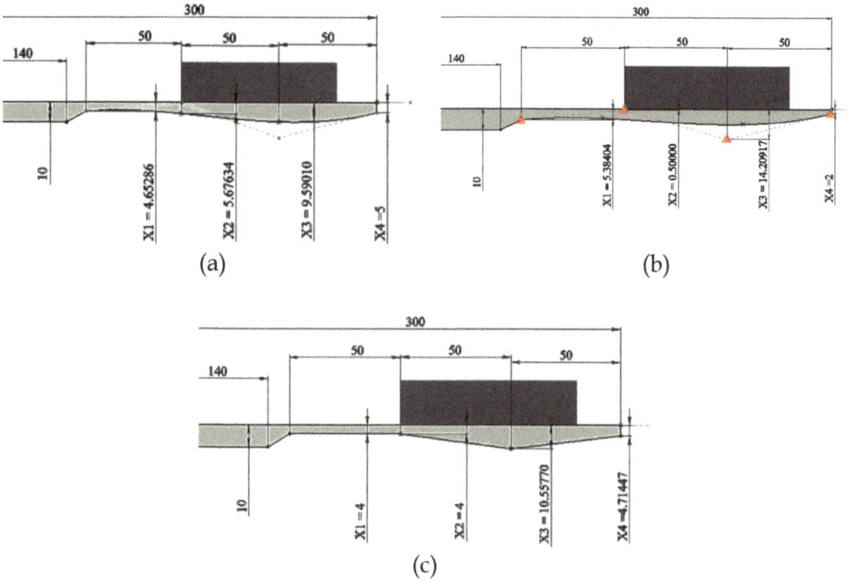

Fig. 12. Results of optimization of ellipsoidal disk. Shape of cross section is defined by (a) NURBS knot points, (b) NURBS polygon points, (c) points that are connected with straight lines.

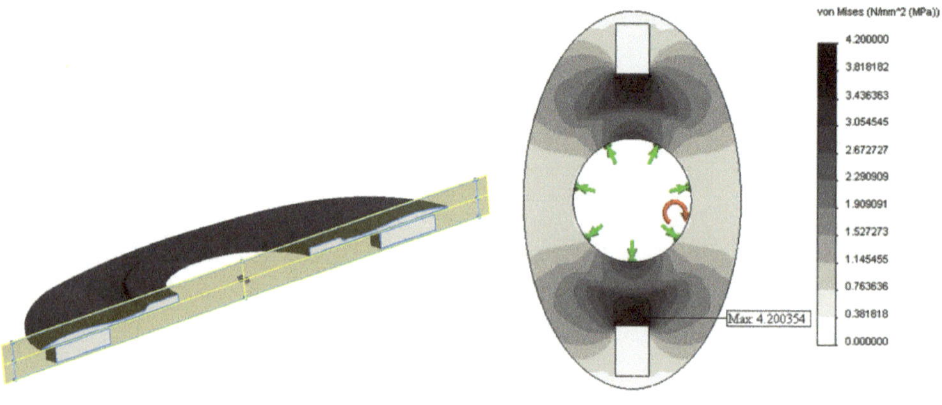

Fig. 13. Half of 3D model of optimal shape disk and real von Mises stresses distribution in it.

3.5 Summary

The proposed equipment allows using the standard wheel pair with removable measurement equipment as a tensometric wheel pair that considerably reduces material and time expenses required for preparing testing. By means of size and shape optimization, the total volume of the mounting disk of railway vehicle measurement system is reduced by ~64 % in comparison with the initial design. The method based on NURBS polygon points gives the shape with at least 3% better objective (volume) than other used methods.

4. Automotive vehicle gage panel

In this section designing the mechanical part of an automotive vehicle gage panel is discussed. Automotive vehicle gage panels (GP) must meet many requirements - such functional characteristics as appropriate stress levels under loads, eigenfrequencies, stiffnesses, weight, accuracy etc. and last but not least they must have minimal environmental pollution during service lifetime. The 3D geometrical models of the gage panel are elaborated using SW. Static and dynamic responses of the gage panel are calculated using SW Simulation and impacts to environment are evaluated using SW Sustainability that include such indices as total energy consumed, carbon footprint, air acidification and water eutrophication. The stationary and transient behaviors of the gage panel under dynamic excitation as well as stress distribution under static loading are investigated. Due to the complexity of the gage panel FEM models, the appropriate metamodels are elaborated based on design of experiments. These metamodels are used for multiobjective optimization using a global search procedure. Partial objectives are aggregated in the complex objective function for optimization purposes. Dynamic behavior of the gage panel is then verified by solution of the full FEM models in case of random vibrations.

4.1 Specific requirements

A constantly pressing problem is the development of safe and environmentally friendly engineering objects with high functional properties, attractive style and competitive price. We should try to take into account not only precisely measurable functional indices, but also such a difficult-to-formalize index as style of GP.

The Industrial Designer Society of America defines industrial design as the professional service of creating and developing concepts and specifications that optimize the function, value, and appearance of products and systems for the mutual benefit of both users and manufacturer. In fact, industrial designers focus their attention upon the form and user interaction of products. There are five critical goals (Ulrich & Eppinger, 2008): 1) Utility: The product`s human interfaces should be safe, easy to use, and intuitive. Each feature should be shaped so that it communicates its function to the user. 2) Appearance: Form, line, proportion, and color are used to integrate the product into a pleasing whole. 3) Easy maintenance: Product must also be designed to communicate how they are to be maintained and repaired. 4) Low costs: Form and features have a large impact on tooling and production costs, so these must be considered jointly by the team. 5) Communication: Product design should communicate the corporate design philosophy and mission through the visual qualities of the products. The practical concept selection methods (Ulrich & Eppinger, 2008) vary in their effectiveness and include the following: 1) External decision: Concepts are turned over to the customer, client, or some other external entity for selection. 2) Product champion: An influential member of the product development team chooses a concept based on personal preference. 3) Intuition: The concept is chosen by its feel. Explicit criteria or trade-offs are not used. The concept just seems better. 4) Multivoting: Each member of the team votes for several concepts. The concept with the most votes is selected. 5) Pros and cons: The team lists the strengths and weaknesses of each concept and makes a choice based upon group opinion. 6) Prototype and test: The organization builds and tests prototypes of each concept, making a selection based upon test data. 7) Decision matrices:

The team rates each concept against pre specified selection criteria, which may be weighted. The concept selection method is built around the use of decision matrices for evaluating each concept with respect to a set of selection criteria. At the same time such formalized methods are elaborated as method of imprecision (Zimmermann, 2001) with non-compensating aggregation and compensating aggregation as well as fuzzy design method with different level interval algorithms.

The GP styles of different cars significantly differ and should be evaluated in context of the specific vehicle. At the same time style determines an arrangement of particular components (distances between gage axes etc.). In Fig. 14 we can see initial styles and the 3D model of the GP designed for new vehicles (Company Amoplant, 2011).

Fig. 14. Frontal view of GP of the initial styles and 3D geometrical model of GP.

4.2 Vehicle GP optimization problem

Now we can try to use the previously discussed approach for multiobjective shape optimization of the GP. The problem is stated as follows:

$$\min_{x} F(x) = [F_1(x), F_2(x), \ldots, F_k(x)]^T \tag{8}$$

subject to $g_j(x) \leq 0, j = 1, 2, \ldots, m$, and $h_l(x) = 0, l = 1, 2, \ldots, e$;
where k is the number of objective functions F_i;
m is the number of inequality constraints;
e is the number of equality constraints;
$x \in E^n$ is a vector of n design variables.

First let us briefly discuss obtaining responses of GP for each particular objective.

4.3 Strength calculation of GP design

Generally the strength of the GP design is checked on special vibrostands. The GP is subjected to different dynamic loads. Vibrostability and vibration strength of the GP are checked on excitations in the frequency domain from 10 to 250 Hz. One of the main natural experiments is a test of shock resistance of the GP design under acceleration level $a = 10g$. Such experiments require significant material and time expenses and for optimization purposes the computer based design check must be used. The 3D geometrical model (Fig. 14) of the GP is created using SW and it consists of 18 parts: 6 deformable bodies and 12 rigid bodies that take into account the inertial characteristics of the internal devices. The deformable parts are made from the ABC 2020 plastic, but for the internal device bodies are assumed as alloy steel. The initial volume of the GP assembly is v_0 = 764674 cm^3 and mass m_0 = 1.02 kg. The 3D model of the GP assembly is used for FEM analysis by SW Simulation to evaluate different responses of the GP. The FE mesh (Fig. 15) is generated with curvature based mesh (max elements size = 9 mm, min element size = 1.8 mm, element size growth ratio = 1.5), that ensures accurate discretization of the complex shape bodies of the GP. The FE mesh consists of ~ 210,000 nodes, ~147,000 elements, ~640,000 DOF.

In the initial design of the GP von Mises stresses from impact loading are shown on Fig. 15. We can see that maximal stresses are concentrated on the bracket's cross-section and it reaches 4 MPa. Other parts of the GP design are stressed considerably less. This implies that the bracket design should be improved.

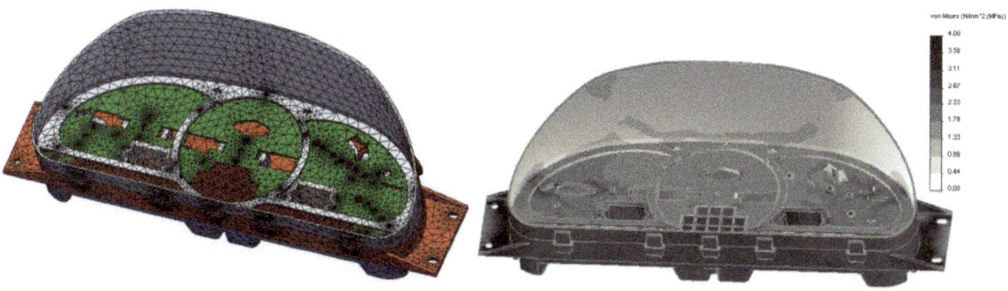

Fig. 15. Meshed 3D model and von Mises stresses distribution in initial design of GP.

4.4 Frequency analysis of GP

Frequency analysis is implemented to find natural frequencies of the GP model and evaluate possible resonance in the case of external excitation. The same FE mesh for model as considered before is used. The contacts between assembly's parts are defined as bonded. The numerical solver FFEPlus of SW is used for calculations.

The obtained results show that the fundamental frequency of the GP is sufficiently high f_1 = 170.47 Hz. The obtained mode shapes for the GP natural frequencies (f_2 = 201.35 Hz, f_3 = 264 Hz, f_4 = 331.85 Hz) are shown on Fig. 16.

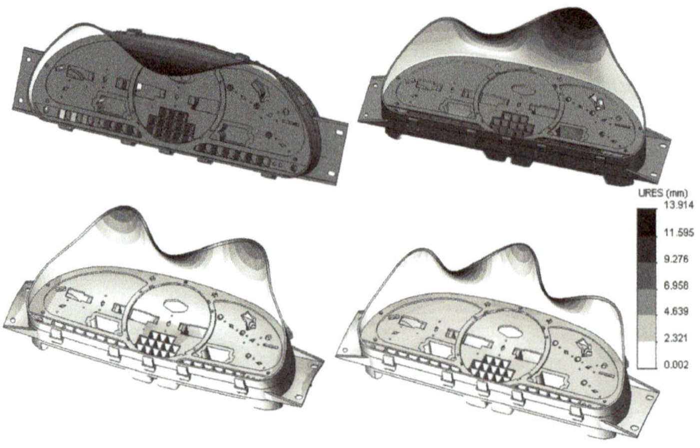

Fig. 16. The mode shapes of the four lower natural frequencies of GP.

4.5 Sustainability analysis of GP

SW Sustainability allows getting immediate feedback on the carbon footprint and other environmental impacts of the GP throughout its entire lifecycle, including material selection, production, transportation, use and end of life (Fig. 17).

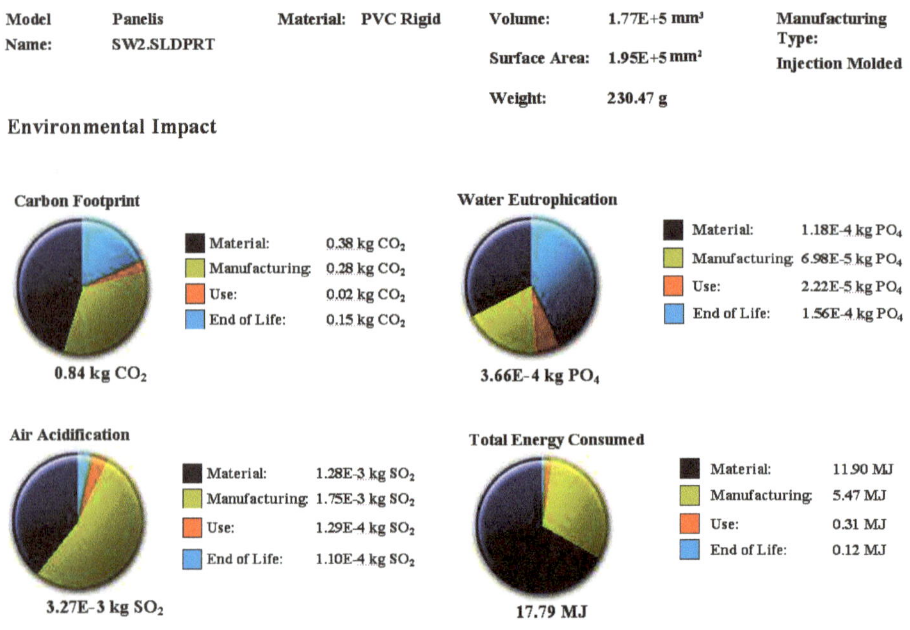

Fig. 17. Environmental impact of the frame component of GP calculated by SW Sustainability.

4.6 Shape optimization of GP

In this specific situation one of the solutions could be increasing the GP bracket cross-section thickness and changing the shape at the most stressed place. The cross-section shape for bracket's strengthening is defined by the 3 knot points (Fig. 18 a) of NURBS. Design parameters are coordinates of the knot points varied in the following ranges: $3 \leq X1 \leq 6$; $2 \leq X2 \leq 5$; $0 \leq X3 \leq 3$. As a cross-section profile is defined, the 3D-shape is created using the path curve (Fig. 18 b). The same shape for strengthening is created on the second bracket of the GP frame component. A maximal von Mises stress in the bracket was minimized with constraint on the GP volume (v <770500 cm^3).

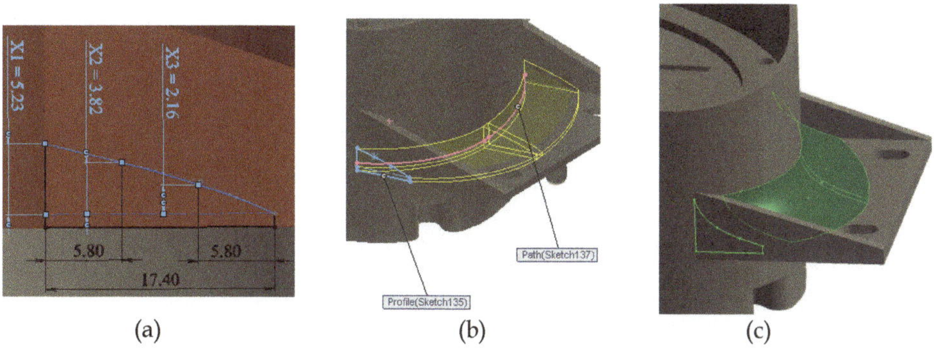

Fig. 18. Shape definition of bracket: (a) cross-section shape; (b) 3D- shape creation through path curve; (c) shape optimization result.

Von Mises stresses are compared in the design of the obtained shape and the initial shape (Fig. 19). There are 6 check points that show von Mises stresses distribution in the most stressed bracket cross-section. Volume of the obtained design is v = 770430 cm^3. Change of

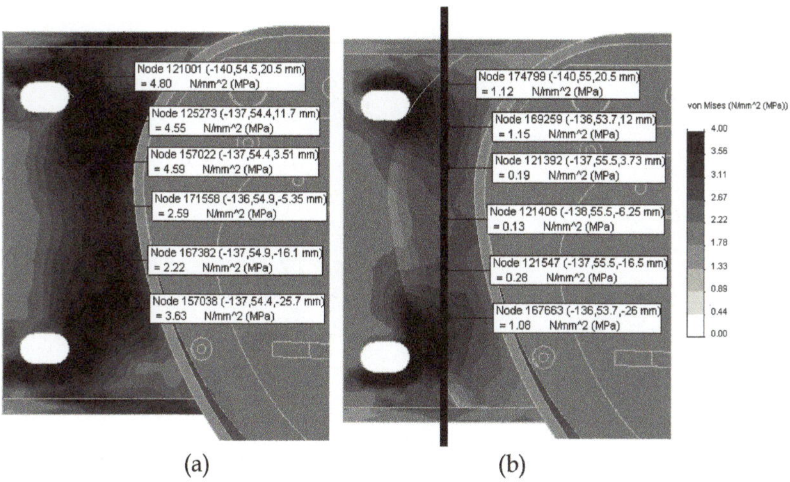

Fig. 19. Von Mises stress distribution in considered cross-section of: (a) initial design of GP and (b) optimized design of GP.

the assembly volume is insignificant, but maximal von Mises stress level is reduced on ~82.4 %. Von Mises stress level changes in the cross-section of the GP bracket for initial and optimized variants are presented in Fig. 20.

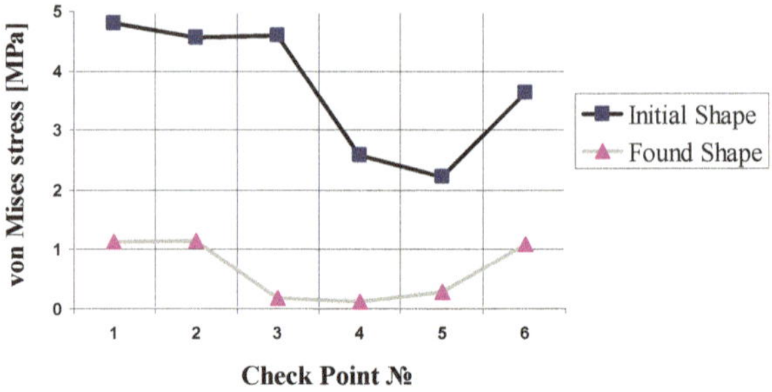

Fig. 20. Von Mises stress distribution in the bracket check points.

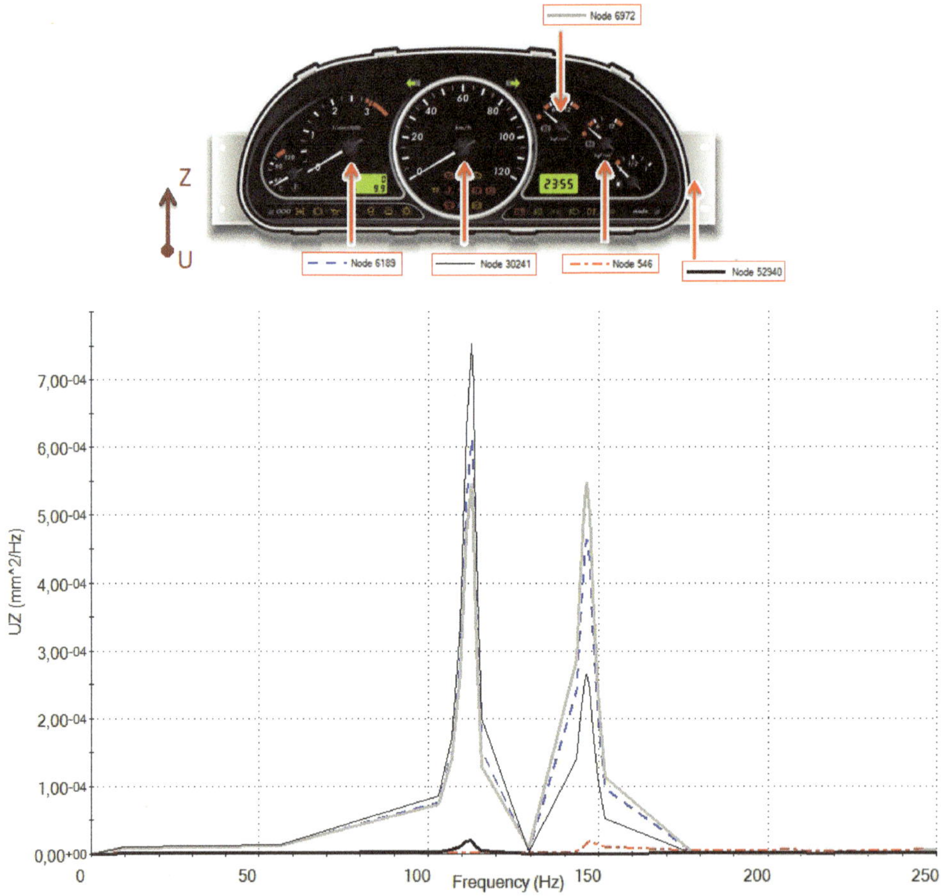

Fig. 21. PSD of vertical displacement in characteristic points of the GP.

Of course, there are possibilities of further optimization of the path curve shape (Fig. 18b) and taking into account simultaneously the additional particular objectives such as maximal von Mises stress in material of the GP from dynamic loading in case of harmonic vibration excitations; styling of the GP using α-cut method (Zimmermann, 2001), the carbon footprint and other environmental impacts of the GP throughout its entire lifecycle calculated by SW Sustainability, as well as natural frequencies of the GP.

For the obtained optimal solution the dynamic behavior of the GP must be verified in case of the uniform base random excitations (see Fig. 21) by analysis of the full FEM model.

5. Conclusion

The results of shape optimization of the mechanical components for two different measurement systems are presented. The described approach allows obtaining smooth shapes that are easy to implement technologically. The jagged forms are excluded from the optimization process and there's no need for excessive computational resources. The most time-consuming step of the current approach is the FEM analysis of the full model for variants defined by design of experiments, the results of which are used for building metamodels of appropriate responses. Then the solution of various single objective problems and the implementation of different aggregation strategies for multiobjective optimization are relatively easy in order to obtain an acceptable final solution.

It will be interesting to compare the effectiveness of the current approach by using the metamodels obtained by kriging and radial basis functions instead of locally weighted polynomial approximations.

6. Acknowledgment

The research work reported here was made possible by partial financial support of Latvian Science Council grant No. 09.1267 and EU Project Filose (ID 231495).

7. References

Arora, J. S. (2004). *Introduction to Optimum Design.* - 2nd ed. - Elsevier

Audze, P. & Eglajs, V. (1977). New Approach for Design of Experiments. *Problems of Dynamics and Strength,* Vol. 35, Riga: Zinatne, RU, pp. 104-107

Auzins, J. Direct Optimization of Experimental Designs. (2004). 10th AIAA/ISSMO *Multidisciplinary Analysis and Optimization Conference,* Albany, NY, 28 Aug.-2 Sep. 2004., AIAA Paper 2004-4578, CD-ROM Number 17, pp. 1-17

Auzins, J. & Janushevskis, A. (2007). *Design of Experiments and Analysis.* Riga, LV

Auzins, J. & Janushevskis, A. (2002). New Experimental Designs for Metamodelling and Optimization, *Proceedings of the Fifth World Congress on Computational Mechanics (WCCM V),* July 7-12, 2002, Vienna, Austria, Editors: Mang, H.A.; Rammerstorfer, F.G.; Eberhardsteiner, J., Publisher: Vienna University of Technology, Austria, ISBN 3-9501554-0-6, pp. 1-10

Auzins, J.; Janushevskis, J.; Janushevskis, A. & Kalnins, K. (2006). Optimisation of designs for natural and numerical experiments, *Extended Abstracts of the 6th Int ASMO-UK/ISSMO conference on Engineering Design Optimization.* Oxford, UK, pp. 118 - 121.

Bendsoe, M. P. & Sigmund, O. (2003). *Topology Optimization: Theory, Methods and Application.* -2nd ed. - Heidelberg (Berlin): Springer

Company Amoplant (2011). Available from http://www.amoplant.lv

Forrester, A. I. J.; Sóbester, A. & Keane, A.J. (2008). *Engineering Design via Surrogate Modelling. A Practical Guide.* Wiley

Grigorov, I. N. (2004). *Transmitting Magnetic Loop Antennas.* Moscow: RadioSoft. RU.

Hart, T. (1986). W5QJR. *The Loop. Small High Efficiency Antennas.* Melbourne: Sage-American Co.

Janushevskis, A.; Akinfiev; T., Auzins, J. & Boyko, A. (2004). A Comparative Analysis of Global Search Procedures. *Proc. Estonian Acad. Sci. Eng.*, Vol.10, No.4, pp. 235-250

Janushevskis, A. & Melnikovs, A. (2010). Shape Optimization of Block. *Scientific Jour. of RTU: Transport and Engineering. Mechanics.* Series 6. Vol. 33, Riga, pp. 89-97

Janushevskis, A.; Melnikovs, A. & Boyko, A. (2010). Shape Optimization of Mounting Disk of Railway Vehicle Measurement System, *Jour. of Vibroengineering*, Vol. 12, Issue 4, pp. 436 - 443

Liang Q.Q.; Xie Y.M. & Steven G.P. (2001) A Performance Index for Topology and Shape Optimization of Plate Bending Problems with Displacement Constraints. *Struct. and Multidisciplinary Optimization*, Berlin, pp. 393-399

Lombard M. (2009). *SolidWorks 2009 Bible.* Indianapolis: Wiley

McKay, M.D.; Beckman, R.J. & Conover, W.J. (1979). A comparison of three methods for selecting values of input variables in the analysis of output from computer code. *Technometrics* 21, pp. 239-245

Mullerschon, H.; Lazarov, N. & Witowski, K. (2010). Application of Topology Optimization for Crash with LC-OPT/Topology. *Proc. 11th Int LS-DYNA Users Conference*, pp. 17-39 - 17-46

Papadrakakis, M.; Lagaros, N.; Thierauf, G. & Cai, J. (1998). Advanced solution methods in structural optimisation using evolution strategies, *Eng. Comp. Jour.*, 15 (1), pp. 12-34

Sacks, J.; Welch, W.J.; Mitchell, T.J. & Wynn, H.P. (1989). Designs and analysis of computer experiments, *Statist. Sci.*, 4, pp. 409–435

Saxena, A. & Sahay, B. (2005). *Computer aided engineering design.* India: Anamaya

Schmit, L. A. & Farcshi, B. (1971). Some Approximation Concepts for Structural Synthesis. *AIAAJ.*,Vol.12, No. 5, pp. 692-699

Sladkowsky, A. & Pogorelov, D. (2008). Investigation of Dynamical Interactions at Rail Wheel Contact in Case of Flat of Wheel. *Bulletin of Ukraine National University*, No. 5, RU pp. 88-95

State Railway Research Institute. (1998). *Codes of Design of Railway Wagons with Wheel Span 1420 mm.* Moscow, RU

Ulrich, K.T. & Eppinger, S. D. (2008). *Product Design and Development.* McGraw - Hill Int.

Vanderplaats, G. N. (2004). *Numerical Optimization Techniques for Engineering Design with Applications.* 4th Ed., Vanderplaats Research & Development, Inc. Colorado Springs, USA

Zimmermann, H. (2001). *Fuzzy Set Theory and its Applications.* Boston: Kluwer Acad. Publ.

Wide Area Measurement Systems

Mohammad Shahraeini[1] and Mohammad Hossein Javidi[2]
[1]*Golestan University,*
[2]*Ferdowsi University of Mashhad,*
Iran

1. Introduction

In last two decades, power industries have been deregulated, restructured and decentralized in order to increase their efficiency, to reduce their operational cost and to free the consumers from their choices of electricity providers (Eshraghnia et al., 2006). As a result of these changes, in comparison with the traditional power systems, new competitive power industries face specific challenges that are related to their generation, operation and planning. As a consequence of these challenges, new intelligent systems should be introduced and established in the power systems in order to tackle such challenges. Wide Area Measurement Systems (WAMS) is a new term, which has been introduced to power system literatures in late 1980s. Recently, they are commercially available in power systems for purposes of monitoring, operation and control.

To be able to monitor, operate and control power systems in wide geographical area, WAMS combines the functions of metering devices (i.e. new and traditional) with the abilities of communication systems (Junce & Zexiang, 2005). The overall capability of this particular combination is that data of the entire system can be obtained at the same time and the same place i.e. the control center. This data, which are obtained from the entire system, can be used by many WAMS functions, effectively. These facts indicate that nowadays, WAMS has been a great opportunity to overcome power systems' challenges related to the restructuring, deregulation and decentralization.

This chapter is allocated to an in-depth survey of WAMS. To carry out this survey, WAMS process is firstly defined and classified into the three main interconnected sub-processes including data acquisition, data transmitting and data processing. These sub-processes are respectively performed by measurement systems, communication systems and WAMS applications.

This chapter is organized as follows. Section 2 provides a basic background and history of WAMS. The definition of the WAMS is given in this section as well. In Section 3, the WAMS process is investigated and divided into three sub-processes. Section 4, 5 and 6 review the pre-mentioned sub-processes, one by one. Finally, this chapter ends with a brief summary and conclusions in Section 7.

2. Background

In this section, a brief background and history of WAMS are provided. Then, the accurate definition of WAMS will be introduced.

2.1 History

Wide Area Measurement System (WAMS) was firstly introduced by Bonneville Power Administration (BPA) in the late 1980s (Taylor, 2006). This was resulted from this fact that the Western System Coordinating Council (WSCC) faced a critical lack of dynamic information throughout the 1980s. As a result of this, in 1990, a general plan to address this problem was formed (Cai et al., 2005). Therefore, the Western Interconnection of the North America power system was the first test-bed for WAMS implementation.

In 1995, the US Department of Energy (DOE) and the Electric Power Research Institute (EPRI) launched the Wide Area Measurement System (WAMS) Project. The aim of this project was to reinforce the Western System Dynamic Information Network called WesDINet. Dynamic information provided by WAMS of WesDINet has been very important and useful for understanding the breakups. This dynamic information can also be used for the purpose of avoiding future disturbances. Furthermore, during deregulation and restructuring process, information resources provided by this WAMS were utilized for maintaining the system reliability (Hauer & Taylor, 1998).

Since 1994, phasor measurement units (PMU) have been used in WAMS and they have provided synchrophasor measurements (Cai et al., 2005). It is noted that a complete survey of PMU will be presented in Section 4. Synchrophasor measurements may contribute previous functions or may introduce some new WAMS functions, which are never achieved previously by conventional measurements. When synchrophasor measurements are used as data resources of a WAMS, such a WAMS will be called PMU based WAMS.

2.2 WAMS definition

There exists a precise and comprehensive definition of WAMS, which has been introduced by Hauer from BPA/Pacific NW National Labs (Taylor, 2006):

"The WAMS effort is a strategic effort to meet critical information needs of the changing power system".

It can be mentioned that a WAMS needs an infrastructure to perform its tasks. This is also defined by Hauer (Taylor, 2006) as follows:

The WAMS infrastructure consists of people, operating practices, negotiated sharing arrangements and all else that are necessary for WAMS facilities to deliver useful information.

In recent years, PMU measurements are commercially available and are widely used in power systems. On the other hand, high speed and low cost communication systems; which are worked based on a layer model, are also well-established in power systems. As a result, the definition of WAMS is slightly different from past. Nowadays, a general definition of WAMS may be presented as follows: The WAMS combines the data provided by synchrophasor and conventional measurements with capability of new communication systems in order to monitor, operate, control and protect power systems in wide geographical area (Junce & Zexiang, 2005).

3. WAMS process

As mentioned at the beginning of this chapter, a WAMS process includes three different interconnected sub-processes (Yan, 2006): data acquisition, data transmitting and data processing. Measurement systems and communication systems together with energy management systems perform these sub-processes, respectively.

In general, a WAMS acquires system data from conventional and new data resources, transmits it through communication system to the control center(s) and processes it. After extracting appropriate information from system data, decisions on operation of power system are made. Occasionally, WAMS may command some actions that are performed by system actuators in remote sites (Shahraeini et al., 2011). All of these facts indicate that WAMS denotes efficient usage of data and data flow to achieve a more secure and a better strategy for the flow of electrical energy. The WAMS process is illustrated in Fig. 1.

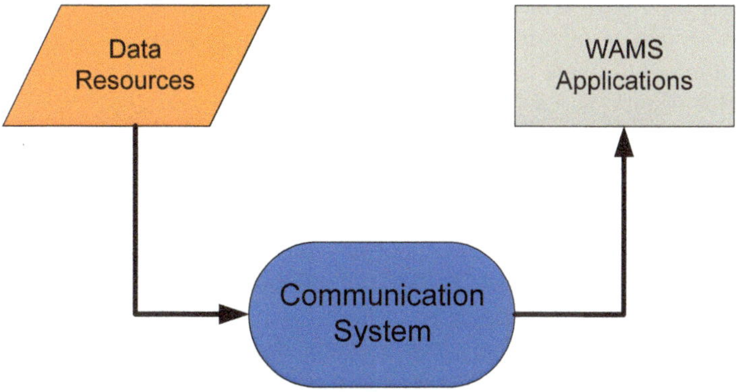

Fig. 1. WAMS Process in Power Systems.

An inspection of the above-mentioned facts together with Fig. 1 indicates that data itself is the fundamental requirement to perform WAMS functions. It can also be concluded that each sub-process has different responsibilities and different tasks that should be performed on system data. Consequently, WAMS main sub-processes should be studied from the data point of view.

In the rest of this chapter, data resources, applications and communication infrastructure of WAMS will be investigated.

4. Data resources of WAMS

Online data and information from the entire system have been essential for secure operation and control of interconnected systems e.g. power systems. In general, system data and information have been provided by data resources of system, which are also called measuring devices (Shahraeini et al., 2010). Data provided by data resources are widely different in terms of their importance, format, volume, sample rate and etc. Thomas et al. (2006) have classified power system data into two main data groups: Operational and Non-Operational data.

The operational data includes the instantaneous measurements of voltages, currents, phasors and breaker statuses that are measured by intelligent devices. Such data is transmitted continuously to the control center(s) through communication systems. Occasionally, they may be used locally for the local decision making.

On the other hand, the non-operational data basically consists of records or logs of multiple events e.g. series of faults, power fluctuations, disturbances and lightning strikes. Typically, the non-operational data is offline data. This means that they are transmitted to the control center(s) either in the specified time intervals (e.g. multiple hours) or when they are requested by the system operator.

There are two main differences between the operational and the non-operational data. The first one is their polling rates. The operational data are normally polled in a regular mode i.e. a continuous stream of data. On the other hand, most of the non-operational data are polled at the defined conditions or they are periodically polled at a specified time intervals. Another major difference is their data format. The operational data is usually transmitted in the form of a stream i.e. stream of numerical variables. While the non-operational data may appear in different formats e.g. waveforms, numerical values, COMTRADE (COMmon format for TRAnsient Data Exchange) format etc. (Thomas et al., 2006).

In this book, the classification of power system data represented by Thomas et al. (2006), is generalized to the data resources of power systems as well. In this section, these two classes of data resources; i.e. operational and non-operational data resources; are summarized in separate subsections. Supervisory Control and Data Acquisition (SCADA) and Synchronized Phasor Measurement System (SPMS) are two operational data resources that will be studied. Alternatively, Digital Fault Recorder (DFR), Digital Protective Relay (DPR) and Circuit Breaker Monitor (CBM) will be investigated as non-operational data resources.

4.1 Supervisory Control and Data Acquisition (SCADA)

SCADA is a generic name for a computerized system, which is capable of gathering and processing data and applying operational controls over long distances. Typical uses of SCADA include power transmission and distribution and pipeline systems (Friedmann, 2003).

In an electrical power system, a SCADA system provides three critical functions in the operation of such a system (Jelatis, 2001):

- Data acquisition
- Supervisory control
- Alarm display and control

In general, a SCADA system consists of both hardware and software. Typically, SCADA hardware may include three parts: Master Terminal Unit (MTU), Remote Terminal Unit (RTU), and Communication System (Stouffer et al., 2008). It should be noted that sometimes Programmable Logic Controllers (PLCs) or Intelligent Electronic Devices (IDEs) may be used as RTU in SCADA systems (Clarke et al., 2004).

The simplest form of SCADA hardware includes: an MTU that located in the control center, one remote field site consisting of either an RTU or a PLC or an IED, and a communication system that provides communication route between remote site and the control center. Fig. 2 shows SCADA sub-systems (Stouffer et al., 2008). Brief descriptions of SCADA sub-systems are as follows:

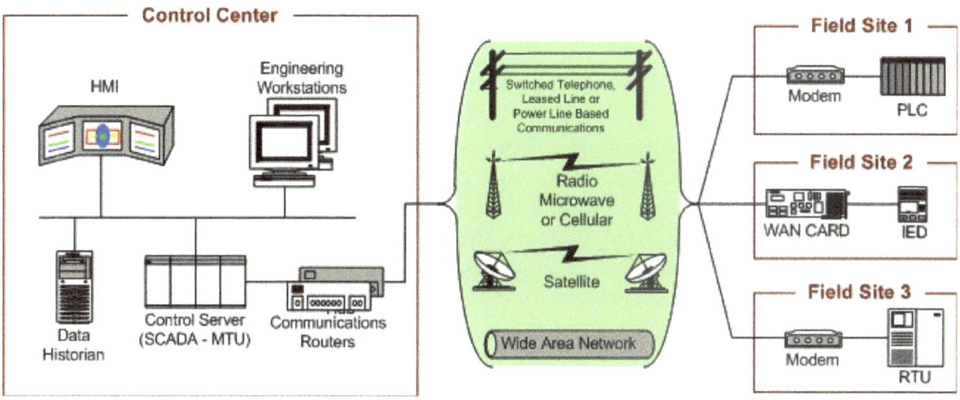

Fig. 2. SCADA sub-systems (Stouffer et al., 2008).

4.1.1 Master Terminal Unit (MTU)

The Master Terminal Unit (sometimes called SCADA center, SCADA server, or master station) may be considered as the heart of a SCADA system. It manages all communications, gathers data of RTUs, stores obtained data and information, sends information to other systems, commands system actuators that are connected to RTUs, and interfaces with operators. Indeed, MTU is a device, which is located in the control center and it acts as master, while RTUs are placed in remote sites and act as slaves (Stouffer et al., 2008).

4.1.2 Remote Terminal Unit (RTU)

The Remote Terminal Unit (RTU) is a stand-alone data acquisition and control unit. RTUs are generally microprocessor based and they monitor and control equipments at remote sites. Their main tasks are twofold: to control and acquire data from process equipments at the remote sites, and to communicate obtained data to a master station (MTU) (Stouffer et al., 2008).

It is useful to mention that traditional RTUs only communicated with a MTU. But nowadays, modern RTUs may also communicate among together. In some cases, an RTU can be configured as a relay. Such an RTU relays data obtained from lower RTUs to an MTU (Clarke et al., 2004).

In general, small-size RTUs include less than 10 to 20 analog and digital signals; medium-size RTUs have 100 digital and 30 to 40 analog inputs. Other RTUs with more inputs are known as large-sized ones (Clarke et al., 2004).

4.1.3 Programmable Logic Controller (PLC)

Programmable Logic Controller (PLC) is a small industrial computer, which is initially designed to perform the logic functions that are carried out by electrical equipments e.g. relays, drum switches, and mechanical timer/counters (Jelatis, 2001). Nowadays, analog control is a standard part of the PLC operation as well (Clarke et al., 2004).

In general, PLCs are modular in nature. They can be expanded to monitor and control additional field devices in remote sites. Since PLCs have built in microprocessor, they can be programmed to function locally even if communication with the master station is lost (Synchrony, 2001).

The PLCs have two main advantages over commercial RTUs. Firstly, they are general-purpose devices and can easily perform variety of different functions. Secondly, PLCs are physically compact and require less space than alternative solutions (Clarke et al., 2004). As a result of these facts, in SCADA systems, PLCs are preferred to special-purpose RTUs because they are more economical, versatile, flexible, and configurable (Stouffer et al., 2008). However, PLCs may not be suitable for specialized requirements e.g. radio telemetry applications (Clarke et al., 2004).

4.1.4 Communication system of SCADA

The communication systems provide communication routes between the master station and the remote sites. This can be done through private transmission media (e.g. fiber optic or leased line) or atmospheric means (wireless or satellite).

There are three main physical communication architectures used in SCADA communications: point-to-point, multipoint and relay station architectures (Clarke et al., 2004).

4.2 Synchronized Phasor Measurement System (SPMS)

The Synchronized Phasor Measurement System (SPMS) was firstly developed and introduced into the power system in mid-1980s. These systems have the ability of measuring currents and voltages, and calculating the angle between them. This ability has been made possible by the availability of Global Positioning System (GPS); on the one hand, and the sampled data processing techniques; on the other hand. In order to synchronize measured angles, SPMS uses time received from GPS as its sampling clock. In addition to measuring angles of voltages and currents, these systems can also measure local frequency and rates of frequency changes, and may be customized to measure harmonics, negative and zero sequence quantities (Phadke & Thorp, 2008).

A SPMS consists of three main parts: Phasor Measurement Unit (PMU), Phasor Data Concentrator (PDC), and communication system. PMUs are normally installed at remote sites. They calculate phasors of voltages and currents and stamp measured phasors with the time received from GPS. A PDC gathers data from several PMUs, rejects bad data, and aligns the time stamps. Communication system of SPMS is responsible for data delivery between PMUs and a PDC or multiple PDCs (Phadke & Thorp, 2008). Fig. 3 shows the hierarchy of SPMS. The next three sub-sections are going to describe each part of SPMS, separately.

4.2.1 Phasor Measurement Unit (PMU)

The Phasor Measurement Unit (PMU) is a microprocessor based device that uses the ability of digital signal processors in order to measure 50/60Hz AC waveforms (voltages and currents) at a typical rate of 48 samples per cycle (2400/2880 samples per second). To do

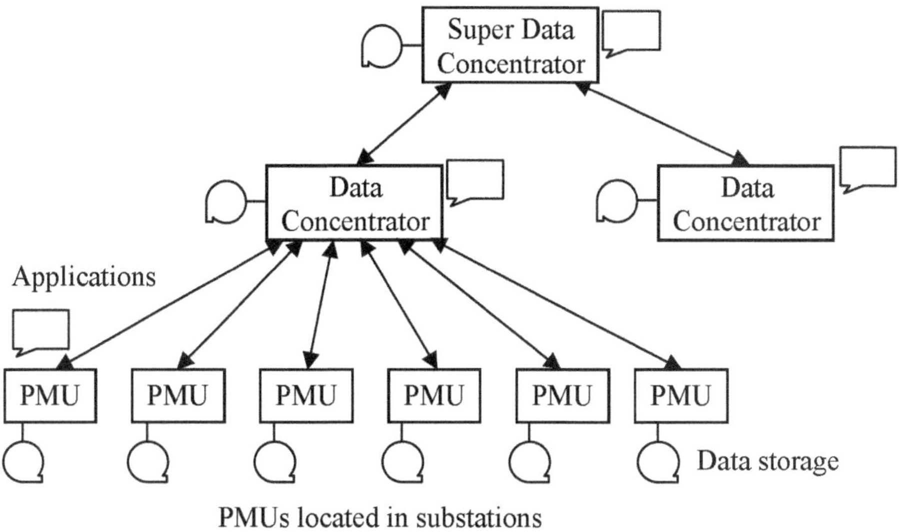

Fig. 3. SPMS sub-systems (Phadke & Thorp, 2008).

this, first, the analog AC waveforms are synchronously sampled by an A/D converter for each phase. In order to provide synchronous clock for the entire system, the time from GPS satellites are used as input for a phase-lock oscillator and thereby, waveforms of the entire system are sampled with 1 microsecond accuracy. In the next step, PMU uses digital signal processing techniques to calculate the voltage and current phasors. Also, line frequencies can be calculated by PMU at each site. By using this technique, a high degree of resolution and accuracy will be achieved. The measured phasors are tagged by GPS time stamps and are transmitted to a PDC at the rates 30-60 samples per second (EPG & CERTS, 2006). Phasor data is formed in COMTRADE format (Phadke & Thorp, 2008).

A study of RTU tasks in SCADA system indicates that PMU and RTU have almost the same tasks in the SPMS and SCADA systems.

4.2.2 Phasor Data Concentrator (PDC)

The main functions of a PDC are: to gather data from several PMUs, to reject bad data, to align the time stamps, and to create a coherent record of simultaneously recorded data. As a consequence, a snap-shot of phasors of the measured area can be obtained (Phadke & Thorp, 2008).

In some cases, a central PDC may concentrate the area data received from other PDCs and may provide phasors of the entire system.

A study of MTU tasks in a SCADA system indicates that the tasks and functions of PDC in SPMS systems are almost the same as those in SCADA systems.

4.2.3 Communication system of SPMS

The communication systems of SPMS may be similar to the SCADA communications in terms of technology, architecture and utilized media. Although these communication systems may be the same, their streamed data are different. The phasor data, which is provided by PMUs, have different nature in comparison with the data of RTUs. Phasor data is continuous and streaming in nature while RTU data is transmitted to the master station either in a specified time intervals or when master station requests it. Another difference between PMU and RTU data is their volume. In general, data of a PMU has more value than data provided by a RTU.

As a result of the above facts, two aspects are of major importance in SPMS communications: communication bandwidth and communication latency (Phadke & Thorp, 2008). High bandwidth communications guarantee that all phasor data can be transmitted to PDCs without any packet drops. On the other hand, low latency communications provide real time streaming between PMUs and PDCs.

4.3 Digital Fault Recorder (DFR)

The Digital Fault Recorder (DFR) acts as the black box of a substation. It records highly accurate waveforms related to faults. The recorded data are huge amount of analog and status data for pre-fault, fault and post-fault conditions (Kezunovic, 2008a). These data may include maximum current, sequence of events, type of fault and the sequence of operation of circuit breakers (Thomas et al., 2006).

The sample rate of the DFR is normally very high and assumed to be 64 to 356 samples per cycle. The DFR data is normally formed in COMTRADE format (IEEE Inc., 2006). Additionally, the data captured by DFRs are offline and they are not used in real time. These data are stored as samples for further offline processing. Generally, the DFRs are installed in the most important substations (Kezunovic, 2008a).

4.4 Digital Protective Relay (DPR)

Historically, different kinds of protective relays have been designed to isolate the area of faults and reduce the impacts of the faults from other parts of the system (Zhang, 2006). The first type of protective relays was electromechanical one. With the introduction of electronic devices e.g. the transistor in the 1950s, electronic protection relays were introduced in the 1960s and 1970s. Recently, digital (microprocessor-based) protective relays have been introduced to the power systems. A DPR uses an advanced microprocessor to analyze voltages and currents of a power system for the purpose of detection of faults in such a system (Hewitson et al., 2004).

Digital protective relays, in addition to performing the traditional relaying tasks, are capable of measuring and recording analog and status data, as well as communicating with a centralized location. They collect current and voltage signals from instrument transformers and digitize them. Due to the fact that relays should act very fast, the accuracy of measured data is not of major concern. Consequently, to speed up A/D conversion, lower sampling rates are normally applied. This implies that data obtained from DPRs are generally less accurate than from the other data resources (Kezunovic, 2008a).

First DPRs sample rates were 4 to 20 samples per cycle, but nowadays, DPRs are available that sample at 64 to 128 samples per cycle (IEEE Inc., 2006).

4.5 Circuit Breaker Monitor (CBM)

The Circuit Breaker Monitor (CBM) is an electronic device that monitors circuit breakers. The CBM captures detailed information about each CB operation in real time; either the operation is initiated manually by the operator or it is initiated automatically by the protection and control equipments (Kezunovic, 2008b). The CBM data is also formed in COMTRADE format.

5. WAMS applications

In general, the information about any system can be extracted from its raw data, which is measured by its data resources. In power systems, this can be achieved by a kind of computer aided tools known as "WAMS Applications" (Shahraeini et al., 2010). Typically, WAMS applications process the raw data measured by data resources and extract usable information for system operator, consumers and customers. Shahraeini et al. (2010) have classified WAMS applications into the three main groups: generation, transmission and distribution applications. Three next sub-sections are going to describe these three groups of applications.

5.1 Generation aplications

Generation applications (GEN): These applications are run in generation level in the way that they acquire and process data of generators in the control center(s). As its consequence, generator information can be obtained in the control center(s) all at once. Generator operation status monitoring and transient angle stability are some examples of such applications (Xiaorong et al., 2006).

In the above mentioned applications, generator status monitoring (GSM) is the most important GEN application since it provides all or part of real time information of generators in the control center. The first kind of GSM was implemented by using DFR as a data recorder (Lee et al., 2000). As DFRs can record the operational and non-operational data with very high sampling rates, they can be used as online recorders in generation sides. If the recorded data is transmitted to the control center in real time, the generator status can be monitored in the control center. After introducing PMUs to the power systems, the information provided by these units can also used for GSM application (Xiaorong et al., 2006). This has been resulted from the fact that PMUs provide phasor data in real time with very high sampling rate (up to 60 samples per second).

5.2 Transmission and Sub-Transmission applications

Transmission and sub-transmission applications (TRAN): In deregulated power industries, some applications are performed at transmission (or sometimes sub-transmission) level by independent system operator (ISO). Historically, these functions are performed by group of computer aided tools called energy management systems (EMS). State estimation (SE), load flow (LF), optimal power flow (OPF), load forecast (LF) and economical dispatch (ED) are some examples of conventional EMS applications (Shahidehpour & Wang, 2003). The aforementioned applications may be considered as conventional WAMS applications since data is fundamental part of all of them.

After introducing phasor measurement units to the power systems, phasor data may contribute conventional WAMS applications or may introduce some new modern WAMS applications. For instance, if a state estimation uses only phasor data as its input, the state equations of the system will be linear. While conventional SEs (that use conventional data) are non-linear and they must use numerical method to solve their equations, iteratively.

Xiaorong et al. (2006) have summarized modern WAMS applications in the power systems. Some of these applications use only phasor data (e.g. Integrated Phasor Data Platform) and some ones may use both phasor and conventional data e.g. hybrid SEs.

Some modern WAMS applications are as follows (Xiaorong et al., 2006): Integrated Phasor Data Platform, Wide-Area Dynamic Monitoring and Analysis, Synchronized Disturbance Record and Replay, Online Low-Frequency Oscillation Analysis, Power Angle Stability Prediction and Alarming, and PMU based State Estimation. Combination of these modern applications with the conventional ones forms a modern EMS in the control center.

In the above mentioned applications, the state estimation is the most important WAMS application and is considered as kernel of EMS (Shahraeini et al., 2011). This has been resulted from the fact that this application extracts creditable data from raw data provided by data resources. Other WAMS applications use obtained creditable data as their input (Shahidehpour & Wang, 2003).

State estimations, based on their utilized data, are classified into the three different types: conventional, PMU based and hybrid state estimations (Shahraeini et al., 2011).

Conventional state estimations use conventional operational data i.e. voltage and current magnitude, active and reactive power flow, and active and reactive power injection.

5.3 Distribution applications

Distribution applications (DIS): In distribution systems, WAMS applications are known as automation applications. According to IEEE definition (Gruenemeyer, 1991), Distribution Automation (DA) systems have been defined as systems that enable a distribution company to monitor, coordinate, and operate distribution components and equipments from remote locations in real time. The DAs aim to reduce costs, to improve service availability, and to provide better consumer services. In general, DA may be classified into three main groups: substation automation, feeder automation, and consumer-side automation (Shahraeini & Alishahi, 2011).

5.3.1 Substation automation

Substation Automation (SA) is the integration of smart sensors with a communication infrastructure to control and monitor substation equipments in real-time (Shahraeini & Alishahi, 2011). The major functions of SA are: service restoration via bus sectionalizing, bus voltage control, substation parallel transformer circulating current control, line drop compensation, and automatic reclosing (Cassel, 1993).

Data resources of SA are located in distribution substations including bus phase voltages, transformer and feeder active and reactive power, feeder currents, statuses of circuit breakers, capacitors and reclosers cut-off switches, load tap changer and voltage regulator positions and statuses, transformer temperatures and relay settings (Cassel, 1993).

Typically, measurement data and status data are measured by pre-mentioned data resources. Then, Remote Terminal Units (RTUs) collect data and send it to SCADA systems. Finally, communication infrastructure has the responsibility of data transmitting from SCADA to the control center(s) (Shahraeini & Alishahi, 2011).

5.3.2 Feeder automation

Nowadays, due to rapid growth in metros, distribution networks have been one of the most extensive infrastructures in metros. In such networks, Feeder Automation (FA) is one of the key elements for efficient management of the power distribution networks. The main purposes of FA are twofold. Firstly, FA aims to automate feeder switching. Secondly, FA controls voltages and active/reactive powers of feeders (Cassel, 1993).

The main data resources and controllable devices of this function are line reclosers, load break switches, sectionalizers, capacitor banks and line regulators. Typically, these data resources are much more than resources of SA and are located at distribution poles (Shahraeini & Alishahi, 2011).

5.3.3 Consumer side automation

Consumer side automation tries to automate the final points of electricity delivery i.e. metering devices of customers. Beyond this, customer equipments may be automated by this application and may be controlled through the control center. Advanced Metering Infrastructure (AMI) and Automatic Meter Reading (AMR) are two main systems utilized to automate consumer sides.

The AMI may be assumed as the central nervous system of the smartgrid architecture in distribution systems. AMI collects consumption data from smart meters and transmits it to the control center. This function of AMI is similar to the purposes of AMR systems. In addition to reading functions, AMI also relays demand signals and pricing information to smart meters in near real-time; and thereby, feedback loop is closed by AMI system (Shahraeini & Alishahi, 2011).

Data resources of AMI /AMR systems are metering devices. But the difference between AMI and AMR is that metering devices of AMI can also be remotely controlled by system operator. Since huge numbers of metering devices are distributed in entire system, the cost of communication system, which is utilized by an AMI/AMR system, is vital. In the view of this fact, low cost communication infrastructure means more automated customers.

6. Communication infrastructure of WAMS

The communication system of WAMS is responsible for data delivery from data resources to the control center(s) and from control center(s) to the system actuators. Indeed, the communication network of WAMS is similar to the neural network of humans. As in case of failure or mal-functioning of neural network paralyzed may happen, failure of communication network may cause huge problems in system operation and control, especially in operation of WAMS applications. Consequently, especial attention should be paid to communication infrastructure, which is as important as electrical infrastructure itself. These two infrastructures (communication and electrical) have become increasingly interdependent so that in the case of failure for each of them, another one may also become out of service (Shahidehpour & Wang, 2003; Lukszo et al., 2010).

New communication systems are designed based on open system interconnection (OSI) layer model. In this architecture, upper layers relay data, assuming that the lower layers work perfectly. In fact, this model is an effective architecture for explanation, designing, implementation, standardization and use of communications networks. The OSI reference model consists of seven layers: physical, data link, network, transport, session, presentation, and application (Shahraeini et al., 2010).

In wide area measurement systems, data resources and WAMS applications normally work at the upper layers of network models. Fig. 4 shows the map between OSI layers and three major blocks of WAMS shown in Fig. 1 i.e. data resources, WAMS applications and communication system (Shahraeini et al., 2010).

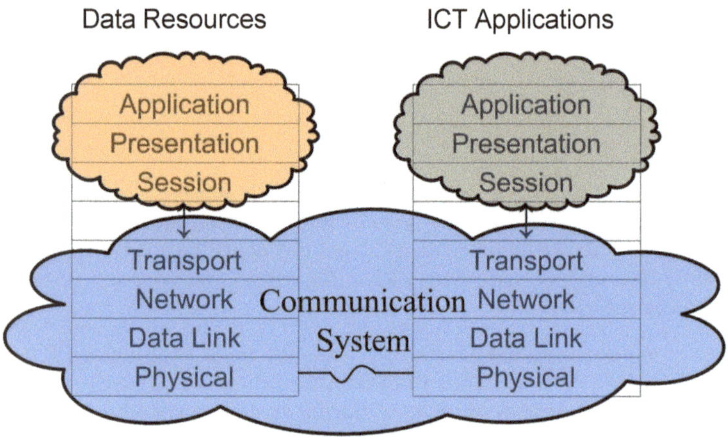

Fig. 4. Layering in WAMS based on OSI reference model (Shahraeini et al., 2010).

The first layer of OSI model, referred to the physical layer, is a kind of medium that establishes the physical connection between transmitter and receiver. The characteristics of the communication systems will become seriously influenced by the characteristics of their media. Therefore, it can be concluded that the characteristics of the transmission media play an important role in communication infrastructure of WAMS. Some major characteristics of a medium are as follows: cost, bandwidth, propagation delay, security and reliability (Shahraeini et al., 2010).

Transmission media, as described below, can be classified as guided and unguided ones (Stallings, 1997). Guided media guides the waves through a solid medium. Twisted pair, coaxial cable, power transmission/distribution line and optical fiber are some examples of guided media. In the case of guided media, the media itself has the most important role in characterizing the limitations of transmission (Stallings, 1997).

On the other hand, unguided media provides a means to transmit electromagnetic waves. However, this media does not guide the signals. The atmosphere and outer space are some examples of this case and usually referred to as wireless communication. Unlike guided media, in the case of unguided media, the signal strength provided by wireless antenna is more important than the media itself (Stallings, 1997). The next sub-sections review these two groups of media.

6.1 Guided media

6.1.1 Power Line Carrier (PLC)

Power line carrier (PLC) has used transmission lines as a medium for communication. This type of transmission media has been one of the first reliable media utilized in power systems for critical communications (Marihart, 2001). This media is also the first guided media commonly utilized in power systems and is a part of power system infrastructure. As a result, failure in power system infrastructure such as line outage causes communication difficulty. PLC systems may be classified as two groups in common, narrow band and broad band PLCs (Shahraeini et al., 2010).

Narrow band PLC usually has low data rates (up to 100kbps). It is used for automation and control applications or few voice channels (Hrasnica et al., 2004). However, due to the fact that the narrowband PLC works in low data rates, this system is very reliable and PLC modems can be installed far from each other (Shahraeini et al., 2010).

Unlike narrowband PLC, broadband PLC establishes a high data rate (beyond 2 Mbps) between two modems (Hrasnica et al., 2004). This kind of communication can be used for multi services such as automation, internet access and telephony at the same time. Broadband PLCs work in high data rates; therefore, distance between two modems is short and modems require more maintenance. This type of communication is not recommended for noisy power lines (Marihart, 2001).

When power lines are used for broadband internet access, power line communication is known as broadband over power line (BPL). BPLs use spread spectrum techniques to deliver data rates previously inaccessible. But because of the fundamental physical constraints, successful data rates will be achieved much above several megabits per second (Nordell, 2008).

6.1.2 Optical fiber

Optical fiber can be used as a medium for communication. Because of its flexibility, fiber optic can be bundled as a cable. As mentioned at the beginning of this section, signals are transmitted through the media by a type of waveform. In fiber cables, the signal is a light wave; either visible or infrared light. Essentially, two types of fiber optic cables including optical power ground wire (OPGW) and all-dielectric self supporting (ADSS) are used in power industries (Shahraeini et al., 2010).

OPGW cable combines the function of grounding and communication. This kind of cable can be used in transmission or distribution lines. In transmission lines, OPGW is replaced with shield wire and is suspended above the lines (Marihart, 2001).

Unlike OPGW, ADSS is a self supporting cable and it does not include any metal component. In fact, they are designed to be fastened into towers or poles underneath the power conductors. Moreover, ADSS is ideal for installation in distribution poles as well as transmission towers, even when live-line installations are required (Marihart, 2001; Nordell, 2008).

6.1.3 Leased line

Historically, leased telephone circuits have been widely used in electric utilities to create a point-to-point or point-to-multipoint communications (Marihart, 2001). The leased lines only provide a share medium for communication and some technologies should be implemented in order to transmit signals through this media. Digital Subscriber Line (DSL) is a group of technologies, which provides digital data transmission over leased telephone circuits. The first version of DSL was defined in 1988 and called ISDN (Integrated Services Digital Network). ISDN provides a maximum of 128 Kbps in both uplink and downlink directions (Hrasnica et al., 2004). Other DSL versions have appeared in different forms, such as high-data-rate DSL (HDSL), single-line DSL (SDSL), asymmetric DSL (ADSL), rate-adaptive DSL (RADSL), and very high-data-rate DSL (VDSL), all of which utilize copper lines. The differences between xDSL technologies are their data rates and directionality of transmission, distances to which those rates can be supported, and the size of the wire.

6.2 Unguided media

Wireless transmission is used when we have several challenges such as environmental or financial limitations for utilization of guided media. However, as transmitted signals using wireless communication can be accessed by anyone, the security of wireless communication is naturally low. On the other hand, various signals, which are transmitted by different sources, may be broadcast on the same frequency and thereby, collision may happen. Thus, it can be concluded that the reliability of wireless communication is less than the reliability of transmission through a guided media. In wireless transmission, signal can take the form of waves in the radio spectrum, including very high frequency (VHF) and microwaves, or it can be light waves including infrared or visible lights such as laser (Shahraeini et al., 2010).

The first important parameter in wireless communication is its range. In accordance with wireless ranges, four wireless types may be defined (Fourty et al., 2005):

- Wireless Personal Area Network (WPAN)
- Wireless Local Area Network (WLAN)
- Wireless Metropolitan Area Network (WMAN)
- Wireless Wide Area Network (WWAN)

6.2.1 Wireless Personal Area Network (WPAN)

Personal networks make a small area networking for a variety of devices. The most popular WPAN has been Bluetooth, which was firstly developed by the Sweden Ericsson. Bluetooth operates in unlicensed 2.4 GHz spectrum, which is also used by Wi-Fi. IEEE 802.15.1 standard for Bluetooth allows data rates up to 3 Mbps and at a range of up to 100 meters.

In addition to the Bluetooth, two industrial technologies, namely UWB (Ultra Wide Band) and Zigbee, make high data rate and low cost WPAN, respectively (Fourty et al., 2005).

UBW, which is standardized under the name IEEE 802.15.3, can use frequencies from 3.1 GHz to 10.6 GHz. UBW allows data rate up to 480 Mbps at the range of several meters and a rate of 110 Mbps at a range of up to 10 meters (Fourty et al., 2005).

Zigbee has been created to become a wireless standard for remote control in industrial fields. It makes very low-cost WPAN for applications that are not too much bandwidth hungry (Fourty et al., 2005). Zigbee allows data rate of 250 Kbps at 2.4 GHz at the range of up to 10 meters (IEEE 802.15.4) and data rate of 20 Kbps at 900 kHz at the range of up to 75 meters (IEEE 802.15.4a). Recently, Zigbee is widely used to create home area communications between smart meters and smart home equipments (Shahraeini et al., 2010).

6.2.2 Wireless Local Area Network (WLAN)

The WLAN technologies connect devices via a wireless distribution method (typically spread-spectrum or OFDM). Wi-Fi is a popular WLAN technology that provides high speed connection on short ranges. In recent years, because of the lack of more suitable metropolitan wireless networks, Wi-Fi has also been used at the metropolitan level. Wi-Fi networks are not suitable for moving devices and they take down in a few kilometers per hour movement. IEEE 802.11 is a set of standards that carry out Wi-Fi (Fourty et al., 2005; Shahraeini et al., 2010). These standards are as follows:

- **IEEE 802.11:** theoretical data rate 2 Mbps - 2.4 GHz unlicensed band (The first standard of the series. It was released in 1997 and clarified in 1999).
- **IEEE 802.11b:** theoretical data rate 11 Mbps - range of 100 meters to a maximum of a few hundred meters - 2.4 GHz unlicensed band.
- **IEEE 802.11a:** theoretical data rate 54 Mbps - range of approximately thirty meters - 5 GHz band.
- **IEEE 802.11g:** theoretical data rate 54 Mbps - range of a hundred meters - 2.4 GHz unlicensed band.
- **IEEE 802.11n:** theoretical data rate 320 Mbps - about thirty meters range - uses two bands 2.4 GHz and 5 GHz.

6.2.3 Wireless Metropolitan Area Network (WMAN)

WiMAX, GPRS, GSM, CDMA and 3G mobile Carrier services are four WMAN technologies, which are used for WMAN communication. The descriptions of these technologies are as follows:

WiMAX: Worldwide Interoperability for Microwave Access (WiMAX) is a communication protocol, which provides fix and fully mobile data networking. WiMAX is based on the IEEE 802.16 standards and its most popular one is 802.16e-2005. Unlike WLAN technologies e.g. Wi-Fi, WiMAX is designed to operate as a WMAN. Various kinds of WiMAX work with both FCC licensed frequencies and unlicensed frequencies. Licensed WiMAX works in the range of 10 to 66 GHz and unlicensed WiMAX works in the range of 2 to 11 GHz. WiMAX theoretical data rate is 70 Mbps with a range of up to a maximum of 50 km with a direct line of sight (LOS). Near line of sight (NLOS) conditions seriously limit their range (Fourty et al., 2005; Shahraeini et al., 2010).

MBWA: Mobile broadband wireless access (MBWA), which is standardized under the name IEEE 802.20, creates mobile metropolitan networks with a speed up to 250 kilometers per hour. It uses licensed frequency band below 3.5 GHz and allows maximal data rates of 1 Mbps for downlink and 300 Kbps for uplink. The maximum range of the cells is 2.5 km. Because of short latency in MBWA technology, this technology is a good choice for mobility data and it can be compared with 3G mobile networks, which focus on the voice (Fourty et al., 2005; Shahraeini et al., 2010).

GPRS: General Packet Radio Service (GPRS or sometimes called 2.5G) is a packet data bearer service for wireless communication over GSM (Global System for Mobile). It applies a packet radio principle to transfer user data packets efficiently between mobile stations and external IP networks. GPRS allows IP-based applications to run on a GSM network (Shahraeini et al., 2010). Using unused channels in the GSM network, it provides moderate speed data transfer. The data speeds can range from 9.6 kbps (using one radio time slot) to 115 kbps (which can be achieved by merging 8 time slots) (Vaishnav et al., 2008).

GSM: It is the most popular second generation standard for mobile telephony systems in the world. There are some differences between GSM and GPRS. GSM is based on circuit-switching technology whereas GPRS makes packet switching network over GSM. GPRS bandwidth is higher than GSM; thus, GPRS has higher data speed toward GSM. In packet switching networks, the bandwidth is used only when a device transmits data. Conversely, connections are "always on" in circuit switching networks. Therefore, GPRS network charges are lower than GSM networks since the billing method is based on data volume and not on call time (Vaishnav et al., 2008).

CDMA: Code Division Multi-Access (CDMA) is another data networking technology for mobile communications. It allows all the users to utilize the entire frequency spectrum for all the time. Multiple simultaneous transmissions are separated by using coding theory. Only users associated with a particular code can understand each other. CDMA can create 64 logical channels whereas 8 channels are available in GPRS (Vaishnav et al., 2008).

3G mobile Carrier services: 3rd Generation networks provide new data carrier services for mobile users. For instance, some networks support High Speed Packet Access (HSPA) data communication with HSDPA standard to provide improved downlink speeds. Furthermore, HSUPA standard is used for uplink speed enhancement. HSDPA provides downlink data rates up to 14.4 Mbps and uplink data rates 384 Kbps. HSUPA provides improved upload data rates of up to 5.76 Mbps (Shahraeini et al., 2010). Another 3G standard for data communication, CDMA2000, allows a maximum theoretical data rate of 2 Mbps (Fourty et al., 2005).

6.2.4 Wireless Wide Area Network (WWAN)

Satellite communications may be used either when a guided media cannot be established between a remote site and the control center or when there is no line-of-sight between such a remote site and pre-installed communication network. Satellites, according to their orbits, may be classified as geostationary, medium, or low earth orbit satellites described below (Fourty et al., 2005):

Geostationary Earth Orbit (GEO) satellites are at an altitude of 35786 kilometers above the equator. GEO rotate around the earth at the same speed of earth rotation; thus, they appear to be fixed from the surface of the earth.

Low Earth Orbit (LEO) satellites rotate between 750 and 1500 kilometers orbit. LEO satellites are widely used for communications. Iridium, Globalstar and Orbcomm are some well-known LEO satellites.

Medium Earth Orbit (MEO) satellites are at altitudes between nearly 10,000 and 20,000 kilometers. From the view point of the earth, MEO rotate slowly in longitude; feel like 6 hours to circle the earth.

These three types of satellites cover surface of the earth almost everywhere, hence; WWAN technologies provide remote sites connections. Although satellites can connect remote sites to the control center, high latency of these connections may create serious problems for some WAMS applications. As a result of this, some critical applications such as WAPS (Wide Area Protection System) should not be implemented under WWAN technologies.

6.3 Comparison of transmission media

To be able to obtain a global comparison among different transmission media used in power system communications, the above mentioned transmission media are compared based on their provided bandwidth, latency and security. This is shown in Table 1.

Media Type	Media	Bandwidth	Latency	Security
Guided	Fiber	High	Low	High
	Power Line	Medium	Low	High
	Leased Line	Medium	Low-Medium	High
Unguided	Wireless			
	WPAN	Low-Medium	Low-Medium	Low
	WLAN	Low-Medium	Medium	Low
	WMAN	Medium	Medium	Low
	WWAN	Low-Medium	High	Low

Table 1. Transmission Media Comparison (Shahraeini et al., 2010).

It should be noted that the above mentioned transmission media may be used either for WAMS communications or for other communications of the system (e.g. SCADA). But the important concept is that the communication infrastructure of WAMS is the most extensive communication infrastructure in a power system. Other communications (e.g. SCADA and SPMS) may be a part of the communication infrastructure of WAMS. Modern communication systems have the ability of physical integration. This means that different communications (e.g. WAMS, SCADA or SPMS) can use the same transmission media and the same routers. In this environment, communication of different applications (e.g. SCADA and SPMS) can virtually be implemented such that communication limitation of each of them is satisfied.

7. Conclusion

Wide Area Measurement Systems (WAMS) is a new opportunity for system operators to monitor, operate, control and protect power systems in wide geographical area. The WAMS combines the data provided by synchrophasor and conventional measurements with the capability of new communication systems in order to obtain dynamic information of the entire system. The WAMS process can be divided into the three interconnected sub-processes; data acquisition, data delivery and data processing. These sub-processes are respectively performed by measurement, communication and energy management sub-systems. Each sub-system has different tasks to perform on system data. As a result, it is definitely important that the functions and the equipments of these sub-systems are deeply investigated from data point of view.

This chapter has extensively reviewed the equipments and the functions of each sub-process, separately. It has been shown that WAMS contributes monitoring systems to shift from the "data acquisition" systems to the "dynamic information" systems. Dynamic information of power systems helps power system operators to overcome generation, operation and planning challenges that may be resulted from system restructuring. Furthermore, it has also been shown that from the big generators to the small home equipments, WAMS systems are capable of monitoring and controlling various functions in real time. It can be concluded that in modern power systems, WAMS is an essential part of power system operation and control.

In particular, this chapter shows that dynamic information of power systems, as a result of WAMS implementation; contributes system operators to make better decisions for system operation and planning. However, in addition to the power systems, dynamic information of any interconnected system (e.g. natural gas pipelines) helps system operators/administrators to reduce operational cost and increase efficiency of such interconnected systems. Consequently, WAMS concepts may be also generalized to other interconnected systems in order to form a dynamic information system and to deliver system data to the related applications in real time.

As a conclusion, it can be stated that although the WAMS was firstly introduced to the power systems in order to obtain dynamic information of such systems, it can also be well established in other critical infrastructures (e.g. natural gas, petroleum, water supply, emergency services, telecommunication and etc.) to operate, monitor and control such infrastructures.

8. References

Cai, J. Y.; Huang, Z.; Hauer, J. & Martin, K. (2005). Current status and experience of WAMS implementation in North America, *Proceeding IEEE/Power Eng. Soc. Transmission and Distribution Conference Exhibition*, Aug. 2005, pp. 1-7

Cassel, W. R. (1993). DISTRIBUTION MANAGEMENT SYSTEMS: FUNCTIONS AND PAYBACK, *IEEE Transactions on Power Systems*, Vol. 8, NO. 3, pp. 796-801

Clarke G.; & Reynders D. (2004). Practical Modern SCADA Protocols: DNP3, 60870.5 and Related Systems. Oxford: Elsevier

EPG & CERTS, (2006). Phasor Technology and Real Time Dynamics Monitoring System (RTDMS) frequently asked question (FAQs), Electric Power Group and CERTS

Eshraghnia, R.; Modir Shanehchi, M. H.; & Rajabi Mashhadi, H. (2006). Generation maintenance scheduling in power market based on genetic algorithm, *Proceeding 2006 IEEE Power Systems Conference and Exposition*, 2006, pp. 1814 - 1819

Fourty, N. Val, T. Fraisse, P. & Mercier, J. J. (2005). Comparative analysis of new high data rate wireless communication technologies "From Wi-Fi to WiMAX"," *Joint*

International Conference on Autonomic and Autonomous Systems and International Conference on Networking and Services - (ICAS/ICNS 2005)

Friedmann, P. G. (2003). The Automation, Systems, and Instrumentation Dictionary, 4th Edition, ISA

Gruenemeyer, D. (1991). Distribution automation: How should it be evaluated?, *Proceeding 35th Annual Rural Electric Power Conference*, Dearborn, MI, C3/1–C31

Hauer, J. F. & Taylor, C. W. (1998). Information, Reliability, and Control in the New Power System, Proceedings of 1998 American Control Conference, Philadelphia, PA., June 24-26, 1998

Hewitson, L.; Brown, M.; & Ramesh B. (2004). Practical Power Systems Protection, Linacre House, Jordan Hill, Oxford, U.K., 2004

Hrasnica, H. Haidine, A. & Lehnert, R. (2004). Broadband Powerline Communications Networks: Network Design, John Wiley & Sons, Ltd, New York

IEEE Inc. (2006). Considerations for use of disturbance recorders, A report to the System Protection Subcommittee of the Power System Relaying Committee of the IEEE Power Engineering Society

Jelatis, G. D. (2001). Information Security Primer, Electric Power Research Institute (EPRI)

Junce, D. & Zexiang, C. (2005). Mixed Measurements State Estimation Based on Wide-Area Measurement System and Analysis, *Proceeding 2005 IEEE/PES Transmission and Distribution Conference & Exhibition*, Asia and Pacific, China, 2005, pp. 1-5

Kezunovic, M. (2006b). Monitoring of power system topology in real-time, *Proceeding 39th Hawaii International Conference System Sciences*

Kezunovic, M. (2008a). Integration of Substation IED Information into EMS Functionality, Final Project Report, *Power Systems Engineering Research Center (PSERC)*

Lee, K. Y. Perakis, M. Sevcik, D. R. Santoso, N. I. Lausterer, G. K. & Samad, T. (2000). Intelligent Distributed Simulation and Control of Power Plant, *IEEE Transactions on Energy Conversion*, vol. 15, no. 1, pp. 116-123

Lukszo, Z. Deconinck, G. & Weijnen M. P. C. (2010). Securing Electricity Supply in the Cyber Age: Exploring the Risks of Information and Communication Technology in Tomorrow's Electricity Infrastructure. New York: Springer

Marihart, D. J. (2001). Communications Technology Guidelines for EMS/SCADA Systems, *IEEE Transactions on Power Delivery*, vol. 16, no. 2, pp. 181 – 188

Nordell, D. E. (2008). Communication Systems for Distribution Automation, *Transmission and Distribution Conference and Exposition*, IEEE/PES, pp. 1 – 14

Phadke, A.G. & Thorp, J.S. (2008). Synchronized Phasor Measurements and Their Applications. New York: Springer

Shahidehpour M. & Wang Y. (2003), Communication and Control in Electric Power Systems. Hoboken, NJ: Wiley-IEEE Press

Shahraeini M. Javidi M. H. & Ghazizadeh M. S. (2010). A New Approach for Classification of Data Transmission Media in Power Systems, *2010 International Conference on Power System Technology (PowerCon 2010)*, China, Oct. 24-28, 2010, pp. 1-7

Shahraeini, M. & Alishahi, S. (2011). A survey on information and communication technology (ICT) applications in distribution systems, *21st International Conference and Exhibition on Electricity Distribution*, Germany

Shahraeini, M. Javidi, M. H. & Ghazizadeh, M. S. (2011). Comparison between Communication Infrastructures of Centralized and Decentralized Wide Area Measurement Systems, IEEE Transaction on Smart Grid, Vol. 2, No. 1, pp. 206-211

Stallings, W. (1997). Data and Computer Communications, fifth edition, Prentice-Hall Inc

Stouffer, K.; Falco, J.; & Kent, K. (2008). Guide to Industrial Control Systems (ICS) Security Recommendations of the National Institute of Standards and Technology, NIST Special Publication, vol. 800

Synchrony (2001). Trends in SCADA for Automated Water Systems

Taylor, C. W. (2006). Wide Area Measurement, Monitoring and Control in Power Systems, Presented at Workshop on Wide Area Measurement, Monitoring and Control in Power Systems, Imperial College, London, 16–17 Mar. 2006

Thomas M. S., Nanda D. & Ali I. (2006). Development of a Data. Warehouse for Non-operational Data in Power Utilities, *Proceeding Power India Conference*, New Dehli, India

Thomas, M. S. Nanda, D. & Ali, I. (2006). Development of a Data Warehouse for Non-Operational Data in Power Utilities, Proc. Power India Conference, New Dehli, India

Vaishnav, R. Gopalakrishnan P. & Thomas, J. (2008). Using public mobile phone networks for distribution automation, *Power and Energy Society General Meeting-Conversion and Delivery of Electrical Energy in the 21st Century*, pp. 1-5

Xiaorong, X., Yaozhong, X. Jinyu, X. Jingtao W. & Yingdao, H. (2006). WAMS applications in Chinese power systems, *IEEE Power Energy Magazine*, Vol. 4, No. 1, pp. 54–63

Yan, D. (2006). Wide-area Protection and Control System with WAMS Based, *2006 International Conference on Power System Technology (PowerCon2006)*, China, Oct. 2006, pp. 1-5

Zhang, P. (2006). Meeting the Protection and Control Challenges of 21st Century, *EPRI Technical Report*

From Conditional Probability Measurements to Global Matrix Representations on Variant Construction – A Particle Model of Intrinsic Quantum Waves for Double Path Experiments

Jeffrey Zheng[1], Christian Zheng[2] and T.L. Kunii[3]
[1]*Yunnan University*
[2]*University of Melbourne*
[3]*University of Tokyo*
[1]*P.R. China*
[2]*Australia*
[3]*Japan*

1. Introduction

1.1 Two types of double slit experiments

Quantum statistics play a key role in Quantum Mechanics QM [Feynman et al. (1965,1989); Penrose (2004)]. Two types of Double Slit Experiment are used to explore the core mysteries of quantum interactive behaviors. These are standard Double Slit Experiments with correlated signals and Single Photon Experiments that use ultra low intensity and lengthy exposures to demonstrate quanta self-interference patterns. The key significance is that intrinsic wave properties are observed in both environments [Barrow et al. (2004); Hawkingand & Mlodinow (2010)].

1.2 Two types of probabilities

Multivariate probabilities acting on multinomial distributions occupy a central role in classical probability theory and its applications. This mechanism has been explored from the early days in the study of modern probability theories [Ash & Doléans-Dade (2000); Durret (2005)]. Conditional probability is a powerful methodology at the heart of classical Bayesian statistics. In the history of probability and statistical developments, there have been long-running debates and a persistent lack of agreement in differentiating between prior distributions and posterior distributions [Ash & Doléans-Dade (2000); Durret (2005)]. It is worthy of note that the uniform distributions or normal distributions of conditional probability are always linked to a relatively large number of probability distributions in non-normal conditions. This points to practical problems with random distributions.**1**

1.3 Advanced single photon experiments

1.3.1 Applying the bohr complementarity principle

The Bohr Complementarity Principle BCP, established back in the 1920s brought us the foundations of QM [Bohr (1949)]. In Bohr's statement:"... we are presented with a choice of either tracing the path of the particle; or observing interference effects ... we have here to do with a typical example of how the complementary phenomena appear under mutually exclusive experimental arrangements." It is significant that BCP provided a powerful intellectual basis for Bohr in key debates in the history of QM and especially in his debates with Einstein [Jammer (1974)].

1.3.2 Testing bell inequality

To help decide between Bohr and Einstein on their approaches to wave and particle issues, Bell proposed a set of Bell-Inequations in the 1960s [Bell (1964)]. In 1969, CHSH proposed a spin measurement approach [Clauser et al. (1969)] and experiments by Aspect in 1982 did not support local realism [Aspect et al. (1982)].

1.3.3 Afshar's measurements

In 2001 Afshar set up an experiment to test the BCP [Afshar (2005)]. This experiment generated strong evidence contradicting the BCP, since both particle and wave distributions can be observed simultaneously. In Afshar's experiments, there are four measurements: ψ_1 - signals via left path, ψ_2 - signals via right path, σ_1 - interactive measurements of $\{\psi_1, \psi_2\}$ on the distance of f , and σ_2 - separate measurements of $\{\psi_1, \psi_2\}$ on the distance of $f + d$ respectively. In this experiment, a measurement quaternion is

$$\langle \psi_1, \psi_2, \sigma_1, \sigma_2 \rangle. \tag{1}$$

1.4 Current situation

From the 1920s through to the start of the 21st century, there was no significant experimental evidence to show that there were problems with the BCP. However, Afshar's 2001 experimental results are clearly not consistent with the BCP and further experimental results have provided solid evidence against the BCP. [Afshar (2005; 2006); Afshar et al. (2007)]. It is interesting to see that neither local realism nor the BCP are validated by the results of modern advanced single photon experiments [Afshar et al. (2007); Aspect (2007)]. It will be a major challenge in this century to redefine the principles on which the quantum approach may now be safely founded.

1.5 Chapter organization

Following on from multivariate probability models, this chapter focusses on a conditional approach to illustrate special properties found in conditional probability measurements via global matrix representations on the variant construction. This chapter is organized into nine sections addressing as follows:

1. general introduction (above)
2. key historical debates on the foundations of QM

3. analysis of key issues of QM
4. conditional construction proposed
5. exemplar results
6. analysis of visual distributions
7. using the variant solution to resolve longstanding puzzles
8. main results
9. final conclusions

2. Wave and particle debates in QM developments

2.1 Heisenberg uncertain principle

The Heisenberg Uncertainty Principle HUP was established in 1927 [Heisenberg (1930)]. The HUP represented a milestone in the early development of quantum theory [Jammer (1974)]. It implies that it is impossible to simultaneously measure the present position of a particle while also determining the future motion of a particle or any system small enough to require a Quantum mechanical treatment. From a mathematical viewpoint, the HUP arises from an equation following the methodology of Fourier analysis for the motion $[Q, P] = QP - PQ = i\hbar$. The later form of HUP is expressed as $\triangle p \cdot \triangle q \approx h$.

This equation shows that the non-commutativity implies that the HUP provides a physical interpretation for the non-commutativity.

2.2 Bohr complementarity principle

The HUP provided Bohr with a new insight into quantum behaviors [Bohr (1958)]. Bohr established the BCP to extend the idea of complementary variables for the HUP to energy and time, and also to particle and wave behaviors. One must choose between a particle model, with localized positions, trajectories and quanta or a wave model, with spreading wave functions, delocalization and interferences [Jammer (1974)].

Under the BCP, complementary descriptions e.g. wave or particle are mutually exclusive within the same mathematical framework because each model excludes the other. However, a conceptual construction allowed the HUP, the BCP and wave functions together with observed results to be integrated to form the Copenhagen Interpretation of QM. In the context of double slit experiments, the BCP dictates that the observation of an interference pattern for waves and the acquisition of directional information for particles are mutually exclusive.

2.3 Bohr-Einstein debates on wave and particle issues

Bohr and Einstein remained lifelong friends despite their differences in opinion regarding QM [Bohr (1949; 1958)]. In 1926 Born proposed a probability theory for QM without any causal explanation. Einstein's reaction is well known from his letter to Born [Born (1971)] in which he said "I, at any rate, am convinced that HE [God] does not throw dice."

Then in 1927 at the Solvay Conference, Heisenberg and Bohr announced that the QM revolution was over with nothing further being required. Einstein was dismayed [Bohr (1949); Bolles (2004)] for he believed that the underlying effects were not yet properly understood.

Perhaps in a spirit of compromise, Bohr then proposed his BCP that emphases the role of the observer over that which is observed. From 1927-1935, Einstein proposed a series of three intellectual challenges to further explore wave and particle issues [Bohr (1935; 1949); Bolles (2004)]:

- First, in 1927 Einstein proposed a double slit experiment on interference properties.
- Second, in 1930 at the sixth Solvay Congress, Einstein proposed weighing a box emitting timed releases of electromagnetic radiation.
- Third, in 1935 the paper "Can Quantum Mechanical Of Description Physical Reality Be Considered Complete?" [Einstein et al. (1935)] by Einstein, Podolsky and Rosen EPR was published in Physical Review.

2.4 EPR claims

The key points of the EPR paper are focused on two aspects: either (1) the description of reality given by the wave function in QM is incomplete or (2); the two quantities P and Q cannot have simultaneous reality.

Both operations: P and Q are applied $PQ - QP = i\hbar$. Such relationships follow the standard Quantum expression.

Property $PQ - QP \neq 0$ implies P and Q operations are related without independent computational properties. Under this condition, it is impossible to execute the two operations simultaneously under extant QM frameworks. From a parallel processing viewpoint, Einstein's view is extremely valuable. As such modern parallel computing theories and practices were only developed in the 1970s [Valiant (1975)] it is remarkable that Einstein pioneered such an approach way back in the 1930s. Modern parallel computing theory and practice support the original EPR paper and the conclusion that a QM description of physical that is expressed only in terms of wave functions is incomplete.

3. Key issues in QM

3.1 Restriction under HUP

For the HUP, different interpretations originate from the equation $[Q, P] = QP - PQ = i\hbar$, and the later HUP form $\triangle q \cdot \triangle p \geq h$. From a mathematical viewpoint, this type of inequality implies $\triangle q, \triangle p \geq h$ too. In other words, a minimal grid of a lattice restricts $\triangle q$ and $\triangle p \to 0$ actions. From the HUP expression, $[Q, P] \neq 0$ indicates the construction with a discrete intrinsic limitation. Such structures cannot directly apply to continuous infinitesimal operations.

Many quantum problems do not extend to the region of Plank constant limitations. Investigation back in the 1930s tended to rely more on theoretical considerations rather than actual experimentation [Bohr (1949)]. Consequently many issues had to wait until the 1980s to become better understood.

Both Q and P are infinite dimensional matrices, the restriction of $[Q, P] = i\hbar$ comes with a clear meaning today on its discrete properties. We cannot apply a continuous approach to make $[Q, P] = 0$. From an operational viewpoint, Einstein correctly identified the root of the matter. Since Q and P cannot exchange, it is not possible to run a simultaneous process on the standard wave functions. Simply extending discrete variations using continuous approaches presents further difficulties for the HUP.

In addition, the following questions need to be addressed in relation to the practical identification of complementary objects.

- What determines when a pair of objects is indeed a complementary pair?
- Is a pair of complex conjugate objects: $a + bi$ and $a - bi$, a pair of complementary objects?
- Is a pair of matrices, a hermit matrix H and its complex conjugate matrix H^*, a pair of complementary objects?
- Why can density matrix operations on infinite dimension be performed without significant errors while a pair of complementary finite matrices must be restricted by the HUP?

In practice, QM computations are mainly applied to wave functions and $[Q, P] = i\hbar$ formula. Intellectual debate on the theoretical considerations is particularly relevant to the HUP. In comparison, the deeper problems of QM cannot be easily explored in the absence of an experimental approach and a viable alternative theoretical construction [Barrow et al. (2004); Jammer (1974)].

3.2 Construction of BCP

Inspired by the HUP, Bohr uses a continuous analogy and a classical logical construction to describe Quantum systems. The BCP extends the HUP to handle different pairs of opposites and to restrict them with exclusive properties [Bohr (1958)].

As there were no well refined critical experiments in these days, all the debates between Bohr and Einstein were based on theoretical considerations alone. Compared with Einstein's open-minded attitudes to QM [Einstein et al. (1935)], Bohr and others insisted on the completeness and consistency of the Copenhagen Interpretation on QM [Bohr (1935)]. Such closed attitudes served to distance Bohr and others of like mind from reasonable suggestions made by Einstein and those expressed in the EPR paper and to lead them to treat such suggestions as if they were attacks on their already strongly held views [Bolles (2004)].

The BCP uses a classical logic framework to support dynamic constructions. Underpinned by the BCP, the HUP and a knowledge of wave functions, the Copenhagen Interpretation played a dominant role in QM from the 1930s on as it had by then been accepted as the orthodox point of view [Jammer (1974)].

Meanwhile, the EPR paper emphases that critical evidence must be obtained by real experiments and measurements. It is in the nature of a priori philosophical considerations that they will run into difficulties when actual experimental results fail to corresponded with their expectations.

3.3 EPR construction

The EPR position [Einstein et al. (1935)] can be re-visited in the light of modern advances in knowledge and computing theory. From a computing viewpoint, simultaneous properties may be the key with which these long-standing mysteries of QM can at last be unlocked. Operators P and Q cannot be exchanged, this indicates operational relevances existing in the lower levels of classical QM construction. In addition, there is a requirement of two systems

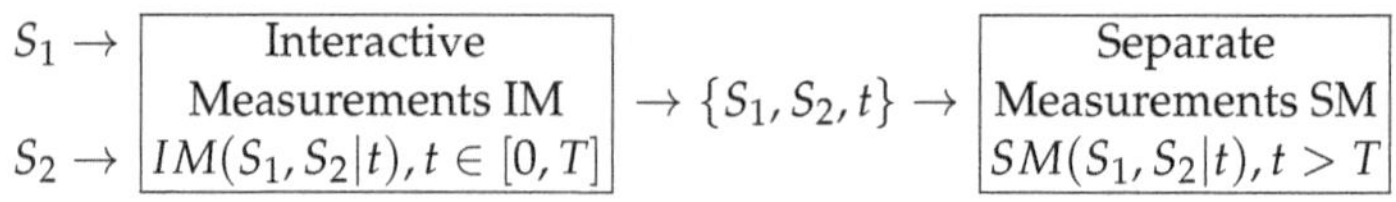

$$\text{ERP Model} = \langle S_1, S_2, IM(S_1, S_2|t)_{t\in[0,T]}, SM(S_1, S_2|t)_{t>T} \rangle$$

Fig. 1. EPR Measurement Quaternion Model on Einstein's Experimental Devices

to have interactive properties in $t \in [0, T]$ and without interactive properties on $t > T$. Such expressions may not be properly formulated by Fourier transformation schemes on wave functions.

However, after 78 years of development in advanced scienctific and ICT technologies, it is now possible to use advanced photonic and optical fiber technologies to implement all the requirements of the experiments proposed by Einstein.

The core EPR model can be shown in Figure 1, listed notations are explained as follows.

Let S_1 be System I, S_2 be System II, $IM(S_1, S_2|t), t \in [0, T]$ be Interactive Measurements IM for S_1 and S_2 on $t \in [0, T]$, $SM(S_1, S_2|t), t > T$ be Separate Measurements SM (non-interactive measurements) for S_1 and S_2 on $t > T$. Einstein's Experimental devices can be described as an EPR measurement quaternion:

$$\langle S_1, S_2, IM(S_1, S_2|t)_{t\in[0,T]}, SM(S_1, S_2|t)_{t>T} \rangle. \tag{2}$$

If an experiment can be expressed in the requisite form for this model, then it can be legitimately claimed as an EPR experiment.

3.4 Afshar experimental device

Afshar's experimental results have shown that it is possible to measure both particle and wave interference properties simultaneously in the same experiment with high accuracy [Afshar (2005; 2006); Afshar et al. (2007)]. Since this set of experiments has produced results that challenge the BCP at its very core it is pertinent to analyze and compare the model with the requirements for valid EPR devices.

In Afshar's experiments, $\{\psi_1, \psi_2\}$ are two signals input through double slits; σ_1 is the location on the distance f to collect interference measurements of $\{\psi_1, \psi_2\}$, and σ_2 is the location on the distance $f + d$ to collect separate measurements of $\{\psi_1, \psi_2\}$. Under this configuration, a 1-1 corresponding map can be established as follows:

$$\begin{cases} \psi_1 \rightarrow S_1; \\ \psi_2 \rightarrow S_2; \\ \sigma_1 \rightarrow IM(S_1, S_2|t), t \rightarrow f; \\ \sigma_2 \rightarrow SM(S_1, S_2|t), t \rightarrow f + d. \end{cases} \tag{3}$$

Using quaternion structures,

$$\langle \psi_1, \psi_2, \sigma_1, \sigma_2 \rangle \rightarrow \langle S_1, S_2, IM(S_1, S_2|f), SM(S_1, S_2|f + d) \rangle. \tag{4}$$

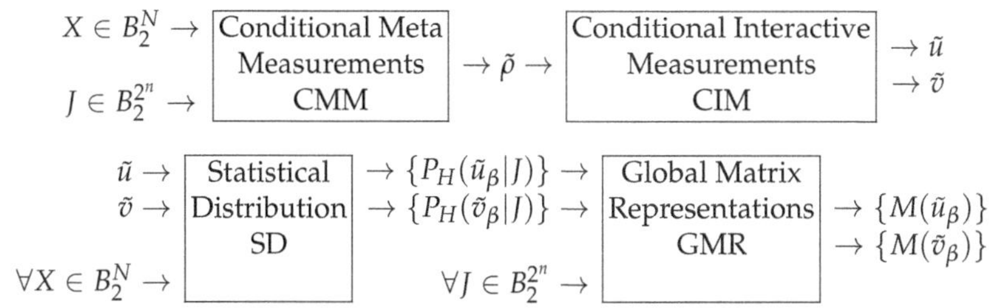

(a) Architecture

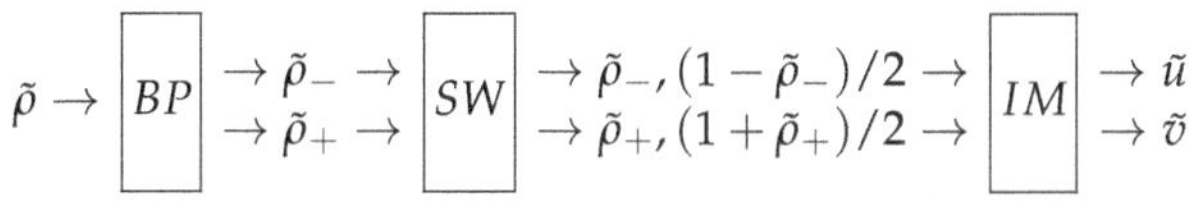

(b) CIM Component

Fig. 2. (a-b) Conditional Variant Simulation and Representation System; (a) System Architecture; (b) Conditional Interactive Measurement CIM Component

Under this correspondence, Afshar experiments are consistent with the EPR model.

4. Conditional variant simulation and representation system

A comprehensive review of the process of variant construction from conditional probability measurements through to global matrix representations is described briefly in this section. It is hoped that this may offer a convenient path for those seeking to devise and carry out experiments to further explore natural mysteries through the application of sound principles of logic and measurement.

Using variant principles described in the following subsections, with a N bit 0-1 vector X and a given logic function f, all N bit vectors are exhausted, variant measures generate two groups of histograms. The variant simulation and representation system is shown in Fig 2 (a-b). The detailed principles and methods are described in Sections 4.2-4.7 respectively. For multivariate probability conditions, please refer to the chapter of "From local interactive measurements to global matrix representations on variant construction" elsewhere in this book for sample cases and group distributions in multivariate probability environments.

4.1 Conditional simulation and representation model

The full measurement and representational architecture as shown in Figure 2(a) has four components: Conditional Meta Measurements CMM, Conditional Interactive Measurements CIM, Statistical Distributions SD and Global Matrix Representations GMR. The key part of the system, the CIM component, is shown in Fig 2(b).

4.1.1 Conditional Meta Measurements

The Conditional Meta Measurement (CMM) component uses N bit 0-1 vector X and a given function $J \in B_2^{2^n}$, CMM transfers N bit 0-1 vector under $J(X)$ to generate four Meta-measures, under a given probability scheme, four conditional probability measurements are generated and output as a quaternion signal $\tilde{\rho}$.

4.1.2 Conditional Interactive Measurements

The Conditional Interactive Measurement (CIM) component is the key location for conditional interactions as shown in Figure 2(b) to transfer a quaternion signal $\tilde{\rho}$ under symmetry / anti-symmetry and synchronous / asynchronous conditions, under four combinations of time effects namely (Left, Right, Double Particle, Double Wave). Two types of additive operations are identified. Each $\{\tilde{u}, \tilde{v}\}$ signal is composed of four distinct signals.

4.1.3 Statistical Distributions

The Statistical Distribution (SD) component performs statistical activities on corresponding signals. It is necessary to exhaust all possible vectors of X with a total of 2^N vectors. Under this construction, each sub-signal of $\{\tilde{u}, \tilde{v}\}$ forms a special histogram with a one dimensional spectrum to indicate the distribution under function J. A total of eight histograms are generated in the probability conditions.

4.1.4 Global Matrix Representations

The Global Matrix Representation (GMR) component uses each statistical distribution of the relevant probability histogram as an element of a matrix composed of a total of 2^{2^n} elements for all possible functions $\{J\}$. In this configuration, C code schemes are applied to form a $2^{2^{n-1}} \times 2^{2^{n-1}}$ matrix to show the selected distribution group.

Unlike the other coding schemes (SL, W, F, ...), only C code schemes provide a regular configuration to clearly differentiate the Left path as exhibiting horizontal actions and the Right path as exhibiting vertical actions . Such clearly polarized outcomes may have the potential to help in the understanding of interactive mechanism(s) between double path for particles and double path for waves properties.

4.2 Variant principle

The variant principle is based on n-variable logic functions [Zheng (2011); Zheng & Zheng (2010; 2011a;b); Zheng et al. (2011)].

4.2.1 Two sets of states

For any n-variables $x = x_{n-1}...x_i...x_0$, $0 \leq i < n, x_i \in \{0,1\} = B_2$ let a position j be the selected bit $0 \leq j < n$, x_j be the selected variable. Let output variable y and n-variable function $f, y = f(x)$, $y \in B_2, x \in B_2^n$. For all states of x, a set $S(n)$ composed of the 2^n states can be divided into two sets: $S_0(n)$ and $S_1(n)$.

$$\begin{cases} S_0(n) = \{x | x_j = 0, \forall x \in B_2^n\} \\ S_1(n) = \{x | x_j = 1, \forall x \in B_2^n\} \\ S(n) = \{S_0(n), S_1(n)\} \end{cases} \tag{5}$$

4.2.2 Four meta functions

For a given logic function f, input and output pair relationships define four meta logic functions $\{f_\perp, f_+, f_-, f_\top\}$.

$$\begin{cases} f_\perp(x) = \{f(x)|x \in S_0(n), y = 0\} \\ f_+(x) = \{f(x)|x \in S_0(n), y = 1\} \\ f_-(x) = \{f(x)|x \in S_1(n), y = 0\} \\ f_\top(x) = \{f(x)|x \in S_1(n), y = 1\} \end{cases} \tag{6}$$

4.2.3 Two polarized functions

Considering two standard logic canonical expressions: the AND-OR form is selected from $\{f_+(x), f_\top(x)\}$ as $y = 1$ items, and the OR-AND form is selected from $\{f_-(x), f_\perp(x)\}$ as $y = 0$ items. Considering $\{f_\top(x), f_\perp(x)\}, x_j = y$ items, they are themselves invariant.

To select $\{f_+(x), f_-(x)\}, x_j \neq y$ in forming a variant logic expression. Let $f(x) = \langle f_+|x|f_-\rangle$ be a variant logic expression. Any logic function can be expressed as a variant logic form. In $\langle f_+|x|f_-\rangle$ structure, f_+ selected 1 items in $S_0(n)$ as the same as the AND-OR standard expression, and f_- selecting relevant parts the same as OR-AND expression 0 items in $S_1(n)$.

4.3 Meta measures and conditional probability measurements

Under variant construction, N bits of 0-1 vector X under a function J produce four Meta measures composed of a measure vector N

$(X : J(X)) \to (N_\perp, N_+, N_-, N_\top), N_0 = N_\perp + N_+, N_1 = N_- + N_\top, N = N_0 + N_1$

Using four Meta measures, relevant probability measurements can be formulated.
$\tilde{\rho} = (\tilde{\rho}_\perp, \tilde{\rho}_+, \tilde{\rho}_-, \tilde{\rho}_\top) = (N_\perp/N_0, N_+/N_0, N_-/N_1, N_\top/N_1), 0 \leq \tilde{\rho}_\perp, \tilde{\rho}_+, \tilde{\rho}_-, \tilde{\rho}_\top \leq 1.$

4.3.1 Variant measure functions

Let Δ be the variant measure function

$$\begin{aligned} \Delta &= \langle \Delta_\perp, \Delta_+, \Delta_-, \Delta_\top \rangle \\ \Delta J(x) &= \langle \Delta_\perp J(x), \Delta_+ J(x), \Delta_- J(x), \Delta_\top J(x) \rangle \\ \Delta_\alpha J(x) &= \begin{cases} 1, & J(x) \in J_\alpha(x), \alpha \in \{\perp, +, -, \top\} \\ 0, & \text{others} \end{cases} \end{aligned} \tag{7}$$

For any given n-variable state there is one position in $\Delta J(x)$ to be 1 and other 3 positions are 0.

4.3.2 Variant measures on vector

For any N bit 0-1 vector $X, X = X_{N-1}...X_j...X_0, 0 \leq j < N, X_j \in B_2, X \in B_2^N$ under n-variable function J, n bit 0-1 output vector Y, $Y = J(X) = \langle J_+|X|J_-\rangle$, $Y = Y_{N-1}...Y_j...Y_0, 0 \leq j < N, Y_j \in B_2, Y \in B_2^N$. For the j-th position $x^j = [...X_j...] \in B_2^n$ to form $Y_j = J(x^j) = \langle J_+|x^j|J_-\rangle$.

Let N bit positions be cyclic linked. Variant measures of $J(X)$ can be decomposed

$$\Delta(X:Y) = \Delta J(X) = \sum_{j=0}^{N-1} \Delta J(x^j) = \langle N_\perp, N_+, N_-, N_\top \rangle \tag{8}$$

as a quaternion $\langle N_\perp, N_+, N_-, N_\top \rangle$, $N = N_\perp + N_+ + N_- + N_\top$.

4.3.3 Example

E.g. $N = 12$, given $J, Y = J(X)$.

$$\begin{aligned} X &= 1\ 0\ 1\ 1\ 1\ 0\ 1\ 1\ 1\ 0\ 0\ 1 \\ Y &= 0\ 0\ 1\ 0\ 1\ 0\ 1\ 0\ 1\ 1\ 0\ 0 \\ \Delta(X:Y) &= -\ \perp\ \top\ -\ \top\ \perp\ \top\ -\ \top\ +\ \perp\ - \end{aligned}$$

$\Delta J(X) = \langle N_\perp, N_+, N_-, N_\top \rangle = \langle 3, 1, 4, 4 \rangle, N = 12.$

Input and output pairs are 0-1 variables for only four combinations. For any given function J, the quantitative relationship of $\{\perp, +, -, \top\}$ is directly derived from the input/output sequences. Four meta measures are determined.

4.4 Four conditional meta measurements

Using variant quaternion, conditional measurements of probability signals are calculated as four meta conditional measurements by following the given equations. For any N bit 0-1 vector X, function J, under Δ measurement: $\Delta J(X) = \langle N_\perp, N_+, N_-, N_\top \rangle$, $N_0 = N_\perp + N_+$, $N_1 = N_- + N_\top$, $N = N_0 + N_1 = N_\perp + N_+ + N_- + N_\top$.

Signal $\tilde{\rho}$ is defined by

$$\begin{cases} \tilde{\rho} = & (\tilde{\rho}_\perp, \tilde{\rho}_+, \tilde{\rho}_-, \tilde{\rho}_\top) \\ \tilde{\rho}_\perp = & \frac{N_\perp}{N_0} \\ \tilde{\rho}_+ = & \frac{N_+}{N_0} \\ \tilde{\rho}_- = & \frac{N_-}{N_1} \\ \tilde{\rho}_\top = & \frac{N_\top}{N_1} \end{cases} \tag{9}$$

4.5 Conditional Interactive Measurements

Conditional Interactive Measurements (CIM) are divided into three stages: BP, SW and SM respectively. The BP stage selects $\{\tilde{\rho}_-, \tilde{\rho}_+\}$ as sub-signals. The SW component extends two signals into four signals with different symmetric properties; The SM component merges different signals to form two sets of eight signals.

Using $\{\tilde{\rho}_+, \tilde{\rho}_-\}$, a pair of signals $\{\tilde{u}, \tilde{v}\}$ are formulated:

$$\begin{cases} \tilde{u} = (\tilde{u}_+, \tilde{u}_-, \tilde{u}_0, \tilde{u}_1) = \{u_\beta\} \\ \tilde{v} = (\tilde{v}_+, \tilde{v}_-, \tilde{v}_0, \tilde{v}_1) = \{v_\beta\} \\ \quad \beta \in \{+, -, 0, 1\} \end{cases} \tag{10}$$

$$\begin{cases} \tilde{u}_+ = \tilde{\rho}_+ \\ \tilde{u}_- = \tilde{\rho}_- \\ \tilde{u}_0 = \tilde{u}_+ \oplus \tilde{u}_- \\ \tilde{u}_1 = (\tilde{u}_+ + \tilde{u}_-)/2 \\ \tilde{v}_+ = \frac{1+\tilde{\rho}_+}{2} \\ \tilde{v}_- = \frac{1-\tilde{\rho}_-}{2} \\ \tilde{v}_0 = \tilde{v}_+ \oplus \tilde{v}_- \\ \tilde{v}_1 = \tilde{v}_+ + \tilde{v}_- - 0.5 \end{cases} \quad (11)$$

where $0 \leq \tilde{u}_\beta, \tilde{v}_\beta \leq 1, \beta \in \{+,-,0,1\}, \oplus$: Asynchronous addition, $+$: Synchronous addition.

4.6 Statistical distributions

The SD component provides a statistical means to accumulate all possible vectors of N bits for a selected signal and generate a histogram. Eight signals correspond to eight histograms respectively. Among these, four histograms exhibit properties of symmetry and the other four histograms exhibit properties of anti-symmetry.

4.6.1 Statistical histograms

For a function J, all measurement signals are collected and the relevant histogram represents a complete statistical distribution.

Using $\tilde{u}$ and $\tilde{v}$ signals, each $\tilde{u}_\beta$ or $\tilde{v}_\beta$ determines a fixed position in the relevant histogram to make vector X on a position. After completing 2^N data sequences, eight symmetry/anti-symmetry histograms of $\{H(\tilde{u}_\beta|J)\}, \{H(\tilde{v}_\beta|J)\}$ are generated.

For a function $J, \beta \in \{+,-,0,1\}$

$$\begin{cases} H(\tilde{u}_\beta|J) = \Sigma_{\forall X \in B_2^N} H(\tilde{u}_\beta|J(X)) \\ H(\tilde{v}_\beta|J) = \Sigma_{\forall X \in B_2^N} H(\tilde{v}_\beta|J(X)), J \in B_2^{2^n} \end{cases} \quad (12)$$

4.6.2 Normalized probability histograms

Let $|H(..)|$ denote the total number in the histogram $H(..)$, a normalized probability histogram $(P_H(..))$ can be expressed as

$$\begin{cases} P_H(\tilde{u}_\beta|J) = \frac{H(\tilde{u}_\beta|J)}{|H(\tilde{u}_\beta|J)|} \\ P_H(\tilde{v}_\beta|J) = \frac{H(\tilde{v}_\beta|J)}{|H(\tilde{v}_\beta|J)|}, J \in B_2^{2^n} \end{cases} \quad (13)$$

Here, all histograms are restricted in $[0,1]^2$ areas.

Distributions are dependant on the data set as a whole and are not sensitive to varying under special sequences. Under this condition, when the data set has been exhaustively listed, then the same distributions are always linked to the given signal set.

The eight histogram distributions provide invariant spectra to represent properties among different interactive conditions.

4.7 Global Matrix Representations

After local interactive measurements and statistical process are undertaken for a given function J, eight histograms are generated. The Global Matrix Representation GMR component performs its operations into two stages. In the first stage, exhausting all possible functions for $\forall J \in B_2^{2^n}$ to generate eight sets, each set contains 2^{2^n} elements and each element is a histogram. In the second stage, arranging all 2^{2^n} elements generated as a matrix by C code scheme. Here, we can see Left and Right path reactions polarized into Horizontal and Vertical relationships respectively.

4.7.1 Matrix and its elements

For a given C scheme, let $C(J) = \langle J^1 | J^0 \rangle$, each element

$$\begin{cases} M_{\langle J^1 | J^0 \rangle}(\tilde{u}_\beta | J) = P_H(\tilde{u}_\beta | J) \\ M_{\langle J^1 | J^0 \rangle}(\tilde{v}_\beta | J) = P_H(\tilde{v}_\beta | J) \\ \qquad J \in B_2^{2^n};\ J^1, J^0 \in B_2^{2^{n-1}} \end{cases} \tag{14}$$

4.7.2 Representation patterns of matrices

For example, using $n = 2, P = (3102), \Delta = (1111)$ conditions, a C code case contains sixteen histograms arranged as a 4×4 matrix.

0	4	1	5
2	6	3	7
8	12	9	13
10	14	11	15

(15)

All matrices in this chapter use this configuration for the matrix pattern representing their elements.

5. Simulation results

For ease of illustration, as different signals have intrinsic random properties, only statistical distributions and global matrix representations are selected in this section.

5.1 Statistical distributions

The simulation provides a series of output results. In this section, $N = \{12, 13\}, n = 2, \{J = 3, J_+ = 11, J_- = 2\}$ are selected. Corresponding to Left path (Left), Right path (Right), Double path for Particles (D-P) and Double path for Waves (D-W) under symmetry and anti-symmetry conditions respectively.

From a given function, a set of histograms can be generated as two groups of eight probability histograms. To show their refined properties, it is necessary to represent them in both odd and even numbers. A total of sixteen histograms are required. For convenience of comparison, sample cases are shown in Figures 3(I-III).

$P_H(\tilde{u}_+\|J)$ (a) Left	$P_H(\tilde{u}_-\|J)$ (b) Right
$P_H(\tilde{u}_0\|J)$ (c) D-P	$P_H(\tilde{u}_1\|J)$ (d) D-W
$P_H(\tilde{v}_+\|J)$ (e) Left	$P_H(\tilde{v}_-\|J)$ (f) Right
$P_H(\tilde{v}_0\|J)$ (g) D-P	$P_H(\tilde{v}_1\|J)$ (h) D-W

(I) Representative patterns of Histograms for function J (a-d) symmetric cases; (e-h) antisymmetric cases

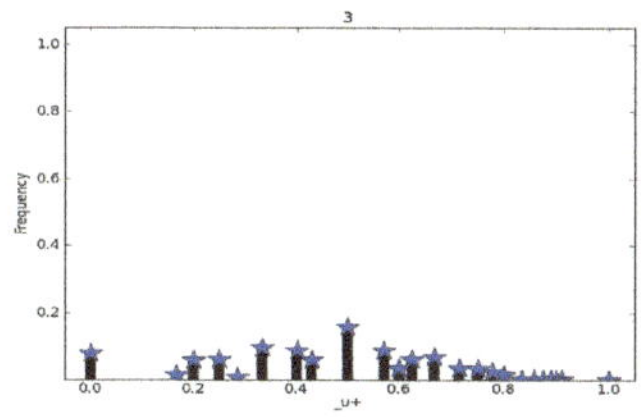

(a) Left

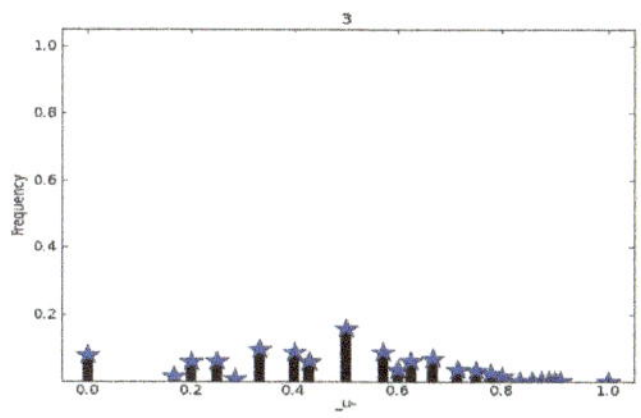

(b) Right

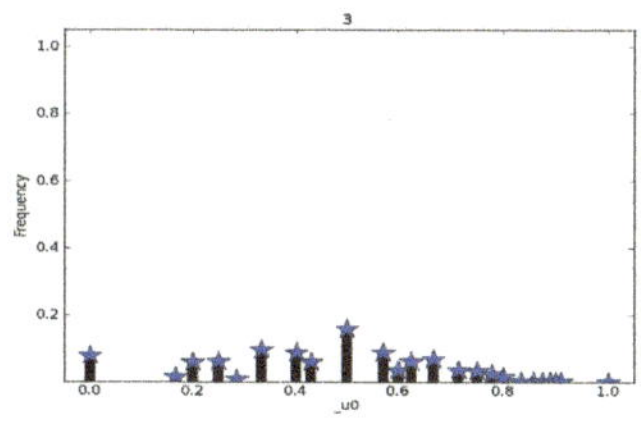

(c) D-P

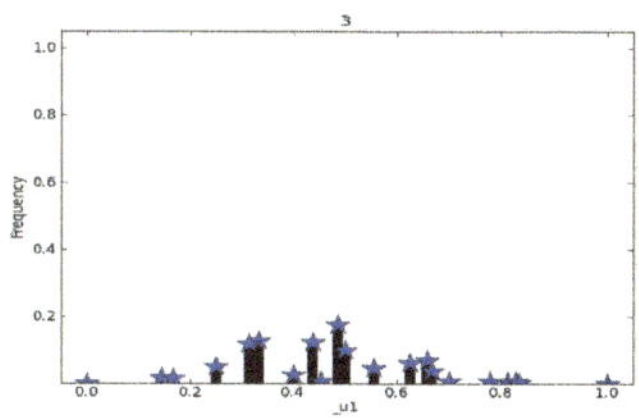

(d) D-W

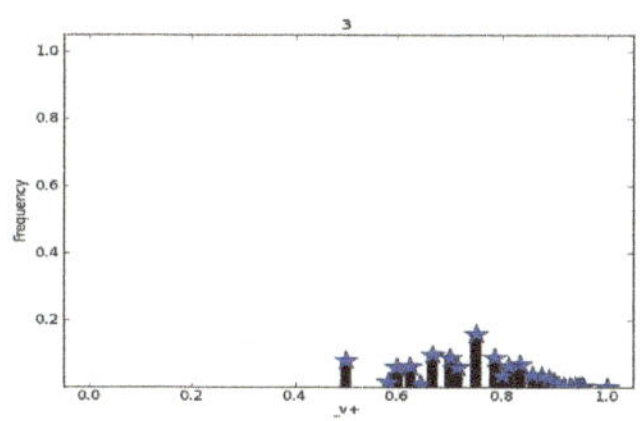

(e) Left

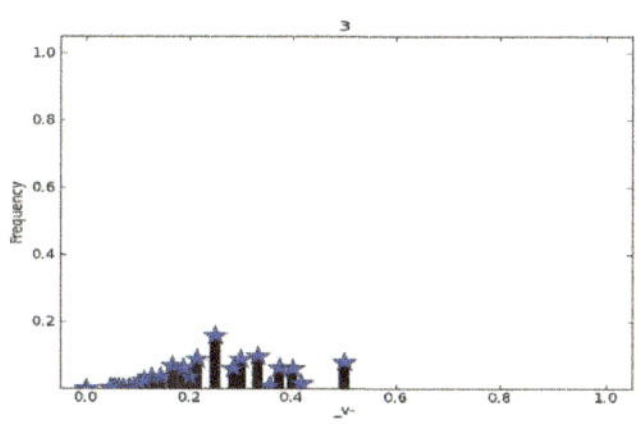

(f) Right

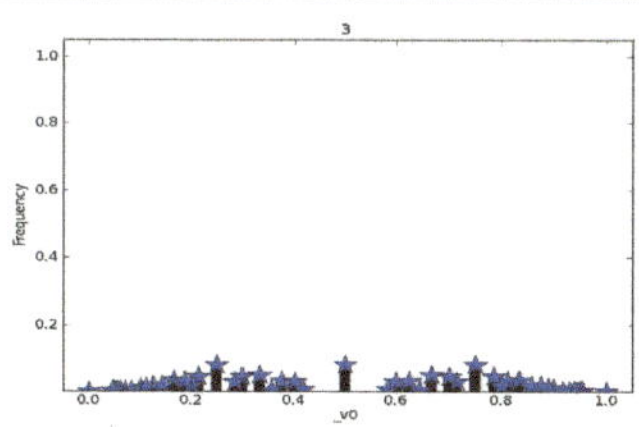

(g) D-P

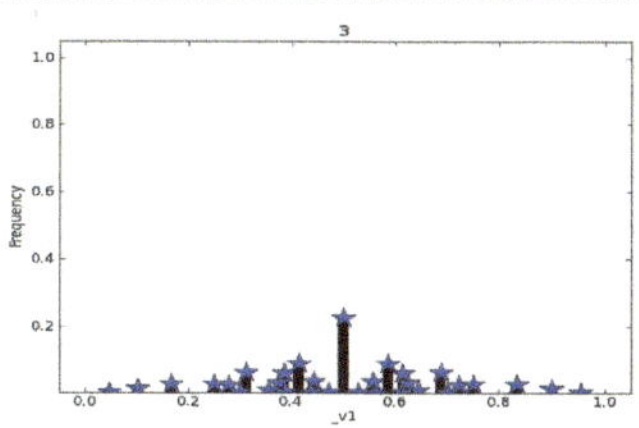

(h) D-W

(II) $N = \{12\}, J = 3$ Two groups of results in eight histograms

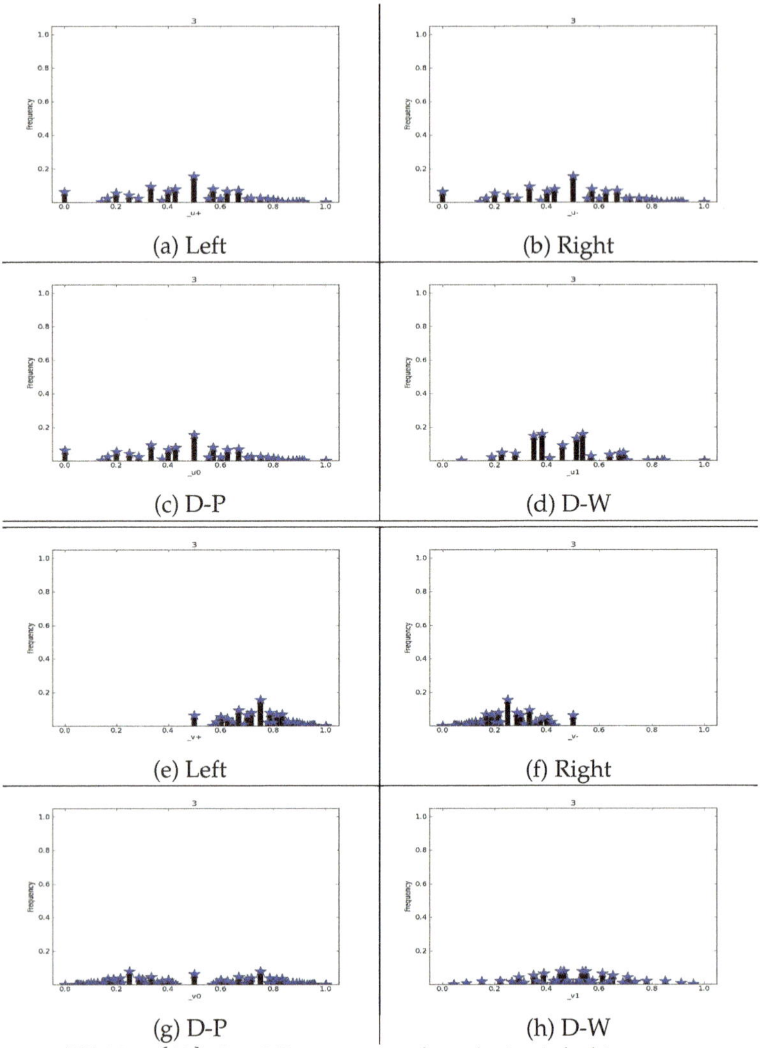

(III) $N = \{13\}, J = 3$ Two groups of results in eight histograms

Fig. 3. (I-III) $N = \{12, 13\}, J = 3$ Simulation results ; (I) Representative Patterns for $P_H(\tilde{u}_+|J) = P_H(\tilde{u}_-|J)$ and $P_H(\tilde{v}_+|J) = P_H(1 - \tilde{v}_-|J)$ conditions; (II) $N = \{12\}, J = 3$ Two groups of eight histograms on conditional probability; (III) $N = \{13\}, J = 3$ Two groups of eight histograms on conditional probability

Representation patterns are illustrated in Fig 3(I). Eight conditional probability histograms of $P_H(\tilde{u}_+|J) = P_H(\tilde{u}_+|J)$ are shown in Fig 3(II) for $N = 12$ to represent four symmetry groups and another eight conditional probability histograms are shown Fig 3(III) for $N = 13$ to represent four anti-symmetry groups respectively.

5.2 Global matrix representations

All possible 2^{2^n} functions are applied. It is convenient to arrange all the histograms generated into a matrix and a C code scheme of variant logic is applied to organize a set of 2^{2^n} histograms into a $2^{2^{n-1}} \times 2^{2^{n-1}}$ matrix.

Applying the C code configuration, any given signal of a function determines a matrix element to represent its histogram. There is one to one correspondence among different configurations.

Using this measurement mechanism, eight types of statistical histograms are systematically illustrated. Each element in the matrix is numbered to indicate its corresponding function and the relevant histogram is shown.

For $n = 2$ cases, sixteen matrices are shown in Figs 5-6 (a-h). Figs 5-6 (a-d) represent Symmetry groups and Figs 5-6 (e-h) represent Anti-symmetry groups. To show odd and even number configurations, Fig 5 (a-h) shows $N = 12$ cases and Fig 6 (a-h) shows $N = 13$ cases respectively.

6. Analysis of results

In the previous section, results of different statistical distributions and their global matrix representations were presented. In this section, plain language is used to explain what various visual effects might be illustrated and to discuss local and global arrangements.

6.1 Statistical distributions for a given function

It is necessary to analyze the differences among the various statistical distributions for a given function.

6.1.1 Symmetry groups for a function

For the selected function $J = 3$, four distributions in symmetry groups are shown in Fig 3 (a-d). (a) $P_H(\tilde{u}_+|J)$ for Left; (b) $P_H(\tilde{u}_-|J)$ for Right; (c) $P_H(\tilde{u}_0|J)$ for D-P; and (d) $P_H(\tilde{u}_1|J)$ for D-W respectively.

Under Symmetry conditions, $P_H(\tilde{u}_+|J) = P_H(\tilde{u}_-|J)$, both Left and Right distributions are the same. $P_H(\tilde{u}_0|J)$ generated with both paths open under asynchronous conditions simulates a D-P. Compared with distributions in (a-b) , it is possible to identify the components from original inputs.

However, for $P_H(\tilde{u}_1|J)$ under synchronous conditions and with the same Left and Right input signals, the simulation shows a D-W exhibiting interferences among the output distributions that are significantly different from the original components.

6.1.2 Anti-symmetry groups for a function

Four distributions are shown in Fig 3 (e-h) as asymmetry groups. A pair of equations $P_H(\tilde{v}_+|J) = P_H(1 - \tilde{v}_-|J)$ shows that one distribution is a mirror image of the other. $P_H(\tilde{v}_+|J)$ distribution is shown in Fig 3 (e) for Left signals and $P_H(\tilde{v}_-|J)$ distribution is shown Fig 3 (f) for Right signals.

$P_H(\tilde{v}_0|J)$ is shown in Fig 3 (g) for both paths open under asynchronous conditions to simulate a D-P. Compared with (e-f) distributions, it is feasible to identify the same components from the original inputs.

However $P_H(\tilde{v}_1|J)$ is shown in Fig 3 (h) under synchronous condition with both path signals as inputs to simulate a D-W exhibiting interferences among the output distributions that are significantly different from the original components.

To differentiate between even and odd numbers, $N = 12$ cases are shown in Fig 3 (II, a-h) and $N = 13$ cases are shown in Fig 3 (III, a-h) respectively.

6.2 Global matrix representations

Sixteen matrices are represented in Fig 4-5 (a-h) with eight signals generating two sets of 16 groups for $N = \{12, 13\}$ respectively.

6.2.1 Symmetry cases

Matrices for the Left in Fig 4-5 (a) show elements in a column with the corresponding histogram showing polarized effects on the vertical.

Matrices for the Right in Fig 4-5 (b) show elements in a row with the corresponding histogram showing polarized effects on the horizontal.

Matrices for D-P in Fig 4-5 (c) provide asynchronous operations combined with both distributions from Fig 4-5 (a-b) to form a unified distribution. From each corresponding position, it is possible to identify each left and right component and the resulting shapes of the histogram.

Matrices for D-W in Fig 4-5 (d) provide synchronous operations combined with both distributions from Fig 4 -5 (a-b) for each element.

Compared with Fig 4-5 (c) and Fig 4-5 (d) respectively, distributions in Fig 4-5 (d) are much simpler with two original distributions especially on the anti-diagonal positions: $J \in \{10, 12, 3, 5\}$. Only less than half the number of spectrum lines are identified.

6.2.2 Anti-symmetry cases

In a similar manner to the symmetry conditions, four anti-symmetry effects can be identified in Fig 4-5 (e-h). Matrices in Fig 4-5 (e) are Left operations for different functions; elements are polarized on the vertical and matrices in Fig 4-5 (f) are Right Operations; elements are polarized on the horizontal. Spectrum lines in Fig 4-5 (e) appear in the right half and spectrum lines in Fig 4-5 (f) appear in the Left half respectively.

Matrices for D-P in Fig 4-5 (g) show additional effects for each distribution according to the relevant position with components that can be identified as corresponding to identifiable inputs in many cases. Anti-symmetry signals are generated in merging conditions.

Matrices for D-W in Fig 4-5 (h) show different properties. In general, only one peak can be observed for each element especially for the $J \in \{10, 12, 3, 5\}$ condition. Spectra appear to be much simpler than the original distributions in Fig 4-5 (e-f), and significant interference properties are observed.

6.3 Four symmetry groups

Pairs of relationships can be checked on symmetry matrices in Figs 4-5 (a-d), four groups are identified.

6.3.1 Left: polarized vertical group

$\{P_H(\tilde{u}_+|J)\}$ elements in Figs 4-5 (a) show (only) four distinct distributions. Each column contains only one distribution. Sixteen elements in the matrix can be classified into four vertical classes: $\{0, 2, 8, 10\}, \{4, 6, 12, 14\}, \{1, 3, 9, 11\}, \{5, 7, 13, 15\}$ respectively. Four meta distributions are given as $\{10, 14, 11, 15\}$.

6.3.2 Right: polarized horizontal group

$\{P_H(\tilde{u}_-|J)\}$ elements in Figs 4-5 (b) show (a further) four distinct distributions. Each row contains only one distribution. Sixteen elements in the matrix can be classified into four horizontal classes: $\{0, 4, 1, 5\}, \{2, 6, 3, 7\}, \{8, 12, 9, 13\}, \{10, 14, 11, 15\}$ respectively. Four meta distributions are given as $\{0, 2, 8, 10\}$.

6.3.3 D-P: particle group

$\{P_H(\tilde{u}_0|J)\}$ elements in Figs 4-5 (c) illustrate symmetry properties. There are six pairs of symmetry elements: $\{8 : 14\}, \{2 : 11\}, \{0 : 15\}, \{6 : 9\}, \{4 : 13\}, \{1 : 7\}$. In addition, four elements on anti-diagonals provide different distributions: $\{10, 12, 3, 5\}$. Under this condition, ten classes of distributions are distinguished.

6.3.4 D-W: wave group

$\{P_H(\tilde{u}_1|J)\}$ elements in Figs 4-5 (d) illustrate symmetry properties. There are six pairs of symmetry elements: $\{8 : 14\}, \{2 : 11\}, \{0 : 15\}, \{6 : 9\}, \{4 : 13\}, \{1 : 7\}$. In addition, four elements on diagonal positions provide the same distribution: $\{0, 6, 9, 15\}$. Two elements on anti-diagonals: $\{12, 3\}$ have the same distribution in Fig 4 (d). Under this condition, nine or ten classes of different distributions can be identified for Fig 4 (d) and Fig 5 (d) respectively.

6.4 Four anti-symmetry groups

Figures 4-5 (e-h) represent anti-symmetry properties, four groups can be identified.

6.4.1 Left: polarized vertical group

$\{P_H(\tilde{v}_+|J)\}$ elements in Figs 4-5 (e) show that (only) four classes can be distinguished. Elements within these groups members are the same as for symmetry groups in Figs 4-5(a). Their distributions fall within the region $[0.5, 1]$.

6.4.2 Right: polarized horizontal group

$\{P_H(\tilde{v}_-|J)\}$ elements in Figs 4-5 (f) show that (only) four classes can be distinguished. Elements within these groups are the same as for symmetry groups in Figs 4-5 (b). Their distributions fall within the region $[0, 0.5]$.

6.4.3 D-P: particle group

$\{P_H(\tilde{v}_0|J)\}$ in Figs 4-5 (g) show six pairs of anti-symmetry distributions: $\{8 \nmid 14\}, \{2 \nmid 11\}, \{0 \nmid 15\}, \{6 \nmid 9\}, \{4 \nmid 13\}, \{1 \nmid 7\}$ four elements are distinguished on the anti-diagonals: $\{10, 12, 3, 5\}$. Under this condition, ten classes can be identified.

6.4.4 D-W: wave group

$\{P_H(\tilde{v}_1|J)\}$ in Figs 4-5 (h) show six pairs of anti-symmetry distributions: $\{8 \nmid 14\}, \{2 \nmid 11\}, \{0 \nmid 15\}, \{6 \nmid 9\}, \{4 \nmid 13\}, \{1 \nmid 7\}$ four pairs of symmetry elements: $\{3 : 5\}, \{10 : 12\}, \{2 : 4\}, \{11 : 13\}$ are distinguished. Under this condition, twelve classes can be identified.

6.5 Odd and even numbers

From a group viewpoint, only D-P and D-W need to be reviewed as different groups in symmetry conditions. Anti-symmetry conditions are unremarkable.

It is reasonable to suggest that anti-symmetry operations will be much easier to distinguish under experimental conditions, since sixteen groups in D-P conditions and twelve groups in D-W conditions will have significant differences. However, under the symmetry conditions (only) minor differences can be identified.

6.5.1 Single and double peaks

Single and Double peaks can be observed in Fig 4(5) (h): $\{3, 5\}$ for even and odd numbers respectively.

For two other members $\{10, 12\}$, (only) single pulse distributions are observed in Figs 4-5 (h) to show the strongest interference results.

6.6 Class numbers in different conditions

To summarize over the different classes, 16 matrices are shown in different numbers of identified classes as follows:

Class No.	Left	Right	D-P	D-W
SE	4	4	10	10
SO	4	4	10	10
AE	4	4	16	14
AO	4	4	16	14

where Left:Left Path, Right: Right Path, D-P: Double Path for Particles, D-W: Double Path for Waves; SE: Symmetry for Even number, SO: Symmetry for Odd number, AE: Anti-symmetry for Even number, AO: Anti-symmetry for Odd number.

6.7 Polarized effects and double path results

In order to contrast the different polarized conditions, it is convenient to compare distributions $\{P_H(\tilde{u}_+|J), P_H(\tilde{u}_-|J\}$ and $\{P_H(\tilde{v}_+|J), P_H(\tilde{v}_-|J\}$ arranged according to the corresponding polarized vertical and horizontal effects. This visual effect is similar to what might be found when using polarized filters in order to separate complex signals into two channels. Different distributions can be observed under synchronous and asynchronous conditions.

6.7.1 Particle distributions and representations

For all symmetry or non-symmetry cases under $\oplus$ asynchronous addition operations, relevant values meet $0 \leq \tilde{u}_0, \tilde{v}_0, \tilde{u}_-, \tilde{v}_-, \tilde{u}_+, \tilde{v}_+ \leq 1$. Checking $\{P_H(\tilde{u}_0|J), P_H(\tilde{v}_0|J)\}$ series, $\{P_H(\tilde{u}_+|J), P_H(\tilde{u}_-|J)\}$ and $\{P_H(\tilde{v}_+|J), P_H(\tilde{v}_-|J)\}$ satisfy following equation.

$$\begin{cases} P_H(\tilde{u}_0|J) = \frac{P_H(\tilde{u}_-|J)+P_H(\tilde{u}_+|J)}{2} \\ P_H(\tilde{v}_0|J) = \frac{P_H(\tilde{v}_-|J)+P_H(\tilde{v}_+|J)}{2} \end{cases} \tag{16}$$

The equation is true for different values of N and n.

6.7.2 Wave distributions and representations

Interference properties are observed in $\{P_H(\tilde{u}_+|J) = P_H(\tilde{u}_-|J)\}$ conditions. Under $+$ synchronous addition operations, relevant values meet $0 \leq \tilde{u}_1, \tilde{v}_1, \tilde{u}_-, \tilde{v}_-, \tilde{u}_+, \tilde{v}_+ \leq 1$. Checking $\{P_H(\tilde{u}_1|J), P_H(\tilde{v}_1|J)\}$ distributions and compared with $\{P_H(\tilde{u}_+|J), P_H(\tilde{u}_-|J)\}$ and $\{P_H(\tilde{v}_+|J), P_H(\tilde{v}_-|J)\}$, non-equations and equations are formulated as follows:

$$\begin{cases} P_H(\tilde{u}_1|J) \neq P_H(\tilde{u}_0|J) \\ P_H(\tilde{v}_1|J) \neq P_H(\tilde{v}_0|J) \end{cases} \tag{17}$$

Spectra in different cases illustrate wave interference properties. Single and double peaks are shown in interference patterns and these are similar to interference effects in classical double slit experiments.

6.7.3 Non-symmetry and non-anti-symmetry

However, for the $\{P_H(\tilde{u}_+|J) \neq P_H(\tilde{u}_-|J)\}$ non-symmetry cases, there are significant differences between $\{P_H(\tilde{u}_0|J), P_H(\tilde{v}_0|J)\}$ and $\{P_H(\tilde{u}_1|J), P_H(\tilde{v}_1|J)\}$. Such cases have interference patterns that exhibit greater symmetry than single path and particle distributions.

Four anti-diagonal positions are linked to symmetry and anti-symmetry pairs, twelve other pairs of functions belong to non-symmetry and non-anti-symmetry conditions. Their meta elements can be identified by the relevant variant expressions.

7. Core debated issues under variant construction

7.1 HUP environment

Under the variant construction, variant measurements can be organized into multiple sets of simultaneous measurements. Each element in a N bit vector provides only a small portion of information, collected measurements are independent of special positions. Under this condition, there is no essential HUP environment for the variant construction. 0-1 groups and their measurements are naturally parallel . They can be processed in simultaneous conditions. Considering these properties, such group measurements do not correspond with the requirements of Heisenberg single particle environments. Viewed as a whole, the system of the variant construction has discrete and separate properties that serve to facilitate complex local interactions for any selected group.

From a measurement viewpoint, the parallel parameters of the variant measurements enable them to exist in different interactive models simultaneously. This set of simultaneous properties exhibits significant differences between the original wave functions and the variant construction.

7.2 Weakness of BCP

The main weakness of the BCP lies deep in the very logic on which it is founded. In his approach to QM, Bohr applied then extant classical principles of logic using static YES/NO approaches to dynamic particle and wave measurements. However, the complex nature of QM phenomena means that such a classical logic framework cannot fully support this quaternion organization or fully model the dynamic systems involved. This is the main reason why the BCP requires the application of exclusive properties to pairs of opposites.

The variant construction provides quaternion measurement groups. This property naturally supports QM-like structures. Useful configurations can be chosen for further development.

The main experimental evidence following Bohr in rejecting particle models are sets of wave interference distributions generated in long duration and very low intensity single photon experiments. These experiments show intrinsic wave interference patterns under many environments. Understandably, such data have long been held to be strongly indicitive of wave properties within even single quanta. Consequently, it has been deemed natural and necessary to apply wave descriptions and analysis tools in the search for QM solutions.

However, evidence residing within the main visual distributions of this chapter, serves to show that statistical distributions under a conditional probability environment naturally link to intrinsic wave properties in the majority of situations. Nearly all interesting distributions show obvious wave properties. Notably, such intrinsic wave distributions may be sufficient to allow a satisfactory alternative explanation of experimental results generated in long duration and very low intensity single photon experiments.

7.3 The BCP for a special subset of QM

We may deduce that there is (only) a special subset of QM for which the BCP is satisfied. Under the variant construction there are six distinct logical configurations that can be used to support 0-1 vectors. Of these six, Bohr's approach is suitable for only the two schemes of pure static YES or NO. Meanwhile, the other four variant, invariant and mixed configurations lie outside the BCP framework. From this viewpoint, Bohr offers insight into important special circumstances of QM rather than provides an all embracing general solution.

Bohr's QM construction is complete and useful in many theoretical and practical environments for static and static-like systems. However, the variant construction provides a more powerful and general mechanism to handle different dynamic systems with variant and invariant properties.

7.4 The EPR contribution on variant construction

From EPR proposed experiments and other theoretical considerations, Einstein demonstrated a depth of understanding of weakness inherent in the foundations of the QM approach. He clearly identified two operators with non-communication properties that failed to support simultaneous operations and recognized that this type of mechanism was still not explained in the Copenhagen interpretation.

Using the variant construction, EPR devices have the following correspondence:

$$\begin{cases} S_1 & \to \{u_\beta, v_\beta, \tilde{u}_\beta, \tilde{v}_\beta ...\}; \\ S_2 & \to \{u_\beta, v_\beta, \tilde{u}_\beta, \tilde{v}_\beta ...\}; \\ IM(S_1, S_2) & \to \{M(u_1), M(v_1), M(\tilde{u}_1), M(\tilde{v}_1) ...\}; \\ SM(S_1, S_2) & \to \{M(u_0), M(v_0), M(\tilde{u}_0), M(\tilde{v}_0) ...\}. \end{cases} \tag{18}$$

$$\begin{cases} \langle S_1, S_2, IM(S_1, S_2), SM(S_1, S_2) \rangle \to \\ \qquad \langle \{u_\beta, v_\beta, \tilde{u}_\beta, \tilde{v}_\beta ...\}, \{u_\beta, v_\beta, \tilde{u}_\beta, \tilde{v}_\beta ...\}, \\ \qquad \{M(u_1), M(v_1), M(\tilde{u}_1), M(\tilde{v}_1) ...\}, \\ \qquad \{M(u_0), M(v_0), M(\tilde{u}_0), M(\tilde{v}_0) ...\} \rangle \end{cases} \tag{19}$$

From this correspondence, many possible configurations of combinations and their subsets are available for future theoretical and experimental exploration.

Using the variant construction, rich configurations can be expressed. From such mapping, it can be seen to be nothing less than astounding that such meta constructions were identified by Einstein as far back as 1935.

7.5 Afshar's experiments on variant construction

Afshar's experiments apply anti-symmetry signals making the following correspondence:

$$\begin{cases} \psi_1 \to \{v_+\}; \\ \psi_2 \to \{v_1\}; \\ \sigma_1 \to \{P_H(v_1|J)\}; \\ \sigma_2 \to \{P_H(v_0|J)\}. \end{cases} \tag{20}$$

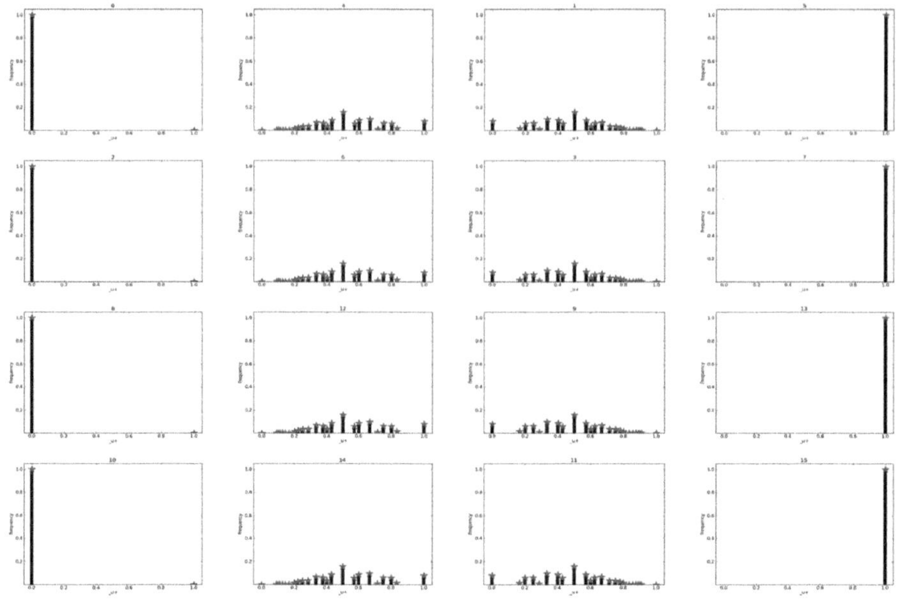

(a) Left

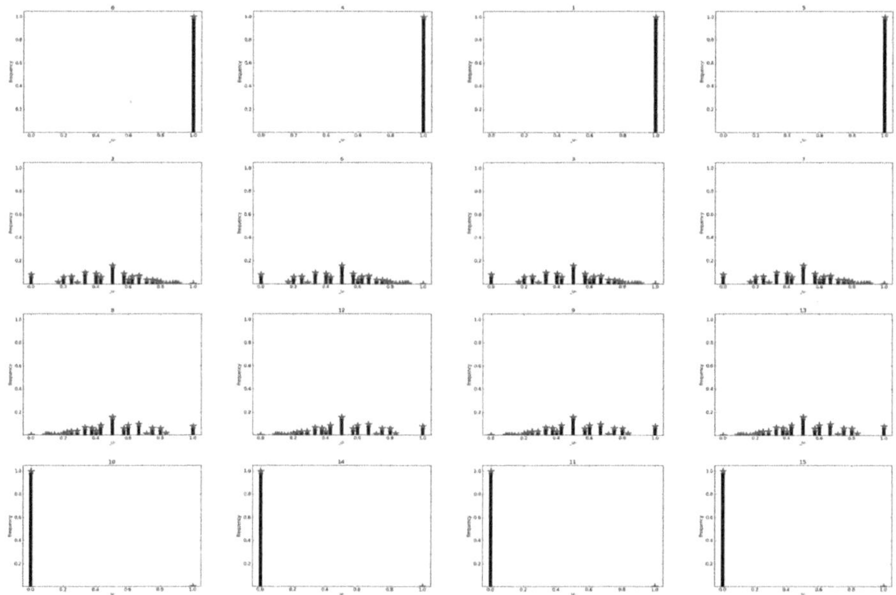

(b) Right

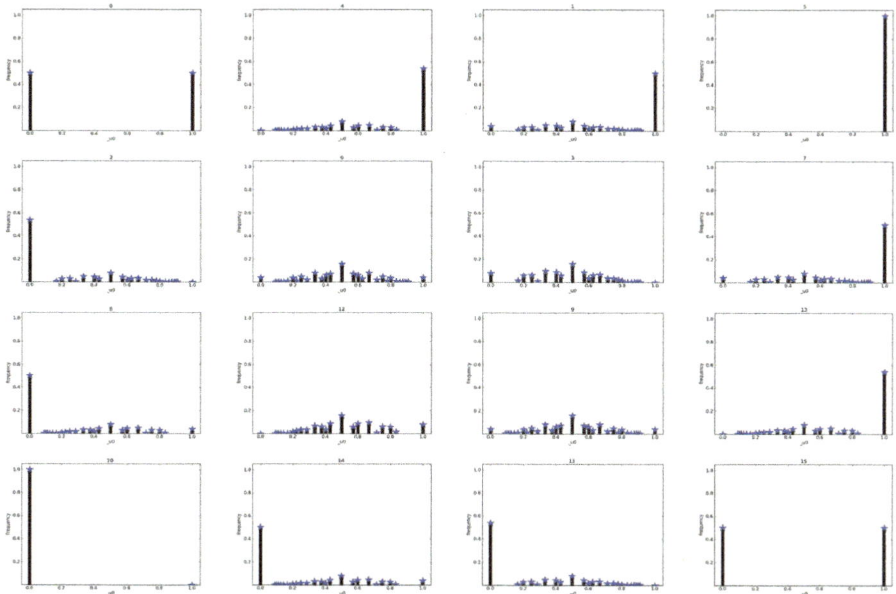

(c) D-P

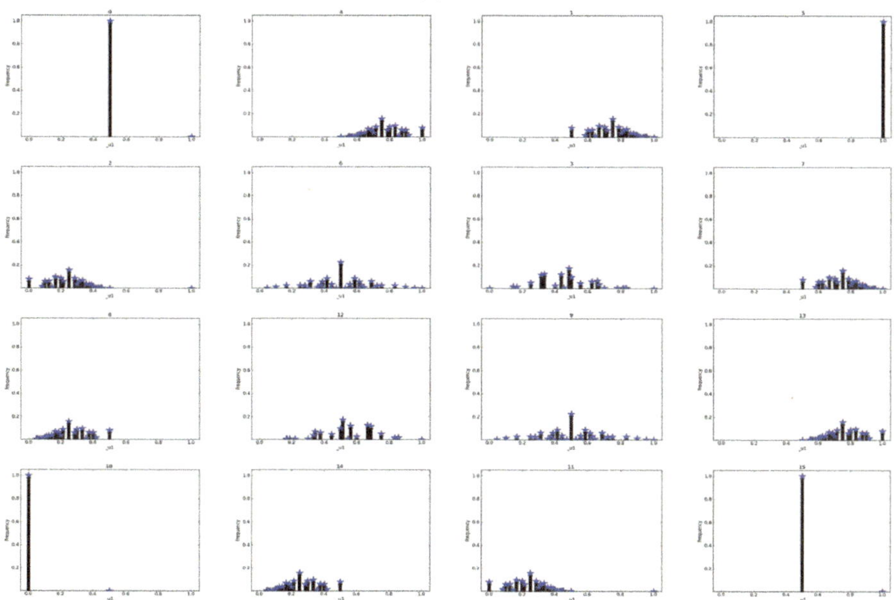

(d) D-W

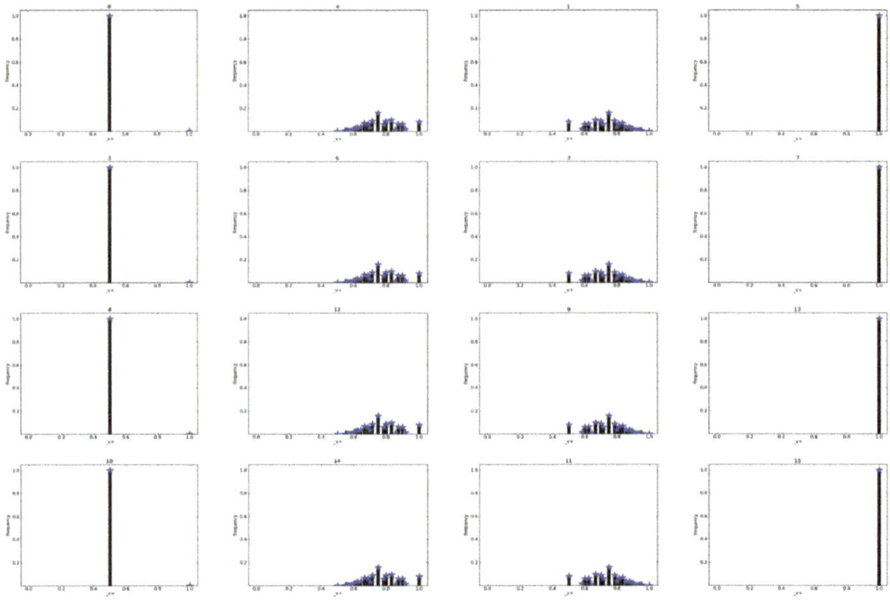

(e) Left

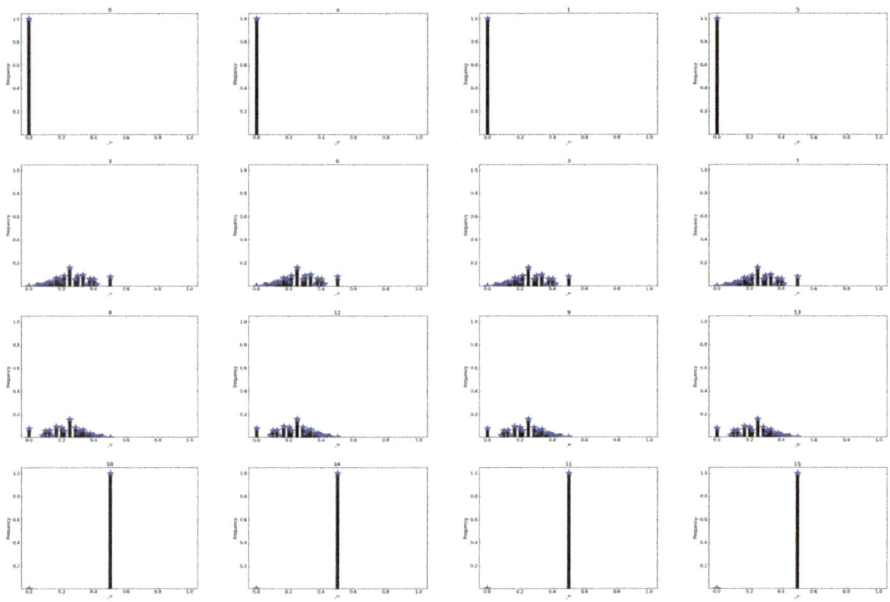

(f) Right

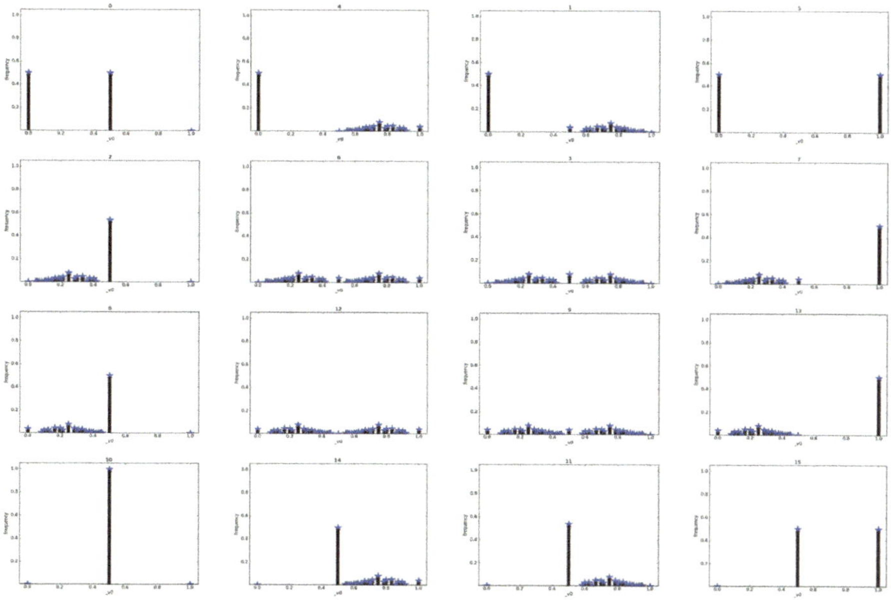

(g) D-P

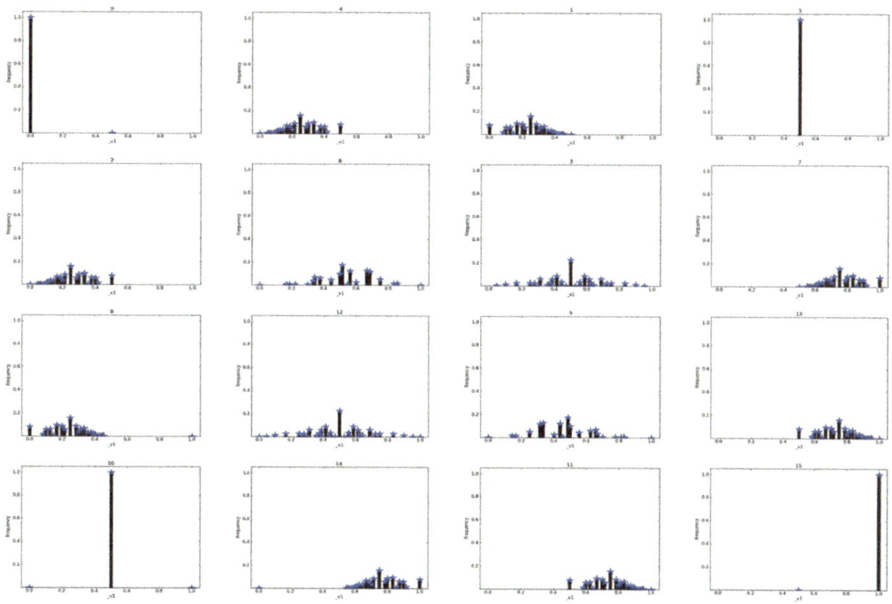

(h) D-W

Fig. 4. (a-h) Even number groups: $N = \{12\}, f \in B_2^4$ Eight Matrices of Global Matrix Representations. (a) Left; (b) Right; (c) D-P; (d)D-W in symmetry conditions; (e) Left; (f) Right; (g) D-P; (h)D-W in anti-symmetry conditions.

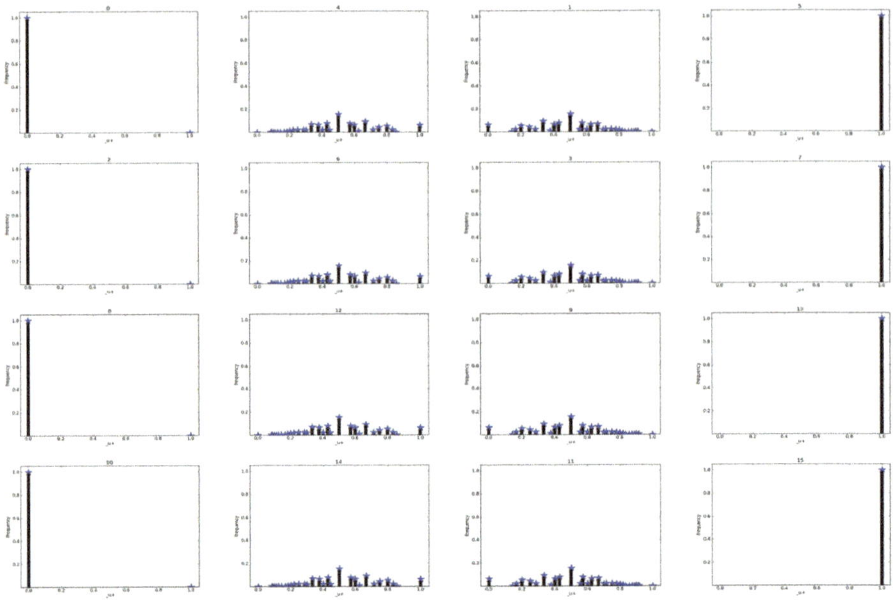

(a) Left

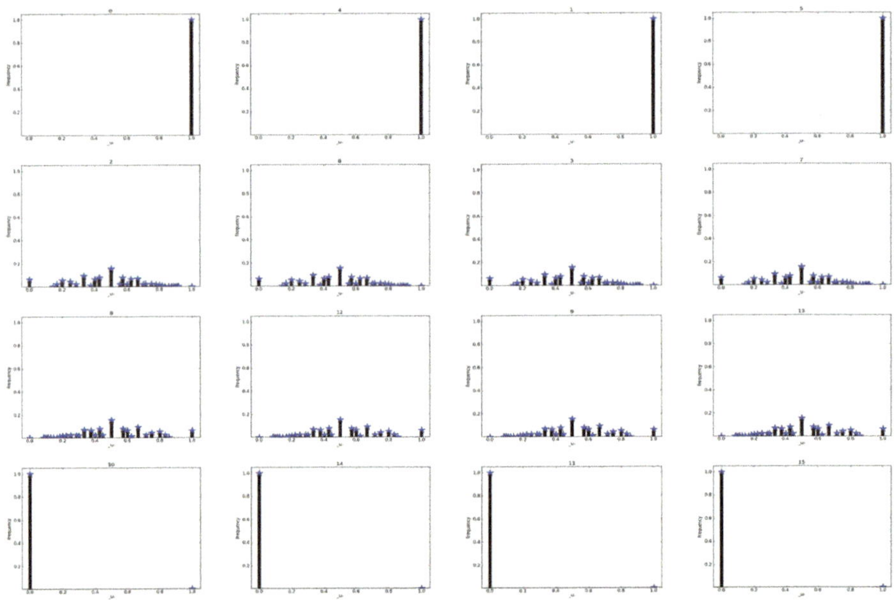

(b) Right

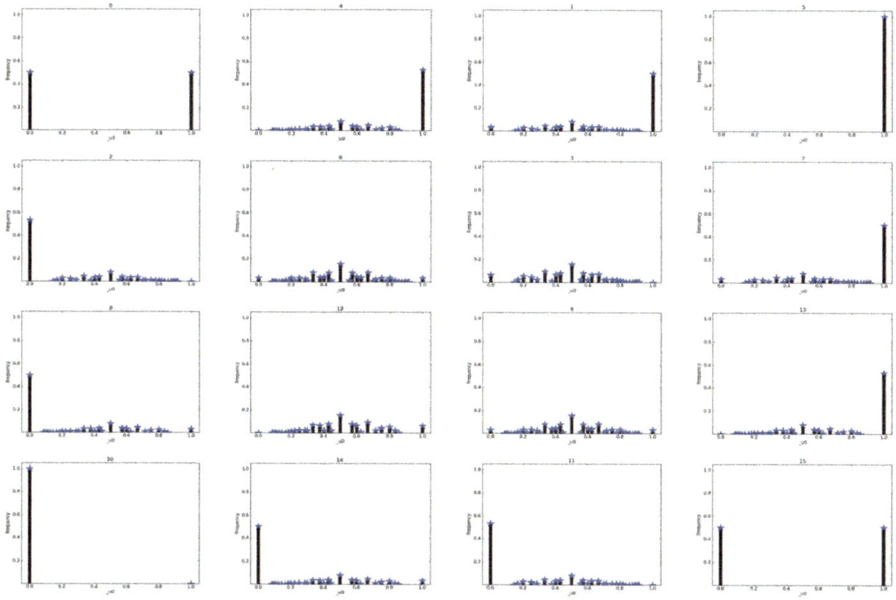

(c) D-P

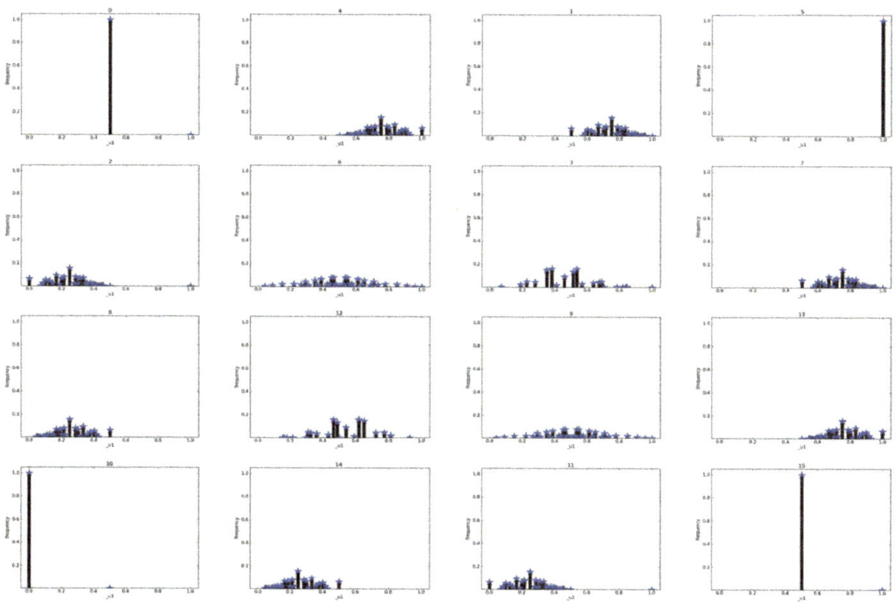

(d) D-W

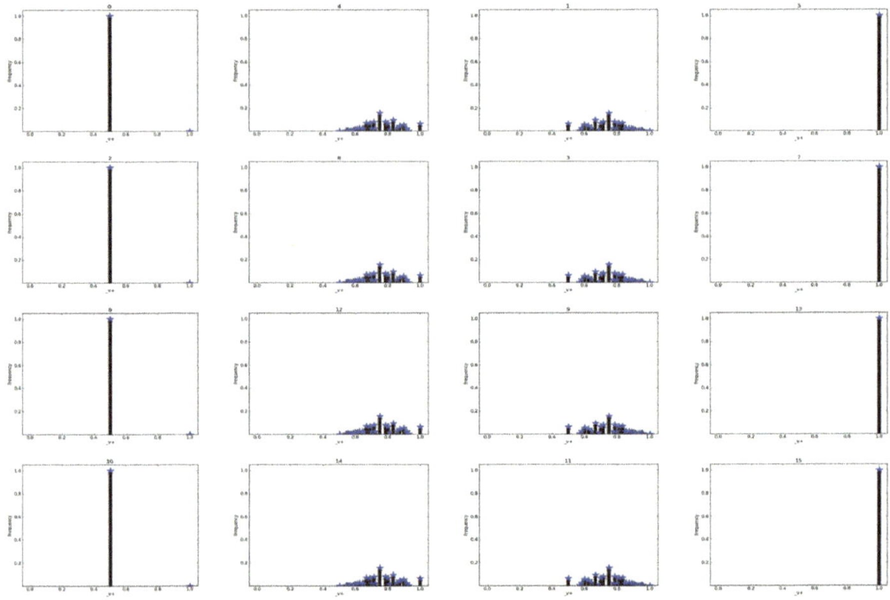

(e) Left

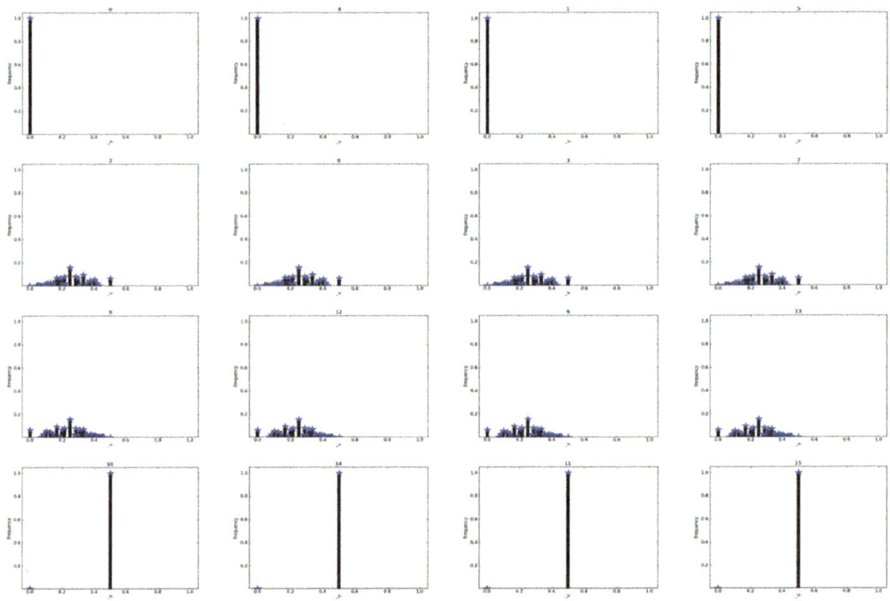

(f) Right

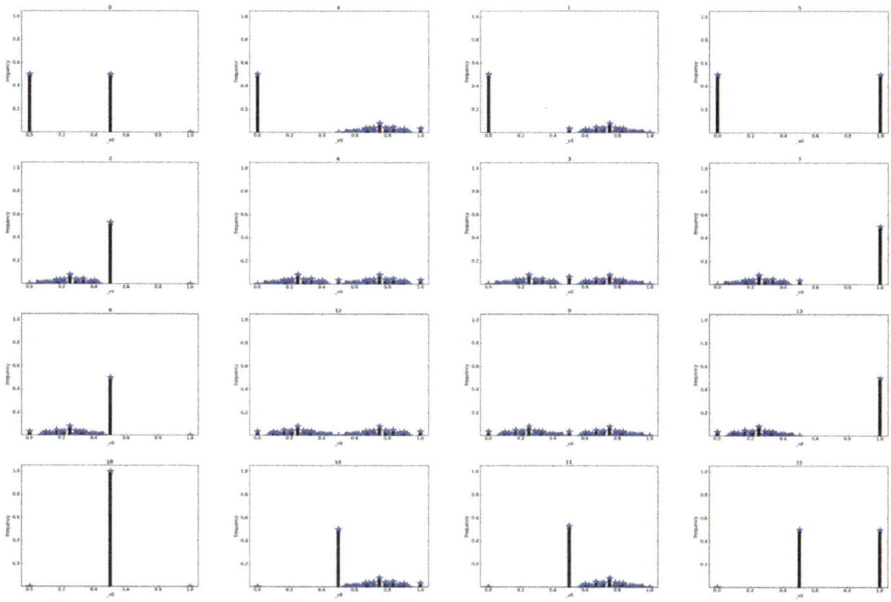

(g) D-P

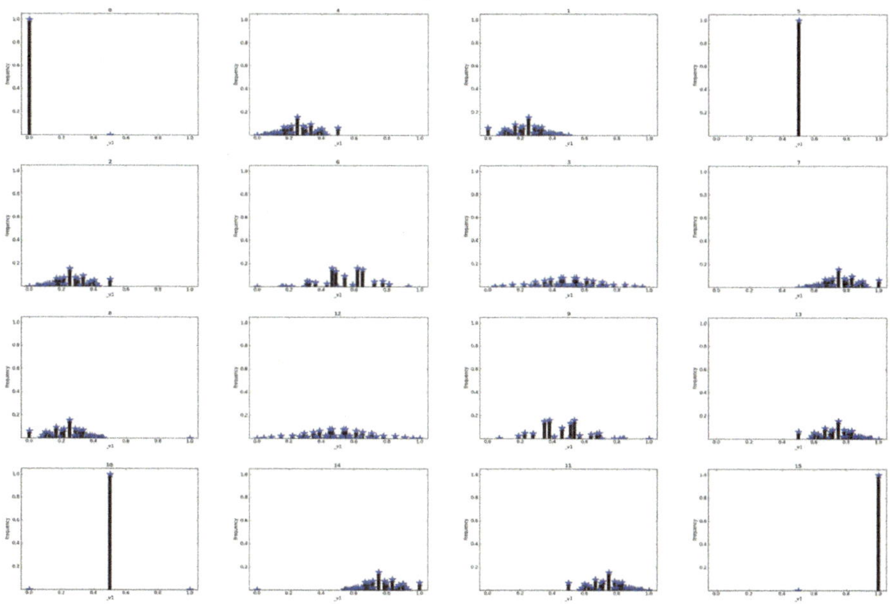

(h) D-W

Fig. 5. (a-h) Odd number groups: $N = \{13\}, f \in B_2^4$ Eight Matrices of Global Matrix Representations. (a) Left; (b) Right; (c) D-P; (d)D-W in symmetry conditions; (e) Left; (f) Right; (g) D-P; (h)D-W in anti-symmetry conditions.

Using quaternion structures,

$$\{ \langle \psi_1, \psi_2, \sigma_1, \sigma_2 \rangle \rightarrow \langle \{v_+\}, \{v_1\}, \{P_H(v_1|J)\}, \{P_H(v_0|J)\} \}\rangle. \tag{21}$$

All Afshar's experiments are a special case of the EPR model.

8. Main results

Presented as predictions and conjectures:

8.1 Predictions

Commensurate with the chapter of local interactive measurements, similar predictions can be described under conditional probability conditions:

Prediction 1: Left distributions have relationships showing polarized vertical behaviors with intrinsic wave properties on conditional environments.

Prediction 2: Right distributions have relationships showing polarized horizontal behaviors with intrinsic wave properties on conditional environments.

Prediction 3: D-P distributions have relationships showing classical particle statistical behaviors with intrinsic wave properties on conditional environments.

Prediction 4: D-W distributions have relationships showing wave interference statistical behaviors with strong wave properties on conditional environments.

Prediction 5: Afshar's experiments are a special case of the EPR model in real photon experimental environments.

Prediction 6: Distributions on conditional environments provide essential evidence to support a series of experimental results on quanta self-interference properties.

8.2 Conjectures

Presented in relation to milestones in the historical debate underpinning the foundations of QM:

Conjecture 1. Einstein may be declared the winner in the Bohr-Einstein debates on QM.

Conjecture 2. EPR construction is a super-powerful model to support different measurements and simulations of quantum behaviors.

Conjecture 3. The variant construction provides a logical measurement based foundation to support the simulation and visualization of quantum behaviors.

Conjecture 4. The next generation of fundamental development in QM will grow out of further theoretical and experimental exploration based on variant construction.

9. Conclusion

Long held views on the wave/particle enigma, especially those investigated through single photon experiments may be founded on a special case rather than a general explanation.

Further insight may be found working from conditional probability measurements to global matrix representation on the variant construction.

Applying conditional probability models on interactive measurements and relevant statistical processes, two groups of parameters $\{\tilde{u}_\beta, \tilde{v}_\beta\}$ describe left path, right path, D-P and D-W conditions with distinguishing symmetry and anti-symmetry properties. $\{P_H(\tilde{u}_\beta|J), P_H(\tilde{v}_\beta|J)\}$ provide eight groups of distributions under symmetry and anti-symmetry forms. In addition, $\{M(\tilde{u}_\beta), M(\tilde{v}_\beta)\}$ provide eight matrices to illustrate global behaviors under conditional environments.

The complexity of n-variable function space has a size of 2^{2^n} and exhaustive vector space has 2^N. Overall simulation complexity is determined by $O(2^{2^n} \times 2^N)$ as ultra exponent productions. How to overcome the limitations imposed by such complexity and how best to compare and contrast such simulations with real world experimentation will be key issues in future work.

Six predictions and four conjectures are offered for testing by further theoretical and experimental work.

10. Acknowledgements

Thanks to Colin W. Campbell for help with the English edition, to The School of Software Engineering, Yunnan University and The Key Laboratory of Yunnan Software Engineering for financial supports to the Information Security research projects (2010EI02, 2010KS06) and sub-CDIO project.

11. References

Afshar, S. (2005). Violation of the principle of complememtarity, and its implications, *Proc. SPIE* 5866. 229-244.

Afshar, S. (2006). Violation of bohr's complememtarity: One slit or both?, *AIP Conf. Proc.* 810. 294-299.

Afshar, S., Flores, E., McDonald, K. & Knoesel, E. (2007). Paradox in wave particle duality, *Found. Phys.* 37. 295-305.

Ash, R. B. & Doléans-Dade, C. A. (2000). *Probability & Measure Theory*, Elsevier.

Aspect, A. (2007). To be or not to be local, *Nature* 446. 866-867.

Aspect, A., Grangier, P. & Roger, G. (1982). Experimental realization of einstein -podolsky -rosen-bohm gedankenexperiment: A new violation of bell's inequalities, *Phys. Rev. Lett.* 49. 91-94.

Barrow, J. D., Davies, P. C. W. & Charles L. Harper, J. E. (2004). *SCIENCE AND ULTIMATE REALITY: Quantum Theory, Cosmology and Complexity*, Cambridge University Press.

Bell, J. S. (1964). On the einstein-podolsky-rosen paradox, *Physics* 1. 195-200.

Bohr, N. (1935). Can quantum-mechanical description of physical reality be considered complete?, *Physical Review* 48. 696-702.

Bohr, N. (1949). *Discussion with Einstein on Epistemological Problems in Atomic Physics*, Evanston. 200-241.

Bohr, N. (1958). *Atomic Physics and Human Knowledge*, Wiley.

Bolles, E. (2004). *Einstein Defiant*, Joseph Henry Press.

Born, M. (1971). *The Born Einstein Letters*, Walker and Company.

Clauser, J., Horne, N., Shimony, A. & Holt, R. (1969). Proposed experiment to test local hidden-variable theories, *PRL* 23. 880-884.

Durret, R. (2005). *Probability:Theory and Examples*, Thomson.

Einstein, A., Podolsky, B. & Rosen, N. (1935). Can quantum-mechanical description of physical reality be considered complete?, *Physical Review* 47. 770-780.

Feynman, R., Leighton, R. & Sands, M. (1965,1989). *The Feynman Lectures on Physics*, Vol. 3, Addison-Wesley, Reading, Mass.

Hawkingand, S. & Mlodinow, L. (2010). *The Grand Design*, Bantam Books.

Heisenberg, W. (1930). *The Physical Principles of Quantum Theory*, Uni. Chicago Press.

Jammer, M. (1974). *The Philosophy of Quantum Mechanics*, Wiley-Interscience Publication.

Penrose, R. (2004). *The Road to Reality*, Vintage Books, London.

Valiant, L. (1975). Parallelism in comparison problems, *SIAM J. Comput.* 4(3). 348-355.

Zheng, J. (2011). Synchronous properties in quantum interferences, *Journal of Computations & Modelling, International Scientific Press* 1(1). 73-90.
URL: *http://www.scienpress.com/upload/JCM/Vol%201_1_6.pdf*

Zheng, J. & Zheng, C. (2010). A framework to express variant and invariant functional spaces for binary logic, *Frontiers of Electrical and Electronic Engineering in China, Higher Education Press and Springer* 5(2): 163–172.
URL: *http://www.springerlink.com/content/91474403127n446u/*

Zheng, J. & Zheng, C. (2011a). Variant measures and visualized statistical distributions, *Acta Photonica Sinica, Science Press* 40(9). 1397-1404.
URL: *http://www. photon.ac.cn/CN/article/downloadArticleFile.do?attachType=PDF&id =15668*

Zheng, J. & Zheng, C. (2011b). Variant simulation system using quaternion structures, *Journal of Modern Optics, Taylor & Francis Group* . iFirst 1-9.
URL: *http://www.tandfonline.com/doi/abs/10.1080/09500340.2011.636152*

Zheng, J., Zheng, C. & Kunii, T. (2011). A framework of variant-logic construction for cellular automata, *Cellular Automata - Innovative Modelling for Science and Engineering edited Dr. Alejandro Salcido, InTech Press*. 325-352, ISBN 978-953-307-172-5.
URL: *http://www.intechopen.com/articles/show/title/a-framework-of-variant-logic--construction-for-cellular-automata*

17

From Local Interactive Measurements to Global Matrix Representations on Variant Construction – A Particle Model of Quantum Interactions for Double Path Experiments

Jeffrey Zheng[1], Christian Zheng[2] and T.L. Kunii[3]
[1]*Yunnan University*
[2]*University of Melbourne*
[3]*University of Tokyo*
[1]*P.R. China*
[2]*Australia*
[3]*Japan*

1. Introduction

1.1 Wave and particle duality in quantum measurements

Right from the introduction of Plank's modern quantum concept, measurement effects have played a central role in both theoretical and experimental considerations [Jammer (1974)]. Einstein (1916) photon effects favor a particle based explanation. de Broglie (1923) proposed wave and particle duality. Heisenberg proposed a matrix approach to handling complex operations based on spectra measurements. Schrödinger established a wave equation for quantum construction extending de Broglie's schemes. von Neumann (1932,1996)'s contribution placed quantum mechanics in Hilbert space to establish a solid mathematical foundation for modern quantum mechanics. Despite developments in the quantum approach spanning more than a century, fundamental measurement problems remain unsolved [Penrose (2004)]. All their lives, Bohr and Einstein engaged in many debates, discussions and arguments trying to reach a common understanding on wave and particle issues [Jammer (1974)]. The EPR (Einstein, Podolsky, Rosen) Paradox [Einstein et al. (1935)] is said to have given Bohr many sleepless nights [Bohr (1935; 1949)].

1.2 Criteria conditions and modern experiments

Quantum measurement puzzles have been explored by [Feynman (1965); Feynman et al. (1965,1989)]. From the 1940s, Feynman emphasized that: "The entire mystery of quantum mechanics is in the double-slit experiment." This experiment establishes an interactive model that can directly illustrate both classical and quantum interactive results. Under single and double slit conditions, dual visual distributions are shown in particle and wave statistical distributions. Both particle probability and wave interactive interference patterns are observed [Barnett (2009); Hawking & Mlodinow (2010); Healey et al. (1998)].

1.3 Modern experiments

Mach-Zehnder interferometers and Stern-Gerlach spin-devices play a key role in Quantum measurement development [Barnett (2009); Barrow et al. (2004); Hawking & Mlodinow (2010); Jammer (1974)]. Wave particle-duality has been demonstrated in larger particles [Arndt et al. (1999)] and advanced optical fibers, communication, computer software, photonics, and integrated technologies have been applied to different quantum media [Barrow et al. (2004); Grangier et al. (1986)].

1.3.1 Bell approaches

In the 1960s, Bell played an important role in exploring the foundations of the quantum approach [Bell (1964)]. Based on the EPR paradox, he proposed inequations for measurable experiments to distinguish between Bohr's Principle of Complementarity and Einstein's EPR paradox under a local realism framework [Aspect et al. (1982); Bell (2004)].

1.3.2 Advanced experiments

By the 1970s, work piloted by [Clauser et al. (1969)], Aspect et al. (1982) was using an experimental approach to test Bell Inequalities and to clearly show a significant gap between Bell Inequalities and real quantum reality.

After 40 years of development, many accurate experiments [Lindner et al. (2005); Zeh (1970); Zeilinger et al. (2005)] have been performed successfully worldwide using Laser, NMRI, large molecular, quantum coding and quantum communication approaches [Afshar et al. (2007); Barrow et al. (2004); Fox (2006); Merali (2007); Schleich et al. (2007)]. Following the application of advanced technologies and simulation methodologies, detailed single and multiple photon detection technologies have been further developed.

1.3.3 Weakness

However it does not matter how successful any single experiment or indeed many experiments might be, those results cannot simply replace the idea experiment of [Feynman et al. (1965,1989); Hawking & Mlodinow (2010); Penrose (2004)]. From a theoretical viewpoint, modern experiments involving Bell Inequations are excellent in illustrating the fundamental differences between a local realism and quantum reality. Since both theoretical and experimental activities focused on supporting or disproving Bell Inequalities cannot on their own provide a full explanation, further investigations are essential to provide a sound foundation on which a full understanding of quantum issues can be constructed.

1.4 From local interactive measurements to global matrix representations

In this chapter, a double path model has been established using the Mach-Zehnder interferometer. Different approaches to quantum measurements taken by Einstein, Stern-Gerlach, CHSH and Aspect are investigated to form quaternion structures. Using multiple-variable logic functions and variant principles, logic functions can be transferred into variant logic expressions as variant measures. Under such conditions, a variant simulation and representation model is proposed.

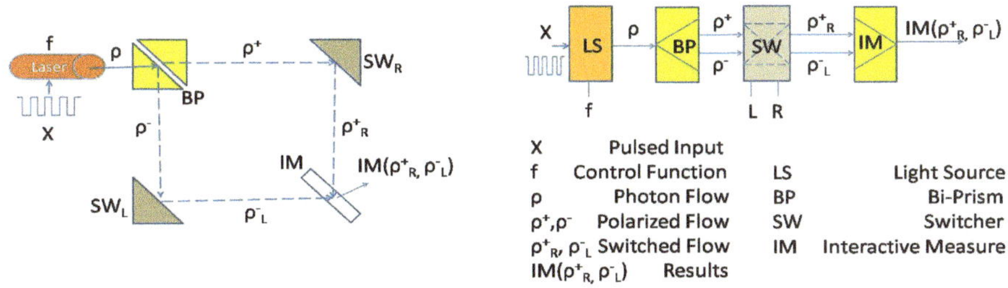

(a) (b)

Fig. 1. (a-b) Double Path Model (a) Mach-Zehnder Double Path Model (b) Description Model

A given logic function f, can be represented as two meta logic functions f_+ and f_- to simulate single and double path conditions. N bits of input vectors are exhausted by 2^N states for measured data, recursive data are organized into eight histograms. Results are determined by symmetry/anti-symmetry properties in histograms. All 2^{2^n} functions are applied to generate a set of histograms. Eight sets of histograms are represented as eight matrices in a selected C code configuration. Under this construction, it is possible to visualize different combinations from symmetry and anti-symmetry categories.

From these results, both additive probability properties in particle condition and wave interference properties with non-addition behaviors are observed. Both types of result are obtained consistently from this model under synchronous/asynchronous conditions. From a simulation viewpoint, this system satisfies all of Feynman's criteria conditions for double slit experiments.

2. Double path model and measurements of quantum interaction

2.1 Mach-Zehnder interferometer model

The Mach-Zehnder interferometer is the most popular device used to support a Young double slit experiment.

In Fig 1(a) a double path interferometer is shown. An input signal X under control function f causes Laser LS to emit the output signal ρ under BP (Bi-polarized filter) operation. The output is in the form of a pair of signals: ρ^+ and ρ^-. Both signals are processed by SW output ρ_L^+ and ρ_R^-, and then IM to generate output signals $IM(\rho_L^+, \rho_R^-)$. In Fig 1(b), a representation model has been described with the same signals being used.

2.1.1 Other devices

A Stern-Gerlach spin measurement device provides equivalent information for double path experiment [Jacques et al. (2008); Jammer (1974)]. This device divides composed signals into vertical $\perp$ and horizontal $\|$ components, in BP part $\rho \to \{\rho^\perp, \rho^\|\}$, through controls and IM output $IM(\rho_L^\perp, \rho_R^\|)$.

2.2 Emission and absorption measurements of quantum interaction

2.2.1 Einstein measurements

Einstein (1916) established the first model to describe atomic interaction with radiation. For two-state systems, Einstein's model is as follows. Let a system have two energy states: the ground state E_1 and the excited state E_2. Let N_1 and N_2 be the average numbers of atoms in the ground and excited states respectively. The number of states are changed from emission state $E_2 \rightarrow E_1$ with a rate $\frac{dN_{21}}{dt}$, and at any point in time, the number of ground states are determined by absorbed energies from $E_1 \rightarrow E_2$ with a rate $\frac{dN_{12}}{dt}$ respectively. For convenience of description, let N_{12} be the number of atoms from E_1 to E_2 and N_{21} be the numbers from E_2 to E_1. In Einstein's model, a measurement quaternion is $\langle N_1, N_2, N_{12}, N_{21} \rangle$.

2.2.2 Spin measurements

Uhlenback and Goudsmit proposed spin using devices devised by Stern-Gerlach [Cohn (1990); Jammer (1974)]. Spin can be represented by $| \uparrow\rangle, | \downarrow\rangle$ in a two-state system. A quaternion $\langle\langle\uparrow | \uparrow\rangle, \langle\uparrow | \downarrow\rangle, \langle\downarrow | \uparrow\rangle, \langle\downarrow | \downarrow\rangle\rangle$ can be established for spin interactions.

CHSH proposed spin measures testing Bell Inequalities [Aspect (2002); Clauser et al. (1969)]. They applied $\perp \rightarrow +$ and $\| \rightarrow -$ to establish a measurement quaternion:
$\langle N_{++}, N_{+-}, N_{-+}, N_{--} \rangle$. CHSH parameters are in the Stern-Gerlach scheme.

2.2.3 Aspect's measurements

Advanced experimental testing of Bell Inequalities for quantum measures were performed by [Aspect (2002); Aspect et al. (1982)]. In this set of experiments, active properties are measured via four measurements: transmission rate N_t, reflection rate N_r, correspondent rate N_c and also the total number N_ω in ω-time period. This set of measurements is a quaternion $\langle N_t, N_r, N_c, N_\omega \rangle$. Among these, N_c is a new data type not found in the Einstein and Stern-Gerlach schemes. As a matched pair of signals, this parameter indicates either single or double path issues. This parameter could be an extension of synchronous/asynchronous time-measurement.

3. Variant simulation and representation system

A comprehensive process of measurement from local interactions through to global matrix representations is described. It is hoped that this may offer a convenient path to assist theorists and experimenters seeking to devise experiments to further explore such natural mysteries through the application of sound principles of logic and measurement.

Using the variant principle described in the next subsections, a N bit 0-1 vector X and a given logic function f, all N bit vectors are exhausted, variant measures generate two groups of histograms. The variant simulation and representation system is shown in Fig 2 (a-b). The detailed principles and methods are described in Sections 3.2-3.7 respectively.

3.1 Simulation and representation model

The full measurement and representation architecture as shown in Figure 2(a) is composed of four components: Meta Measurements MM, Local Interactive Measurements LIM, Statistical

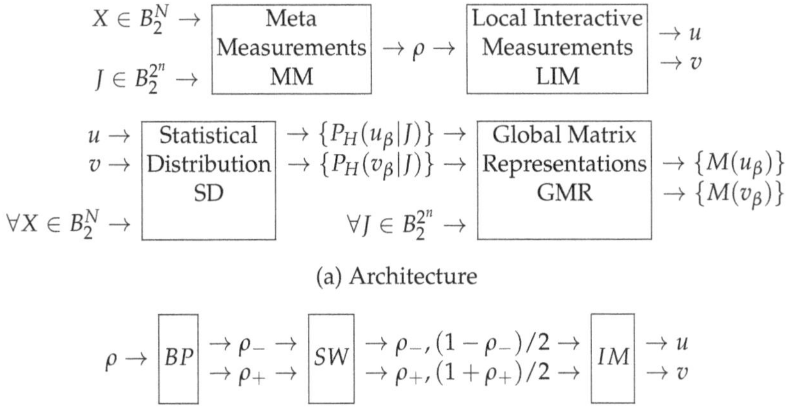

(b) LIM Component

Fig. 2. (a-b) Variant Simulation and Representation System; (a) System Architecture; (b) Local Interactive Measurement LIM Component

Distributions SD, Global Matrix Representations GMR respectively. The key part of the system: the LIM component is shown in Fig 2(b).

3.1.1 Meta Measurements

The Meta Measurement (MM) component uses N bit 0-1 vector X and a given function $J \in B_2^{2^n}$, MM transfers N bit 0-1 vector under $J(X)$ to generate four Meta-measures, under a given Probability scheme, four probability measurements are generated to output as a quaternion signal ρ.

3.1.2 Local Interactive Measurements

The Local Interactive Measurement (LIM) component is the key location for local interactions as shown in Figure 2(b) to transfer quaternion signal ρ under symmetry / anti-symmetry and synchronous / asynchronous conditions, in relation to four combination of time effects as (Left, Right, Double Particle, Double Wave) respectively. Two types of additive operations are identified. Each $\{u, v\}$ signal is composed of four distinct signals.

3.1.3 Statistical Distributions

The Statistical Distribution (SD) component performs statistical activities on corresponding signals. It is necessary to exhaust all possible vectors of X with a total of 2^N vectors. Under this construction, each sub-signal of $\{u, v\}$ forms a special histogram with a one dimensional spectrum to indicate the distribution under function J. A total of eight histograms are generated in the probability conditions.

3.1.4 Global Matrix Representations

The Global Matrix Representation (GMR) component uses each statistical distribution of the relevant probability histogram as an element of a matrix composed of a total of 2^{2^n} elements for all possible functions $\{J\}$. In this configuration, C code schemes are applied to form a $2^{2^{n-1}} \times 2^{2^{n-1}}$ matrix to show the selected distribution group.

Unlike the other coding schemes (SL, W, F, ...), only C code schemes provide a regular configuration to clearly differentiate the Left path as exhibiting horizontal actions and the Right path as exhibiting vertical actions . Such clearly polarized outcomes may have the potential to help in the understanding of interactive mechanism(s) between double path for particles and double path for waves properties.

3.2 Variant principle

The variant principle is based on n-variable logic functions [Zheng (2011); Zheng & Zheng (2010; 2011a;b); Zheng et al. (2011)].

3.2.1 Two sets of states

For any n-variables $x = x_{n-1}...x_i...x_0$, $0 \leq i < n, x_i \in \{0,1\} = B_2$ let a position j be the selected bit $0 \leq j < n$, x_j be the selected variable. Let output variable y and n-variable function $f, y = f(x)$, $y \in B_2, x \in B_2^n$. For all states of x, a set $S(n)$ composed of the 2^n states can be divided into two sets: $S_0(n)$ and $S_1(n)$.

$$\begin{cases} S_0(n) = \{x | x_j = 0, \forall x \in B_2^n\} \\ S_1(n) = \{x | x_j = 1, \forall x \in B_2^n\} \\ S(n) = \{S_0(n), S_1(n)\} \end{cases} \tag{1}$$

3.2.2 Four meta functions

For a given logic function f, input and output pair relationships define four meta logic functions $\{f_\perp, f_+, f_-, f_\top\}$.

$$\begin{cases} f_\perp(x) = \{f(x) | x \in S_0(n), y = 0\} \\ f_+(x) = \{f(x) | x \in S_0(n), y = 1\} \\ f_-(x) = \{f(x) | x \in S_1(n), y = 0\} \\ f_\top(x) = \{f(x) | x \in S_1(n), y = 1\} \end{cases} \tag{2}$$

3.2.3 Two polarized functions

Considering two standard logic canonical expressions: AND-OR form is selected from $\{f_+(x), f_\top(x)\}$ as $y = 1$ items, and OR-AND form is selected from $\{f_-(x), f_\perp(x)\}$ as $y = 0$ items. Considering $\{f_\top(x), f_\perp(x)\}, x_j = y$ items, they are invariant themselves.

To select $\{f_+(x), f_-(x)\}, x_j \neq y$ in forming variant logic expression. Let $f(x) = \langle f_+ | x | f_- \rangle$ be a variant logic expression. Any logic function can be expressed as a variant logic form. In $\langle f_+ | x | f_- \rangle$ structure, f_+ selected 1 items in $S_0(n)$ as same as the AND-OR standard expression, and f_- selecting relevant parts the same as OR-AND expression 0 items in $S_1(n)$.

3.2.4 $n = 2$ representation

For a convenient understanding of the variant representation, 2-variable logic structures are illustrated for its 16 functions as follows.

f No.	$f \in S(n)$	3 11	2 10	1 01	0 00	$f_+ \in S_0(n)$	3^0 11^0	2^1 10^1	1^0 01^0	0^1 00^1	$f_- \in S_1(n)$
0	$\{\varnothing\}$	0	0	0	0	$\langle\varnothing\|$	1	0	1	0	$\|3,1\rangle$
1	$\{0\}$	0	0	0	1	$\langle 0\|$	1	0	1	1	$\|3,1\rangle$
2	$\{\varnothing\}$	0	0	1	0	$\langle\varnothing\|$	1	0	0	0	$\|3\rangle$
3	$\{1,0\}$	0	0	1	1	$\langle 0\|$	1	0	0	1	$\|3\rangle$
4	$\{2\}$	0	1	0	0	$\langle 2\|$	1	1	1	0	$\|3,1\rangle$
5	$\{2,0\}$	0	1	0	1	$\langle 2,0\|$	1	1	1	1	$\|3,1\rangle$
6	$\{2,1\}$	0	1	1	0	$\langle 2\|$	1	1	0	0	$\|3\rangle$
7	$\{2,1,0\}$	0	1	1	1	$\langle 2,0\|$	1	1	0	1	$\|3\rangle$
8	$\{3\}$	1	0	0	0	$\langle\varnothing\|$	0	0	1	0	$\|1\rangle$
9	$\{3,0\}$	1	0	0	1	$\langle 0\|$	0	0	1	1	$\|1\rangle$
10	$\{3,1\}$	1	0	1	0	$\langle\varnothing\|$	0	0	0	0	$\|\varnothing\rangle$
11	$\{3,1,0\}$	1	0	1	1	$\langle 0\|$	0	0	0	1	$\|\varnothing\rangle$
12	$\{3,2\}$	1	1	0	0	$\langle 2\|$	0	1	1	0	$\|1\rangle$
13	$\{3,2,0\}$	1	1	0	1	$\langle 2,0\|$	0	1	1	1	$\|1\rangle$
14	$\{3,2,1\}$	1	1	1	0	$\langle 2\|$	0	1	0	0	$\|\varnothing\rangle$
15	$\{3,2,1,0\}$	1	1	1	1	$\langle 2,0\|$	0	1	0	1	$\|\varnothing\rangle$

(3)

Checking two functions $f = 3$ and $f = 6$ respectively.
$\{f = 3 := \langle 0|3\rangle, f_+ = 11 := \langle 0|\varnothing\rangle, f_- = 2 := \langle\varnothing|3\rangle\}$;
$\{f = 6 := \langle 2|3\rangle, f_+ = 14 := \langle 2|\varnothing\rangle, f_- = 2 := \langle\varnothing|3\rangle\}$.

3.3 Meta measures

Under variant construction, N bits of 0-1 vector X under a function J produce four Meta measures composed of a measure vector N

$$(X : J(X)) \to (N_\perp, N_+, N_-, N_\top), N = N_\perp + N_+ + N_- + N_\top$$

Using four Meta measures, relevant probability measurements can be formulated.
$\rho = (\rho_\perp, \rho_+, \rho_-, \rho_\top) = (N_\perp/N, N_+/N, N_-/N, N_\top/N), 0 \le \rho_\perp, \rho_+, \rho_-, \rho_\top \le 1$.

From a methodological viewpoint, this set of probability parameters belongs to *multivariate probability measurements*.

3.3.1 Variant measure functions

Let Δ be the variant measure function

$$\begin{aligned} \Delta &= \langle \Delta_\perp, \Delta_+, \Delta_-, \Delta_\top \rangle \\ \Delta J(x) &= \langle \Delta_\perp J(x), \Delta_+ J(x), \Delta_- J(x), \Delta_\top J(x) \rangle \\ \Delta_\alpha J(x) &= \begin{cases} 1, & J(x) \in J_\alpha(x), \alpha \in \{\perp, +, -, \top\} \\ 0, & \text{others} \end{cases} \end{aligned} \tag{4}$$

For any given n-variable state there is one position in $\Delta J(x)$ to be 1 and other 3 positions are 0.

3.3.2 Variant measures on vector

For any N bit 0-1 vector $X, X = X_{N-1}...X_j...X_0, 0 \le j < N, X_j \in B_2, X \in B_2^N$ under n-variable function J, n bit 0-1 output vector Y, $Y = J(X) = \langle J_+|X|J_-\rangle$, $Y = Y_{N-1}...Y_j...Y_0, 0 \le j <$

$N, Y_j \in B_2, Y \in B_2^N$. For the j-th position $x^j = [...X_j...] \in B_2^n$ to form $Y_j = J(x^j) = \langle J_+|x^j|J_-\rangle$. Let N bit positions be cyclic linked. Variant measures of $J(X)$ can be decomposed

$$\Delta(X:Y) = \Delta J(X) = \sum_{j=0}^{N-1} \Delta J(x^j) = \langle N_\perp, N_+, N_-, N_\top \rangle \quad (5)$$

as a quaternion $\langle N_\perp, N_+, N_-, N_\top \rangle$, $N = N_\perp + N_+ + N_- + N_\top$.

3.3.3 Example

E.g. $N = 12$, given $J, Y = J(X)$.

$$\begin{aligned} X &= 1\ 0\ 1\ 1\ 1\ 0\ 1\ 1\ 1\ 0\ 0\ 1 \\ Y &= 0\ 0\ 1\ 0\ 1\ 0\ 1\ 0\ 1\ 1\ 0\ 0 \\ \Delta(X:Y) &= -\ \perp\ \top\ -\ \top\ \perp\ \top\ -\ \top\ +\ \perp\ - \end{aligned}$$

$\Delta J(X) = \langle N_\perp, N_+, N_-, N_\top \rangle = \langle 3, 1, 4, 4 \rangle, N = 12$.

Input and output pairs are 0-1 variables for only four combinations. For any given function J, the quantitative relationship of $\{\perp, +, -, \top\}$ is directly derived from the input/output sequences. Four meta measures are determined.

3.4 Four meta measurements

Using variant quaternion, local measurements of probability signals are calculated as four meta measurements by following the given equations. For any N bit 0-1 vector X, function J, under Δ measurement: $\Delta J(X) = \langle N_\perp, N_+, N_-, N_\top \rangle$, $N = N_\perp + N_+ + N_- + N_\top$

Signal ρ is defined by

$$\begin{cases} \rho = \frac{\Delta J(X)}{N} = (\rho_\perp, \rho_+, \rho_-, \rho_\top) \\ \rho_\alpha = \frac{N_\alpha}{N}, \alpha \in \{\perp, +, -, \top\} \end{cases} \quad (6)$$

The four meta measurements are core components in the *multivariate probability framework*.

3.5 Local Interactive Measurements

Local Interactive Measurements (LIM) are divided into three stages: BP, SW and SM respectively. The BP stage selects $\{\rho_-, \rho_+\}$ as sub-signals. The SW component extends two signals into four signals with different symmetric properties; The SM component merges different signals to form two sets of eight signals.

Using $\{\rho_+, \rho_-\}$, a pair of signals $\{u, v\}$ are formulated:

$$\begin{cases} u = (u_+, u_-, u_0, u_1) = \{u_\beta\} \\ v = (v_+, v_-, v_0, v_1) = \{v_\beta\} \\ \quad \beta \in \{+, -, 0, 1\} \end{cases} \quad (7)$$

$$\begin{cases} u_+ = \rho_+ \\ u_- = \rho_- \\ u_0 = u_+ \oplus u_- \\ u_1 = u_+ + u_- \\ v_+ = \frac{1+\rho_+}{2} \\ v_- = \frac{1-\rho_-}{2} \\ v_0 = v_+ \oplus v_- \\ v_1 = v_+ + v_- - 0.5 \end{cases} \tag{8}$$

where $0 \leq u_\beta, v_\beta \leq 1, \beta \in \{+,-,0,1\}, \oplus$: Asynchronous addition, $+$: Synchronous addition.

3.6 Statistical distributions

The SD component provides a statistical means to accumulate all possible vectors of N bits for a selected signal and generate a histogram. Eight signals correspond to eight histograms respectively. Among these, four histograms exhibit properties of symmetry and another four histograms exhibit properties of anti-symmetry.

3.6.1 Statistical histograms

For a function J, all measurement signals are collected and the relevant histogram represents a complete statistical distribution.

Using u and v signals, each u_β or v_β determines a fixed position in the relevant histogram to make vector X on a position. After completing 2^N data sequences, eight symmetry/anti-symmetry histograms of $\{H(u_\beta|J)\}, \{H(v_\beta|J)\}$ are generated.

For a function $J, \beta \in \{+,-,0,1\}$

$$\begin{cases} H(u_\beta|J) = \sum_{\forall X \in B_2^N} H(u_\beta|J(X)) \\ H(v_\beta|J) = \sum_{\forall X \in B_2^N} H(v_\beta|J(X)), J \in B_2^{2^n} \end{cases} \tag{9}$$

3.6.2 Probability histograms

Let $|H(..)|$ denote the total number in the histogram $H(..)$, a normalized Probability histogram $(P_H(..))$ can be expressed as

$$\begin{cases} P_H(u_\beta|J) = \frac{H(u_\beta|J)}{|H(u_\beta|J)|} \\ P_H(v_\beta|J) = \frac{H(v_\beta|J)}{|H(v_\beta|J)|}, J \in B_2^{2^n} \end{cases} \tag{10}$$

Here, all histograms are restricted in $[0,1]^2$ areas respectively.

Distributions are dependant on the data set as a whole and are not sensitive to varying under special sequences. Under this condition, when the data set has been exhaustively listed, then the same distributions are always linked to the given signal set.

The eight histogram distributions provide invariant spectrum to represent properties among different interactive conditions.

3.7 Global Matrix Representations

After local interactive measurements and statistical process are undertaken for a given function J, eight histograms are generated. The Global Matrix Representation GMR component performs its operations into two stages. In the first stage, exhausting all possible functions for $\forall J \in B_2^{2^n}$ to generate eight sets, each set contains 2^{2^n} elements and each element is a histogram. In the second stage, arranging all 2^{2^n} elements generated as a matrix by C code scheme. Here, we can see Left and Right path reactions polarized into Horizontal and Vertical relationships respectively.

3.7.1 Matrix and Its elements

For a given C scheme, let $C(J) = \langle J^1|J^0\rangle$, each element

$$\begin{cases} M_{\langle J^1|J^0\rangle}(u_\beta|J) = P_H(u_\beta|J) \\ M_{\langle J^1|J^0\rangle}(v_\beta|J) = P_H(v_\beta|J) \\ \quad J \in B_2^{2^n};\ J^1, J^0 \in B_2^{2^{n-1}} \end{cases} \tag{11}$$

3.7.2 Representation patterns of matrices

For example, using $n = 2, P = (3102), \Delta = (1111)$ conditions, a C code case contains sixteen histograms arranged as a 4×4 matrix.

0	4	1	5
2	6	3	7
8	12	9	13
10	14	11	15

(12)

All matrices in this chapter use this configuration for the matrix pattern to represent their elements.

4. Simulation results

For ease of illustration, as different signals have intrinsic random properties only statistical distributions and global matrix representations are selected in this section.

4.1 Statistical distributions

The simulation provides a series of output results. In this section, $N = \{12, 13\}, n = 2, \{J = 3, J_+ = 11, J_- = 2\}$ are selected. Corresponding to Left path (Left), Right path (Right), Double path for Particles (D-P) and Double path for Waves (D-W) under symmetry and anti-symmetry conditions respectively.

From a given function, a set of histograms can be generated as two groups of eight probability histograms. To show their refined properties, it is necessary to represent them in both odd and even numbers. A total of sixteen histograms are required. For convenience of comparison, sample cases are shown in Figures 3(I-III).

$P_H(u_+\|J)$ (a) Left	$P_H(u_-\|J)$ (b) Right
$P_H(u_0\|J)$ (c) D-P	$P_H(u_1\|J)$ (d) D-W
$P_H(v_+\|J)$ (e) Left	$P_H(v_-\|J)$ (f) Right
$P_H(v_0\|J)$ (g) D-P	$P_H(v_1\|J)$ (h) D-W

(I) Representative patterns of Histograms for function J (a-d) symmetric cases; (e-h) antisymmetric cases

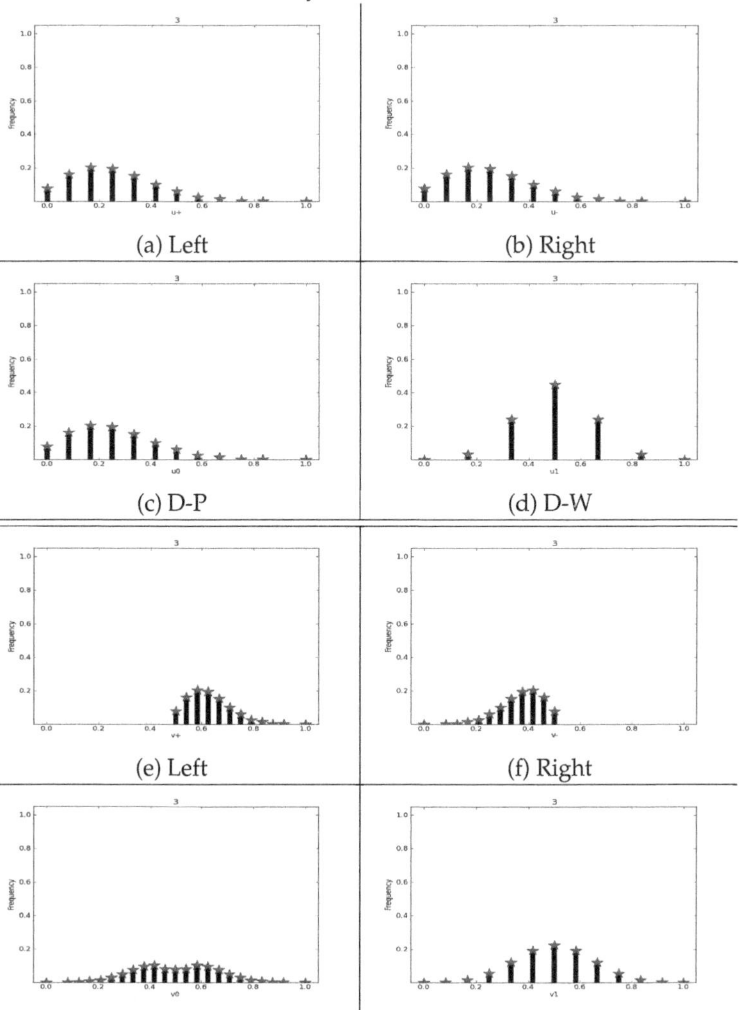

(a) Left (b) Right

(c) D-P (d) D-W

(e) Left (f) Right

(g) D-P (h) D-W

(II) $N = \{12\}, J = 3$ Two groups of results in eight histograms

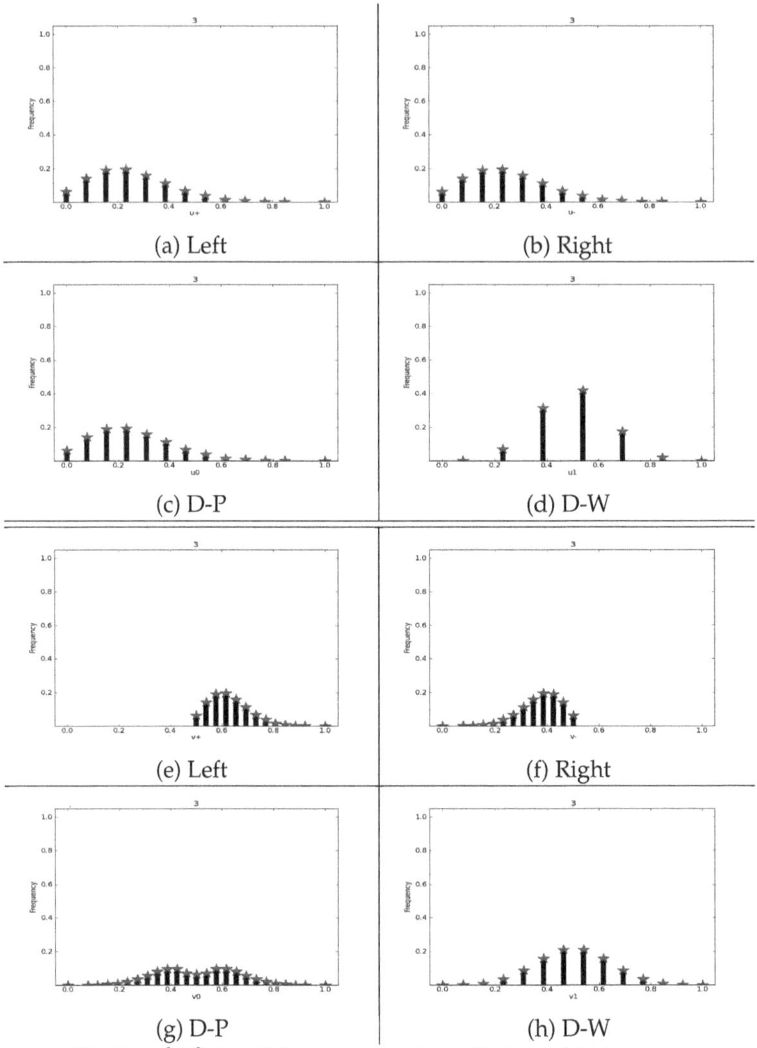

(a) Left (b) Right

(c) D-P (d) D-W

(e) Left (f) Right

(g) D-P (h) D-W

(III) $N = \{13\}, J = 3$ Two groups of results in eight histograms

Fig. 3. (I-III) $N = \{12, 13\}, J = 3$ Simulation results ; (I) Representative Patterns for $P_H(u_+|J) = P_H(u_-|J)$ and $P_H(v_+|J) = P_H(1 - v_-|J)$ conditions; (II) $N = \{12\}, J = 3$ Two groups of eight probability histograms; (III) $N = \{13\}, J = 3$ Two groups of eight probability histograms

Representation patterns are illustrated in Fig 3(I). Eight probability histograms of $P_H(u_+|J) = P_H(u_+|J)$ are shown in Fig 3(II) for $N = 12$ to represent four symmetry groups and another eight probability histograms are shown Fig 3(III) for $N = 13$ to represent four anti-symmetry groups respectively.

4.2 Global matrix representations

All possible 2^{2^n} functions are applied. It is convenient to arrange all generated histograms as a matrix, a C code scheme of variant logic applied to organize a set of 2^{2^n} histograms into a $2^{2^{n-1}} \times 2^{2^{n-1}}$ matrix.

Applying the C code configuration, a given signal of a function determines an element on a matrix to represent its histogram. There is one to one correspondence among different configurations.

Using this measurement mechanism, eight types of statistical histograms are systematically illustrated. Each element in the matrix is numbered to indicate its corresponding function and also the relevant histogram will be put on the position.

For $n = 2$ cases, sixteen matrices are shown in Figs 5-6 (a-h). Figs 5-6 (a-d) represent Symmetry groups and Figs 5-6 (e-h) represent Anti-symmetry groups. To show odd and even number configurations, Fig 5 (a-h) shows $N = 12$ cases and Fig 6 (a-h) shows $N = 13$ cases respectively.

5. Analysis of results

In the previous section, results of different statistical distributions and their global matrix representations were presented. In this section, plain language is used to explain what various visual effects might be illustrated and to discuss local and global arrangements.

5.1 Statistical distributions for a given function

It is essential to analyze differences among various statistical distributions for a given function.

5.1.1 Symmetry groups for a function

For the selected function $J = 3$, four distributions in symmetry groups are shown in Fig 3 (a-d). (a) $P_H(u_+|J)$ for Left; (b) $P_H(u_-|J)$ for Right; (c) $P_H(u_0|J)$ for D-P; and (d) $P_H(u_1|J)$ for D-W respectively.

Under Symmetry conditions, $P_H(u_+|J) = P_H(u_-|J)$, both Left and Right distributions are the same. $P_H(u_0|J)$ generated with both paths open under asynchronous conditions simulates D-P. Compared with distributions in (a-b) , it is feasible to identify the same components from original inputs.

However, for $P_H(u_1|J)$ under synchronous conditions and with the same Left and Right input signals, the simulation shows D-W exhibiting interferences among the output distributions that are significantly different from the original components.

5.1.2 Anti-symmetry groups for a function

Four distributions are shown in Fig 3 (e-h) as asymmetry groups. A pair of equation $P_H(v_+|J) = P_H(1 - v_-|J)$ shows that one distribution is a mirror image of another one. $P_H(v_+|J)$ distribution is shown in Fig 3 (e) for Left signals and $P_H(v_-|J)$ distribution is shown Fig 3 (f) for Right signals.

$P_H(v_0|J)$ is shown in Fig 3 (g) for both paths open under asynchronous conditions to simulate D-P. Compared with (e-f) distributions, it is feasible to identify the same components from the original inputs.

However $P_H(v_1|J)$ is shown in Fig 3 (h) under synchronous condition with both path signals as inputs to simulate D-W exhibiting interferences among the output distributions that are significantly different from the original components.

To show even and odd number's differences, $N = 12$ cases are shown in Fig 3 (II, a-h) and $N = 13$ cases are shown in Fig 3 (III, a-h) respectively.

5.2 Global matrix representations

Sixteen matrices are represented in Fig 4-5 (a-h) with eight signals generating two sets of 16 groups for $N = \{12, 13\}$ respectively.

5.2.1 Symmetry cases

Matrices for the Left in Fig 4-5 (a) show elements in a column with the corresponding histogram showing polarized effects on the vertical.

Matrices for the Right in Fig 4-5 (b) show elements in a row with the corresponding histogram showing polarized effects on the horizontal.

Matrices for D-P in Fig 4-5 (c) provide asynchronous operations combined with both distributions from Fig 4-5 (a-b) to form a unified distribution. From each corresponding position, it is possible to identify each left and right component and the resulting shapes of the histogram.

Matrices for D-W in Fig 4-5 (d) provide synchronous operations combined with both distributions from Fig 4-5 (a-b) for each element. Compared with Fig 4-5 (c) and Fig 4-5 (d) respectively, distributions in Fig 4-5 (d) are much simpler with two original distributions especially on the anti-diagonal positions: $J \in \{10, 12, 3, 5\}$. Only less than half the number of spectrum lines are identified.

5.2.2 Anti-symmetry cases

In a similar manner to the symmetry conditions, four anti-symmetry effects can be identified in Fig 4-5 (e-h). Matrices in Fig 4-5 (e) are Left operations for different functions, elements are polarized on the vertical and matrices in Fig 4-5 (f) are Right operations, elements are polarized on the horizontal. Spectrum lines in Fig 4-5 (e) appear in the right half and spectrum lines in Fig 4-5 (f) are appeared in the left half respectively.

Matrices for D-P in Fig 4-5 (g) show additional effects for each distribution according to the relevant position with components that can be identified as corresponding to identifiable inputs in many cases. Anti-symmetry signals are generated in merging conditions.

Matrices for D-W in Fig 4-5 (h) show different properties. In general, only one peak can be observed for each element especially for the $J \in \{10, 12, 3, 5\}$ condition. Spectra appear to be much simpler than the original distributions in Fig 4-5 (e-f), and significant interference properties are observed.

5.3 Four symmetry groups

Pairs of relationships can be checked on symmetry matrices in Figs 4-5 (a-d), four groups are identified.

5.3.1 Left: polarized vertical group

$\{P_H(u_+|J)\}$ elements in Figs 4-5 (a) show (only) four distinct distributions. Each column contains only one distribution. Sixteen elements in the matrix can be classified into four vertical classes: $\{0, 2, 8, 10\}, \{4, 6, 12, 14\}, \{1, 3, 9, 11\}, \{5, 7, 13, 15\}$ respectively. Four meta distributions are given as $\{10, 14, 11, 15\}$.

5.3.2 Right: polarized horizontal group

$\{P_H(u_-|J)\}$ elements in Figs 4-5 (b) show (a further) four distinct distributions. Each row contains only one distribution. Sixteen elements in the matrix can be classified into four horizontal classes: $\{0, 4, 1, 5\}, \{2, 6, 3, 7\}, \{8, 12, 9, 13\}, \{10, 14, 11, 15\}$ respectively. Four meta distributions are given as $\{0, 2, 8, 10\}$.

5.3.3 D-P: particle group

$\{P_H(u_0|J)\}$ elements in Figs 4-5 (c) illustrate symmetry properties. There are six pairs of symmetry elements: $\{8 : 14\}, \{2 : 11\}, \{0 : 15\}, \{6 : 9\}, \{4 : 13\}, \{1 : 7\}$. In addition, four elements on anti-diagonals provide different distributions: $\{10, 12, 3, 5\}$. Under this condition, ten classes of distributions are distinguished.

5.3.4 D-W: wave group

$\{P_H(u_1|J)\}$ elements in Figs 4-5 (d) illustrate symmetry properties. There are six pairs of symmetry elements: $\{8 : 14\}, \{2 : 11\}, \{0 : 15\}, \{6 : 9\}, \{4 : 13\}, \{1 : 7\}$. In addition, four elements on diagonal positions provide same distribution: $\{0, 6, 9, 15\}$. Two elements on anti-diagonals: $\{12, 3\}$ have the same distribution in Fig 4 (d). Under this condition, nine or ten classes of different distributions can be identified for Fig 4 (d) and Fig 5 (d) respectively.

5.4 Four anti-symmetry groups

Figures 4-5 (e-h) represent anti-symmetry properties, four groups can be identified.

5.4.1 Left: polarized vertical group

$\{P_H(v_+|J)\}$ elements in Figs 4-5 (e) show that (only) four classes can be distinguished. Elements within these groups members are the same as for symmetry groups in Figs 4-5(a). Their distributions fall within the region $[0.5, 1]$.

5.4.2 Right: polarized horizontal group

$\{P_H(v_-|J)\}$ elements in Figs 4-5 (f) show that (only) four classes can be distinguished. Elements within these groups are the same as for symmetry groups in Figs 4-5 (b). Their distributions fall within the region $[0, 0.5]$.

5.4.3 D-P: particle group

$\{P_H(v_0|J)\}$ in Figs 4-5 (g) show six pairs of anti-symmetry distributions: $\{8 \nmid 14\}, \{2 \nmid 11\}, \{0 \nmid 15\}, \{6 \nmid 9\}, \{4 \nmid 13\}, \{1 \nmid 7\}$ four elements are distinguished on the anti-diagonals: $\{10, 12, 3, 5\}$. Under this condition, ten classes can be identified.

5.4.4 D-W: wave group

$\{P_H(v_1|J)\}$ in Figs 4-5 (h) show six pairs of anti-symmetry distributions: $\{8 \nmid 14\}, \{2 \nmid 11\}, \{0 \nmid 15\}, \{6 \nmid 9\}, \{4 \nmid 13\}, \{1 \nmid 7\}$ four pairs of symmetry elements: $\{3 : 5\}, \{10 : 12\}, \{2 : 4\}, \{11 : 13\}$ are distinguished. Under this condition, twelve classes can be identified.

5.5 Odd and even numbers

From a group view point, only D-P and D-W need to be reviewed as different groups in symmetry conditions. Anti-symmetry conditions are unremarkable.

It is reasonable to suggest that anti-symmetry operations will be much easier to distinguish under experimental conditions, since sixteen groups in D-P conditions and twelve groups in D-W conditions will have significant differences. However, under the symmetry conditions (only) minor differences can be identified.

5.5.1 Single and double peaks

Single and Double peaks can be observed in Fig 4(5) (h): $\{3, 5\}$ for even and odd numbers respectively.

For two other members $\{10, 12\}$, (only) single pulse distributions are observed in Figs 4-5 (h) to show the strongest interference results.

5.6 Class numbers in different conditions

To summarize over the different classes, 16 matrices are shown in different numbers of identified classes as follows:

Class No.	Left	Right	D-P	D-W
SE	4	4	10	9
SO	4	4	10	10
AE	4	4	16	12
AO	4	4	16	12

where Left:Left Path, Right: Right Path, D-P: Double Path for Particles, D-W: Double Path for Waves; SE: Symmetry for Even number, SO: Symmetry for Odd number, AE: Anti-symmetry for Even number, AO: Anti-symmetry for Odd number.

5.7 Polarized effects and double path results

In order to contrast the different polarized conditions, it is convenient to compare distributions $\{P_H(u_+|J), P_H(u_-|J\}$ and $\{P_H(v_+|J), P_H(v_-|J\}$ arranged according to the corresponding polarized vertical and horizontal effects. This visual effect is similar to what might be found when using polarized filters in order to separate complex signals into two channels. Different distributions can be observed under synchronous and asynchronous conditions.

5.7.1 Particle distributions and representations

For all symmetry or non-symmetry cases under $\oplus$ asynchronous addition operations, relevant values meet $0 \leq u_0, v_0, u_-, v_-, u_+, v_+ \leq 1$. Checking $\{P_H(u_0|J), P_H(v_0|J)\}$ series, $\{P_H(u_+|J), P_H(u_-|J)\}$ and $\{P_H(v_+|J), P_H(v_-|J)\}$ satisfy following equation.

$$\begin{cases} P_H(u_0|J) = \frac{P_H(u_-|J)+P_H(u_+|J)}{2} \\ P_H(v_0|J) = \frac{P_H(v_-|J)+P_H(v_+|J)}{2} \end{cases} \tag{13}$$

The equation is true for different values of N and n.

5.7.2 Wave distributions and representations

Interference properties are observed in $\{P_H(u_+|J) = P_H(u_-|J)\}$ conditions. Under $+$ synchronous addition operations, relevant values meet $0 \leq u_1, v_1, u_-, v_-, u_+, v_+ \leq 1$. Checking $\{P_H(u_1|J), P_H(v_1|J)\}$ distributions and compared with $\{P_H(u_+|J), P_H(u_-|J)\}$ and $\{P_H(v_+|J), P_H(v_-|J)\}$, non-equations and equations are formulated as follows:

$$\begin{cases} P_H(u_1|J) \neq P_H(u_0|J) \\ P_H(v_1|J) \neq P_H(v_0|J) \end{cases} \tag{14}$$

Spectra in different cases illustrate wave interference properties. Single and double peaks are shown in interference patterns similar to interference effects in classical double slit experiments.

5.7.3 Non-symmetry and non-anti-symmetry

However, for the $\{P_H(u_+|J) \neq P_H(u_-|J)\}$ non-symmetry cases, there are significant differences between $\{P_H(u_0|J), P_H(v_0|J)\}$ and $\{P_H(u_1|J), P_H(v_1|J)\}$. Such cases have interference patterns with more symmetric properties than single path and particle distributions.

Four anti-diagonal positions are linked to symmetry and anti-symmetry pairs, twelve other pairs of functions belong to non-symmetry and non-anti-symmetry conditions. Their meta elements can be identified by the relevant variant expressions.

6. Other relevant measurements and properties

6.1 Quaternion measurements

It is interesting to note the relationship between the variant quaternion and other quaternion measurements.

6.1.1 Variant quaternion

In the variant quaternion, $\Delta f(X) = (N_\perp, N_+, N_-, N_\top), N = N_\perp + N_+ + N_- + N_\top$.

6.1.2 Einstein quaternion

Einstein's two-state system of interaction $(N_1, N_2, N_{12}, N_{21})$ allows the following equations to be established.

$$\begin{cases} N_1 = & N_\perp + N_+ \\ N_2 = & N_- + N_\top \\ N_{12} = N_+ \\ N_{21} = N_- \\ N = & N_1 + N_2 \end{cases} \tag{15}$$

From the equations, the measured pair $\{N_{21}, N_{12}\}$ has a 1-1 correspondence to $\{N_-, N_+\}$.

6.1.3 CHSH quaternion

Selecting $+ \rightarrow 1, - \rightarrow 0$, CHSH's $N_{\pm,\pm}(a,b)$ measurements meet

$$\begin{cases} N_{+,+}(a,b) \rightarrow N_\top \\ N_{+,-}(a,b) \rightarrow N_- \\ N_{-,+}(a,b) \rightarrow N_+ \\ N_{-,-}(a,b) \rightarrow N_\perp \end{cases} \tag{16}$$

$(N_{++}, N_{+-}, N_{-+}, N_{--}) \rightarrow (N_\top, N_-, N_+, N_\perp)$, let $N = N_{++} + N_{+-} + N_{-+} + N_{--}$. CHSH quaternion is a permutation of the variant quaternion.

6.1.4 Aspect quaternion

Aspect's quaternion $(N_t, N_r, N_c, N_\omega)$ have following corresponding:

$$\begin{cases} N_t \rightarrow N_- \\ N_r \rightarrow N_+ \\ N_\omega \rightarrow N \end{cases} \tag{17}$$

For N_c, there is no parameter in the variant quaternion for parameter N_c. N_c indicates joined action numbers to distinguish single and double paths, corresponding to $\{u_1, v_1\}$ times.

This parameter is of significance in an actual experiment. In a simulated system, the parameter provides a control coefficient that separates two types of paths $\{u_0, v_0\}$ and $\{u_1, v_1\}$ that would be measured in real experiments.

6.2 Different particle models

From Newton's particles to Young's Double slit experiments, the question of how to distinguish particle and wave measurements has a long history [Hawking & Mlodinow (2010); Penrose (2004)]. From a measurement viewpoint, recent activities testing Bell Inequations can be seen to be consistent with historical viewpoints [Jammer (1974)].

The fundamental assumptions of Bell Inequations are based on a local realism [Eberhard (1978); Fine (1999)]. A key condition of measure theory can be seen in a review of authoritative definitions of local realism [SEP (2009)].

6.2.1 Independent conditions in probability

Kolmogorov developed modern probability construction [Ash & Doléans-Dade (2000)] to use measure theory approaches to handle probability measurements. Modern expressions of Bell Inequalities have many forms [SEP (2009)], all of these are based on the conceptual framework of locality which is understood as the conjunction of independent conditions on probability measurements.

For any independent events A, B,

$$\begin{array}{l} P(A \cap B) = P(A)P(B) \\ P(A \cup B) = P(A) + P(B) - P(A \cap B), 0 \leq P(A), P(B) \leq 1 \\ P(A \cup B) \leq P(A) + P(B) \end{array} \tag{18}$$

Probability measurement expressions play the core role in Bell Inequalities. In real single photon experiments, people found that $P(A \cup B) \leq P(A) + P(B)$ did not hold true.

In quantum reality environment, testing measurements could be $\tilde{P}(A \cup B) > \tilde{P}(A) + \tilde{P}(B)$ under specific conditions.

6.2.2 Bell inequalities and Newton-Einstein-Feynman particle distributions

From a measurement viewpoint, measurements of local realism correspond to a real number construction that links to Kolmogorov probability [Ash & Doléans-Dade (2000)]. von Neumann (1932,1996)'s mathematical foundation of quantum mechanics is based on a complex number construction. By their nature, these measurement constructions reveal significant differences between the classical and complex probability framework.

Probability deductions under local realism must be restricted to real number systems. Under the independent condition, $P(A \cup B) \leq P(A) + P(B)$ is always true.

6.3 Further predictions

Observing modern experiments to test Bell Inequations, it is necessary to measure the events in synchronous conditions to create multiple pairs of photons. Different time conditions indicate asynchronous and synchronous conditions playing a critical role in distinguishing between classical and quantum activities. Experimental evidence and case study results are not sufficient at this time to permit firm propositions. However, a summary of predictions for the measurement construction of variant frameworks which can be extrapolated from the simulations is provided below.

Prediction 1: Left distributions have relationships showing polarized vertical behaviors.

Prediction 2: Right distributions have relationships showing polarized horizontal behaviors.

Prediction 3: D-P distributions have relationships showing classical particle statistical behaviors.

Prediction 4: D-W distributions have relationships showing wave interference statistical behaviors.

Prediction 5: Under the same conditions of Bell Inequations, it will be possible to design and implement experiments to distinguish D-P and D-W distributions in real photon experimental environments.

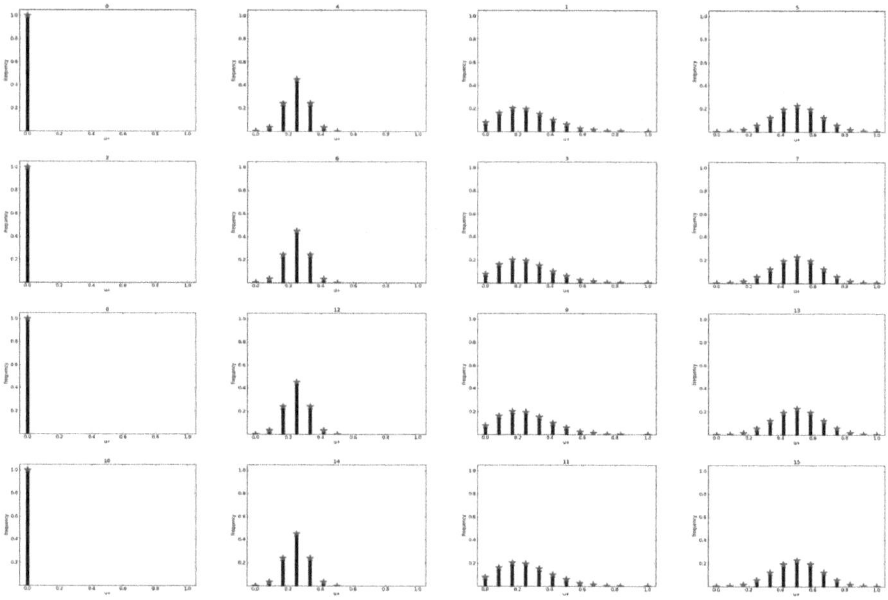

(a) Left

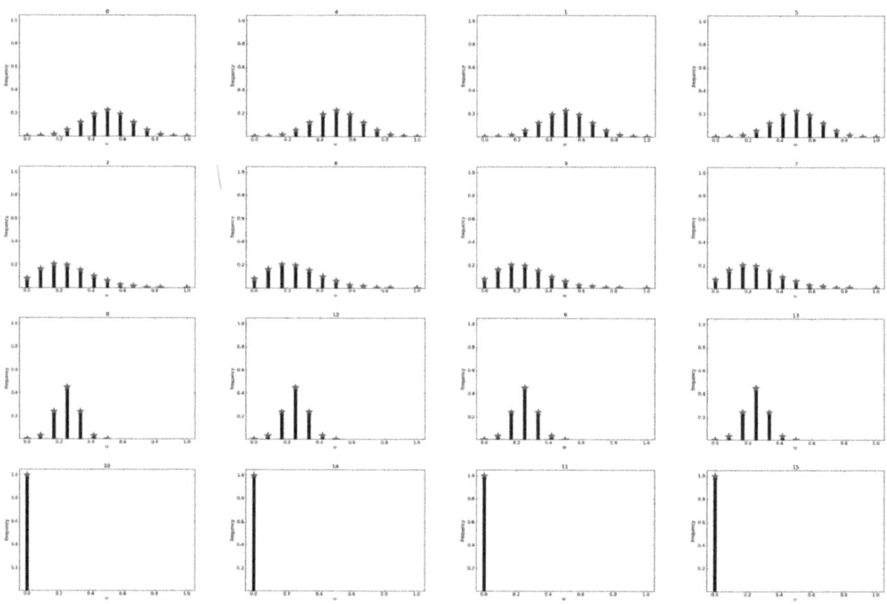

(b) Right

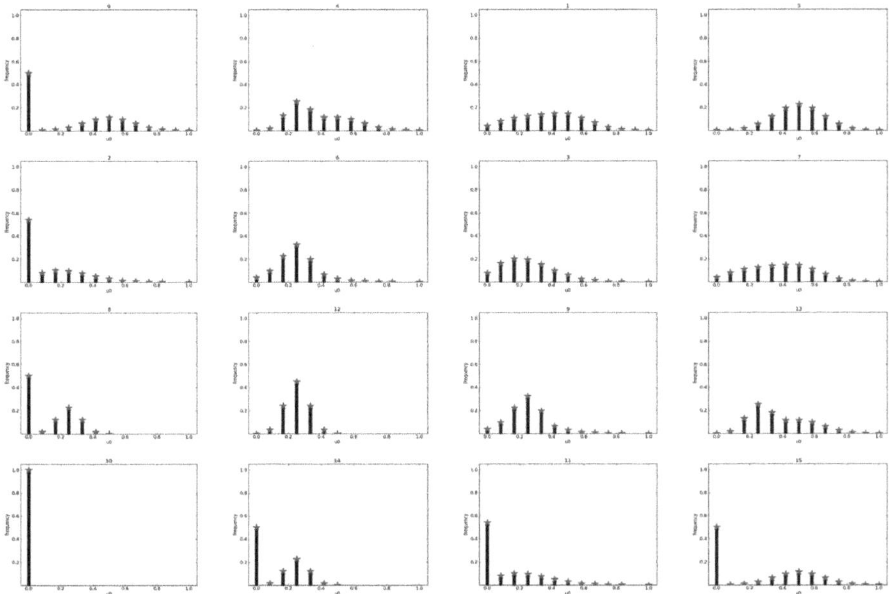

(c) D-P

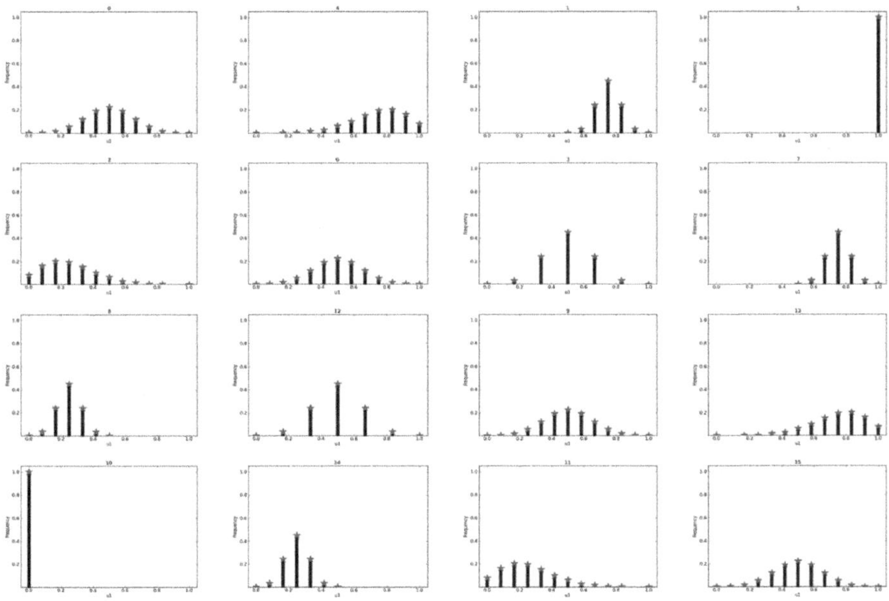

(d) D-W

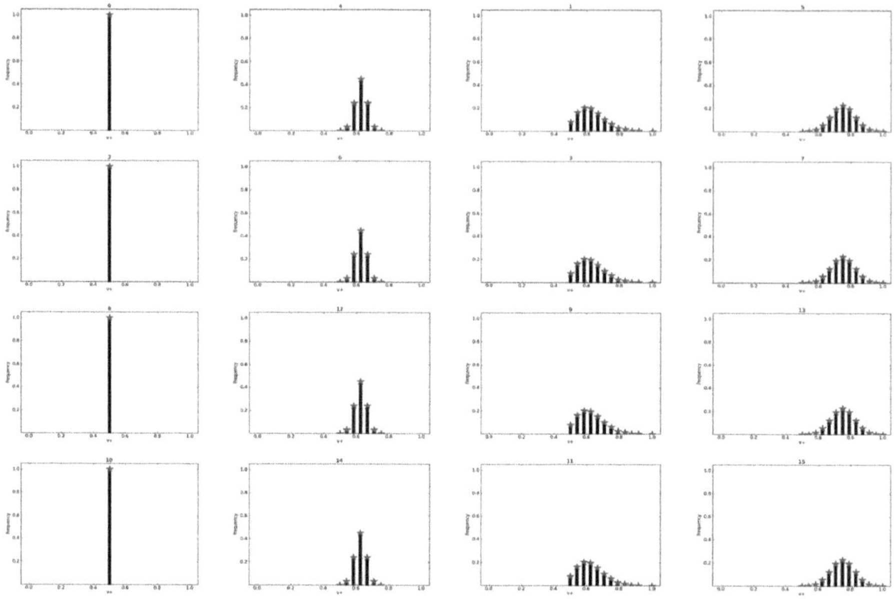

(e) Left

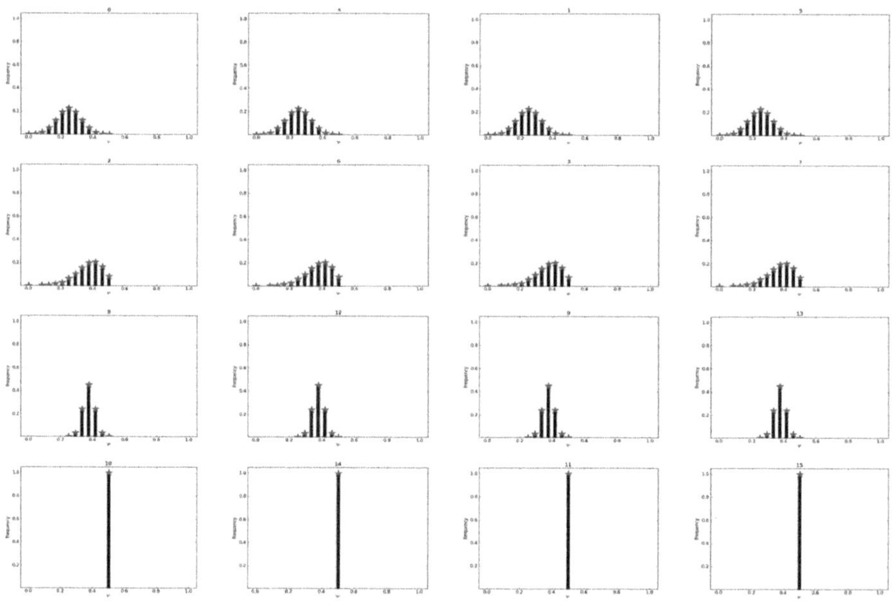

(f) Right

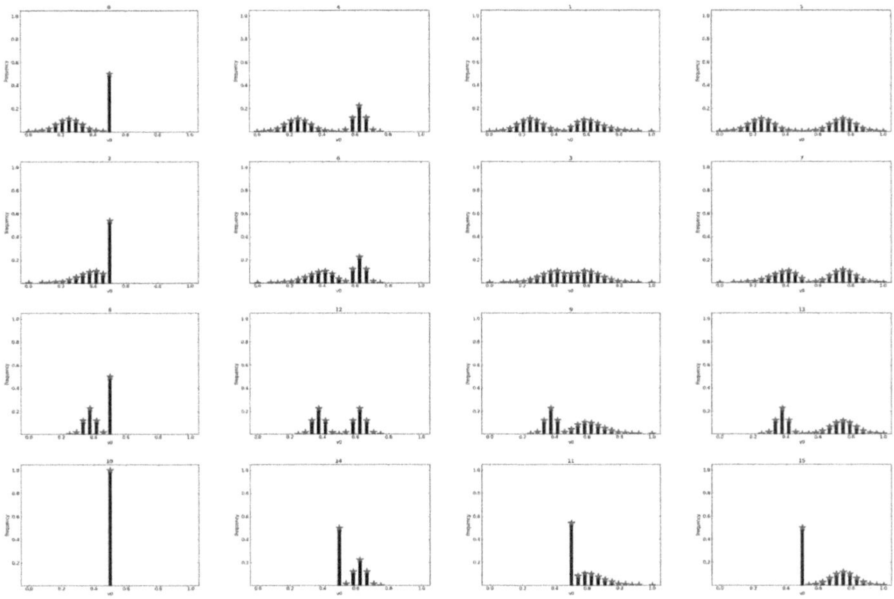

(g) D-P

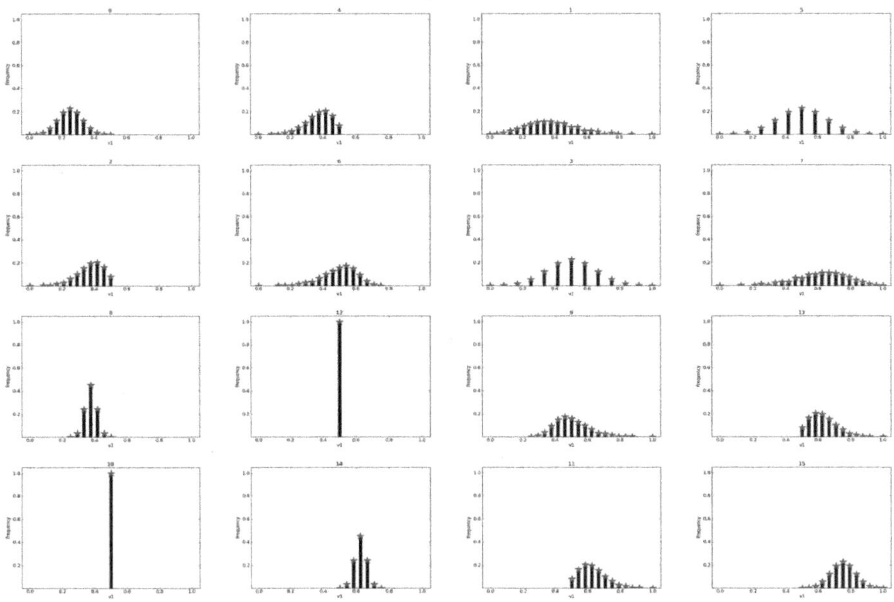

(h) D-W

Fig. 4. (a-h) Even number groups: $N = \{12\}, f \in B_2^4$ Eight Matrices of Global Matrix Representations. (a) Left; (b) Right; (c) D-P; (d)D-W in symmetry conditions; (e) Left; (f) Right; (g) D-P; (h)D-W in anti-symmetry conditions.

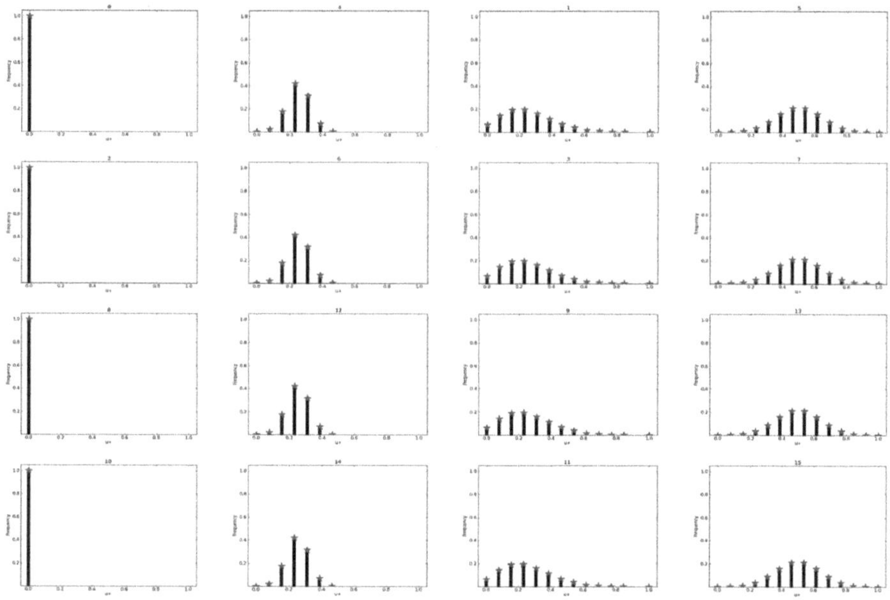

(a) Left

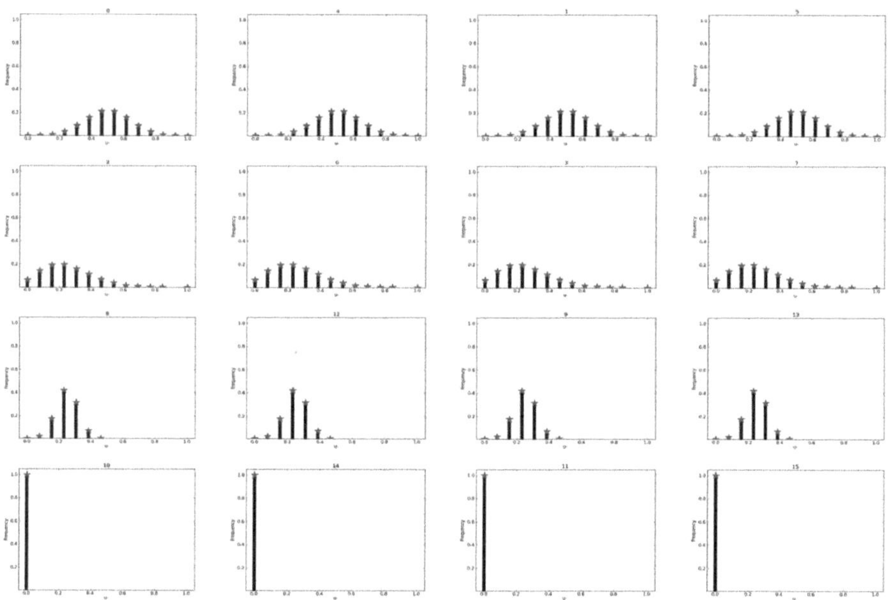

(b) Right

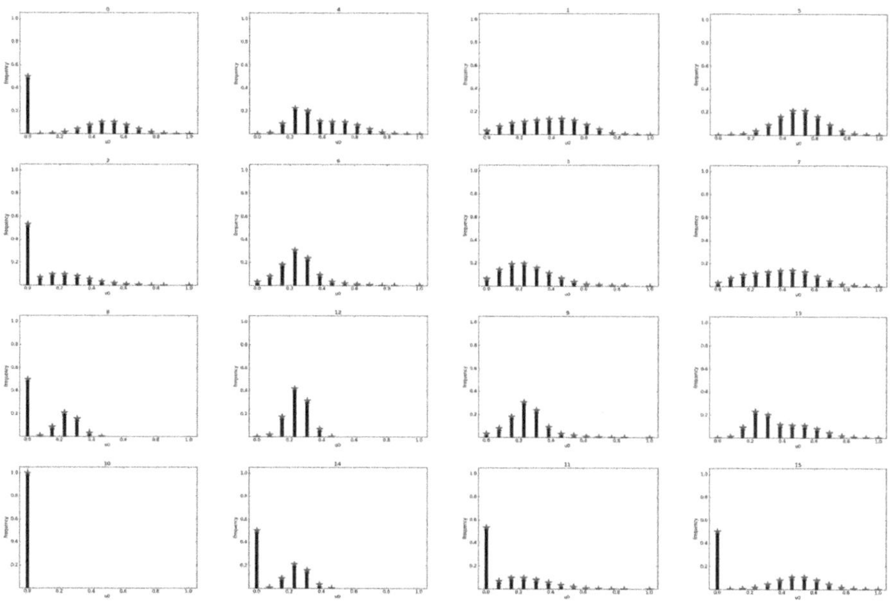

(c) D-P

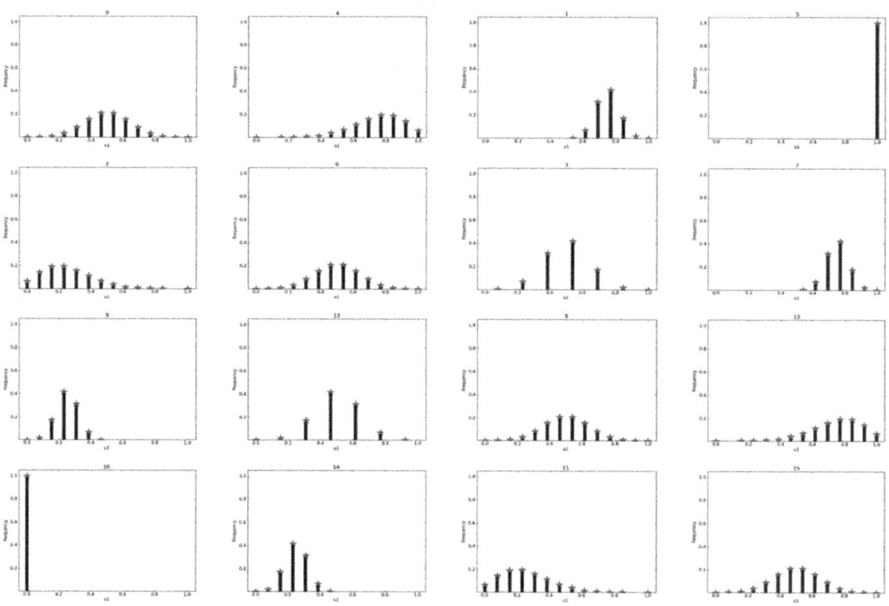

(d) D-W

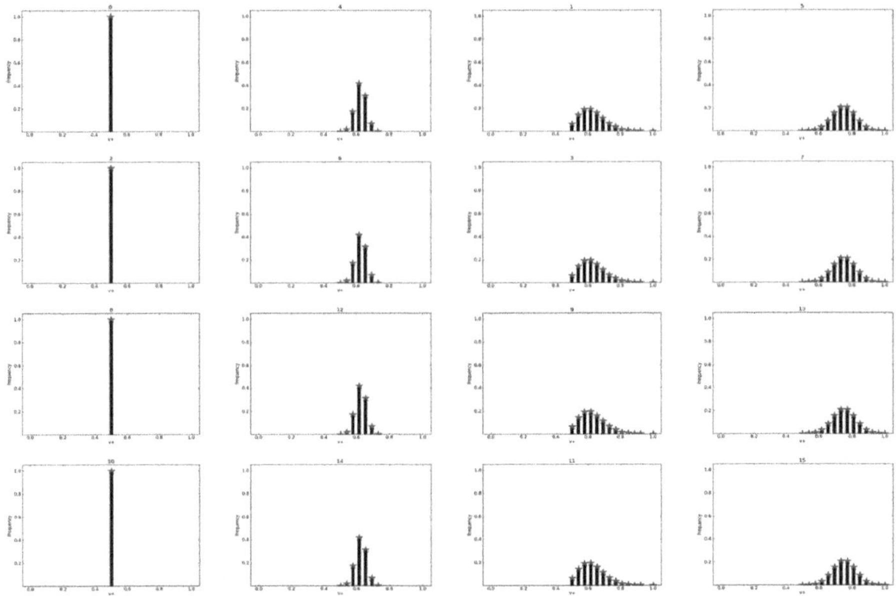

(e) Left

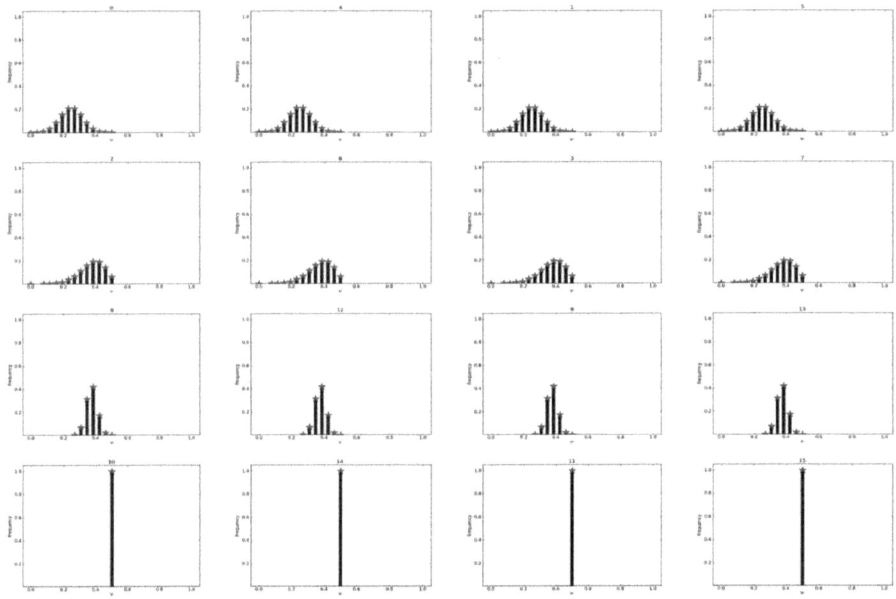

(f) Right

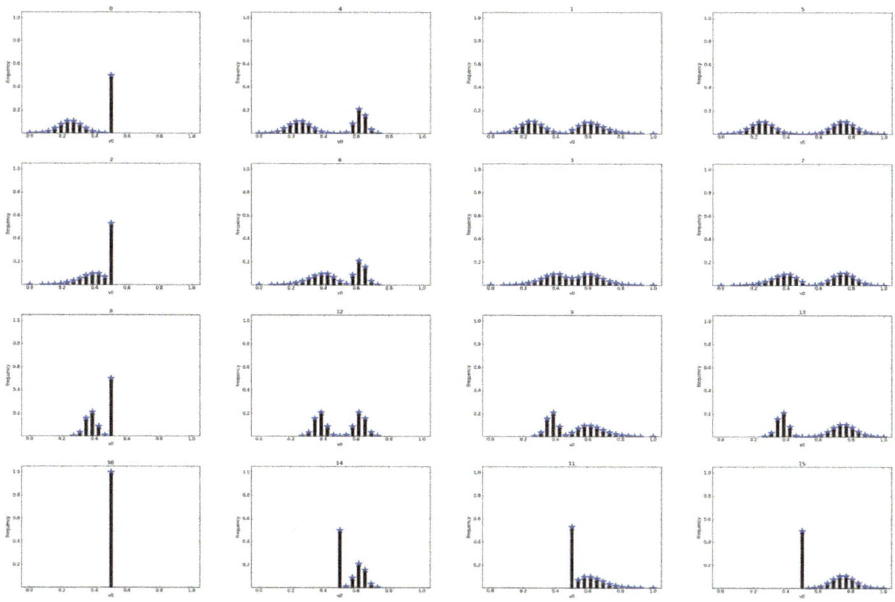

(g) D-P

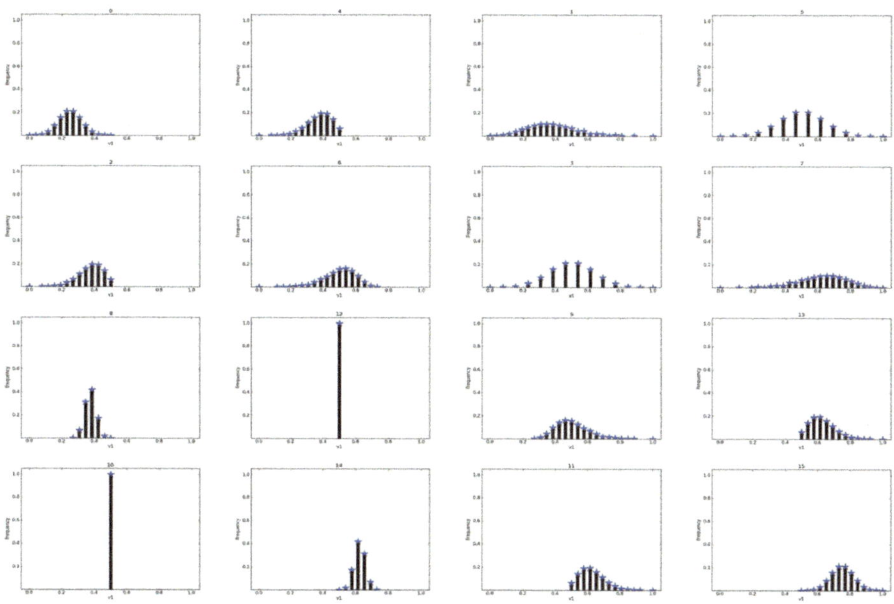

(h) D-W

Fig. 5. (a-h) Odd number groups: $N = \{13\}, f \in B_2^4$ Eight Matrices of Global Matrix Representations. (a) Left; (b) Right; (c) D-P; (d)D-W in symmetry conditions; (e) Left; (f) Right; (g) D-P; (h)D-W in anti-symmetry conditions.

Prediction 6: It will be much easier to design and implement key experiments to distinguish D-P and D-W behaviors in asynchronous conditions than in synchronous conditions.

In other words, under proposed variant measurements, the simplest effects are polarized properties in Left and Right matrices. Both D-P and D-W distributions are generated from pairs of polarized signals in general cases. In addition, significant differences can be observed between D-P and D-W distributions in asynchronous conditions. This set of theoretical predictions could help experimenters to design and implement effective experiments to check variant measurements under real quantum environments.

6.4 Two conjectures

Back to Young's waves and Newton's particles, Bohr's complementarity, EPR and Feynman's particle and wave conditions [Hawking & Mlodinow (2010); Jammer (1974); Penrose (2004)], it is essential to list two conjectures to summarize our results as follows:

Conjecture 1. Measurement results of Newton-Einstein-Feynman particles and Variant D-P models must obey Bell Inequations.

This conjecture could be approved from listed models satisfied independent conditions. From this viewpoint, Newton-Einstein-Feynman particle models and Variant D-P models could satisfy Bell Inequalities. Bell Inequations at most could provide only a logical foundation for different particle models.

Conjecture 2. Measurement results of Young-Bohr-Feynman waves and Variant D-W models satisfy the same types of entanglement conditions.

Since the Local Realism cannot be supported by quantum construction, a solid foundations is required to validate this conjecture using complex-probability conditions for different entanglements in real quantum environments.

7. Conclusion

Analyzing a N bit 0-1 vector and its exhaustive sequences for variant measurement, from a double path experiment viewpoint, this system simulates double path interference properties through different accurate distributions from local interactive measurements to global matrix representations. Using this model, two groups of parameters $\{u_\beta\}$ and $\{v_\beta\}$ describe left path, right path, and double paths for particles and double paths for waves with distinguishing symmetry and anti-symmetry properties. $\{P_H(u_\beta|J), P_H(v_\beta|J)\}$ provide eight groups of distributions under symmetry and anti-symmetry forms. In addition, $\{M(u_\beta), M(v_\beta)\}$ provide eight matrices to illustrate global behaviors under complex conditions.

Compared with the variant quaternion and other quaternion measurements, it is helpful to understand the usefulness and limitations of variant simulation properties.

The complexity of n-variable function space has a size of 2^{2^n} and exhaustive vector space has 2^N. Whole simulation complexity is determined by $O(2^{2^n} \times 2^N)$ as ultra exponent productions. How to overcome the limitations imposed by such complexity and how best to compare and contrast such simulations with real world experimentation will be key issues in future work.

Six predictions and two conjectures are summarized in this chapter to guide further theoretical and experimental exploration.

In addition, real world experiments are expected to be designed and implemented in the near future to test results given in this chapter.

8. Acknowledgements

Thanks to Colin W. Campbell for help with the English edition, to The School of Software Engineering, Yunnan University and The Key Laboratory of Yunnan Software Engineering for financial supports to the Information Security research projects (2010EI02, 2010KS06) and sub-CDIO project.

9. References

Afshar, S., Flores, E., McDonald, K. & Knoesel, E. (2007). Paradox in wave particle duality, *Found. Phys.* 37. 295-305.

Arndt, M., Nairz, O., Vos-Andreae, J., Keller, C., van der Zouw, G. & Zeilinger, A. (1999). Wave-particle duality of C60 molecules, *Nature* 401. 680-682.

Ash, R. B. & Doléans-Dade, C. A. (2000). *Probability & Measure Theory*, Elsevier.

Aspect, A. (2002). Bell's theorem: The naive view of an experimentalist, *Quatum [Un]speakables - From Bell to Quantum Information, Ed. Bertlmann and Zeilinger, Springer* .

Aspect, A., Grangier, P. & Roger, G. (1982). Experimental realization of einstein-podolsky-rosen-bohm gedankenexperiment: A new violation of bell's inequalities, *Phys. Rev. Lett.* 49. 91-94.

Barnett, S. M. (2009). *Quantum Information*, Oxford Uni. Press.

Barrow, J. D., Davies, P. C. W. & Charles L. Harper, J. E. (2004). *SCIENCE AND ULTIMATE REALITY: Quantum Theory, Cosmology and Complexity*, Cambridge University Press.

Bell, J. S. (1964). On the einstein-podolsky-rosen paradox, *Physics* 1. 195-200.

Bell, J. S. (2004). *Speakable and Unspeakable in Quantum Mechanics*, Cambridge Univ. Press.

Bohr, N. (1935). Can quantum-mechanical description of physical reality be considered complete?, *Physical Review* 48. 696-702.

Bohr, N. (1949). *Discussion with Einstein on Epistemological Problems in Atomic Physics*, Evanston. 200-241.

Clauser, J., Horne, N., Shimony, A. & Holt, R. (1969). Proposed experiment to test local hidden-variable theories, *PRL* 23. 880-884.

Cohn, E. G. D. (1990). George e. uhlenbeck and statistical mechanics, *Amer. J. Phys.* 58(7). 619-625.

de Broglie, L. (1923). Radiation, waves and quanta, *Comptes rendus* 177. 507-510.

Eberhard, P. (1978). Bell's theorem and the different concepts of locality, *Nuovo Cimento* 46B. 392-419.

Einstein, A. (1916). Strahlungs-emission und -absorption nach der Quantentheorie, *Verhandlungen der Deutschen Physikalischen Gesellschaft* 18. 318-323.

Einstein, A., Podolsky, B. & Rosen, N. (1935). Can quantum-mechanical description of physical reality be considered complete?, *Physical Review* 47. 770-780.

Feynman, R. (1965). *The Character of Physical Law*, MIT Press.

Feynman, R., Leighton, R. & Sands, M. (1965,1989). *The Feynman Lectures on Physics*, Vol. 3, Addison-Wesley, Reading, Mass.

Fine, A. (1999). Locality and the hardy theorem, *in From Physics to Philosophy, Cambridge University Press* .

Fox, M. (2006). *Quantum Optics*, Oxford Uni. Press.

Grangier, P., Roger, G. & Aspect, A. (1986). Experimental evidence for a photon anticorrelation effect on a beam splitter: A new light on single-photon interferences, *Europhys. Lett.* 1. 173-179.

Hawkingand, S. & Mlodinow, L. (2010). *The Grand Design*, Bantam Books.

Healey, R., Hellman, G. & Edited. (1998). *Quantum Measurement: Beyond Paradox*, Uni. Minnesota Press.

Jacques, V., Lai, N., Dréau, A., Zheng, D., Chauvat, D., Treussart, F., Grangier, P. & Roch, J. (2008). Illustration of quantum complementarity using single photons interfering on a grating, *New J. Phys.* 10. 123009, arXiv:0807.5079.

Jammer, M. (1974). *The Philosophy of Quantum Mechanics*, Wiley-Interscience Publication.

Lindner, F., Schätzel, M. G., Walther, H., Baltuska, A., Goulielmakis, E., Krausz, F., Milosevic, D. B., Bauer, D., Becker, W. & Paulus, G. G. (2005). Attosecond double-slit experiment, *Physical Review Letters* 95. 040401.

Merali, Z. (2007). Parallel universes make quantum sense, *New Scientist* 2622. http://space.newscientist.com/article/mg19526223.700-parallel-universes-make-quantum-sense.html.

Penrose, R. (2004). *The Road to Reality*, Vintage Books, London.

Schleich, W. P., Walther, H. & Edited (2007). *Elements of Quantum Information*, Wiley-VCH Verlag GmbH & Co KGaA Weinheim.

SEP (2009). Bell's theorem, *Stanford Encyclopedia of Philosophy* .
URL: *http://plato.stanford.edu/entries/bell-theorem/*

von Neumann, J. (1932,1996). *Mathematical Foundations of Quantum Mechanics*, Princeton Univ. Press.

Zeh, H. D. (1970). On the interpretation of measurement in quantum theory, *Foundation of Physics* 1. 69-76.

Zeilinger, A., Weihs, G., Jennewein, T. & Aspelmeyer, M. (2005). Happy centenary, photon, *Nature* 433. 230-238.

Zheng, J. (2011). Synchronous properties in quantum interferences, *Journal of Computations & Modelling, International Scientific Press* 1(1). 73-90.
URL: *http://www.scienpress.com/upload/JCM/Vol%201_1_6.pdf*

Zheng, J. & Zheng, C. (2010). A framework to express variant and invariant functional spaces for binary logic, *Frontiers of Electrical and Electronic Engineering in China, Higher Education Press and Springer* 5(2): 163–172.
URL: *http://www.springerlink.com/content/91474403127n446u/*

Zheng, J. & Zheng, C. (2011a). Variant measures and visualized statistical distributions, *Acta Photonica Sinica, Science Press* 40(9). 1397-1404. URL: *http://www.photon.ac.cn/CN /article/downloadArticleFile.do?attachType=PDF&id=15668*

Zheng, J. & Zheng, C. (2011b). Variant simulation system using quaternion structures, *Journal of Modern Optics, Taylor & Francis Group*. iFirst 1-9.
URL: *http://dx.doi.org/10.1080/09500340.2011.636152*

Zheng, J., Zheng, C. & Kunii, T. (2011). A framework of variant-logic construction for cellular automata, *Cellular Automata - Innovative Modelling for Science and Engineering edited Dr. A. Salcido, InTech Press*. 325-352. URL: *http://www.intechopen.com/articles/show/title/a-framework-of-variant-logic-construction-for-cellular-automata*

18

Dynamic State Estimator Based on Wide Area Measurement System During Power System Electromechanical Transient Process

Xiaohui Qin, Baiqing Li and Nan Liu
Power System Department, China Electric Power Research Institute, The People's Republic of China

1. Introduction

The wide area measurement system (WAMS) developed rapidly in recent years [1-3]. It has been applied to the studies on many topics in monitoring and control of power systems. But as a kind of measurement system, WAMS has the measurement error and bad data unavoidably. The steady measurement errors of WAMS have been prescribed in corresponding IEEE standard [4], but the dynamic measurement errors now become the focus of discussion [5-6] and attract the attention of PSRC workgroup H11. If the dynamic raw data is applied directly, the unpredictable consequence will be resulted in, which will do a lot of damage to power systems. Therefore, the dynamic estimation for the state variables during electromechanical transient process is the backbone for WAMS based dynamic applications and real-time control.

There was no effective means to measure the power system dynamic process before WAMS come forth; therefore, the dynamic estimation for power system state variables during electromechanical transient process was not feasible. In reference [7], a dynamic estimator for generator flux state variables during transient process is proposed, but the dimension of the flux state variables is relative high, and the accurate values of parameters can not be achieved easily. In reference [8], a non-linear dynamic state observer for generator rotor angle during electromechanical transient process is proposed, but the method is only applicable to one machine infinite bus system (OMIB), and the fault scenarios is required to satisfy the preset mode.

After WAMS come forth, many references focused on the steady state estimation with PMU measurements and had many achievements [9-11]. Comparing with the steady state estimation, the traditional dynamic state estimation [12-15] aims at the relative slow load fluctuation, which is different with the proposed dynamic state estimation during electromechanical transient process. The traditional dynamic state estimation employs the measurement equations based on the network constraints, and predicts the state variables using exponential smoothing techniques. But during the power system fault stage and consequent dynamic process, the network topology is changed and can not be acquired in time; the bus voltage phase angles has jump discontinuities and are not easy to be predicted. Thereby, the centralized dynamic estimation which adopts the measurement equations based on network constraints and regards the bus complex voltages as the state variables is not feasible for the electromechanical dynamic process.

In this paper, a novel WAMS based dynamic state estimator during power system electromechanical transient process is proposed. The estimator chooses the generator rotor angle and electrical angular velocity as the state variables to estimate. The generator output power measured by PMU is used to decouple the generator rotor movement equation and the outer network equation. And the linear Kalman filter based dynamic state estimator mathematical model is presented. The WAMS measurement noise and dynamic model noise are analyzed in detail. The total flow chart and the bad data detection and elimination approach are given as well. The numerical simulation is carried out on the IEEE 9-bus test system and a real generator in North China power grid. The simulation results indicate that the proposed real time dynamic estimator can estimate the generator rotor angles accurately; therefore, which can serve the power system dynamic monitoring and control system better.

2. Kalman filter

The time-invariant linear system model can be written as [16]

$$\begin{cases} \dot{\mathbf{x}} = \mathbf{A}\mathbf{x} + \mathbf{B}\mathbf{u} \\ \mathbf{y} = \mathbf{C}\mathbf{x} \end{cases} \tag{1}$$

The discrete form of equation (1) is

$$\begin{cases} \mathbf{x}(t+T) = \mathbf{\Phi}(T)\mathbf{x}(t) + \mathbf{\Gamma}(T)\mathbf{u}(t) \\ \quad \mathbf{y}(t) = \mathbf{C}\mathbf{x}(t) \end{cases} \tag{2}$$

where
T is the sampling period, $t=kT$, $k=0, 1, 2, \ldots$,
$\boldsymbol{\Phi}(T)$ is the state transition matrix, and

$\boldsymbol{\Phi}(T)=\exp(AT)=\sum_{i=0}^{\infty}\mathbf{A}^i T^i/(i\,!)$, $A^0=\boldsymbol{I}$; $\mathbf{\Gamma}(T)=\left[\int_0^T \mathbf{\Phi}(\alpha)d(\alpha)\right]\mathbf{B}$,and I is the identity matrix.

When subject to a random plant and measurement noise, the sampled-data system can be expressed as

$$\begin{cases} \mathbf{x}(k+1) = \mathbf{\Phi}\mathbf{x}(k) + \mathbf{\Gamma}\mathbf{u}(k) + \mathbf{\Gamma}\mathbf{w}(k) \\ \mathbf{y}(k) = \mathbf{C}\mathbf{x}(k) + \mathbf{v}(k) \end{cases} \tag{3}$$

The plant noise sequence $w(k)$ and the measurement noise sequence $v(k)$ are assumed to be Gaussian stationary white-noise sequence with zero means and covariance:

$$\begin{aligned} &E[\mathbf{v}(k)] = E[\mathbf{w}(k)] = \mathbf{0} \\ &E[\mathbf{w}(k)\mathbf{w}^T(j)] = \mathbf{Q}\boldsymbol{\delta}_{kj} \quad \boldsymbol{\delta}_{kj} = \begin{cases} 1 & k=j \\ 0 & k \neq j \end{cases} \\ &E[\mathbf{v}(k)\mathbf{v}^T(j)] = \mathbf{R}\boldsymbol{\delta}_{kj} \end{aligned}$$

where, Q is the plant noise covariance matrix, R is the measurement noise covariance matrix.

The initial state $x(0)$ is assumed to be uncorrelated with the plant and measurement noise sequences and is Gaussian with mean covariance:

$$E[x(0)]=\hat{x}(0)\,;\; E[(x(0)-\hat{x}(0))(x(0)-\hat{x}(0))^T]=\boldsymbol{P}_0$$

The Kalman filter formulas for time-invariant linear model can be largely divided into two steps: prediction step and filtering step.

The prediction step is:

$$\begin{cases} \overline{\mathbf{x}}(k+1)=\mathbf{\Phi}\mathbf{x}(k)+\mathbf{\Gamma}\mathbf{u}(k) \\ \mathbf{P}'_{k+1}=\mathbf{\Phi}\mathbf{P}_k\mathbf{\Phi}^T+\mathbf{\Gamma}\mathbf{Q}\mathbf{\Gamma}^T \end{cases} \tag{4}$$

where, $\overline{\mathbf{x}}(k+1)$ is the predictive value of state variables x at time step k+1, $\mathbf{P}'_{k+1}$ is the predictive error covariance matrix (pre-measurement covariance matrix) at time step k+1, and $\boldsymbol{P}_k$ is the estimated error matrix at time step k.

The filtering step is:

$$\begin{cases} \mathbf{K}_{k+1}=\mathbf{P}'_{k+1}\mathbf{C}^T(\mathbf{C}\mathbf{P}'_{k+1}\mathbf{C}^T+\mathbf{R})^{-1} \\ \hat{\mathbf{x}}(k+1)=\overline{\mathbf{x}}(k+1)+\mathbf{K}_{k+1}(\mathbf{y}(k+1)-\mathbf{C}\overline{\mathbf{x}}(k+1)) \\ \mathbf{P}_{k+1}=(\mathbf{I}-\mathbf{K}_{k+1}\mathbf{C})\mathbf{P}'_{k+1} \end{cases} \tag{5}$$

where, $\boldsymbol{K}_{k+1}$ is the Kalman gain matrix at time step k, $\hat{\mathbf{x}}(k+1)$ is the estimated value of state variables x at time step k+1, and $\boldsymbol{P}_{k+1}$ is the estimated error matrix at time step k+1 (post-measurement covariance matrix).

3. The proposed dynamic estimator model

3.1 The proposed dynamic estimator model for generator variables

During the power system fault stage and consequent dynamic process, the network topology is changed and can not be acquired in time; the bus voltage phase angles has discontinuities and are not easy to be predicted. Therefore, the generator rotor angle and electrical angular velocity are regarded as the estimated state variables which can not mutate suddenly and obey the rotor motion equation. Moreover, the generator angle trajectories implicate abundant dynamic information; thereby, acquiring accurate generator rotor trajectories is of great importance to power system real time control.

The rotor motion equation is written as follows:

$$\begin{cases} \dfrac{d\delta}{dt}=(\omega-1)\omega_0 \\ \dfrac{d\omega}{dt}=\dfrac{1}{T_J}(T_m-T_e-D\omega)=\dfrac{1}{T_J}(\dfrac{P_m}{\omega}-\dfrac{P_e}{\omega}-D\omega) \end{cases} \tag{6}$$

where, δ is the generator rotor angle (rad), ω is the generator electrical angular velocity; T_m and T_e are the mechanical torque and electrical torque on generator shaft respectively; P_m and P_e are the mechanical input power and electrical output powert respectively; T_J is the moment of inertia of the machine rotor and D is the damping coefficient.

It can be seen form equation (6) that if the electrical torque (power) and mechanical torque (power) at any time step are known, the generator motion equation will be decoupled from outer network [17]. Therefore, the generator rotor motion becomes single rigid body motion in two-dimension state space (displacement and velocity). If the generator mechanical torque is assumed constant, when only the electrical torque curve in time domain is known, the rotor motion equation is decoupled from outer network.

Equation (6) is written as follow form:

$$\begin{bmatrix} \dot{\delta} \\ \dot{\omega} \end{bmatrix} = \begin{bmatrix} 0 & \omega_0 \\ 0 & -\dfrac{D}{T_J} \end{bmatrix} \begin{bmatrix} \delta \\ \omega \end{bmatrix} + \begin{bmatrix} -\omega_0 \\ \dfrac{T_m - T_e}{T_J} \end{bmatrix} \tag{7}$$

It is noted that equation (7) is a standard time-invariant linear system which is qualified to linear Kalman filter well.

The torque can not be measured easily, hence the second term (controlling variable vector) of the right side of equation (7) is written as the form of P_m and P_e, then the equation (7) is written as follow:

$$\begin{bmatrix} \dot{\delta} \\ \dot{\omega} \end{bmatrix} = \begin{bmatrix} 0 & \dfrac{180}{\pi}\omega_0 \\ 0 & -\dfrac{D}{T_J} \end{bmatrix} \begin{bmatrix} \delta \\ \omega \end{bmatrix} + \begin{bmatrix} 1 & 0 \\ 0 & \dfrac{1}{T_J} \end{bmatrix} \begin{bmatrix} -\dfrac{180}{\pi}\omega_0 \\ (P_m - P_e)/\omega \end{bmatrix} \tag{8}$$

where, the unit of δ is degree.
Comparing equation (8) with equation (1), we have

$$\mathbf{x} = \begin{bmatrix} \delta \\ \omega \end{bmatrix} \quad \mathbf{A} = \begin{bmatrix} 0 & \dfrac{180}{\pi}\omega_0 \\ 0 & -\dfrac{D}{T_J} \end{bmatrix} \quad \mathbf{B} = \begin{bmatrix} 1 & 0 \\ 0 & \dfrac{1}{T_J} \end{bmatrix} \quad \mathbf{u} = \begin{bmatrix} -\dfrac{180}{\pi}\omega_0 \\ \dfrac{P_m - P_e}{\omega} \end{bmatrix} \tag{9}$$

It should be pointed out that even the sate variable ω appears in controlling variable vector u of formula (8), thereby, formula (8) is not a strict state equation, but it does not affect the application of Kalman filter. The reasons are:

1. The state differential equation only adopted in predictive step to calculate the state variable at next time step. Even though $u(k)$ is the function of $\omega(k)$, $\omega(k)$ has been estimated by Kalman filter at last time step, so it can be regarded as a known variable to substitute in equation (8)

2. During electromechanical transient process, the value of ω is about 1 (p.u.) and the off nominal range of ω is about from parts per thousand to 2 percent. It can be seen form equation (8) that the impact of ω fluctuation upon the controlling variable is so small that can be covered by the dynamic plant noise.

δ can be measured by PMU synchronistically (direct measurement or inferred by generator terminal electrical variables). If the pulse sequences of the rotor angular velocity meter are synchronized by GPS, the electrical angular velocity ω also can be measured by PMU. Therefore,

$$\mathbf{y} = \begin{bmatrix} \delta \\ \omega \end{bmatrix} \quad \mathbf{C} = \begin{bmatrix} 1 & 0 \\ 0 & 1 \end{bmatrix} \tag{10}$$

If only δ can be measured by PMU, then:

$$\mathbf{y} = [\delta] \quad \mathbf{C} = [1 \quad 0] \tag{11}$$

Calculating the state transition matrix (equation (2)) in engineering, only the three former terms are taken into account, which can achieve sufficient accuracy.

$$\mathbf{\Phi}(T) \approx \mathbf{I} + \mathbf{A}T + \mathbf{A}^2T^2 / 2 = \begin{bmatrix} 1 & \frac{180}{\pi}\omega_0 T(1 - \frac{DT}{2T_J}) \\ 0 & 1 - \frac{DT}{T_J}(1 - \frac{DT}{2T_J}) \end{bmatrix}$$

$$\mathbf{\Gamma}(T) = \left[\int_0^T \mathbf{\Phi}(\alpha) d(\alpha) \right] \mathbf{B} = \begin{bmatrix} T & \frac{180}{\pi}\omega_0 \frac{T^2}{2T_J}(1 - \frac{DT}{3T_J}) \\ 0 & \frac{T}{T_J} - \frac{DT^2}{2T_J^2}(1 - \frac{DT}{3T_J}) \end{bmatrix}$$

Now, the generator dynamic equation qualified to Kalman filter (equation (2)) is built up, but to realize the predictive step (equation (4)) and filtering step (equation (5)) of Kalman filter algorithm, the exact value of plant noise covariance Q and measurement noise R. the following error analysis analyzes the problem in detail.

3.2 Error and noise analysis

The measurement noise (error) covariance matrix $\boldsymbol{R}$ and plant noise covariance matrix $\boldsymbol{Q}$ are analyzed respectively in terms of the rotor angle measurement mode (direct measurement mode or indirect measurement mode).

3.2.1 The direct measurement mode of rotor angle and electrical angular velocity

The direct measurement method of generator rotor angle using rotor position sensor has relative higher accuracy. The method assumes that the electrical angular velocity is constant during one cycle, but in fact, the electrical angular velocity varies about parts per thousand during one cycle. According to analysis, the measurement error due to this is about 1~2°. Moreover, the detection precision of rotor position pulse also affects the measurement error. Considering all above factors, the standard deviation of the rotor angle direct measurement error can be set as 2°, and the corresponding variance is 4°.

The measurement of electrical angular velocity is equivalent to the measurement of rotor angular velocity (electrical angular velocity=number of pole pairs× rotor angular velocity). In modern power systems, the rotor angular velocity can be measured by velocity meter for turbine generator and hydro generator. The principle of velocity meter is described below. There is a 60-tooth gear installed in the rotor shaft and the detection circuit of velocity meter detects the pulse generated by each tooth of the gear. Therefore, 6° divided by the time interval between two pluses is the value of instantaneous angular velocity. If the pulse sequences generated by rotor angular velocity meter are synchronized by GPS, the electrical angular velocity ω can also be measured by PMU. The synchronized accuracy and the pulse detection accuracy both affect the measurement error, thereby, the standard deviation of the direct measurement error of rotor electrical angular velocity can be set as 0.001(p.u., i.e. 0.05~0.06 Hz), and the corresponding variance is1e-6.

It can be seen that when both the rotor angle and the electrical angular velocity can be measured by PMU directly, the corresponding measurement noise covariance matrix is as follows:

$$\mathbf{R} = \begin{bmatrix} 4 & 0 \\ 0 & 10^{-6} \end{bmatrix} \tag{12}$$

3.2.2 The indirect measurement mode of rotor angle

The direct measurement method of generator rotor angle is required to install the rotor position sensor, but some old style generators do not satisfy the installation condition. The direct synchronized measurement of electrical angular velocity is also required to install the additional GPS receiver and carry out necessary alteration. Thereby, these generators can not adopt the direct measurement mode and have to infer the rotor angle using the generator terminal electrical variables measured by PMU. The inferred rotor angle is regarded as the indirect measurement value of rotor angle and used in Kalman filter to estimate the rotor angle and electrical angular velocity.

If the generator terminal voltage phasor $\dot{U}_t$ and current phasor $\dot{I}_t$ is measured, the virtual internal voltage $\dot{E}_Q$ which is used to fix on the q axis position is written as,

$$\dot{E}_Q = \dot{U}_t + \dot{I}_t(R + X_q) \tag{13}$$

where, R is stator resistance, X_q is q axis synchronous reactance, the angle of $\dot{E}_Q$ is rotor angle δ.

An alternative equation is:

$$\delta' = a\tan\frac{PX_q - QR}{U_t^2 + PR + QX_q} \tag{14}$$

where, δ' is the rotor angle with reference to terminal voltage phase. P, Q are the generator output active power and reactive power, respectively.

It is assumed that the voltage phase of the generator terminal bus is θ, therefore, the generator rotor angle δ is:

$$\delta = \theta + \delta' \tag{15}$$

where, δ' is the rotor angle with reference to terminal voltage phase. P, Q are the generator output active power and reactive power, respectively.

It should be kept in mind that equation (13) and (14) are both derived with the assumption that there is no damping current in rotor amortisseur, thereby, the equations have sufficient precision in steady state, but considerable errors may be resulted in during dynamic process. Especially in fault duration, the inferred rotor angle will produce an obvious discontinuity.

Strictly speaking, the error propagation theory should be used to calculate the indirect measurement error variance of the inferred rotor angle accurately, whereas, the formula is very complex and hard to calculate due to the factors such as measurement variation, damping current, time-variant parameters and iron saturation, etc. therefore, according to the errors variance of PMU direct measurements (terminal voltage, terminal current and output power) and considering all factors mentioned above synthetically and simply, the standard deviation of the indirect measurement error of rotor angle can be set as 3°, and the corresponding variance is 9°.

Therefore, when only the rotor angle can be inferred by the PMU generator terminal measurements indirectly, the corresponding measurement noise covariance matrix is as follows:

$$\mathbf{R} = [9] \tag{16}$$

Since the phasor calculation has time delay, the sampling value with synchronous time stamp is adopted to calculate the instantaneous output active and reactive power directly.

$$\begin{aligned} P &= u_a i_a + u_b i_b + u_c i_c \\ Q &= \frac{1}{\sqrt{3}}(u_a(i_c - i_b) + u_b(i_a - i_c) + u_c(i_b - i_a)) \end{aligned} \tag{17}$$

where, u_a, i_a; u_b, i_b and u_c, i_c are the instantaneous sampling value of phase a, phase b and phase c generator terminal voltage and current, respectively.

3.2.3 The dynamic plant noise analysis

The dynamic plant noise represents the errors of model and parameters. It can be seen form equation (8) that the involved parameters are generator inertial constant T_J and damping coefficient D. T_J can be acquired exactly normally; D is very small and only reflects the mechanical friction and the windage since the electrical damping has been covered by the measured output active power. Therefore, the dynamic plant noise mainly roots in the measurement error of electrical output active power and the variation of mechanical input power P_m. Since P_m is hard to measure accurately, it is assumed constant and its variation due to governor action is regarded as dynamic plant noise when only governor operates and generator reject and fast valving are not triggered.

In terms of relative standards [4], the standard deviation of electrical active power measurement error varies between 1%~2%, considering the variation of P_m, the plant noise covariance matrix Q is set as:

$$\mathbf{Q} = \begin{bmatrix} 0 & 0 \\ 0 & 0.0004P_{e0} + 0.0001 \end{bmatrix} \quad (18)$$

where, P_{e0} is the generator electrical output active power in pre-fault steady state.

It should be pointed out that Q varies with the measured active power P_e, but considering the impact of computational cost and bad data, equation (18) can satisfy the precision requirement generally.

To avoid the impact of the time delay of phasor calculation, the instantaneous sampling value should be employed to calculate the instantaneous electrical active power P_e.

$$P_e = P + I_t^2 R = u_a i_a + u_b i_b + u_c i_c + I_t^2 R \quad (19)$$

where, It^2R is the copper loss of generator stator armature.

4. Flow chart and implementation

The proposed WAMS based dynamic state estimator during electromechanical process is described as in Fig. 1.

The problems which should be noted in the implementation of the proposed estimator are described as follows:

4.1 Startup criterion

The steady state estimation aims at the power system steady state, whereas the proposed dynamic state estimator aims at the power system electromechanical transient process. Therefore, the startup criterion is needed for the proposed dynamic state estimator. The general startup criterion of the micro-computer based protection can be adopted as the startup criterion in the proposed dynamic estimator. The criterion triggers startup if three sequential instantaneous sampling values of generator terminal voltage and current all exceed the preset threshold.

After startup, a appropriate length of historical data of rotor angle, electrical angular velocity and electrical output power are backdated to ensure the convergence of Kalman filter before the startup time (power system fault occurring time).

4.2 Bad data identification and elimination

The rotor angle and electrical angular velocity have no discontinuous variation; therefore, the absolute residuals related with the bad measurement data will be increased anomaly. This feature can be used to identify and eliminate the bad data.

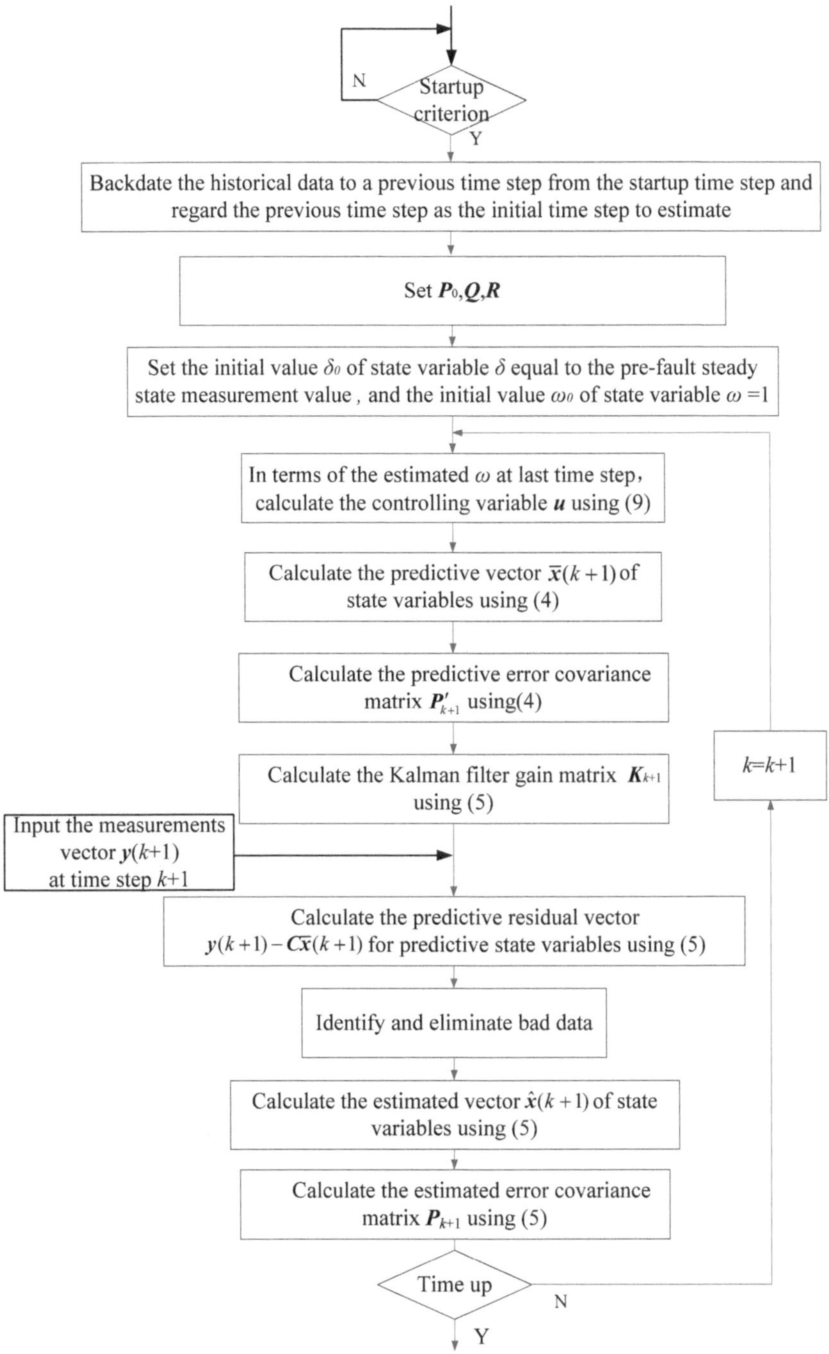

Fig. 1. Flow chart of the proposed dynamic state estimator.

For the example of rotor angle direct measurement, $\Delta\delta$(k+1) denotes the predictive residual at time step k+1, and it can be calculated as follows:

$$\Delta\delta(k+1) = \delta_m(k+1) - \bar{\delta}(k+1) \tag{20}$$

where, δ_m(k+1) denotes the measurement value at time step k+1, $\bar{\delta}(k+1)$ denotes the predictive value at time step k+1.

The average of three previous sequential predictive absolute residual $\Delta\delta absmean=(|\Delta\delta(k)|+|\Delta\delta(k-1)|+|\Delta\delta(k-2)|)/3$, so, the bad data identification criterion at time step k+1 is described below:

$$|\Delta\delta(k+1)| > 5\times\Delta\delta_{\mathrm{absmean}} \tag{21}$$

If equation (21) holds, the corresponding measurement is identified as bad data. To eliminate its adverse impact, the absolute predictive residual $|\Delta\delta(k+1)|$ at time step k+1 is replaced by $\Delta\delta_{absmean}$, but its sign is reserved. After that, the filtering step of Kalman filter and the estimation at next time step are continued.

Equation (21) aims at the single bad data point, but does not work for the serial bad data points. For instance, there are serial bad data points with negative residuals in measurement form time step $k+1 \sim k+n$, if the criterion is applied, the absolute predictive residuals at time step $k+1 \sim k+n$ will be all replaced by $\Delta\delta_{absmean}$, and the negative sign will be reserved, thus, the predictive residuals at time step $k+1 \sim k+n$ tend to reach a negative constant, which breaks the randomness of predictive residuals. In this situation, the divergence of Kalman filter may be resulted in and the sequent normal measurement points may be identified the bad data wrongly.

To avid this, the function sin(k) is used to recovery the randomness of predictive residuals at bad data points. In general, The value of function sin(k), $k=1,\cdots n$ (rad) obey the random distribution rule in the range of [-1,+1], thereby, if there is bad data at time step k+1, the substitute of the predictive residuals at the time step is

$$\Delta\delta(k+1) = \sin(k+1)\cdot\Delta\delta_{\mathrm{absmean}} \tag{22}$$

As mentioned before, the inferred rotor angle using equation (13) and (14) will produce an obvious discontinuity in fault duration. But in fact, the real rotor angle has no discontinuous variation; therefore, the discontinuity during fault stage can be regarded as a series of bad data. Applying the above bad data identification and elimination method, the more accurate rotor angle will be estimated.

5. Numerical study

IEEE 9-bus test system shown in Fig.2 and a real generator in North China power grid are chosen to carry out the numerical simulation.

In IEEE 9-bus test system, Gen1 and Gen2 are chosen as the generator to estimate. Gen1 and Gen2 are both equipped with governors and voltage regulators, and Gen2 is equipped with PSS. All the generators adopt the sixth-order detail model. A three-phase metal short-circuit fault is set on the beginning end of line BusB-Bus1 at the 50th cycles, and the circuit breakers

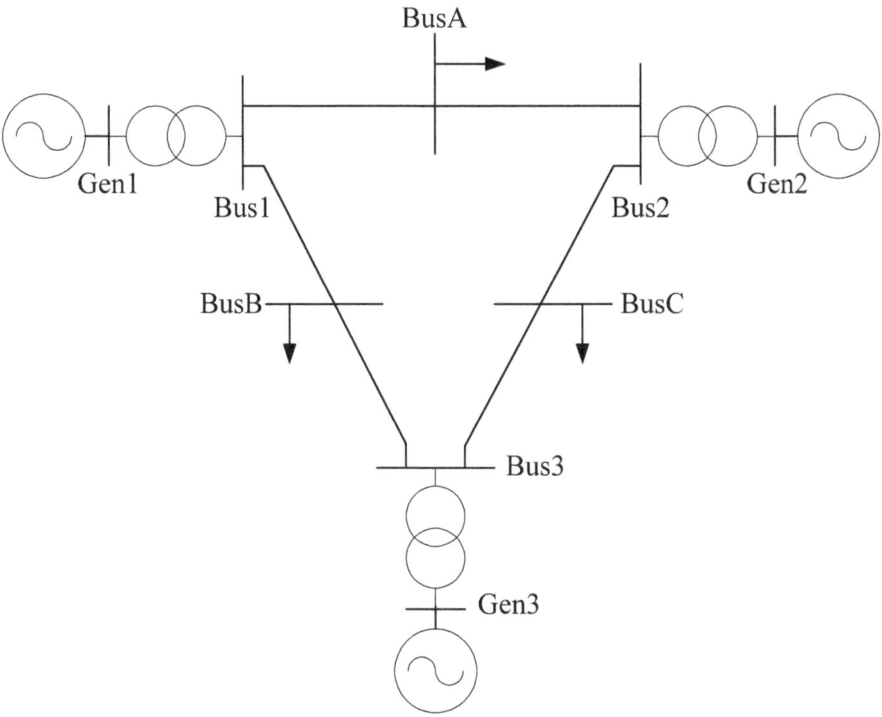

Fig. 2. IEEE 9-bus test system.

of both ends trip to clear fault at the 56th cycles. The true values of rotor angles, electrical angular velocities and electrical output power are acquired by commercial power system simulation software BPA, and the time step length is 2 cycles. The measurement values consist of true values, additional measurement errors and bad data. The errors of all types of measurements are assumed to follow the normal distribution with zero mean and the standard deviations which are given in Section III.B.

Since the measurement precision of rotor angle in steady state is relative high, the steady measurement value can be set as the initial value of state variable δ, and the initial value of state variable ω is set to 1 (p.u.). the reliability of initial value is relative high, the initial covariance matrix P0 is set to zero matrix for convenience. The startup time step is the 50th cycle, the initial time step to estimate is the 26th cycle and the end time step is the 600th cycle. Fig. 3 gives the dynamic estimation effect of Gen 2 with δ and ω direct measurement. Fig. 4 shows the dynamic estimation effect of Gen 2 when only δ can be measured indirectly.

It can be seen form Fig.3 that the proposed dynamic estimator has favorable filter effect and eliminates the measurement noise and bad data effectively for δ and ω direct measurement. The subgraph (c) which zooms in the interested part of estimation effect for δ shows clearly that the proposed estimator eliminates the adverse impact of serial bad data points successfully.

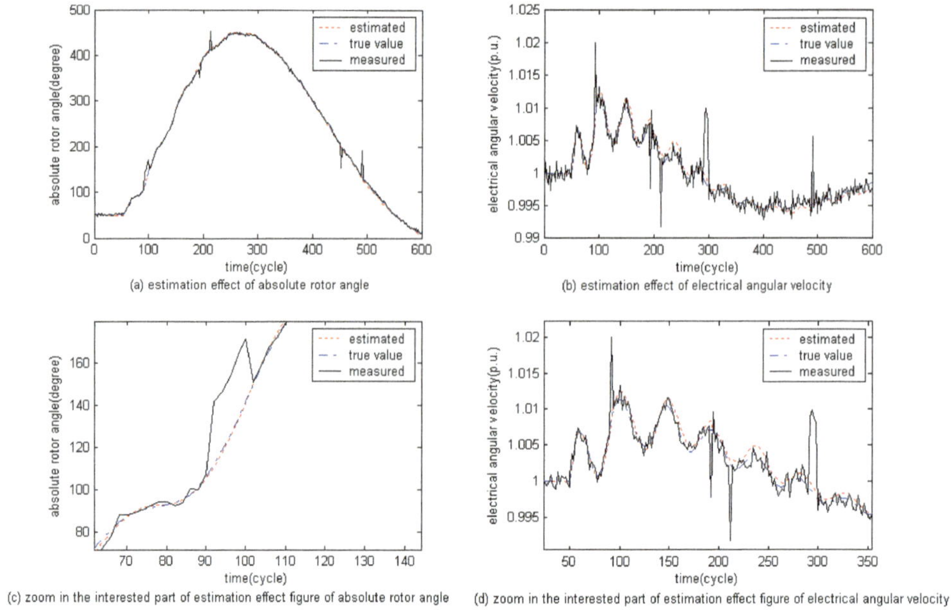

Fig. 3. The estimation effect of Gen2 with δ and ω direct measurement.

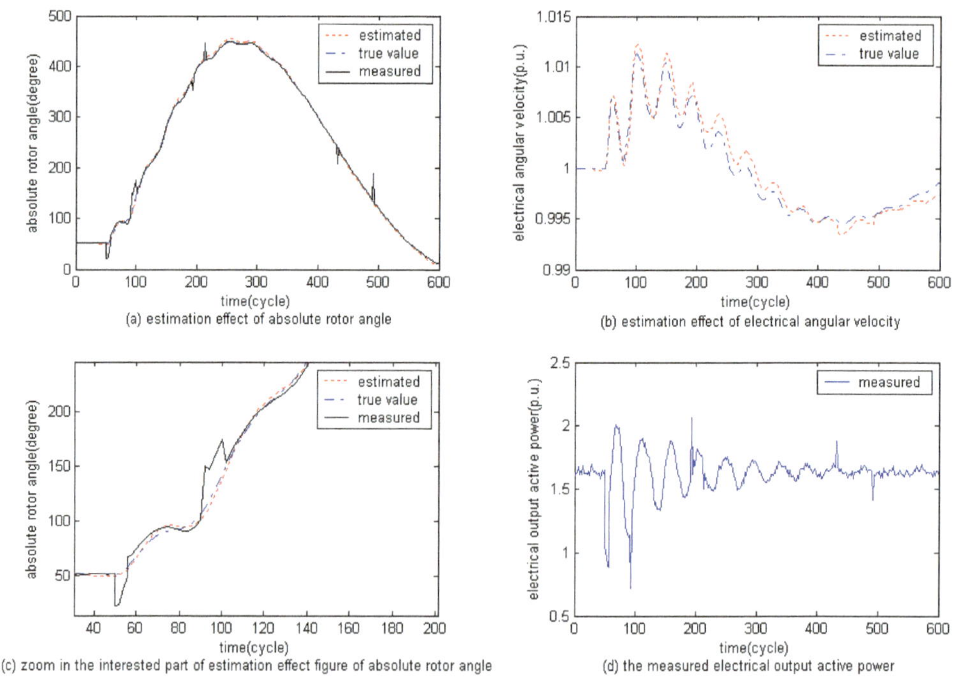

Fig. 4. The estimation effect of Gen2 when only δ can be measured indirectly.

The similar conclusion can be drawn form Fig. 4, Moreover, the subgraph (a) and (c) show that the indirect measurement of rotor angle δ has an obvious sag in fault duration. Applying the proposed bad data elimination method for serial bad data, the sag is made up well, which smoothes the estimated curve for δ and achieves excellent filter effect. Since there is no measurements of electrical angular velocity, the estimation effect for ω is not as good as that in Fig.3, but is still acceptable.

Whether the state variables δ and ω are measured directly or indirectly, the considerable measurement noises and bad data are added on the electrical output active power introduced as the controlling variable (Fig. 4(d)), but the estimation effect is almost not affected. The reason is that the proposed dynamic estimator is an integral process of the electrical output active power substantially, and the integral itself has better antinoise performance.

To acquire the quantified estimation index, the filter effect ρ and index ε[13] are defined as:

$$\rho = \frac{\sum_{i=1}^{n} (\hat{x}_i - x_i^+)^2}{\sum_{i=1}^{n} (x_i^M - x_i^+)^2} \tag{23}$$

$$\varepsilon = \frac{\sum_{i=1}^{n} \left| \frac{\hat{x}_i - x_i^+}{x_i^+} \right| \times 100\%}{n} \tag{24}$$

where, i indicates the sequence number of time step, n indicate the total number of time steps. $\hat{x}_i$,x_i^M and x_i^+ indicate the estimated value, measured value and true value of state variable x (δ or ω) at time step i respectively. The filter effect ρ and index ε of δ and ω are calculated respectively for the difference between their quantities is very large.

Table 1 gives the estimation indices of Gen1 and Gen2 with different measurement modes, and all results in Table I are the average values over 100 runs of Monte Carlo. It can be seen from Table I that the proposed dynamic estimator achieves good estimation effects with different measurement modes, and the estimation time is about 0.1 ms, which can satisfy the requirements of real-time applications. The estimation results of Gen 1 and Gen 2 with direct measurement of δ and ω is higher than those with only indirect measurement of δ. Moreover, the estimation precisions of Gen1 are higher than those of Gen2, because that the inertia constant of Gen1 is bigger than of Gen2, which makes the estimation of Gen1 is more immune to the dynamic plant noise.

A real generator in North China power grid is also chosen as the estimated generator, a three-phase metal short-circuit fault is put on a 500 kV line in the grid to excite the electromechanical transient, and the corresponding circuit breakers are tripped to clear fault after 5 cycles. All the simulation conditions are same as those of IEEE-9 test system except that the time step length is 1 cycle. The simulation results is given in Table 2.

The similar conclusion can be drawn form Table II; moreover, the estimation precisions are higher than those in IEEE 9 test-system. Since the generator is decoupled with outer network in the proposed dynamic estimator, the parallel dynamic estimation of different generators in large power system can be executed simultaneously. Therefore, the fast speed of the proposed dynamic estimator is not affected by the increasing numbers of generators in large power systems.

Generator	Measurement mode		State variable	Estimation indices ρ	Estimation indices ε	Estimation time(ms)
Gen1	1	direct	δ	0.0539	1.20	0.1247
		direct	ω	0.0132	0.0157	
	2	indirect	δ	0.1663	1.65	0.1047
		No	ω	No	0.0178	
Gen2	1	direct	δ	0.1226	1.47	0.1273
		direct	ω	0.1554	0.0619	
	2	indirect	δ	0.2620	2.20	0.1067
		No	ω	No	0.0803	

Table 1. Estimation results of G1 and G2 in IEEE 9-bus system.

Generator	Measurement mode		State variable	Estimation indices ρ	Estimation indices ε	Estimation time(ms)
Generator	1	direct	δ	0.0282	0.84	0.1240
		direct	ω	0.0078	0.0091	
	2	indirect	δ	0.0594	1.12	0.1160
		No	ω	No	0.0119	

Table 2. Estimation results of one generator in practical grid.

6. Conclusion

A WAMS based distributed dynamic state estimator for generator rotor angle and electrical angular velocity during power system electromechanical transient process is proposed in this paper. The WAMS measurement noise and dynamic model noise are analyzed concretely, and the bad data detection and elimination approach are given as well. The simulation results indicate that the proposed dynamic estimator has high estimation precision and fast computational speed; which satisfies the real-time requirements. The estimated generator rotor angle and electrical angular velocity eliminate the adverse impact of measurement noises and bad data; therefore, the proposed dynamic estimator can serve the power system dynamic monitoring and control system better.

7. References

[1] A. G. Phadke, "Synchronized phasor measurements—a historical overview," in *Proc. IEEE Power Eng. Soc. Asia Pacific Transmission Distribution Conf. Exhib*, Print ISBN: 0-7803-7525-4, Yokohama, Japan, 6–10, Oct., 2002, vol. 1, pp. 476–479.

[2] T. W. Cease and B. Feldhaus, "Real-time monitoring of the TVA power system," *IEEE Comput. Appl. Power*, vol. 7, no. 3, pp. 47–51, Jul. 1999. ISSN : 0895-0156

[3] A. G. Phadke, "Synchronized Phasor Measurements in Power Systems", *Comput. Appl. Power*, vol. 6, no.2, pp. 10-15, 1993, ISSN : 0895-0156

[4] *Standard for Synchrophasors for Power Systems*, IEEE Standard C37.118-2005m, 2005. ISBN: 0-7381-4820-2 SH95382

[5] Zhenyu Huang, John F. Hauer, and Kenneth E. Martin. "Evaluation of PMU dynamic performance in both lab environments and under field operating conditions," *IEEE Power Engineering Society General Meeting*, ISSN: 1932-5517, Tampa, 24-28, Jun, 2007, pp. 1-6

[6] J. F. Hauer, K. E. Martin, and Harry Lee, "Evaluating the dynamic performance of phasor measurement units: experience in the western power system," WECC Disturbance Monitoring Work Group, June 15, 2004.

[7] William L. Miller and John B. Lewis. "Dynamic state estimation in power systems," *IEEE Trans. Automatic Control*, vol. 16, no. 6, pp. 841-846, Dec. 1971, ISSN : 0018-9286

[8] Jaewon Chang, Glauco N. Taranto, and Joe H. Chow. "Dynamic State Estimation in Power System using a Gain-Scheduled Nonlinear Observer," in *Proceedings of the 4th IEEE Conference on Control Applications*, Print ISBN: 0-7803-2550-8, Albany, NY,USA,28-29,Sep, 1995, pp. 221-226

[9] A. G. Phadke, J. S. Thorp, and K. J. Karimi, "State estimation with phasor measurements," *IEEE Trans. Power Syst.*, vol. 1, no. 1, pp. 233–240, Feb,1986. ISSN : 0272-1724

[10] J. S. Thorp, A. G. Phadke, and K. J. Karimi, "Real time voltage-phasor measurements for static state estimation," *IEEE Trans. Power App. Syst.*, vol. PAS-104, no. 11, pp. 3098–3104, Nov. 1985. ISSN : 0272-1724

[11] R. Zivanovic and C. Cairns, Implementation of PMU technology in state estimation: an overview, in *Proc. IEEE 4th AFRICON,* Print ISBN: 0-7803-3019-6, Stellenbosch , South Africa , 24–27, Sep, 1996, vol. 2, pp. 1006–1011.

[12] A. S. Debs and R. E. Larson. "A dynamic estimator for tracking the state of the power system," *IEEE Trans. Power Apparat. Syst.*, vol. 89, no. 7, pp. 1670–1678, 1970, ISSN : 0018-9510

[13] A. Silva, M. Filho, and J. Cautera. "An efficient dynamic state estimation algorithm including bad data processing". *IEEE Trans. Power Syst.* vol 2, no. 4 , pp. 1050–1058, 1987, ISSN : 0885-8950

[14] Durgaprasad G., and Thakur S.S. "Robust dynamic state estimation of power system based on M-estimation and realistic modeling of system dynamics". *IEEE Trans. Power Syst.*, vol. 13, no. 4, pp. 1331-1336, 1998, ISSN : 0885-8950

[15] Kuang-Rong Shih and Shyh-Jier Huang. "Application of a Robust Algorithm for Dynamic State Estimation of a Power System". IEEE Trans. Power Syst., vol 17, no.1, pp. 141-147, 2002, ISSN : 0885-8950

[16] Shande Shen, power system parameters identification, Beijing. Hydro and Power Press, 1993

[17] Zhenyu Huang, Dmitry Kosterev and Ross Guttromson, etc. Model Validation with Hybrid Dynamic Simulation, IEEE 2006 Power Engineering Society General Meeting, Print ISBN: 1-4244-0493-2, Montreal, Canada, 18-22, Jun,2006,pp.1-9,

Permissions

The contributors of this book come from diverse backgrounds, making this book a truly international effort. This book will bring forth new frontiers with its revolutionizing research information and detailed analysis of the nascent developments around the world.

We would like to thank Prof. Md. Zahurul Haq, for lending his expertise to make the book truly unique. He has played a crucial role in the development of this book. Without his invaluable contribution this book wouldn't have been possible. He has made vital efforts to compile up to date information on the varied aspects of this subject to make this book a valuable addition to the collection of many professionals and students.

This book was conceptualized with the vision of imparting up-to-date information and advanced data in this field. To ensure the same, a matchless editorial board was set up. Every individual on the board went through rigorous rounds of assessment to prove their worth. After which they invested a large part of their time researching and compiling the most relevant data for our readers. Conferences and sessions were held from time to time between the editorial board and the contributing authors to present the data in the most comprehensive form. The editorial team has worked tirelessly to provide valuable and valid information to help people across the globe.

Every chapter published in this book has been scrutinized by our experts. Their significance has been extensively debated. The topics covered herein carry significant findings which will fuel the growth of the discipline. They may even be implemented as practical applications or may be referred to as a beginning point for another development.

The editorial board has been involved in producing this book since its inception. They have spent rigorous hours researching and exploring the diverse topics which have resulted in the successful publishing of this book. They have passed on their knowledge of decades through this book. To expedite this challenging task, the publisher supported the team at every step. A small team of assistant editors was also appointed to further simplify the editing procedure and attain best results for the readers.

Our editorial team has been hand-picked from every corner of the world. Their multi-ethnicity adds dynamic inputs to the discussions which result in innovative outcomes. These outcomes are then further discussed with the researchers and contributors who give their valuable feedback and opinion regarding the same. The feedback is then collaborated with the researches and they are edited in a comprehensive manner to aid the understanding of the subject.

Apart from the editorial board, the designing team has also invested a significant amount of their time in understanding the subject and creating the most relevant covers. They scrutinized every image to scout for the most suitable representation of the subject and create an appropriate cover for the book.

The publishing team has been involved in this book since its early stages. They were actively engaged in every process, be it collecting the data, connecting with the contributors or procuring relevant information. The team has been an ardent support to the editorial, designing and production team. Their endless efforts to recruit the best for this project, has resulted in the accomplishment of this book. They are a veteran in the field of academics and their pool of knowledge is as vast as their experience in printing. Their expertise and guidance has proved useful at every step. Their uncompromising quality standards have made this book an exceptional effort. Their encouragement from time to time has been an inspiration for everyone.

The publisher and the editorial board hope that this book will prove to be a valuable piece of knowledge for researchers, students, practitioners and scholars across the globe.

List of Contributors

Yoshitake Yamamoto
Himeji Dokkyo University, Japan

Kazuaki Jikuya
Kawasaki University of Medical Welfare, Japan

Toshimasa Kusuhara, Takao Nakamura and Takuji Okamoto
Okayama University, Japan

Hiroyuki Michinishi
Okayama University of Science, Japan

Koji Nakanishi and Toshiaki Ohta
The SR Center, Ritsumeikan University, Japan

Laurent Lamaignère
CEA, DAM, CESTA, BP N°2, F-33114 Le Barp, France

Andrzej B. Dobrucki, Przemysław Plaskota and Piotr Pruchnicki
Wroclaw University of Technology, Poland

Karel Janák
Mendel University Brno, Czech Republic

L. E. Toledo and F. J. Cuevas
Centro de Investigaciones en Optica A.C., Dept. of Computer Vision and Artificial Intelligence, Optical Division, Leon, Mexico

J.F. Jimenez Vielma and J. H. Sossa
Center for Computing Research, National Polytehnical Institute, Artificial Inteligence Laboratory, Mexico

Takuya Kawamura, Kazuo Tani and Hironao Yamada
Department of Human and Information Systems, Gifu University, Japan

Manuel Sierra-Castañer, Alfonso Muñoz-Acevedo, Francisco Cano-Fácila and Sara Burgos
Technical University of Madrid (Universidad Politécnica de Madrid - UPM), Spain

Abdullah Beyaz
Ankara University, Faculty of Agriculture, Department of Agricultural Machinery, Turkey

James Baker-Jarvis
Electromagnetics Division, National Institute of Standards and Technology, Boulder, CO, USA

Seon-Gyoo Kim
Kangwon National University, South Korea

Diogo Acatauassu, Igor Almeida, Francisco Muller and Aldebaro Klautau
Federal University of Pará, Brazil

Chenguang Lu, Klas Ericson and Boris Dortschy
Ericsson AB, Sweden

R˘azvan Deaconescu
University POLITEHNICA of Bucharest, Romania

Alexander Janushevskis, Janis Auzins, Anatoly Melnikovs and Anita Gerina-Ancane
Riga Technical University, Latvia

Mohammad Shahraeini
Golestan University, Iran

Mohammad Hossein Javidi
Ferdowsi University of Mashhad, Iran

Jeffrey Zheng
Yunnan University, P.R. China

Christian Zheng
University of Melbourne, Australia

T.L. Kunii
University of Tokyo, Japan

Xiaohui Qin, Baiqing Li and Nan Liu
Power System Department, China Electric Power Research Institute, The People's Republic of China